U0284911

乐嘉陵院士

乐嘉陵简介

乐嘉陵，男，中国工程院院士，原中国人民解放军总装备部科学技术委员会兼职委员、顾问，原总装备部空气动力学专业组长，中国空气动力学学会副理事长。现任中国空气动力研究与发展中心研究员，《实验流体力学》杂志主编。

乐嘉陵院士长期从事高超声速地面试验设备和试验技术、弹头突防气动物理、运载火箭空气动力、高超声速推进与飞行器等领域的理论和试验研究，是我国高超声速空气动力学、再入气动物理、火箭空气动力学和吸气式高超声速技术的学科带头人之一。近 20 多年来领导了近百名中国空气动力研究与发展中心（CARDC）吸气式高超声速技术研究人员组成的团队，开展了高超声速飞行器和超燃冲压发动机关键技术研究，并取得重大突破，包括研制 2.4 米口径大型脉冲燃烧风洞，开展相应的飞行试验研究，研制基于超级计算的超声速燃烧流动三维大规模并行数值软件及相应的试验诊断研究，为我国吸气式高超声速技术及其战略发展做出了重要贡献。

乐嘉陵院士曾获何梁何利基金科学与技术进步奖，国家科技进步奖二等奖 1 项、三等奖 1 项，军队科技进步奖一等奖 7 项、二等奖 7 项，在国内外学术期刊和会议上发表论文 310 余篇。

2002 年 6 月，安县晓坝镇牛头山下，初期建立的流量 1kg/s 双管式脉冲燃烧风洞，
开展氢和液态碳氢燃料超声速燃烧流动研究。

乐嘉陵院士和爱人(曹履晖)，2013年11月在酒泉观看模型飞行试验。

2017 年 3 月，乐嘉陵院士领导团队部分青年研究人员参加厦门高超声速流动国际会议

乐嘉陵院士的最早培养一批博士生和初期吸气式高超研究小组部分人员于2019年7月在绵阳CARDC会议中心钱学森塑像前合影留念。（前排左起）王兰，贺伟，曾来荣，刘伟雄，乐嘉陵，曾学军(1984年入学)，白菡尘，贺元元，赵慧勇。（后排左起）焦伟，倪鸿礼，董维中，李向东，王晓东，吴东升，杨顺华，贺旭照，邢建文。

乐嘉陵院士研究
论文选集

曾学军　倪鸿礼　白菡尘　焦　伟　张若凌　编

科学出版社

北京

内 容 简 介

本书选编了乐嘉陵院士及其研究团队多年来在高超声速空气动力学方面具有代表性的论文，主要涉及高超声速非平衡流动、激波与流动显示和高超声速技术等研究内容。前言简要介绍了乐嘉陵院士及其带领的团队开展研究工作的情况。文集分为三个部分。第一部分主要是乐嘉陵院士在高超声速非平衡流动方面发表的文章，这与中国空气动力研究与发展中心早期规划的研究内容相关。第二部分是激波与流动显示研究论文，研究的是运动激波与(运动和静止)物体间的相互作用。第三部分高超声速技术研究论文，选编了自 1995 年以来中国空气动力研究与发展中心高超中心开展的燃烧空气动力学研究工作。附录是乐嘉陵院士于 2017 年 12 月前发表论文的统计。

本书可供从事高超声速空气动力学的研究人员、高校相关专业学生和教师阅读，也可作为航空航天相关领域工程技术人员的工具书和参考书。

图书在版编目(CIP)数据

乐嘉陵院士研究论文选集/曾学军等编. —北京: 科学出版社, 2019.9
ISBN 978-7-03-059032-9

Ⅰ. ①乐⋯　Ⅱ. ①曾⋯　Ⅲ. ①空气动力学–文集　Ⅳ. ①V211-53

中国版本图书馆 CIP 数据核字 (2018) 第 224632 号

责任编辑:赵敬伟 / 责任校对: 胡庆家
责任印制:肖　兴 / 封面设计: 耕者工作室

科学出版社 出版
北京东黄城根北街 16 号
邮政编码:100717
http://www.sciencep.com

中国科学院印刷厂印刷
科学出版社发行　各地新华书店经销
*

2019 年 11 月第 一 版　开本: 720×1000 1/16
2019 年 11 月第一次印刷　印张: 24 1/4
字数: 471 000

定价:228.00 元
(如有印装质量问题, 我社负责调换)

序　言

　　乐嘉陵院士，1954 年考入北京航空学院，1956 年进入空气动力学专业，1964 年研究生毕业后在北京空气动力研究所工作，1971 年赴四川绵阳参加中国空气动力研究与发展中心建设与相关研究至今。

　　乐嘉陵院士的主要研究领域为高超声速空气动力学，包括再入物理、各类脉冲设备、CFD 及相关测量，近 20 余年主要从事吸气式高超声速和超声速燃烧研究等。本论文集共 44 篇文章，分为三个部分，即高超声速非平衡流动研究、激波与流动显示研究和高超声速技术研究，大致反映了不同时期乐嘉陵院士在中国空气动力研究与发展中心关注的高超声速领域取得的研究进展。

　　高超声速非平衡流动是中国空气动力研究与发展中心早期规划的研究内容。20 世纪 60–70 年代美国阿波罗登月、苏联再入弹头和人造卫星发射，以及国内外高超声速热障攻关与突破，大大推动了高超声速空气动力学和气动物理的发展。我国在钱学森等的领导下亦制订了高超声速技术(包括探月返回再入辐射传热等)研究和人才发展规划。1964 年左右我国就开始进行气动物理研究，1970 年航空航天工业部和中国科学院力学研究所利用中国空气动力研究与发展中心的激波管、电弧风洞等实验装置开展了实验和理论研究，1980 年起，航空航天工业部开始组织气动物理研究与交流。1999-2003 年在国家自然科学基金重点项目支持下，乐嘉陵院士继续主持 "高温高焓气体非平衡特性研究" 的项目研究工作。文集中高超声速非平衡流动研究部分中有几篇论文涉及了辐射传热、强激波后电离非平衡，另外在弹道靶中直径为 12mm、飞行 $M_\infty=15$ 圆球模型的非平衡流动，其研究目的是验证大量非平衡流动计算方法，当前这一领域仍然值得关注。

　　激波与流动显示研究部分，研究的是运动激波与(运动和静止)物体间的相互作用，主要实验工具是小尺寸光学激波管($40\text{mm} \times 40\text{mm}$)。通过各类密度场测量与计算流动显示(Computational Flow Imaging, CFI)对比，以验证 CFD 计算方法。乐嘉陵院士独立发展的彩色干涉条纹技术对清晰判断强激波前后的定量密度场发挥了重大作用。此外，这方面的研究亦为靶场安全评估发挥了作用。在分析 1996 年 2 月 15 日发生的某火箭爆炸时，采用了一套经实验验证的数值软件系统，及时评估了爆炸波在复杂山区 1.5km 外对建筑物的冲击载荷和损毁效应。

　　高超声速技术研究部分，选编了自 1995 年以来在中国空气动力研究与发展中心高超中心开展的燃烧空气动力学研究工作。其中包括乐嘉陵院士根据当时中国国情提出的实施国家高超声速飞行器计划的马赫数、航程和小型飞行器尺寸等建议，以及确定了机体/推进一体化研究的具体参数和所需风洞试验参数。为开展这方面研究，乐嘉陵团队做出了包括2.4米喷管直径在内的系列脉冲燃烧风洞研制及其在超燃发动机中应用等开创性工作。与传统的连续式风洞进行燃烧试验研究不同，首创建立的工作时间仅几百毫秒的脉冲燃烧风洞，最大贡献在于能以最经济高效方式获得从基础到实际应用的结果，包括开展带动力(液态煤油燃料)三维机体/推进一体化飞行器整体气动性能及超燃发动机脉冲和连续风洞的对比试验。通过长达十多年研究，最终获得了国内外认可。此外，高超中心建立的计算工具AHL3D是国内最早独立开发用于燃烧流动的三维大规模并行数值模拟平台，经过了两相流和光学诊断(OH-PLIF、自发光高速摄影、差分干涉、阴影)和反应动力学的算例考核，在AHL3D的基础上发展了较完整的再生冷却热结构设计计算工具，已经为内流道热环境研究和液态燃料再生冷却通道设计发挥了重要作用。三维大规模并行数值软件也用于了复杂构型复杂化学反应的单头部航发燃烧室计算，并进行了实验验证，表明它可用于更复杂构型的计算与分析。

再入物理(或气动物理)和燃烧空气动力学是高超声速空气动力学两个重要的发展分支，亦是空气动力学与物理化学学科进一步交叉与发展。虽然论文集中的内容仅仅涉及一小部分，但从学科和工程方面国内外正在从深度和广度上发展。

最后需要说明的是，选编的论文是乐嘉陵院士和合作者(包括研究生)共同辛勤劳动和潜心研究的成果，这些成果的取得，得益于中国空气动力研究与发展中心和上级机关各级领导的组织和推动。乐嘉陵院士在此向他们表示衷心感谢!

编　者
2019年7月

目 录

高超声速非平衡流动研究

高温空气总辐射强度的测定

乐嘉陵

高温空气辐射问题，无论是对高速物体再入大气层的辐射传热还是对强爆炸在大气中的能量传输，都是一项重要的研究课题。而高温空气总辐射强度，则是上述研究课中的一个基本的物理量。虽然这个物理量可以用理论的方法进行计算，但理论和实验之间差别很大，因此从实验上确切测定这个物理量显得很重要。国外在 64 年前对温度低于 $10,000°K$ 的空气辐射已经作了充分的研究 [1][2][3]，近期来主要是研究温度高于 $10,0000°K$ 类似于文献 [4] 的金星再入及具有强烈真空紫外、自吸现象 [5] 的气体辐射问题。本文简略介绍在激波管装置中开展高温平衡空风总辐射强度测定的有关实验工作。

一、 高温空气总辐射强度测定的基本原理

设模型置于高速气流中 (见图 1)，模型前端面装有石英窗口，控测器置于模型内于窗口之后，当高速气流以速度 V_∞ 向模型运动时，模型前形成脱体弓形激波，激波与模型之间形成高温高压区域。如果区域 2 为等温等压区，并且吻合光学薄平面平行激波层理论 [6] 的话，则此区域对探测器的平衡辐射热流为：

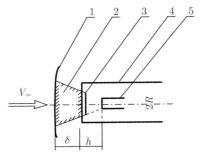

1. 脱体激波

2. 高温高压区

3. 石英窗口

4. 模型

5. 探测器

图 1 辐射测量原理

$$q = \frac{1}{2}E \cdot \delta \cdot F \tag{1}$$

乐嘉陵. 高温气体总辐射强度的测定. 气动研究与发展, 1979, 3: 1-10.

上式中,

$$2\pi F = \int_0^\Omega d\Omega$$

Ω —— 探测器所能接受的辐射气体立体角;

δ —— 脱体距离;

E —— 空气的总辐射强度,是气体状态参数 P、T 的函数。其物理意义是单位时间内单位容积的、高温气体在 4π 立体角内发射的总能量,即:

$$E = 4\pi j = 4\pi \int_0^\infty J_v dv \qquad (2)$$

其中 J_v —— 高温气体在单位立体角内 v 频率下的发射强度。

如果考虑到窗口的透射率 B_W 和探测器的吸收系数 α_G,则探测器所吸收的辐射热流为:

$$q = \frac{1}{2} E \cdot \delta \cdot F \cdot \alpha_G \cdot B_W \qquad (3)$$

$$E = q \Big/ \frac{1}{2} \delta \cdot F \cdot \alpha_G \cdot B_W \qquad (4)$$

上式中,q —— 探测器实际所接受的能量,也是探测器的主要任务;

δ —— 脱体距离,可由计算或实验确定。

B_W —— 窗口的透射率,可由分光光度计测定。

F —— 几何因子,纯属几何关系,可由计算给出。

解决了 q、δ、B_W、α_G、F,由式(4)可推得空气总辐射强度 E 值,或者说已经确定了一定温度、压力状态下的 $E(P,T)$ 值。这样就可以用 E 值估算飞行器驻点在飞行速度为 8 公里/秒再入大气层时的辐射传热 [6]。由文献 [7] 可知,当气体温度小于 $10,000°$K 时,气体的辐射只有黑体辐射的百分之几,此时气体的自吸影响可以不加考虑。因此采用式(1)的方案取得的 E 值是可靠的。由于上述实验中探测器所吸收的气体辐射是经越窗口的,石英窗口的截止频率为 2000Å 左右,因此本文总辐射的积分范围是 2000Å~1.5μm。

二、 实验装置与试验方案

辐射测量中,高速气流由激波管装置产生。激波管装置见图 2。高压段与低压段间用铝膜隔开。高压段充 H_2、O_2、He 气体。点火方式为轴向点火,用 $\phi = 0.1\text{mm}$ 钨丝引燃。低压段可充空气或其它试验气体。低压段在充试验气体前先用 FW–140 型涡轮分子泵抽到 $10^{-3} \sim 10^{-4}$ 托以下,并经过过滤干燥器再充以试验气体。沿激波管低压段的轴向每隔一米就装有电离探头,激波经越探头的时间由十五通道计数器给出，由此就可获得激波速度。试验时高压段点火，使 H_2、O_2、He 气体瞬间引燃造成 150atm 和 2500～3000°K 的压力和温度，高低压段之间铝膜就突然破裂，于是处于静止的低压段试验气体就被激波加速为高速气流并冲击在模型上，形成模型前的弓形脱体激波与高温高压区域。试验过程中要经常清洗激波管以防止杂质污染。通过气体光谱的摄谱，可以观察到用这种办法获得高温气体辐射没有杂质的污染。

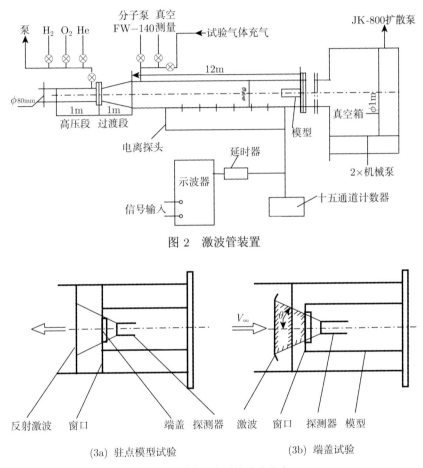

图 2　激波管装置

(3a) 驻点模型试验　　　　(3b) 端盖试验

图 3　辐射测量两种试验方案

　　本文试验采用两种方案，即驻点模型与端盖模型两种方案（图 3a、3b）。驻点模型与一般风洞试验没有区别。而端盖模型实际上就是测定激波反射后形成的高温气体对探测器的辐射传热。对于驻点模型方案，要保证高温气体区域为平面激波的等温等压区、并增大脱体距离 δ，可以采用平头模型；同时探测器所接收的应该是在驻点附近的极小部分的气体辐射[8]，即是说 θ 角应该限制在很小的范围，只有在这种条件下才能用式（1）处理实验数据，所取得的 E 值可以说是处在驻点的 P、T 状态下的总辐射强度，才是有意义的。例如文献 [3]，$\theta \approx 8°$，探测器元件的长度 $l = 3.3\text{mm}$。根据上述原则，本文的 $\theta \approx 12°$，探测器元件长度 $l = 3\text{mm}$。对于端盖模型方案，由于反射激波后的流场比较复杂，反射激波的速度也并非均匀。据文献 [9] 和本文在端盖处用光敏管和薄膜电阻辐射探测器测量的结果（见图 4），激波在端盖处反射后的 $20 \sim 25\mu s$ 范围内，探测器所接收的辐射强度可以近似认为随时间的变化是线性的，因此本文对端盖模型方案，取 $t \approx 20 \sim 25\mu s$。

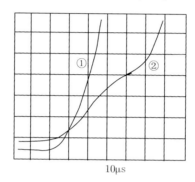

10μs

a 薄膜探测器与光敏二极管讯号示波图：①光敏管讯号；②薄膜探测器讯号.

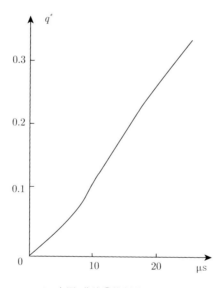

b. 由图a曲线②推得的q^*(无因次)

图 4　端盖试验 No. 211, $P_1 = 1\text{mmHg}$, $V_s = 4.88\text{Km/sec}$

三、 探测器的结构及其使用

从式（4）可知，在总辐射强度测量中，探测器所接收的辐射热流 q 的测定是问题的关键。由于波激管中试验时间小于 100μs，因此对探测器来说首先要有足够快的几微秒的响应时间，其次对于本文测定的 2000Å～1.5μm 的光谱范围内探测器亦应有平坦的光谱吸收特性。

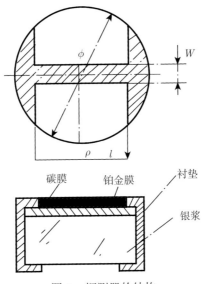

图 5　探测器的结构

本文使用的探测元件是铂金膜电阻，图 5 是它的结构，顶层是炭黑体，第二层是铂金膜，再下面就是玻璃衬底，两侧是引线。铂金膜是用真空溅射法溅射在清洗干净的玻璃基片上，铂金膜的形状是直线式的；同时在两端引线处涂上银浆作为铂金膜的引线。此后就送入高温炉内加热 600℃左右烘烤以老化铂金膜，使其阻值稳定；炭膜是蒸发在铂金膜上。为保证几微秒的响应时间，文献 [3][5] 要求铂金膜厚度为 300 ～ 500Å，炭膜厚度小于 1500Å。

本文使用探测器的三种规格如下（见图 5）：

ϕ/mm	l/mm	w/mm	R/Ω
15	12	2	40～60
10	6	1	40～50
5	3	1	15～20

本文使用的铂金膜较文献 [10] 形状简单，尺寸也小，制作方便，这样就可以在 $\theta \approx 12°$ 情况下利用高灵敏度示波器测量比文献 [9] 低一个量级的辐射讯号，从而

获得正确测量的 E 值。

探测器铂金膜厚度为 $300 \sim 450$Å，炭膜厚度为 1200Å、1500Å 两种，都是由样品用西德的莱茨双光束干涉显微镜测定。炭膜蒸发是由科学院生物物理所协作完成。整个探测器的光谱吸收数可由日本的 MPS–5000 分光光度计测定。铂金膜的透射率测定结果小于 0.05，可以忽略不计。所以探测器的吸收系数

$$\alpha_G = 1 - 反射系数$$

此项工作由北京玻璃研究所、北京工业学院、上海水产研究所协作完成。几个样品的探测器吸收系数见图 6，其中 2# 探测器的吸收特性比较均匀平坦。衬底的材料本文使用 GG–17 玻璃。

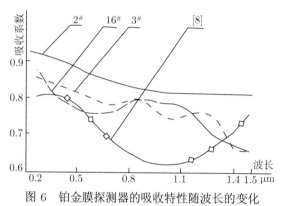

图 6　铂金膜探测器的吸收特性随波长的变化

四、　辐射热流测量

图 7，是辐射热流测量原理。详细的推演见文献 [11]。当铂金膜受辐射热流时有一个温升

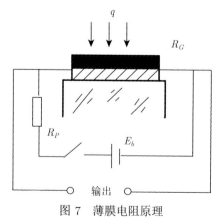

图 7　薄膜电阻原理

$$\Delta T = \frac{1}{\sqrt{\pi K \rho C}} \int_0^t \frac{q(\lambda)}{(t-\lambda)^{1/2}} d\lambda \tag{5}$$

上式中 K、ρ、C 是衬底的导热系数、密度、比热。或者说

$$\alpha \Delta T = \frac{1}{2 \left(\frac{\sqrt{\pi K \rho C}}{2\alpha} \right)} \int_0^t \frac{q(\lambda)}{(t-\lambda)^{1/2}} d\lambda \tag{6}$$

上式中 α 是电阻温度系数，上式说明辐射热流 $q(t)$ 与膜电阻的变化 $\alpha \Delta T(t)$ 建立了一一的对应关系；如果对铂膜通以恒电流，则铂膜电阻变化引起的电压

$$e = \frac{R_P \cdot R_G}{R_P + R_G} I_G \alpha \Delta T = \frac{1}{2K} \int_0^t \frac{q(\lambda)}{(t-\lambda)^{1/2}} d\lambda \tag{7}$$

上式中

$$K = \frac{\sqrt{\pi K \rho C}}{2\alpha} \frac{R_P + R_G}{I_G R_P R_G}, I_G = \frac{E_t}{R_P + R_G}.$$

由式（7）可知输入辐射热流 $q(t)$ 与输出电压是一一对应的。有了输出电压 $e(t)$，就能用不同的方法确定 $q(t)$ 值。实际的线路见图 8，对于小 θ 角情况还必须解决 100 ～ 200μV 的小讯号接收问题。本文除了使用高灵敏度示波器外，还采取了相应的隔离与屏蔽措施以防止外界的干扰，从而获得了几百微伏的讯号接收。

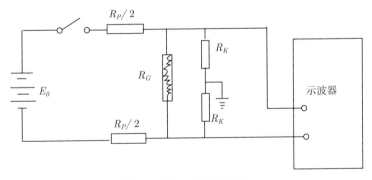

图 8 辐射热流测量路线

五、 试验情况与数据处理

高压段用 H_2、O_2、He 混合驱动，He 气比例为 70%～ 75%，低压段充空气 $P_1 = 0.1 \sim 12\text{mmHg}$，获得气体的温度、压力和最大模拟速度为：$T = 5900 \sim 8000°\text{K}$，$P = 0.74 \sim 20\text{atm}$，$V_{\max} = 4.96 \sim 8.65\text{km/sec}$。模型尺寸 $\phi = 50\text{mm}$ 窗口是 $\phi = 10\text{mm}$ 的石英玻璃，探测器离窗口的距离是：

$$h = 22.8\text{mm}, \quad \theta = 12.4°, \quad F = 0.0233(\text{驻点模型})$$
$$h = 18\text{mm}, \quad \quad \theta = 15.5°, \quad F = 0.0365(\text{端盖模型})$$

图 9　石英窗口光谱透射率

石英窗口的透射率 $B_W \approx 0.85$，见图 9，探测器的吸收特性 $\alpha_G = 0.8$，测定的铂金膜电阻温度系数 $\alpha = 2.78 \times 10^{-3}/°\text{K}$，由文献 [12] 测定 GG–17 衬底玻璃的

$$\sqrt{\pi K \rho C} = 0.274\text{J}/\text{CM}^{2°}\text{K}(\text{sec})^{1/2}, R_G = 50\Omega, R_P = 200\Omega, E_b = 11\text{V},$$

对驻点模型，脱体距离 δ 由有关文献或图 10 得

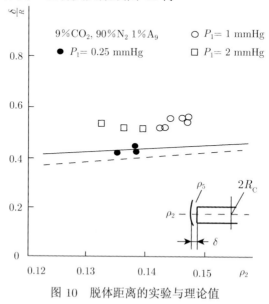

图 10　脱体距离的实验与理论值

$$\delta \approx 0.45R$$

对端盖试验情况

$$\delta = W_R \cdot t$$

W_{R-} 反射激波速度，$t-$ 激波反射后的工作时间。曲型的讯号输出波型见图 11，12。由输出讯号 $e(t)$，通过数值解可以获得 q 值。q 值的取得也可以通过模拟线路 RC 网络等办法获得。

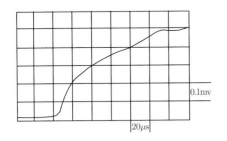

0.1mv

20μs

No. 188 P_1=5mmHg V_S=3.81km/sec

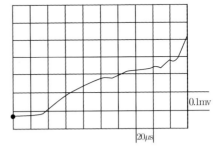

0.1mv

20μs

No. 199 P_1= 1mmHg V_S=4.95km/sec

图 11　驻点试验示波图

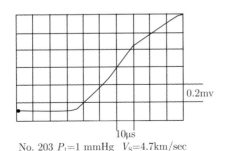

0.2mv

10μs

No. 203 P_1=1 mmHg V_S=4.7km/sec

图 12　端盖试验示波图

六、　试验结果与进一步工作

本文试验获得了驻点模型与反射激波条件下总辐射强度 E 值的测定。典型的驻点与端盖试验结果见下表：

试验结果与国外实验和理论数据作了比较，见图 13，图 13 中曲线是一经验曲线，试验证明这一曲线是有价值的。各种状态下的 E 值可以利用这一曲线，并可估算高速物体再入时的辐射传热。

试验结果说明探测器的输出压只取决于 q 值和 R_G 值而与探测器尺寸无关，因此探测器可以制成 $\phi 5mm$ 尺寸以提高其测量精度。

端盖试验结果

No.	$P_1/(mmHg)$	$V_S/(km/sec)$	ρ_5/ρ_0	$T_5/(°K)$	$V_{max}/(km/sec)$	$E/(W/cm^3)$	$E/(\rho_5/\rho_0)^{1.7}$
77—115	6	3.84	0.505	6670	5.58	170	5.4×10^2
77—116	1	5.08	0.115	7420	7.32	70	2.8×10^3
77—119	0.1	6.17	0.0168	7950	8.65	30.2	3×10^4
77—121	12	3.4	0.82	5920	4.96	10.9	1.53×10^2

驻点试验结果

No.	$P_1/(mmHg)$	$V_S/(km/sec)$	ρ_5/ρ_0	$T_5/(°K)$	$V_{max}/(km/sec)$	$E/(W/cm^3)$	$E/(\rho_5/\rho_0)^{1.7}$
78—199	1	4.95	0.105	6960	6.7	75	3.34×10^3
78—188	5	3.81	0.355	6030	5.1	128	7.1×10^2

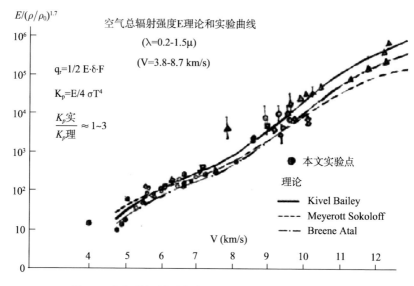

图 13　空气总辐射强度实验曲线及与理论值的比较

本文 E 值测定中需要测量 α、K、ρ、C、α_G、B_W 等几个参数，因此就很难提高辐射测量精度。最直接的办法是在激波管试验之前，用标准脉冲氙灯对每个探测器作标定与鉴定，这是下一步的工作方向。

本文试验结果还是初步的，对于数据的理论分析，气体温度更高时的测量与试验技术的改进尚有待一步的努力。

参 考 文 献

[1] J. Keck, B. Kivel and T. Wentink, Emissivity of High Temperature Air. Annals of

Physics.

[2] R. M. Nerem and G. H. Stikford, Shock-Tube Studies of Equilibrium Air Radiation. AIAA J. 1967. No 3. p 517.

[3] H. Hoshizaki, Equilibrium Total Radiation Measurements in Air at Superorbital Entry Velocities. Third Hypervelocity Techniques Symposium 1964, p 245.

[4] J. E. Nealy and K. V. Haggard, A Shock-Tube Study of Radiation Behind Shock Waves in CO_2. Proceedings of the Ninth International Shock Tube Symposium. 1973. p 330.

[5] H. Hoshizaki, A. D. Wood, J. C. Andrems, Radiant Energy Transfer Measurements in Air. NASA-CR-585, 1966.

[6] Bradford H. Wick: Radiative Heating of Vehicles Earths Atmosphere. The High Temperature Aspects of Hypersonic Flow. AGARD-68.

[7] Kennet. H and Strak. S. L, Stagnation Point Radiation Transfer ARS. J31. 1961. p 370.

[8] G. M. Thomas, W. A. Menard: Experimental Measurements of Nonequilibrium and Equilibrium Radiation from Planetary Atmosphere, AIAA. J. 1966. No 2. p 227.

[9] J. C. Keck: Radiation from Hot Air. Annals of Physics. Vol.7 1959. p 1-38.

[10] 竺乃宜等, 薄膜辐射热探测器, 力学学报, 1978 年第 3 期.

[11] D. L. Schuttz and T. V. Jone, Heat-Transfer Measurements in Short-Duration Hypersonic Facilities. AGARD-168.

[12] 王曾亚, 薄膜电阻温度计衬底材料 $(K\rho C)^{1/2}$ 的测定。中国人民解放军 89955 部队, 1978 年.

Studies of nonequilibrium effects in nozzle flow under high reservoir conditions

Le Jialing and Dong Weizhong

China Aerodynamics Research and Development Center, Mianyang, Sichuan, China

Abstract. A set of equations is presented, governing quasi-one-dimensional chemical, vibrational and ionization nonequilibrium nozzle flow in high reservoir conditions, with v-d coupling and electron excitation effects. In order to remove the problem of the small space marching step size under high reservoir pressure, the method of chemically frozen, thermal equilibrium flow with an equilibrium species correction is introduced. Numerical solutions were obtained for an arc-jet wind-tunnel and a shock wind-tunnel. The influence of different reservoir pressures and of various chemical models, v-d coupling effects and electron excitation effects is studied.

Key words: Nozzle flow, Nonequilibrium flow, High-enthalpy flow

1. Introduction

With the development of flight vehicles, the stagnation enthalpy encountered in hypersonic flight results in flowfield temperatures becoming high enough to dissociate chemical species in air. In China, new hypersonic testing facilities are being designed and built in order to simulate nonequilibrium phenomena such as dissociation, ionization, vibrational excitation and electron excitation of the air around flight vehicles, and to validate the methods of calculation for nonequilibrium flow. In addition to high stagnation enthalpy, extreme reservoir pressures are needed to simulate hypersonic flow. Also, in the nozzle expansion, the test gas remains in nonequilibrium. Presently, the composition of the air in the test section is still uncertain because of the nonequilibrium effects in the nozzle flow, so it is quite difficult to explain the data and the results of aerodynamic force and heat-transfer measurements obtained in the tests, and to know the properties of test gas. Therefore, it is very important to study nonequilibrium effects for nozzle flows under high reservoir conditions.

Table 1. Nozzle data

Nozzle	L_1 (cm)	L_2 (cm)	A^* (cm^2)	θ_1 (°)	θ_2 (°)	D_e (cm)	T_0 (K)	P_0 (MPa)
#1	3.5	21.5	1.0	42	11.3	8.6	4800	5
#2	3.5	296.5	1.0	42	11.3	120	8000	100, 200

2. Numerical method

2.1. Governing equations

For a steady quasi-one-dimensional flow, the nondimensional governing equations for chemical, vibrational and ionization nonequilibrium flow with v-d coupling and electron excitation are as follows:

$$\frac{d\nu_j}{dx} = (\rho v)^{-1} S_{cj}, \qquad (j = 1, \ldots s, s+1) \tag{1}$$

$$\frac{de_{vj}}{dx} = v^{-1} S_{vj} \tag{2}$$

Le Jialing, Dong Weizhong. Studies of nonequilibrium effects in nozzle flow under high reservoir conditions. Proceedings of the 19th International symposium on shock waves. Marseille, France, 1993: 307-312.

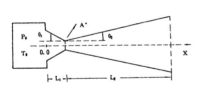

Fig. 1. Nozzle configuration

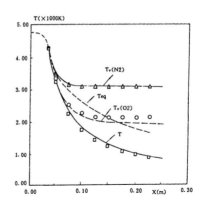

Fig. 2. Temperatures; $P_0 = 5$ MPa, $T_0 = 4800$ K. Moss model; points are data of Sagnier (1991)

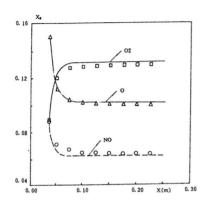

Fig. 3. Species molar fraction;
$P_0 = 200$ MPa, $T_0 = 8000$ K. Moss model;
points are data of Sagnier (1991)

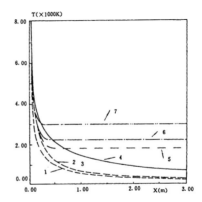

Fig. 4. Temperatures; $P_0 = 200$ MPa, $T_0 = 8000$ K. Dunn-Kang model; lines 1 to 6 plot respectively the values of $T_{f.eq}, T, T_e, T_{eq}, T_v(O_2), T_v(NO), T_v(N_2)$

$$\frac{dT_e}{dx} = -\frac{2T_e}{3v}\frac{dv}{dx} + \frac{2}{3\rho v \nu_e}(Q_{el} + Q_{in}) \tag{3}$$

$$\frac{dv}{dx} = \frac{1}{1 - M_f^2}\frac{D_v}{D} \tag{4}$$

$$\frac{dp}{dx} = -\rho v \frac{dv}{dx} \tag{5}$$

$$\frac{dT}{dx} = \left[\left(\frac{p}{v} - \rho v\right)\frac{dv}{dx} + \frac{p}{A}\frac{dA}{dx} - \rho \nu_e \frac{dT_e}{dx} - \frac{1}{v}\left(T\sum_{j=1}^{s} S_{cj} + T_e S_{ce} \right) \right] \bigg/ \left(\rho \sum_{j=1}^{s} \nu_j \right) \tag{6}$$

where

$$M_f = V/a_f \tag{7}$$

$$a_f^2 = \frac{\sum\limits_{j=1}^{s}\left[\frac{5+2(n_j-1)}{2} + C_{ej}\right]\nu_j}{\sum\limits_{j=1}^{s}\left[\frac{3+2(n_j-1)}{2} + C_{ej}\right]\nu_j} T \sum_{j=1}^{s}\nu_j + \frac{5}{3}T_e\nu_e \tag{8}$$

$$D_v = \frac{1}{\rho} \left\{ \sum_{j=1}^{s} \left[\frac{5 + 2(n_j - 1)}{2} + C_{ej} \right] \nu_j \right\} \left(T \sum_{j=1}^{s} S_{cj} + T_e S_{ce} - \frac{pv}{A} \frac{dA}{dx} \right)$$

$$- \frac{1}{\rho} \left(\sum_{j=1}^{s} \nu_j \right) \left\{ T \sum_{j=1}^{s} \left[\frac{5 + 2(n_j - 1)}{2} S_{cj} \right] + \frac{5}{2} T_e S_{ce} + \sum_{j=1}^{s} [(n_j - 1)(e_{vj} S_{cj} + \rho \nu_j S_{vj}) \right.$$

$$\left. + S_{cj} \left(\Delta h_j^0 + e_j \right)] \right\} + \frac{2}{3\rho} \left[\sum_{j=1}^{s} (n_j - 1 + C_{ej}) \nu_j \right] (Q_{el} + Q_{in}) \qquad (9)$$

$$D = a_f^2 \left\{ \sum_{j=1}^{s} \left[\frac{3 + 2(n_j - 1)}{2} + C_{ej} \right] \nu_j \right\} \quad \text{and} \quad p = \rho \left(T \sum_{j=1}^{s} \nu_j + T_e \nu_e \right) \qquad (10)$$

where ν_j is the molar number of species j per unit mass, e_j the excitation energy of the electron of species j and $C_{ej} = \partial e_j / \partial T$ (Clarke et al. 1964), S_{cj} the chemical source term, S_{vj} the vibrational source term, Q_{el} and Q_{in} the production terms of free electron energy due to elastic and inelastic collisions of free electrons with heavy particles (Matsuzaki 1988). The solution of Eqs.1-6 gives $P, T, T_e, V, T_{vj}(e_{vj})$ and ν_j. The results are obtained by using the standard Runge-Kutta-Gill method.

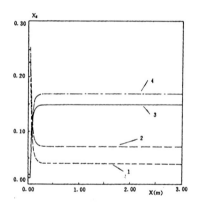

Fig. 5. Species molar fraction, Dunn-Kang model. Lines 2 and 3: x_O and x_{O_2} at $P_0 = 100$ MPa; lines 1 and 4: at 200 MPa

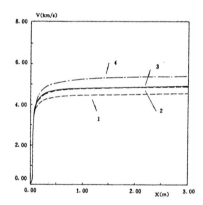

Fig. 6. Velocity; lines 1, 2 and 4: $V_{f.eq}, V$ and V_{eq} at $P_0 = 200$ MPa; line 3: V at 100 MPa

2.2. Chemical and $v - d$ coupling model

In this study, the high-temperature air is assumed to consist of the following 11 species: O_2, N_2, NO, O, N, O_2^+, N_2^+, NO^+, O^+, N^+ and e. The rate processes considered for the above species are taken from the Moss, Park, Gardiner and Dunn-Kang model (Sagnier et al.1991). The real reaction constants (with the v-d coupling nonpreferential model) are given by:

$$k_{fi} = k_{fi}^{\infty} \prod_{j=1}^{s} V_j^{A_{ij}}, \qquad k_{bi} = k_{fi}^{\infty}/K_i \quad \text{and} \quad k_{fi}^{\infty} = a_i T^{bi} \exp\left(-c_i/T\right) \qquad (11)$$

where V_j is the v-d coupling factor, A_{ij}=0 or 1 according to whether or not the i-th reaction is affected by the coupling process. K_i is the equilibrium constant of the i-th reaction, which is calculated by the partition function of species (Vincenti 1965).

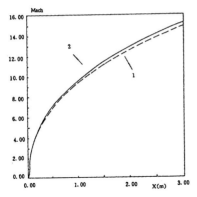

Fig. 7. Mach number; line 1, $P_0 = 200$ MPa; line 2, $P_0 = 100$ MPa

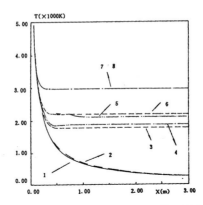

Fig. 8. Model effect on temperatures; $P_0 = 200$ MPa. Lines 1, 4, 5, 7: $T, T_v(O_2), T_v(NO), T_v(N_2)$, Park; Lines 2, 3, 6, 8: $T, T_v(O_2), T_v(NO), T_v(N_2)$, Dunn-Kang

The nonpreferential model is employed for the v-d coupling effect (Sagnier 1991):

$$S_{vj} = \frac{e_{vj}^{\infty} - e_{vj}}{\tau_j} - \left(\frac{d\nu_j}{dt}\right)_f \frac{E_j(T, T_{vj}) - e_{vj}}{\nu_j} + \left(\frac{d\nu_j}{dt}\right)_b \frac{G_j(T) - e_{vj}}{\nu_j} \tag{12}$$

where $\left(\frac{d\nu_j}{dt}\right)_f$ and $\left(\frac{d\nu_j}{dt}\right)_b$ are production and depletion terms, $E_j(T, T_{vj})$ and $G_j(T)$ are the average vibrational energies of, respectively, a dissociating molecule j and of a molecule j immediately after recombination.

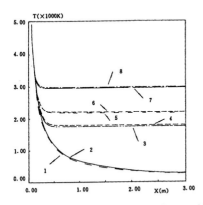

Fig. 9. Model effect on temperatures; $P_0 = 200$ MPa. Lines 1, 3, 5, 7: $T, T_v(O_2), T_v(NO), T_v(N_2)$, Gardiner Lines 2, 4, 6, 8: $T, T_v(O_2), T_v(NO), T_v(N_2)$, Dunn-Kang

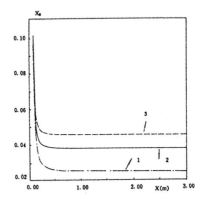

Fig. 10. Chemical model effect on molar fraction of oxygen atom; $P_0 = 200$ MPa. Lines 1, 2, 3: Park, Dunn-Kang, Gardiner models

2.3. The method of chemically-frozen - thermal equilibrium flow with an equilibrium species correction

Under high reservoir pressure, from the reservoir up to the throat, the expansion is weak enough for the flow to stay in a near-equilibrium state (Sagnier et al. 1991), and a small space marching

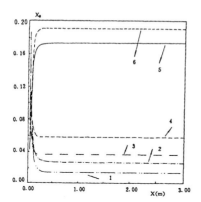

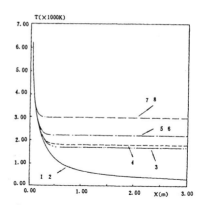

Fig. 11. Electronic excitation effect on molar fractions; $P_0 = 200$ MPa, Park model. Lines 1,3,6: x_{NO}, x_O, x_{O_2}, without electron excitation; lines 2, 4, 5: with electron excitation

Fig. 12. $v - d$ coupling effect on temperatures; $P_0 = 200$ MPa, Dunn-Kang model. Lines 1, 3, 5, 7: $T, T_v(O_2), T_v(NO), T_v(N_2)$, without coupling; lines 2, 4, 6, 8: with coupling

step size must be taken. There are two methods of calculation for equilibrium flow, the first is an empirical formula method, however possibly inducing large errors for high reservoir conditions. The second is to solve the equations directly according to the properties of thermochemistry, however, practically it is not possible to find analytic expressions for the derivatives of the unknown ν_j (Anderson 1989). In the present study, first we solved the governing equations in the assumption of chemicallly frozen-thermal equilibrium flow, then chemical species were solved alone according to the local values of $p(x)$ and $T(x)$ by the equilibrium method (Vincenti et al. 1965).

3. Numerical results

3.1. The arc-jet wind-tunnel case
In order to verify the numerical methods, first the nozzle flow in the arc-jet wind-tunnel was computed (nozzle #1, see Table 1 and Fig.1). Figs.2 and 3 show temperatures and molar fractions of species. Results compare pretty well with those of Sagnier et al. (1991). The nonequilibrium temperatures (T, T_{vj}) deviate quickly from the equilibrium temperature T_{eq}. At the nozzle exit, there are great differences between T and T_{eq}, or T and T_{vj} (shown in Fig.2). The freezing of O_2, NO and O is observed at about 7 cm (Fig.3).

3.2. The shock wind-tunnel case
After the code was carefully checked, we have made an investigation for a high-enthalpy, high reservoir pressure nozzle flow (nozzle #2, see Table 1 and Fig.1). The diameter of the exit is about 120 cm. Equilibrium values were calculated by using the data of Srinivasan et al. (1987). Hereafter some of the results will be given. Nonequilibrium computation was started with the initialization of our method 1 mm downstream of throat.

3.2.1. Influence of different reservoir pressures
Fig.4 shows the temperatures along the nozzle ($P_0 = 200$ MPa, $T_0 = 8000$ K). $T_{f.eq}$ is the value for chemicallly-frozen - thermal equilibrium flow with an equilibrium species correction. Downstream of the throat, nonequilibrium temperatures (T, T_e, T_{vj}) are, for a certain distance, close to $T_{f.eq}$, which means that the present method of calculation for the region upstream of the throat is reliable and successful. With an increase in P_0 from 100 MPa to 200 MPa, T and T_e increase, but the vibrational temperature T_{vj} decreases. The molar fractions of O_2 and O change from 0.1462

and 0.0685 to 0.1659 and 0.0374 respectively at the exit (as shown in Figs.4 and 5). It means that the high reservoir presure can slightly suppress the nonequilibrium of the nozzle flow. Figs. 6 and 7 show the velocity and frozen Mach number along the nozzle. The velocity keeps at first equilibrium values, then detaches from them, and finally lies between the equilibrium value and $V_{f.eq}$. At the exit, the frozen Mach number is about 15.

3.2.2. Influence of different chemical models

Figs.8-10 show the temperature and molar fractions of the species. The model that is closest to equilibrium is Park's, the farthest one is Gardiner's. The molar fraction of atomic oxygen can vary from 0.025 to 0.045, depending on the chemical models, at the exit. The effect on $T_v(O_2)$ and $T_v(NO)$ is more important than on $T_v(N_2)$ because the O_2 and NO chemistry is more active. The situation is reversed for the vibrational temperatures because the chemical energy storage is high and so there is less energy available for vibration, and the lower translational temperature for the Landau-Teller law imply lower frozen vibrational temperatures.

3.2.3. Influence of electron excitation

It appears that the influence is weak on temperatures, but important on the molar fractions of O, O_2 and NO (shown in Fig.11). The results show that the recombination rate decreases because of electron excitation.

3.2.4. Influence of the $v - d$ coupling effect

$v - d$ coupling effect has a certain influence on $T_v(O_2)$ because O_2 chemistry is more active (shown in Fig.12).

4. Conclusions

The present numerical method of calculation for nozzle flows in high reservoir conditions is successful. In order to determine accurately the gas compositions at the nozzle exit, consideration of nonequilibrium effects and selection of the thermochemistry model are important. The high reservoir pressure can slightly reduce the nonequilibrium of the nozzle flow. If the reservoir pressure is lower, or the reservoir temperature is higher, $v - d$ coupling effect would have a stronger influence on flow properties.

References

Anderson JD (1989) Hypersonic and high temperature gas dynamics. McGraw-Hill

Clarke JF, McChesney M (1964) The dynamics of real gases. Butterworths, London

Mastsuzaki R (1988) Quasi-one-dimensional aerodynamics with chemical, vibrational and thermal nonequilibrium. Trans. Japan Soc. Aero. Space Sci. 243-258

Sagnier P, Marraffa L (1991) Parametric study of thermal and chemical nonequilibrium nozzle flow. AIAA J. 334-343

Srinivasan S, Tannehill JC, Weilmuenster KJ (1987) Simplified curve fits for the thermodynamic properties of equilibrium air. NASA-RP-1181

Vincenti WG, Kruger CH (1965) Introduction to physical gas dynamics. John Wiley

$2H_2+O_2$ 爆轰波后壁面热交换测量

乐嘉陵　　杜锡鑫

(中国气动力研究与发展中心)

摘要：实验测量了 $2H_2+O_2$ 爆轰波后壁面热交换率。所用传感器为镀有厚约 2000Å 硅层的铂金薄膜电阻温度计。实验结果与数种理论计算做了比较。实验曲线的两头与层流附面层理论相吻合；但从 30 微秒到 150 微秒内，实验数据却与湍流附面层理论相一致。值得指出的是，2 大气压下 $2H_2+O_2$ 气体爆轰波后诱导引起的附面层可能经历着从层流向湍流的转捩，而后成为湍流附面层。

一、实 验 概 述

本实脸中所用的爆轰管是内径为 80 毫米，长为 5 米的不锈钢管。沿管壁装有测压和测温传感器，传感器端面安装与管内壁并齐，用来测量爆轰波速度和热交换率。爆轰管的一端装有电火花塞，用来点燃 $2H_2+O_2$ 混合气体。电火花塞放电能量在 2 至 100 焦耳范围内可以调节，但在大多数实验中放电能量为 100 焦耳。

爆轰波的传播速度用几个陶瓷压电晶体传感器测量，它们的讯号送进一台 TCH-1000 型的记录仪记录。在距离点火位置分别为 2.02 米和 3.74 米二个测量位置间所测得的平均爆轰波速度为 2.93 公里/秒，比 C-J 爆轰理论值 2.84 公里/秒稍大一点。

用薄膜温度计测量了离点火位置 2.2 米测点上的壁面热交换率。薄膜计是用真空溅射技术制成的，膜厚约为 700Å，通常镀有厚约 2000Å 的硅层，以防止电离效应和机械损坏。薄膜计的衬底为 GG-17 玻璃，它的热特性，用脉冲放电法测得为 0.036-0.037 卡/（厘米 ^{2}K 秒 $^{1/2}$）。薄膜计的频响时间为 1~2 微秒[1]。由于爆轰波扫过而引起的温度升高，产生了相应的电阻增量，使薄膜计因而输出一个电压讯号(因为膜上事前通有一个恒定的电流)，由 TCH-1000 记录仪记录，所用的时间分辨率为 0.5 微秒。

乐嘉陵, 杜锡鑫. $2H_2+O_2$ 爆轰波后壁面热交换测量. 空气动力学学报, 1984, 3: 71-74.

二、热交换数据处理

薄膜计的热交换率可以用半无限体热传导理论计算[2]：

$$q_w = \sqrt{\frac{\rho k c}{\pi}} \frac{1}{\alpha E_f} \left[\frac{E(t)}{\sqrt{t}} + \frac{1}{2} \int_0^t \frac{E(t) - E(\tau)}{(t-\tau)^{1/2}} d\tau \right]$$

其中 q_w——热交换率；$\rho k c$——薄膜计的电阻温度系数；E_f——作用在薄膜计两端的初始电压；$E(t)$——薄膜计在时刻 t 的输出电压。

如果 $E(t)$ 在 Δt 间隔内能用一直线段来代替，方程(1)则可以改为[2]：

$$q_w(t) = \sqrt{\frac{\rho k c}{\pi}} \frac{1}{\alpha E_f} \left\{ \frac{E(t_n)}{\sqrt{t_n}} + \sum_{i=1}^{n-1} \left[\frac{E(t_n) - E(t_i)}{(t_n - t_i)^{1/2}} - \frac{E(t) - E(t_{i-1})}{(t_n - t_{i-1})^{1/2}} \right. \right.$$

$$\left. \left. + 2 \frac{E(t_i) - E(t_{i-1})}{(t_n - t_i)^{1/2} + (t_n - t_{i-1})^{1/2}} \right] + \frac{E(t_n) - E(t_{n-1})}{\Delta t^{1/2}} \right\}$$

因为 $i=0$，$E(t)=0$，实际上方程（2）还可以进一步简化为计算起来省机时的简单式：

$$q_w(t_n) = 2 \sqrt{\frac{\rho k c}{\pi \Delta t}} \frac{1}{\alpha E_f} \left[E(t_n) + \sum_{i=1}^{n-1} E(t_i) \left(\sqrt{n-i+1} - 2\sqrt{n-i} + \sqrt{n-i-1} \right) \right]$$

因为薄膜计表面经受着十分严重的热流率(高达 5~10 千瓦/（厘米²）)，在很短的时间内温升可达 400℃，因此衬底材料的热特性可能经受很大的变化，给数据处理带来困难和造成误差。由于问题复杂，这一误差的大小尚难估计。

三、结 果 与 讨 论

测得的温升曲线示于图 1。根据该温升曲线求得的壁面热交换率则示于图 2 和图 3。同示于图 2 和图 3 的还有根据文献[3]和[4]的理论曲线，以作比较。可以看出，当 20 微秒>t>150 微秒时，实验结果与文献[3]的理论值相吻合。而当 20 微秒<t<150 微秒时，实验结果却与文献[4]的理论值相一致。文献[3]是层流附面层理论。它比较全面地考虑了爆轰波后自由流的速度、压力和密度的快速衰减效应，而当自由流速度变得很小时，附面层方程简化为热传导方程。文献[4]是湍流附面层理论。但是在该理论中 Sichel 等作了一个比较粗糙的假设：在爆轰

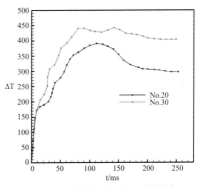

图 1　薄膜温度计温升的随
时间（微秒）变化

波后的压力、温度和速度保持不变并等于 C-J 值。此外，Sichel 等还应用了壁面热交换 q_w 与剪应力 τ_w 之间的雷诺比拟。通过以上分析，不难看出实验曲线的变化是由于：(1)当 t<20 微秒时，爆轰波后的附面层为层流型的，所以与文献[3]的层流理论相一致。但随着自由流雷诺数的增加，附面层有可能产生转捩，由层流变为紊流。所以实验值又趋向于 Sichel 的湍流理论值。随着时间的消逝，测点处气流速度越来越小而渐渐趋于静止，因而强迫对流热交换的作用，为传导热交换所代替，所以实验值又向文献[3]的理论值靠拢。在图 2 中，同时也示出了自由流雷诺数随时间的变化。可以看出当 t=20 微秒时，Re=10^6，这已进入了转捩雷诺数的范围。文献[5]和[6]的激波管实验研究表明，当激波马赫数大于 2 时，转捩雷诺数的范围为。从图 1 可以看出，对应着爆轰波扫过后的 25 微秒时刻，确实存在着一个明显的温度跃升，它可能标明了转捩的开始，指示出了一个热交换率的巨大变化。这个有趣的现象，将在今后的实验中进一步加以探讨。

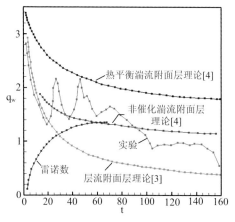

图 2　热交换率(千卡/(厘米)^{2}秒)随时间 t（微秒）的变化以及实验结果与理论的比较

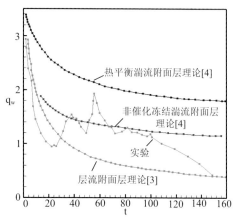

图 3　热交换率(千卡/(厘米)^2·秒)随时间 t（微秒）的变化以及实验结果与理论的比较

四、结　束　语

通过本实验结果与数种理论计算的比较，可以看出，Sichel 等的冻结湍流理论与实验结果相当一致，从而证明了该理论的可用性。

但是，值得指出的另一点是，正如本文实验结果文献[3]和[7]所指出的那样，在紧挨爆轰波的后面，可能存在着层流附面层，以及随后的转捩。这似乎与爆轰波的胞格结构理论相矛盾。根据文献[8]White 的理论，爆轰波后 的流动是湍流结构。说明对爆轰波机理的认识有待于进一步深化。

参 考 文 献

[1] Hanson,R.K. 8th International Shock Tube Symposium

[2] Cook, W.J. and Felderman. E. J., AIAA J., Vol.4. No.3, 1966,pp.561.

[3] Du, X. X. Liu, W. S. and Glass. I.I., TIAS Report No.259.June.1982.

[4] Sichel, M. And David, T. S. AIAA J. Vol. 4, No. 6, 1966.pp. 1089.

[5] Hartunian, R. A., J. Aerospace Sci. Vol. 27, No.8, 1960, pp. 587.

[6] Futhrer. R. G., 7th International shock Tubes Symposium.

[7] Paillard, C. et al, Int. Gasdynamics of Detonations and Explosions (ed. J.R. Bowen), 1981 pp. 134-140. New York, AIAA.

[8] White, D.R., The Physics of Fluids, 1961, pp.165.

HEAT TRANSFER MEASUREMENT ON THE WALL
BEHIND DETONATION WAVES

Le jialing, Du xixin

China Aerodynamic Research & Development Center

Abstract

Heat transfer rates were measured on the wall behind detonation waves. The transducers used were the platinum thin film resistance gauges, painted with a thin layer of silicon about 2000Å thickness.

The experimental data were compared with several analytical results. Good agreement is found in the comparisons during early and later times between the experimental data and the laminar boundary layer theory. From 30μs to 150μs. however. the experimental results agree well with the turbulent boundry layer analytical values. It shows that the boundary layer induced   behind a 2 atm $2H_2+O_2$ mixture detonation wave may undergo a transition from laminar to turbulent and then turn out a turbulent boundary layer.

驻点壁面催化速率常数确定的研究

董维中, 乐嘉陵, 刘伟雄

（中国空气动力研究与发展中心，四川　绵阳 621000）

摘要：以平衡流动作为热环境估算的依据，提出了用数值求解非平衡 Navier-Stokes 方程和实验测量热流值确定模型表面材料催化速率常数的方法。用 5 组分 17 个化学反应 Dunn-Kang 空气化学模型和轴对称热化学非平衡 Navier-Stokes 方程，对激波管中球头和平头圆柱模型绕流流场进行了数值模拟，给出了驻点热流随催化速率常数变化的分布，并根据激波管实验测量的热流值确定了表面材料 Pt、SiO_2、Ni 和某种飞船材料的催化速率常数，建立了数值分析高焓流动边界层催化特性的软件。

关　键　词：催化速率常数；热化学非平衡；激波管
中图分类号：TQ032.3；V211.22　　　**文献标识码**：A

The determination of catalytic rate constant of surface materials of testing model in the shock tube

DONG Wei-zhong, LE Jia-ling, LIU Wei-xiong

（China Aerodynamics Research and Development Center，Mianyang 62100，China）

Abstract：It is introduced that the catalytic rate constant of surface materials of testing model is determined by solving non-equilibrium Navier-Stokes equations and measuring heat transfer rate of the stagnation point of testing model. The flow-field around a hemisphere and a flat-nosed cylinder in the shock tube is simulated numerically by using non-equilibrium Navier-Stokes equations with an air chemical model of 5 species 17 reactions, and the catalytic rate constant of surface materials such as Pt, SiO_2, Ni and a flight vehicle materials is determined by using the heat transfer rate of testing in the shock tube and the distribution of heat transfer rate along with the catalytic rate constants.

Key words：catalytic rate constant；thermal and chemical non-equilibrium；shock tube

0　引　言

高超声速非平衡绕流的物面热环境是十分重要的，其中材料表面催化速率常数则是

董维中, 乐嘉陵, 刘伟雄. 驻点壁面催化速率常数确定的研究. 流体力学实验与测量, 2000, 14(3): 1-6.

一个最基本的影响参数,也是当前国内外非平衡流动研究中的一个突出的问题。由于没有掌握这一方面确切的数据,因而在热环境的估算上,常常以平衡流动作为设计依据,从而使物面热流和防热结构重量大大增加。这是因为在平衡条件下原子瞬时复合发生在整个边界层流场中而不是发生在壁面处,其结果是原子释放的化学能必然使边界层流场中焓值增加从而导致壁面热流的增加,也就是说在这种情况下壁面热流并不受壁面材料催化特性的影响。在非平衡流动条件下原子的复合发生在壁面,其复合程度或者说壁面不同程度的热将明显受到不同壁面材料催化特性的影响。由于壁面材料对原子复合过程的机理十分复杂,通常都采用宏观的办法,通过壁面的热流测量间接确定壁面材料的催化特性。在工程试验方面,由于壁面温度要有一定范围的变动,因而多数是(美国[1]、德国[2]、俄罗斯)采用电弧风洞进行驻点和平板的试验,并通过材料表面测温和测热流的方法来获得表面催化率特性,对这样的试验,要求电弧风洞气流的品质十分干净、非平衡来流试验气体的状态要十分清楚(这两方面的技术要求是十分严格的)。另一方面,激波管是十分经济和理想的高焓流动实验装置,气流参数比较确切也很少受污染,用脉冲加热提高壁温的技术也正在发展之中,因而是一项有前景的试验技术,各个国家也都在进行研究和开发[3,4]。为了使壁面催化率特性对热流有最大的灵敏度,试验的状态都控制在非平衡流动中的冻结边界层状态。但存在的问题是,安置在激波管中入射激波后、模型前的流动气体已经处于离解状态,模型的绕流将十分复杂,与物体再入大气时的流动有很大不同,在这种情况下非平衡流动的冻结状态将很难用定性的方法来确定[3]。笔者通过非均匀来流高焓非平衡钝体绕流的层流 N-S 方程的数值方法,较严格地确定了划分非平衡流动的 Damkohler 数,从而确定了绕钝体模型驻点边界层非平衡冻结流动状态,并通过驻点热流的实验测量和数值计算结果初步确定了材料催化速率常数。

1 数值方法

在计算坐标系下,无量纲化的热化学非平衡的轴对称或二维 Navier-Stokes 控制方程为[2]:

$$\frac{\partial Q}{\partial \tau} + \frac{\partial F}{\partial \xi} + \frac{\partial G}{\partial \eta} + \delta \cdot H = \frac{1}{Re}\left(\frac{\partial F_v}{\partial \xi} + \frac{\partial G_v}{\partial \eta} + \delta \cdot H_v\right) + W \tag{1}$$

$$Q = J(\rho_i, \rho E_v, \rho, \rho u, \rho v, \rho E)^T \quad (i = 1, 2, \cdots, N_s - 1)$$

$\delta = 0$ 或 1 分别是二维或轴对称情况。化学模型为 5 组分 17 个化学反应的 Dunn-Kang 空气化学模型。化学反应速率常数和热化学非平衡源项可参见文献[5]。方程(1)用全隐式 TVD 格式差分[5]。

2 壁面催化条件

对于 5 组分空气化学模型,一般有如下假设:NO 组分是非催化,而且在壁面主要是原子组分发生复合反应。因此壁面催化条件有如下形式[6,7]:

$$\frac{\partial C_O}{\partial n} = K_O C_O, \qquad \frac{\partial C_{O_2}}{\partial n} = -K_O C_O \tag{2a}$$

$$\frac{\partial C_N}{\partial n} = K_N C_N, \qquad \frac{\partial C_{N_2}}{\partial n} = - K_N C_N \tag{2b}$$

$$\frac{\partial C_O}{\partial n} = 0 \tag{2c}$$

这里 $K_{O,N} = K_W/D_{O,N}$，$D_{O,N}$ 是原子 O 和 N 的扩散系数。其中假设：对于原子 O 和 N，催化速率常数 K_W 是一样的。对于完全非催化情况，$K_W = 0$；对于完全催化情况，原子组分的质量分数等于零。

3 热流计算公式

在非平衡流动中，壁面热流由热传导（含输运项和振动项）和组分扩散两部分组成：

$$q_w = k_{tr} \frac{\partial T}{\partial n} + k_v \frac{\partial T_v}{\partial n} + \sum_{i=1}^{N_S} \rho D_i h_i \frac{\partial C_i}{\partial n} \tag{3}$$

这里 T_v 和 k_v 是振动温度和振动热传导系数，D_i 和 h_i 是组分扩散系数和焓。

4 流动的热力学特性

描述一个特别的热化学过程的流动，一般用 Damkohler 数 Da，对于化学反应情况定义如下[6,7]：

$$Da = \tau_{flow}/\tau_{chem} \tag{4}$$

这里 τ_{flow} 和 τ_{chem} 分别是流动和化学反应的特征时间。严格地讲，要描述流动的热力学特征，必需考虑所有非平衡过程的 Damkohler 数。然而，对于钝体离解流动，基本的非平衡过程是离解-复合反应。因此，可考虑下述双原子分子 A_2（N_2, O_2）的反应：

$$A_2 + M = 2A + M$$

这里 M 代表碰撞体。这个化学反应的化学速率 Q_A 为：

$$Q_A = \sum_m (k_{bm} x_A^2 x_m - k_{fm} x_{A2} x_m)$$

这样，可以定义化学反应的特征时间 τ_{chem} 为：

$$\tau_{chem} = x_T/Q_A \tag{5}$$

这里 $x_T = \sum x_i$，$x_i = \rho_i/M_i$。在边界层，流动特征时间可以用边界层边缘的速度梯度表示：

$$\tau_{flow} = \left(\frac{\partial u}{\partial x}\right)_e^{-1} \cong R_N \left(\frac{\rho}{2p}\right)_\delta^{\frac{1}{2}} \tag{6}$$

由（5）和（6）可得 Damkohler 数：

$$Da = \frac{Q_A}{x_T} R_N \left(\frac{\rho}{2p}\right)_\delta^{\frac{1}{2}} \tag{7}$$

当 Da 是负值时，流动是离解占主要的流动；当 Da 是正值时，流动是复合占主要的流动。当 $|Da| \leqslant 0.01$ 时，流动是冻结流。

5 结果与分析

5.1 实验和计算的来流条件

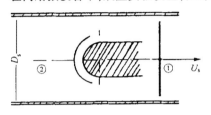

图 1 激波管流动示意图

Fig.1 Model in the shock tube

在高焓激波管中,放置实验模型,如球头、平头圆柱(如图 1 所示),测量其驻点或其它点热流,再通过数值计算,确定实验模型表面材料催化速率。笔者实验的激波管直径 D_s =150mm,实验气体为空气,初始激波管压力 P_1 =6.5Pa,初始激波温度 T_1 =300K,激波速度 U_s =5.3 km/s。计算的来流条件是入射运动激波后的参数,可根据平衡气体条件求得。计算来流条件如表 1 所示。

表 1 计算来流条件

$H_0/(\mathrm{MJ \cdot kg^{-1}})$	$U_\infty/(\mathrm{m \cdot s^{-1}})$	p_∞/Pa	$\rho_\infty/(\mathrm{kg \cdot m^{-3}})$	T_∞/K	$T_{v\infty}/\mathrm{K}$
28.16	4880	1958	9.45×10^{-4}	5211	5211

T_W/K	$C_{\mathrm{N_2}}$	$C_{\mathrm{O_2}}$	C_{NO}	C_{O}	C_{N}
300	0.5922	2.96×10^{-5}	2.373×10^{-3}	0.2316	0.1738

5.2 球头模型催化特性研究

球头半径 R_N =25mm ,计算的催化条件:完全非催化(NCW)、有限催化(K_w =1.0, 4.0, 8.0, 16.0, 32.0)和完全催化(FCW)。图 2 给出了完全非催化时的密度等值线。图

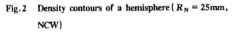

图 2 球头(R_N =25mm)密度等值线(NCW)

Fig.2 Density contours of a hemisphere(R_N =25mm, NCW)

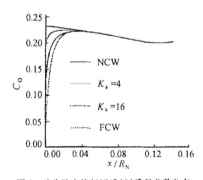

图 3 球头驻点线氧原子(O)质量分数分布

Fig.3 Mass fraction of O at the stagnation line of a hemisphere(R_N =25mm)

3 给出了球头驻点线上不同催化速率常数的氧原子质量分数分布。图 4 给出了非催化时驻点线边界层内 Damkohler 数的分布。结果表明:在所计算的条件下,边界层内 $|Da| \leqslant$ 0.01,所以边界层处于化学冻结状态,可以通过实验和理论分析研究边界层有限催化特

性。图 5 给出了驻点热流随催化速率 K_W 变化的分布图。

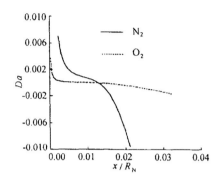

图 4 球头驻点线 N_2 和 O_2 的 Damkohler 数分布

Fig.4 Damkohler number at the stagnation line of a hemisphere(R_N = 25mm)

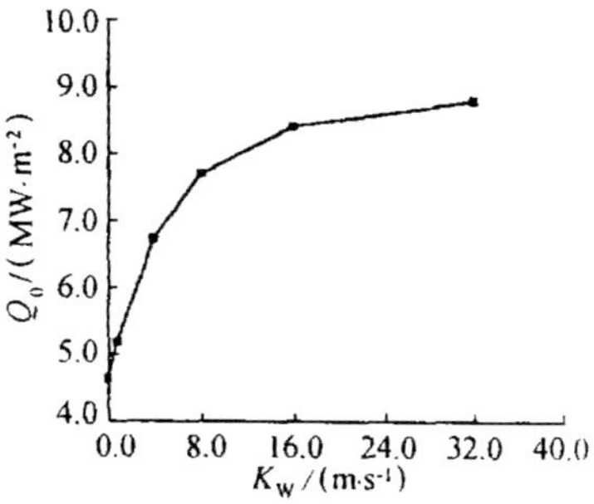

图 5 球头(R_N = 25mm)驻点热流随催化速率常数的分布

Fig.5 Heat transfer rate at the stagnation point of a hemisphere(R_N = 25mm)

另外,为了与实验相结合,还计算了球头半径 R_N = 15mm 的情况,计算的催化条件:完全非催化(NCW)、有限催化(K_W = 1.0, 8.0, 16.0, 24.0)。图 6 给出了驻点热流随催化速率 K_W 变化的分布图,同时也给出了实验测量的壁面材料为 Ni 的驻点热流值 Q_0 = 11.89MW/m^2。这样,用图 6 插值可得材料 Ni 的催化速率 $K_W \approx 15$。

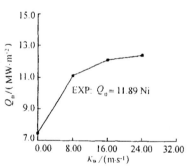

图 6 球头(R_N = 15mm)驻点热流随催化速率常数的分布

Fig.6 Heat transfer rate at the stagnation point of a hemisphere(R_N = 15mm)

图 7 平头圆柱密度等值线

Fig.7 Density contours of a flat -nosed cylinder

5.3 平头圆柱模型催化特性研究

为了提高驻点流动的均匀性,也采用了平头圆柱半径 R_N = 20mm 的实验模型,计算的催化条件:完全非催化(NCW)、有限催化(K_W = 1.0, 2.0, 3.0, 4.0, 8.0, 16.0)和完全催化(FCW)。图 7 给出了完全非催化时的密度等值线。图 8 给出了驻点热流随催化速率 K_W 变化的分布图,同时也给出了实验测量的壁面材料为 Pt、SiO$_2$ 和某种飞船材料 X 的驻点热流值 Q_0 = 7.99、4.93、5.27MW/m^2。这样,用图 8 插值可得 Pt、SiO$_2$ 和 X 材料的催化速率 $K_W \approx 12, 0.7, 1.0$。

6 结束语

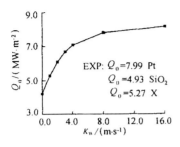

图 8 平头圆柱驻点热流随催化速率常数的分布

Fig. 8 Heat transfer rate at the stagnation point of

a flat-nosed cylinder(R_N = 20mm)

（1）笔者确定了钝体驻点边界层划分平衡和非平衡流动的 Damkohler 数和冻结流动的实验参数。

（2）建立了通过实验测量热流并获得壁面催化特性的数值分析方法。

（3）通过激波管实验初步给出了壁面温度 T_w = 300K 条件的 Pt、SiO$_2$、Ni 和某种飞船材料的催化速率常数。

（4）进一步的技术发展将是改变壁面温度和更严格的热流测量。

参考文献：

[1] STEWART D A. Predicting material surface catalytic efficiency using arc-jet tests. AIAA-95-2013.

[2] STOCKLE T,KURTZ M A. Material catalysis in high enthalpy air flows. AIAA-96-1904.

[3] McCAFFREY B J, EAST R A. Non-equilibrium stagnation point heat transfer measurements to catalytic surfaces in shock heated air . 10th ISSW, 1975.

[4] MAUL J. Shock tube measurements for aerothermodynamic gas /wall interaction. 20th ISSW, 1995.

[5] 董维中. 热化学非平衡效应对高超声速流动影响的数值计算与分析:[博士学位论文]. 北京航空航天大学, 1996.

[6] DONG Wei-zhong, LE Jia-ling. Numerical studies of non-equilibrium heat transfer testing in the shock tube and extrapolation to flight. 2nd Asia Workshop on CFD, Tokyo, 1996,219-225.

[7] GOKCEN T. Effects of flowfield non-equilibrium on convective heat transfer to a blunt body. AIAA-96-0352.

The calculations of aerodynamic heating and viscous friction forces on the surface of hypersonic flight vehicle

LE Jia-ling[1], V L Ganimedov[2], M I Muchnaja[2], V N Vetlutsky[2]

(1. China Aerodynamics Research and Development Center; 2. ITAM, Russia)

Abstract: In this paper the results of calculations of the coefficients of skin friction and heat transfer (Stanton number) on the surface of hypersonic flight vehicle are presented. Two methods, namely "The plane sections method" i. e. 2D boundary layer approximation and "The engineering method", are used for calculation of viscous effects at hypersonic flow around the flight vehicle. Calculation results of both methods are presented and compared in detail. It can be seen from this article that the two methods have demonstrated the working capacity for calculation of frictional force and thermal load on surface of a hypersonic winged flight vehicle, which moves in dense atmospheric layers on a trajectory with stationary values of velocity head. They can successfully be applied to parametric calculations for flow around bodies of complex form.

Key words: hypersonic flight vehicle; skin friction; heat transfer; numerical computation; engineering computation

高超声速飞行器表面气动热和粘性摩擦力计算

乐嘉陵[1]，詹妮迈德芙 V L[2]，曼彻娜娅 M I[2]，维特拉斯基 V N[2]

(1. 中国空气动力研究与发展中心，四川 绵阳 621000; 2. 俄罗斯理论和应用力学研究所)

·摘要: 本文给出了高超声速飞行器表面摩阻和传热系数(斯坦顿数)的计算结果。采用两种方法平面切面法亦即二维边界层近似法和工程方法计算了飞行器高超声速绕流的粘性效应，并对两种方法的计算结果作了仔细的比较。由文可见，对于在稠密大气层内，沿轨道运行头速度恒定的高超声速有翼飞行器，能够用本文所采用的两种方法计算其表面摩阻和热载荷。此二法可成功地应用于绕复杂形状物体的流动参数计算。

关键词: 高超声速飞行器; 表面摩阻; 传热; 数值计算; 工程计算

Le Jialing, V L Ganimedov, M I Muchnaja, V N Vetlutsky. The calculations of aerodynamic heating and viscous friction forces on the surface of hypersonic flight vehicle. Experiments and Measurements in Fluid Mechanics, 2002, 16(1): 8-20.

0 Introduction

In this paper the results of calculations of the coefficients of skin friction and heat transfer
(Stanton number) on the surface of hypersonic flight vehicle are presented. The shape of the vehicle
shown in Fig. 1 is taken from[1]. This shape has been obtained on base of the analysis of parame-

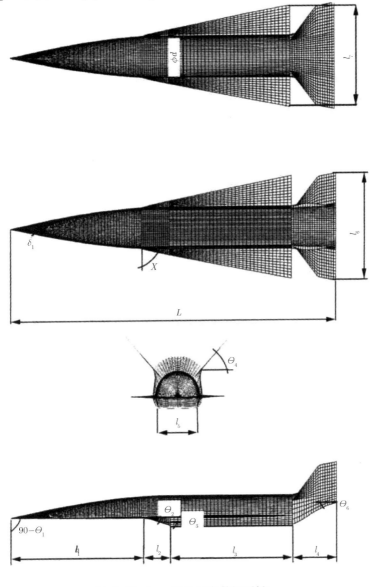

Fig. 1 The shape of hypersonic flight vehicle

图 1 高超声速飞行器外形图

tric calculations of 3-D Euler equations. The following requirements are key for the shape selection (at the given base parameters): (1) high lift-to-drag ratio; (2) longitudinal static stability of the vehicle; (3) providing of uniform flow at entrance of engine combustor.

The top part of the vehicle nose has the form of half-ellipses in each cross-sections. This form smoothly mates with the main cylindrical part. The maximum angle of the nose is $11°$. The lower surface of the nose has the smooth, slightly convex form, which is close to a triangular plate. This surface is followed by two compression wedges of inlet. The wedges are at the angles $7.5°$ and $17.5°$ to a building axis of the vehicle respectively. Further there is the engine box which is close to rectangular form. The engine bottom is located slightly below than the back boundary of the second compression wedge. This gap simulates an entrance in the engine. In the central part of fuselage the thin triangular wings are located. The wings have relative thickness 3% and the angle of sweep is $80°$. X-coordinates of the beginning of cylindrical part of fuselage, inlet and wings coincide. Also X-coordinates of the end of fuselage, engine and wings coincide. The flow around the tail part of the vehicle is not considered in the present work. The length of the nose is equal to 2m, the length of the vehicle without a tail part is 4.5m.

The body surface is given by rather detailed grid in rectangular coordinates X, Y, Z, where the axis X is aligned with building axis of the vehicle. The origin of coordinates coincides with the apex of the body.

The parameters of inviscid flow around body are specified for each grid point with coordinates x, y, z. There are velocity components u_e, v_e, w_e, pressure p_e, density ρ_e, temperature T_e, Mach number M_e. The field of inviscid flow is taken from the calculations[1]. In addition the flow parameters at infinity (designated by a subscript "∞"), the angle of attack (and wall temperature T_w are set. In the present work this parameters are used to solve the problems of determining of viscous effects.

1　Mathematical models

Two mathematical models are used for calculation of viscous effects at hypersonic flow around the flight vehicle. The brief description of both models is presented below.

1.1　The plane sections method (two—dimensional boundary layer approximation)

For solving of boundary layer equations the coordinate system (ξ, η) is used which is tied with a body. The surface coordinate ξ is measured along body axis and is calculated from Cartesian coordinates of the body. The η—coordinate is a normal to the surface.

Let some plane cross the vehicle surface. The line of intersection is taken as the origin of normal lines of body surface, these perpendiculars generate the surface. It's assumed that the transverve flows hardly cross this surface. Under this assumption, many similar normal surface can be formed. For example they can be based on intersections of fuselage with the planes that are parallel to the axis x and passed through body tip. Also this surfaces can be based on intersections of engine outer surface with the planes that are parallel to free stream, and so on (see Fig. 1). Then the 3-D boundary layer equations[2] are reduced to 2-D system[3, 4]. The equations of a compressible boundary layer in non-dimensional form are:

$$\frac{\partial}{\partial \xi}\bigg| \rho u \sqrt{\frac{g}{g_{11}}}\bigg| + \sqrt{g}\frac{\partial \rho v}{\partial \eta} = 0$$

$$\frac{\rho u}{\sqrt{g_{11}}}\frac{\partial u}{\partial \xi} + \rho v\frac{\partial u}{\partial \eta} - Q - \frac{1}{Re}\frac{\partial}{\partial \eta}\bigg|\mu\frac{\partial u}{\partial \eta}\bigg| = 0 \tag{1}$$

$$c_p\bigg| \frac{\rho u}{\sqrt{g_{11}}}\frac{\partial T}{\partial \xi} + \rho v\frac{\partial T}{\partial \eta}\bigg| - \frac{1}{Pr}\frac{1}{Re}\frac{\partial}{\partial \eta}\bigg|\lambda\frac{\partial T}{\partial \eta}\bigg| + uR - (\gamma - 1)M_\infty^2\frac{1}{Re}\mu\bigg|\frac{\partial u}{\partial \eta}\bigg|^2 = 0$$

$$\rho = \gamma M_\infty^2 p / T$$

Here u, v, T, p, ρ are velocity components, temperature, pressure and density correspondingly; μ is the coefficient of viscosity, λ is heat conductivity coefficient, c_p is specific heat at constant pressure, γ is ratio of specific heats. All parameters are normalized with the body length L and those in free stream. Re is Reynolds number, and Pr is Prandtl number.

The quantities g and g_{11} are the metric coefficients of the surface, and the quantities Q and R are known functions of the flow parameters. For example,

$$R = -\frac{\rho_e c_p}{\sqrt{g_{11}}}\frac{\partial T_e}{\partial \xi}.$$

The expression for quantities Q is more bulky.

It is assumed that the body surface in the beginning (near the body tip, at the origin of engine, near leading edge of a wing) can be approximated by a conical surface. In this case the boundary—layer equations allow a self-similar solution, which depends on one variable $\eta / \sqrt{\xi}$. This solution can be used as the initial data for the problem as a whole. On the edge of the boundary layer the parameters u_e, w_e, T_e, p_e are taken from the inviscid flow around the body, previously calculated by the Euler equations in the Cartesian coordinates. The boundary condition at the body surface consists of a no-slip condition and a specific wall temperature.

The new independent variables t, n were introduced in order to increase the accuracy of the calculations obtained by difference scheme with constant steps. The change of variables is made thus: $\xi = t$, $\eta = \eta(t, n)$. The function $\eta(t, n)$ is obtained from the solution of the differential equation:

$$\frac{\partial \eta}{\partial t} = \frac{\partial}{\partial n}\bigg| d\frac{\partial \eta}{\partial n}\bigg|, \quad d\bigg|\frac{\partial u}{\partial \eta}\bigg|^\omega + \varepsilon$$

With the help of these variables the stretching of the region with large gradients was realized. The two-layer implicit difference weighted scheme of Lax-Wendroff type is used. The scheme has the accuracy $O((\theta - 1/2)\Delta t + \Delta t^2 + \Delta n^2)$. It is absolutely stable for $u > 0$ and the weighting coefficient $0.5 \leqslant \theta \leqslant 1.0$.

At the boundary layer edge the given parameters of the inviscid flow are over-calculated for this grid by means of the cubic spline.

For the calculation of the laminar, transitional and turbulent flows the equations contain the global viscosity and thermal conductivity coefficients. They equal the proper laminar coefficient plus the turbulent coefficient multiplied by a certain transition factor Γ. Four eddy viscosity models of turbulence are used for turbulent viscosity. They give almost the same results. In the present work the calculations are carried out with Michel's model[5].

The two-dimensional boundary layer on the body surface is calculated by marching technique along the t-variable. The profiles of the velocity, temperature, local skin-friction coefficient C_f and the local Stanton number St are obtained in every cross section:

$$C_f = \frac{\tau_\xi}{0.5\rho_\infty U^2_\infty}, \quad St = \frac{q_w}{\rho_\infty U_\infty(h_0 - h_w)} \tag{2}$$

Here ρ_∞, U_∞ are the density and the velocity at the infinity, h_0, h_w are the total enthalpy and the air enthalpy at the wall, τ_ξ is the skin-friction components along the ξ-coordinate, and q_w is the wall heat flux.

The boundary layer calculations are carried out on the difference grid with 80 mesh points along a normal to a streamlined surface. The physical value of the step is less in 50 times near body surface, than the value for boundary layer edge. The step along the ξ-coordinate is equal to 0.01 meter.

1.2 The engineering method

The second used approach is borrowed from[6] and[7]. The skin friction coefficients and the Stanton numbers are found in every grid point of body surface given in Cartesian coordinates. First of all a regime of boundary-layer flow is determined: is it laminar or turbulent. For this purpose the Reynolds number of transition $Re_{tr} = C_{tr}(M_e)Re_1^{0.6}$ is determined. Here Re_1 is unit Reynolds number estimated by parameters of free flow. The factor C_{tr} depends on a local Mach number. On the basis of a great body of different experimental data the following approximating relation can be obtained for the range $0 < M_e < 8$:

$$C_{tr} = 480 - 80M_e \qquad \text{at } M_e \leqslant 5$$
$$C_{tr} = -220 - 60M_e \qquad \text{at } M_e > 5.$$

If local Reynolds number $Re_x < Re_{tr}$ then a flow is laminar. If $Re_x \geqslant Re_{tr}$ then a flow is turbulent.

For the laminar boundary layer the local skin-friction coefficient is determined with the formula:

$$C_{f\delta} = \tau_w/(0.5\rho_e u_e^2) = \sqrt{C(1+2j)/Re_x}$$

Here $C = \dfrac{(1+110.4/T_e)\sqrt{h_e}}{h_e + 110.4/T_e}$, where $h_e = 0.4605 + 0.595h_w + 0.0268M_e^2$, $h_w = h_w/h_e$ $\approx (0.23T_w + 0.0002T_w^2)/(0.24T_e)$.

In the above formulas u_e, ρ_e, T_e, and M_e are local flow parameters on boundary layer edge, T_w is wall temperature. The parameter j equals zero for flat flows, $j = 1$ for spatial flows. The flow is considered as flat flow if the condition $\bar{r} > 1.43$ is satisfied, where $\bar{r}$ is local radius of cross-section curvature referred to a semi-range of local cross-section of the body. If $\bar{r} < 1.43$ then the flow is considered as spatial (conical). In other words, the flow is considered as flat flow if the local curvature is not too large.

In the case of turbulent boundary layer the following formulas are used: for flat flow

$C_{f\,\delta} = C_{f\,\mathit{pl}} = 1/(3.3\,a\lg Re_x - 2.83\,b)^2,$

and for conical flow $C_{f\,\delta} = 1.15\,C_{f\,\mathit{pl}}.$

Here $Ma - a、b$ relation is shown in Fig.2.

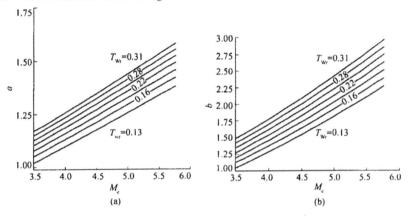

Fig.2　Curves of Ma vs a、b
图 2　马赫数随 a, b 的变化曲线

$$T_{wr} = T_w/T_r, \qquad T_r = T_e(1 + 0.178M_e^2).$$

Here T_r is approximate value of the recovery temperature in a boundary layer.

The value of the local heat transfer coefficient $St_{\delta} = q_w/[\,\rho_e u_e(h_r - h_w)\,]$ is connected to the local coefficient of skin-friction by the ratio of Reynolds analogy between friction and heat transfer (q_w is specific heat flux, h_r and h_w are the values of enthalpy at recovery temperature and on a wall correspondingly).

For a laminar boundary layer $\quad St_{\delta} = C_{f\,\delta}/1.6016,$

and for a turbulent boundary layer $St_{\delta} = \dfrac{C_{f\,\delta}}{2(9 - d\sqrt{C_{f\,\delta}})}.$

where $d = 0.885\sqrt{T_{wr}(1 + 0.175M_e^2)},$

And finally the values of skin-friction and heat transfer coefficients are referring to the free flow parameters:

$$C_f = C_{f\,\delta}(\rho_e u_e^2)/(\rho_\infty U_\infty^2), \qquad St = St_{\delta}(\rho_e u_e)/(\rho_\infty U_\infty) \qquad (3)$$

It is necessary to notice that the value of the skin-friction coefficient C_f, which is obtaining in this method, is the same as C_f which has been determined in the previous approach. But the values of Stanton number St are different here and in the previous method. Here St is determined from the recovery temperature T_r, whereas it is determined from the stagnation temperature in the 2-D approximation.

2 The comparison of calculation results of both methods

The solution with the help of 3-D equations is more precise, but at the same time this approach manages to be applied to narrower class of surfaces. For comparison there are the symmetric plane at the lee side (Fig.3) and the plane $\phi=58°$ at the wind side (Fig.4). The angle ϕ is meridional angle, $\phi=0°$ corresponds to the symmetry plane at the windward side, $\phi=180°$ corresponds to the symmetry plane at the lee side. The abscissa is the distance along the axis of the vehicle from its nose.

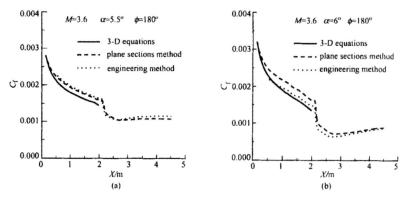

Fig.3 Comparison of calculation results at lee side

图 3 在背风面计算结果的比较

The comparison has been made only for the skin friction coefficient because Stanton number has different definition in those methods as it was noted earlier. The values of skin friction coefficient C_f are compared in two points of base trajectory of the vehicle: 1) $M=3.6$, $\alpha=5.5°$; 2) $M=6$, $\alpha=6°$. The unit Reynolds number estimated by parameters of free flow is $Re_1=9.1\times10^6/m$ for the first variant and $Re_1=5.1\times10^6/m$ for the second one. The temperature of a surface is set constant and equal to 400 ℃. The comparison is for the regions with developed turbulent regime. The founding of position and length of a zone of transition from a laminar regime to turbulent one is not stated as the independent task, and they have some arbitrariness. Therefore results for the flow in the vicinity of the fuselage nose are not presented on the graphs.

As one can see in the Fig.3, the results of calculation of the coefficient C_f ₈obtained by various algorithms agree well with each other for $\phi=180°$. In the end of vehicle nose part ($x=2m$) the difference in values between distinct models does not exceed 15%. The results of the 3-D calculation are below than others. It depends upon the fact that the 3-D model takes into account gas mass transferring to symmetric plane by cross flows. It's should be noted that the angles of attack are not large on the base trajectory. Consequently the intensity of inviscid flowing to the symmetric plane is relatively small, and all three models reach a sufficiently good agreement.

In the plane $\phi=58°$ the curves for distinct methods are practically equidistant. At $M=3.6$ the maximum difference between calculation results is 15% (Fig. 4a). At $M=6$ difference for 3-D and plane sections models is equal to 11%, whereas the difference between results of 3-D and engineering models reaches 30% (Fig. 4b).

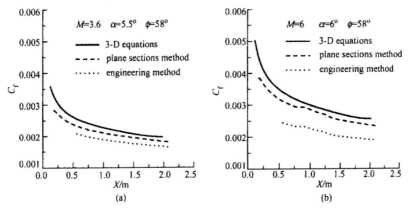

Fug. 4   Comparison of calculation results at wind side
图 4   在迎风面计算结果的比较

The results of three-dimensional calculation for the plane $\phi=58°$ are above those yielded by other methods. This is due to the fact that 3-D model takes into account the cross flows driving of gas masses away from this area.

Nevertheless it is possible to consider that the present comparisons show satisfactory agreement between different approaches, and the method of plane sections and the engineering one are rather well suited for parametric calculations.

3   The results of calculations of viscosity effects on vehicle surface

The results have been received by the plane sections method. The values of determining parameters of the problem have been taken from the base trajectory (five points on the trajectory). The temperature of vehicle surface is considered constant and equal to 400 ℃. The results of parametric calculations are presented in Figs. 5~8.

The coefficients of skin friction and heat transfer in the plane of symmetry at lee side are shown in Fig. 5 (a, b). The calculation is carried out up to a tail part of the vehicle only ($x\leqslant4.5m$). The shock induced by stabilizing flare does not allow to continue the solution of boundary layer equations on the flare.

All curves for the coefficient C_f at five points of the based trajectory (Fig. 5a) are located closely to each other and have similar character. In the beginning there is a sharp growth, which corresponds to transition zone of flow. Then in a developed turbulent boundary layer the coefficient C_f decreases monotonously on the nose part. At transition to a cylindrical part there is a sharp decreasing of C_f, which connected with sharp pressure decreasing. Further, on account of the effect of pressure, the decreasing of the coefficient C_f value becomes quieter and passes smoothly in feeble increase. The close analysis shows that the increase of angle of attack results in decreasing of the values of C_f. With increasing of Mach number the coefficient C_f grows on the nose and decreases on a cylindrical part.

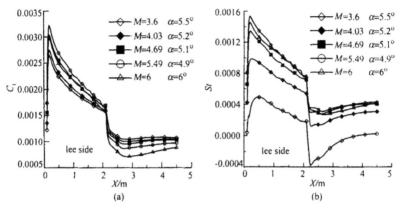

Fig. 5   Curves of C_f, St vs x at lee side

图5   在对称面背风侧表面摩阻系数 C_f 和热传导系数 St 数与 x 的关系

The curves for the Stanton number (Fig. 5b) have similar character with curves for the skin—friction coefficient. A part of distribution of the function $St = F(x)$ is negative for the point of the trajectory with $M=3.6$. It is explained by that circumstance that at $M=3.6$ and $\alpha=5.5°$ the given surface temperature is higher than local recovery temperature but is less than stagnation tempera-

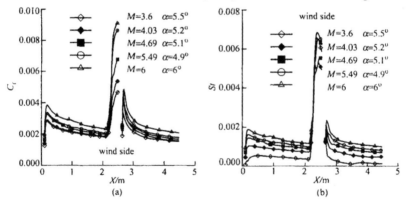

Fig. 6   The distribution of C_f and St at wind side

图6   在对称面迎风侧表面摩阻系数 C_f 和热传导系数 St 的分布

ture of flow. The effect of negative value of Stanton number would not appear if it had been normalized on equilibrium temperature. However equilibrium temperature is a unknown quantity for algorithm of the plane sections method.

In Fig. 6 the distribution of skin friction coefficients C_f and heat transfer coefficients St in the symmetry plane at wind side along axis X are shown. The gap of function is corresponded to the place of entrance of external flow into engine combustion chamber. The function after the gap corresponds to distribution of parameters on bottom engine cowling. The growth of the functions at the beginning of the nose and the engine corresponds to zone of transition. As it is visible from these graphs, on a windward side the coefficients C_f and St monotonously increase with increase of Mach number and angle of attack.

The distribution of the coefficient C_f on lateral surface of engine cowling (in its middle) is presented in Fig. 7. This surface segment represents the flat panel. The flow about this panel is determined in many respects by influence of wedges of compression of inlet. The distributions of functions depend on a Mach number monotonously.

In Fig. 8 the distribution of the coefficient C_f on the vehicle surface is shown at various value of the angle ϕ for the variant $M=6$, $\alpha=6°$. On an axis of abscissas the axial distance of the vehicle is plotted. The distributions of the parameters for $\phi=0°$ and $82°$ are presented up to inlet entrance only. The results for $\phi=0°$ and $\phi=58°$ practically coincide, and therefore the curve for $\phi=58°$ is not shown.

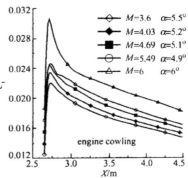

Fig. 7　The distribution of C_f on lateral surface of engine cowl

图 7　在发动机罩表面的摩阻系数分布

As an illustration of friction distribution on wing surface, the projection to the wing of the straight line is chosen which connected the forward point of onboard chord of the wing with the middle of trailing edge of the wing. The distribution of the coefficient C_f on this line is presented in Fig. 9. The results are presented for the cruise regime of flight.

4　The results of calculations by the engineering method

The engineering method is allowed for the complicated task due to its simplicity and speed. Usually the temperature of a wall is set very roughly. In parallel with calculation of coefficients C_f and St with use of given wall temperature T_w as in the previous method, also this coefficients are calculated using the local equilibrium temperature of wall T_{we}. To do this, the recalculation of $C_{f\delta}$ and St_δ values is made by successive approximation procedure, using the initial value T_w as the first approximation. The equilibrium temperature T_{we} is found as a result of solution of the thermal bal-

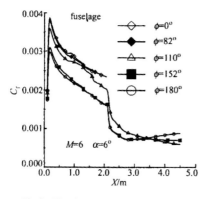

Fig. 8   The distribution of C_f on the vehicle
surface at various value of ϕ
图 8   在不同 ϕ 角下机身表面的 C_f 分布

Fig. 9   The distribution of C_f on wing surface
图 9   在机翼表面的 C_f 分布

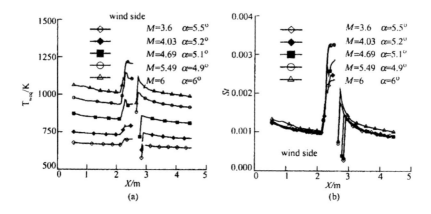

Fig. 10   The computed results of equilibrium temperature T_w and Stanton number St
in planes of symmetry at wing side
图 10   对称面上迎风侧平衡温度和 St 数的计算结果

ance equation

$$\alpha(T_r - T_{we}) = \varepsilon\sigma T_{we}^4$$

where $\alpha = St\ c_p\rho_e u_e$, c_p — specific heat, ε-the factor of surface blackness, σ-Stefan-Boltzmann factor. In the present work $\varepsilon = 0.8$, and the equation of thermal balance was solved by the successive
approximation method of Newton. It is needed 3-5 iterations for convergence. As soon as iterations
converge, the coefficients C_f and St are found from formulas (3). Further the local equilibrium
temperature of a surface is used for the Stanton numbers calculation.

The calculations of equilibrium temperature and Stanton number in planes of symmetry are carried out for five points of the trajectory and are shown in Figs. 10 ~ 11. In the graphs on the axis of
abscissas the distance along the vehicle axis is laid which is read out from its nose.

As seen in the graphs the maximum values of temperature and Stanton number are on wedges of

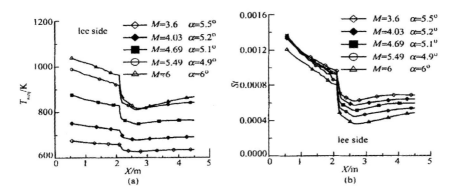

Fig. 11   The computed results of equilibrium temperature T_w and St in planes of symmetry at lee side

图 11   在对称面背风侧平衡温度 T_w 和 St 的计算结果

compression of inlet where the pressure is maximum. The oscillations in distribution of temperature stem from the oscillations in the inviscid solution. The increase of Mach number results in increase of equilibrium temperature of a surface, despite of a constancy of dynamic head on the trajectory. On the trajectory the maximum equilibrium temperature of a surface is equal approximately 1200K if not to take into consideration leading edges and neighborhood of fuselage nose.

In Fig. 12 the circumferential distribution of Stanton number and the equilibrium temperature are presented for the cross-section $x = 2m$ that corresponds to the beginning of inlet. On an axis of abscissas the polar angle ϕ is laid. The results are presented for two points on the trajectory. In circumferential distribution the maximum of the surface temperature is on a lateral side where the pressure and velocity gradient are maximum.

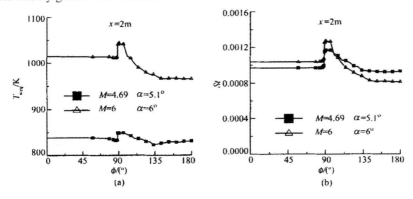

Fig. 12   The distribution of St and T_w for the cross-section $x = 2m$

图 12   在机身 $x = 2m$ 剖面处的 St 和 T_w 分布

The distribution of Stanton number and the equilibrium temperature on a wing surface is presented for the cross—section $x = 4m$ in Fig. 13. On an axis of abscissas the dimensionless coordinate $\bar{z}$ along wing span is laid ($\bar{z} = 0$ is correspond to airborne chord of wing). The maximum value of temperature on a wing surface does not exceed 1000K.

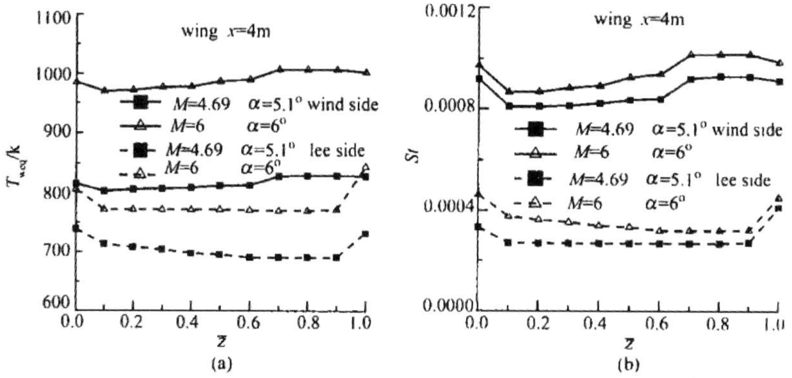

Fig. 13　The distribution of St and T_w on wing surface for the croos-section $x=4$m.

图 13　在机翼表面 $x=4$m 剖面处的 St 和 T_w 分布

5　In summary

It can be noticed that two used methods have shown the working-capacity for calculation of frictional force and thermal loads on a surface of a hypersonic winged flight vehicle, which moves in dense atmospheric layers on a trajectory with stationary values of velocity head. They can be successfully applied to parametric calculations for flow around bodies of complex form.

This research was supported by the China Aerodynamics and Development Center under contract number 99XSDC-3-33.

Rererences:

[1]　GANIMEDOV V L, KOROTAEVA T A, MUCHNAJA MIA etc. Investigation of the flow around vehicle // Report of ITAM SB RAS, August 2000, Novosibirsk.

[2]　VETLUTSKY V N, KRAUSE E. Calculation of three—dimensional compressible boundary layers on pointed bodies and comparison with experiments // Z. Flugwiss., 1992, N.16, pp.308—316.

[3]　FANNELOP T K. A method of solving the three—dimensional laminar boundary—layer equations with application to a lifting re—entry body // AIAA Journal, 1968, v.6, N.6, pp.1075—1083.

[4]　BRADLEY R G. Approximate solution for compressible turbulent boundary layers in three—dimensional flow // AIAA Journal, 1968, v.6, N.5, pp.859—864.

[5]　MICHELI R, QUEMARD C, DURANT R. Hypotheses on the mixing length and application to the calculation of the turbulent boundary layers. In: Keine S. J. et. al. (Eds.): Proceedings Computation of Turbulent Boundary Layers. AFOSR—IFP—Stanford Conference, 1968, v.1, pp.195—207.

[6]　Engineering methods and programs of aerodynamic computer calculation of characteristics of flight vehicle at hypersonic velocities (paper's collection) // Trudy TsAGI, issue 1580, 1974 (in Russian).

[7]　Materials for calculation of friction drag and heart transfer of different bodies at hypersonic velocities of flight // Trudy TsAGI, issue 937, 1964 (in Russian).

高超声速圆球模型飞行流场的数值模拟和实验验证

柳 军[1]，乐嘉陵[2]，杨 辉[2]

（1. 国防科技大学航天与材料工程学院，湖南 长沙 410073；2. 中国空气动力研究与发展中心，四川 绵阳 621000）

摘要：对高超声速圆球模型飞行流场进行数值模拟，分别采用空气完全气体模型、平衡气体模型以及热化学非平衡 11 组元气体模型求解非定常轴对称 N—S 方程组。使用有限差分时间相关法捕捉激波，得到了定常流场的解。差分方程隐式部分采用了 LU—SGS 方法以避免矩阵运算，对化学反应和振动能量源项采用预处理矩阵以解决刚性问题。由计算结果处理得到的阴影图和干涉条纹图与再入物理弹道靶实验照片进行了对比分析，验证了实验中圆球飞行流场大部分区域接近于平衡状态。

关键词：数值计算；弹道靶；热化学非平衡；平衡气体；干涉图；阴影图

中图分类号：V211.7　　　文献标识码：A

Numerical simulation of hypersonic flowfield around sphere model and experimental verification

LIU Jun[1], LE Jia-ling[2], YANG Hui[2]

（1. College of Aerospace and Material engineering，NUDT Changsha 410073，China；2. China Aero-dynamics Research and Development Center，Mianyang 621000，China）

Abstract：The hypersonic flowfields around the sphere model are simulated numerically by solving axisymmetric Navier—Stokes equations with various thermal and chemical models，such as perfect gas model，equilibrium gas model，and non—equilibrium gas model of two temperatures. The time—dependent finite difference scheme is used to capture the shock. The LU—SGS method is used in implicit part of the difference scheme to avoid matrix operations. The preconditioning technique is used for solving stiffness problem. The computational interferograms and shadowgraphs of equilibrium gas model are good compared with the experimental photos.

Key words：numerical calculation；ballistic range；thermochemical nonequilibrium；equilibrium gas；interferogram；shadowgraph

柳军，乐嘉陵，杨辉. 高超声速圆球模型飞行流场的数值模拟和实验验证. 流体力学实验与测量，2002，16(1): 67-79.

0 引 言

高超声速飞行器在再入大气层的过程中,由于粘性滞止和激波压缩,使得其周围空气的温度增加,分子振动激发并产生离解和电离,同时引起了电子激发和光辐射。空气组分发生化学反应,改变了激波层中的组分浓度、电子密度及流场特性。高空由于密度较低,非平衡效应占据主要地位。研究高温气体非平衡流动是高超声速飞行器设计中的重要一环。

高温气体非平衡流动十分复杂,与设计有关的工程问题的解决,多数要采用飞行试验、地面实验和数值计算三结合的方法。飞行试验代价昂贵,因而大量的研究性问题主要依赖于地面实验和数值计算。随着计算机和 CFD 技术的发展,数值计算不仅在研究方面发挥越来越大的作用,而且也能为实验设计发挥重要指导作用,特别是由于高超声速地面实验和测量的困难,更需要用数值计算分析和预测实验的可能结果从而使实验更为经济有效。

高温气体流动研究中普遍采用高焓激波风洞和弹道靶这两类地面设备。高焓激波风洞由于喷管流动中的快速膨胀导致空气中的氧、氮分子处于热化学冻结状态,因而风洞中实验来流处于典型的有部分组元离解、振动温度冻结的非平衡状态,从而使实验结果的分析更加困难。弹道靶实验中由于来流气体处于静止的大气状态,数据分析相对而言比较可靠。虽然弹道靶中模型尺寸不能很大(一般模型直径为几十毫米),测量也很困难,但目前仍然是高温气体非平衡研究中普遍采用的实验装置。

弹道靶中多数用圆球体模型进行研究,这是因为高温气体(分子的振动、离解、电离)对圆球体的激波脱体距离非常敏感,因此许多作者在这方面进行了研究[1~4],但由于实验的种种困难,发表的数据极其有限,流动显示多数是阴影照片,流场定量密度场的显示与测量极少[5]。本文进行了弹道靶中速度为 5.1km/s 的圆球实验,获得了同一次实验流场的清晰的阴影图和全息有限干涉图。圆球模型弹道靶实验在中国空气动力研究与发展中心超高速所进行,实验设备为超高速所再入物理弹道靶。

与此同时,本文用数值计算分析了不同气体模型对激波脱体距离和流场的影响并且十分成功地进行了数值干涉图与实验干涉图的比较。数值计算中采用了完全气体模型、平衡气体模型、热化学非平衡 11 组元气体模型等三种空气模型,对轴对称 Navier-Stokes 方程进行求解。流场作层流假设,采用有限差分时间相关法捕捉激波,差分方程隐式部分采用 LU-SGS[6]方法以避免矩阵运算,提高了计算速度。对化学反应和振动能量源项采用预处理矩阵以解决刚性问题。

1 理论模型与计算方法

1.1 控制方程

在计算坐标系(τ, ξ, η)中,轴对称守恒形式无量纲化的 Navier-Stokes 方程为:

$$\frac{\partial U}{\partial \tau} + \frac{\partial F}{\partial \xi} + \frac{\partial G}{\partial \eta} + H = \frac{1}{Re}\left|\frac{\partial F_v}{\partial \xi} + \frac{\partial G_v}{\partial \eta} + H_v\right| + W \tag{1}$$

其中,对完全气体模型和平衡气体模型有

$$U = J(\rho, \rho u, \rho v, \rho E)^T \tag{2}$$

$$W = J(0, 0, 0, 0)^T \tag{3}$$

对两温度热化学非平衡十一组元气体模型有

$$U = J(\rho_i, \rho u, \rho v, \rho E, \rho E_v)^T \tag{4}$$

$$W = J(\omega_i, 0, 0, 0, S_v)^T \tag{5}$$

这里 ρ, ρ_i, E, E_v 分别是混合气体密度、组元密度、单位质量总内能和单位质量振动能。ω_i 为各组元的化学反应质量生成率，S_v 为振动能量的生成率。其余项见文献[7]。

热化学非平衡气体的组元粘性系数 μ_i、热传导系数 k_i、k_{vi} 以及混合气体粘性系数 μ、热传导系数 k 的计算均采用拟和公式和经验公式，具体公式参见文献[8]。

1.2 状态方程和热力学关系式

在各种气体模型中，单位质量总内能为

$$E = e + \frac{1}{2}(u^2 + v^2) \tag{6}$$

各种气体模型的状态方程和能量关系式如下：

（1）完全气体模型

$$p = \rho RT, \quad e = C_V T, \quad h = C_p T$$

（2）平衡气体模型采用了 W. Neuenschwander 的热力学显式关系式[9]

$$p = Z\rho RT \tag{7}$$

$$C_{pK} = f(Z, p), \quad h_D, h_I = f(Z)$$

$$T = \frac{h - h_D - h_I}{C_{pK}} \tag{8}$$

（3）双温度热化学非平衡气体模型[10]

$$p = \frac{R}{M}\rho T, \quad e = \sum_i C_i e_i, \quad e_i = e_{i,t} + e_{i,r} + e_{i,v} + e_{i,el} + \Delta h_i^0$$

$$h = \sum_i C_i h_i, \quad h_i = e_i + \frac{p_i}{\rho_i}$$

1.3 热化学模型

（1）振动松弛模型

双原子分子或离子要考虑振动能量非平衡，组元（N_2、O_2、NO、NO^+、N_2^+、O_2^+）振动松弛时间 τ_i 确定为 Millikan—White 松弛时间 τ_i^{MW} 与有限碰撞时间 τ_i^P 之和

$$\tau_i = \tau_i^{MW} + \tau_i^P \tag{9}$$

对于温度从 300K 到 8000K 的空气，Millikan—White 给出了振动松弛时间的拟合公式[11]，对于 8000K 以上的温度，Park 给出振动松弛的形式[12]。

（2）化学反应模型

化学反应采用 Dunn &Kang 考虑 11 组元 32 个化学反应式的空气化学反应模型[13]，组元为 N_2、O_2、N、O、NO、NO^+、N_2^+、O_2^+、N^+、O^+、e^-，其反应式及质量生成率 ω_i 的计算公

式见文献[10]。

(3)化学—振动耦合模型

在双温度热化学非平衡模型下,空气离解反应的特征时间和振动松弛的特征时间是可比的,因此需建立相应的化学—振动耦合模型。不同的化学反应有着不同的控制温度:对于电子碰撞电离反应,正向反应由振动温度控制,反向反应由平动温度控制;离解反应其正向反应由混合控制温度控制,反向反应由平动温度控制;置换反应全由平动温度控制;化合电离反应正向反应由混合控制温度控制,反向反应由振动温度单独控制。混合控制温度采用

$$T_d = (TT_v)^{1/2} \tag{10}$$

进行计算[10]。

热化学模型种类较多,对计算结果影响不同,优劣难分,有待进一步研究。

1.4 计算方法

笔者用有限差分方法进行流场的数值模拟,对控制方程(1)采用通量差分分裂的NND格式[14]进行离散,粘性项采用中心差分离散,隐式部分采用LU—SGS方法。热化学非平衡气体模型计算中对流体力学和热化学动力学问题采用全耦合解法,对化学反应和振动源项采用时间预处理以解决数值计算中的stiff问题。差分方程隐式部分分裂为两步处理

$$\lambda \Delta U_{i,j}^* - A_{i-1,j}^+ \Delta U_{i-1,j}^* - B_{i,j-1}^+ \Delta U_{i,j-1}^* = RHS \tag{11}$$

$$\lambda \Delta U_{i,j} + A_{i+1,j}^- \Delta U_{i+1,j} + B_{i,j+1}^- \Delta U_{i,j+1} = \lambda \Delta U_{i,j}^* \tag{12}$$

$$\lambda = \left| \frac{1}{\alpha t} + \lambda_{max}^A + \lambda_{max}^B - \frac{\partial W}{\partial U} \right|, \quad U_{i,j}^{n+1} = U_{i,j}^n + \Delta U_{i,j}$$

这里 RHS 为差分方程右端项, A、B 为Jacobian 矩阵, λ_{max} 为Jacobian 矩阵特征值的绝对值最大者。化学反应和振动能量源项Jacobian 矩阵只保留对角项作为时间预处理矩阵。

计算初场由自由流条件给出,壁面给出无滑移条件和等温壁条件。热化学非平衡流计算中壁面组元给出完全催化壁条件。

2 结果分析

2.1 实验和计算条件

a. 实验条件:

圆球直径 $\phi = 12mm$, $U = 5100m/s$, $p = 11195Pa(84mmHg)$, $T = 300K$, $\rho R = 7.8 \times 10^{-4}$ kg/m²;

b. 计算条件:

A: $U_\infty = 5420m/s$, $p_\infty = 1173Pa(8.8mmHg)$, $T_\infty = 288K$, $\rho R = 8.5 \times 10^{-5} kg/m^2$;

B: $U_\infty = 5100m/s$, $p_\infty = 11195Pa(84mmHg)$, $T_\infty = 300K$, $\rho R = 7.8 \times 10^{-4} kg/m^2$。

算例 B 是根据实测的实验状态所确定的计算条件。由于飞行过程中圆球受热,壁温要增加,初步取定 $T_W = 1000K$。计算网格节点数为 121×101,为了精确捕捉激波,在实验照片基础上对激波附近区域进行了加密处理。

2.2 结果分析

图1为条件A时热化学非平衡气体模型计算结果处理得到的干涉条纹,图中I区为无限干涉条纹,II区为有限干涉条纹。由于环境密度较低,图中的激波虽然能够识别,但比较模糊。如果在该状态下进行实验,拍照测量的难度很大,所以试验状态选择了有较高密度环境的条件B。以后各图中的计算结果也均对应于算例B条件下的计算数据。

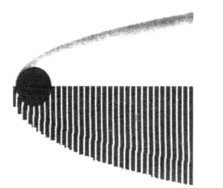 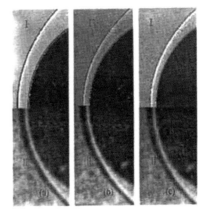

图1 计算干涉条纹(算例A)

Fig.1 Computational interferograms(Case A)

图2 阴影图局部放大

Fig.2 Zoom up of shadowgraphs

表1为圆球头部轴线上激波脱体距离的计算值和实验测量值比较。实验测量值与热非平衡气体模型计算值更为接近,说明实验中在圆球头部的局部区域内流动状态接近于非平衡。图2为圆球头部计算阴影图和试验阴影图对比的局部放大,(a)、(b)、(c)中计算结果分别对应完全气体、热化学非平衡气体和平衡气体模型,I区为计算结果,II区为实验照片。从图中可以直观地看到,平衡气体模型头部激波脱体距离最小,实验激波位置确实更接近于热非平衡气体模型计算结果。

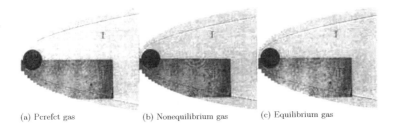

(a) Pcrefct gas   (b) Nonequilibrium gas   (c) Equilibrium gas

图3 计算与实验阴影图对比

Fig.3 Comparison of experimental and computational shadowgraphs

图3为计算结果阴影图与实验阴影照片的对比,同图2一样,I区为计算结果,II区为实验窗口范围内拍摄到的圆球模型飞行流场阴影照片。由图3可见,平衡气体的弓形激波形状与实验一致,实验照片中远端激波的发展趋势与计算结果吻合。图4为激波位置的计算结果与实验值对比,实验值是由阴影照片扫描后放大处理得到。图中坐标原点位于圆球头部顶点,X轴沿圆球轴线指向圆球飞行的相反方向,R为圆球半径。从图中可

以看到,完全气体模型计算的激波位置明显偏大,与其头部激波脱体距离最大的趋势一致。而热非平衡气体模型与平衡气体模型计算的激波位置则在大部分区域内一致,并与实验激波位置相符。本文实验的环境密度相当于高空18km处的大气密度,此时高超声速飞行器的绕流场近于平衡态。本文平衡气体模型计算的激波位置与实验相符说明圆球飞行流场的大部分区域应该接近于平衡态。高温高超声速流动研究中有用热化学非平衡气体模型程序计算平衡状态流动的方法,图4中这两种气体模型计算结果一致可以理解。

表1 圆球头部轴线上激波脱体距离

Table 1 The detached shock wave distance at sphere nose

	计算结果(算例B)			实验
	完全气体 $\gamma=1.4$	热化学非平衡气体	平衡气体	
激波脱体距离 $\triangle R$	0.1358	0.0815	0.0710	0.0805

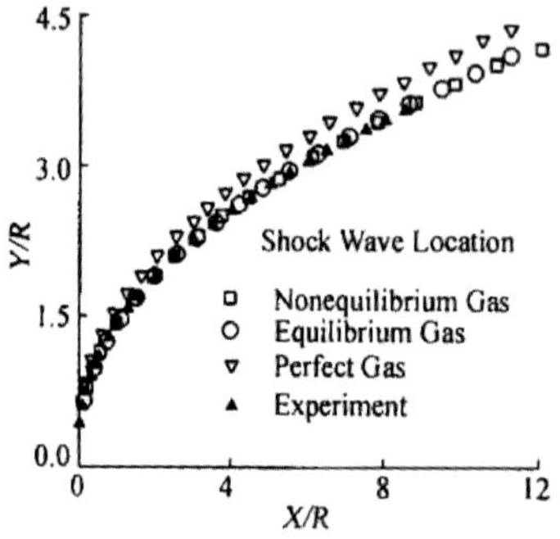

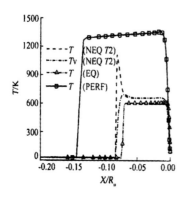

图4 激波位置对比

Fig.4 Comparison of shock wave location

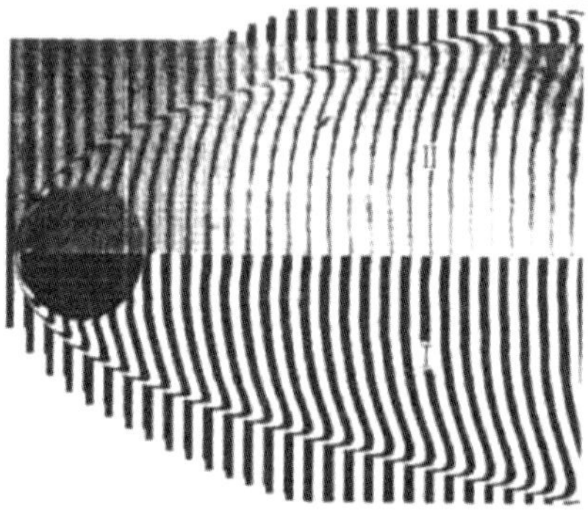

图5 热化学非平衡流干涉条纹与实验对比

Fig.5 Comparison of computational (thermochemical nonequilibrium)and experimental interferograms

图5为热化学非平衡气体计算有限干涉条纹与实验全息有限干涉条纹照片的对比,I区为计算结果,II区为实验照片,计算干涉图中条纹间隙与实验一致。计算干涉条纹是根据本文计算数据处理得到,计算干涉图的制作原理和方法见文献[17]。比较说明,在实验照片

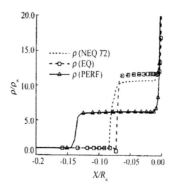

图6 驻点线上温度分布(算例B)

Fig.6 Temperatures at stagnation line(Case B)

图7 驻点线上密度分布(算例B)

Fig.7 Density at stagnation line(Case B)

所拍摄到的可视流场区域内,试验条纹与计算条纹吻合良好。根据干涉条纹图可以确定圆球模型飞行流场的密度,实现高超声速流场的定量测量。

图6为算例B条件下圆球驻点线上计算温度分布,图7为驻点线上计算密度分布,图中 Rn 为圆球半径,$X=0$ 处为壁面。热化学非平衡气体模型在激波处由于振动能级冻结,其平动温度较高,而激波层内平动温度与振动温度则几乎一致。由于部分热能完全用于化学反应和振动激发,平衡气体模型的温度最低。不同气体模型对激波后压力分布影响并不显著,根据状态方程和温度分布情况可知,激波后平衡气体模型的密度值最高,热化学非平衡气体模型次之,图7符合上述分析。

3   结束语

(1)本文采用不同气体模型计算了高超声速圆球模型的飞行流场,并与实验结果进行对比分析。验证了实验流场头部小区域接近于非平衡状态,其余大部分区域接近于平衡状态。

(2)本文热化学非平衡气体模型和平衡气体模型计算结果处理为阴影图,与实验阴影照片进行了对比。实验圆球模型头部激波脱体距离与非平衡气体模型结果接近,除头部区域外的弓形激波位置与计算吻合。

(3)本文得到的计算有限干涉条纹图与实验有限干涉条纹照片吻合良好。

致谢   本文实验得到了气动中心超高速所杨贵山的大力协助,计算阴影图和干涉图的制作使用了吴颖川博士生编写的图像生成软件,在此一并表示衷心的感谢。

参考文献:

[1] SAWADA K, DENDOU E. Validation of hypersonic equilibrium flow calculations using ballistic — range data. AIAA 97 — 0344, Jan. 1997.

[2] FURUDATE M, NONAKA S, SAWADA K. Behavior of two-temperature model in intermediate hypersonic regime. AIAA 99 — 0223, Jan. 1999.

[3] LOWRY H, STEPANEK C, et al. Shock structure of a spherical projectile in weakly ionized air. AIAA 99 — 0600, Jan. 1999.

[4] LOBB R K. Experimental measurement of shock detachment distance on sphere fired in air at hypervelocities. The high Temperature Aspects of Hypersonic Flow, Nelson, W. C., ed., Pergamon press, New York, 1964, pp519 — 527.

[5] NONAKA S, MIZUNO H, HASHIMOTO T. Density measurement over a sphere in ballistic range. AIAA 2000 — 0837, Jan. 2000.

[6] JAMESON A, YOON S. LU implicit scheme with multiple grids for the Euler equations. AIAA 86 — 0105, 1986.

[7] 黄   华等. 飞船轴对称热化学非平衡流场数值求解. 国防科技大学学报. 1999.21(3).

[8] CANDLER G V, MACCORMACK R W. The computation of hypersonic ionized flows in chemical and thermal non — equilibrium. AIAA — 88 — 0511, 1988.

[9] NEUENSCHWANDER W. Explicit relationships for the thermodynamic properties of molecular and dissociated air mixtures to 25,000 degree R. AIAA 89 — 1735, 1989.

[10] GNOFFO P A, GUPTA R N, SHINN J L. Conservation equations and physical models for hypersonic air flows in thermal and chemical nonequilibrium. NASA TP — 2867.

(下转第79页)

[11]   MILLIKAN R C, WHITE D R. Systematics of vibrational relaxation. J. Chem. Phys., 1963, 39(12): 3029~3213.

[12]   PARK C. Problems of rate chemistry in the flight regimes of aeroassisted orbital transfer vehicles. Progress in Astronautics and Aeronautics, 1985, 96(期): 511~537.

[13]   DUNN M G, KAND S W. Theoretical and experimental studies of reentry plasmas. NASA CR—2232, 1973.

[14]   沈   清. 一种高精度、高分辨率激波捕捉的迎风型 NND 格式. 第七届全国计算流体力学会议论文集, 1994.

[15]   NONAKA S, TAKAYAMA K, PARK C. Measurement of shock stand—off distance for sphere in ballistic range. AIAA 97—0563, Jan. 1997.

[16]   NONAKA S, MIZUNO H, TAKAYAMA K. Ballistic range measurement of shock shapes in intermediate hypersonic regime. AIAA 99—1025, Jan. 1999.

[17]   LE Jia—ling, WU Ying—chuan, NI Hong-Li et al. Computational interferometry for three—dimension flow. Experiments and Measurements in Fluid Mechanics, 2001, 15(2): 1~9.

基于 CFD 的电磁散射数值模拟

许　勇, 乐嘉陵

(中国空气动力研究与发展中心,四川 绵阳　621000)

摘　要:计算流体力学(CFD)技术中的时域有限体积法(FVTD)被推广到计算电磁学中,直接数值求解时变麦克斯韦方程组。FVTD 使用电磁场矢量共置于网格单元中心、散射场积分形式的麦克斯韦方程组守恒形,空间离散使用基于特征值的近似黎曼解构建网格单元边界通量,时间推进采用四阶龙格– 库塔法。TM 和 TE 波极化下几种二维完全导电体的表面诱导电流密度和雷达散射截面(RCS)计算验证表明,时域有限体积法是一种高精度有效的时域方法。

关键词:麦克斯韦方程组;计算流体力学;时域有限体积法;雷达散射截面

中图分类号:V211.3　　　**文献标识码**:A

0 引　言

电磁波传播、散射以及衍射等时变电磁现象由电磁学中麦克斯韦方程组控制,结合初值和边界值构成具有唯一确定电磁场解的偏微分方程系统。

电磁场计算方法包括频域方法和时域方法。频域方法主要包括高频渐进方法和积分方程方法。高频渐进方法是一种射线跟踪方法,把散射和衍射近似于一种局部现象,其应用范围受到各种限制。积分方程法中的矩量法(MOM)虽然计算精度高,但是因为包含矩阵求逆运算,所以局限于特征尺度不大于 10 倍波长的三维物体电磁散射。时域方法由于直接求解时变麦克斯韦方程组,因此入射电磁波既可为简谐波也可为宽频脉冲波,对脉冲波在一次计算中可给出多个频率下的雷达截面,这不同于传统频域方法一次仅能计算单频简谐波散射。

最著名的时域方法是 1966 年 Yee[1]提出的时域有限差分法,采用电场、磁场矢量空间、时间交错放置的笛卡儿直角网格,计算方法是二阶中心差分格式(蛙跳格式),在模拟复杂外形物体散射时存在阶梯效应,因此 Holland[2]进一步研究了非正交曲线坐标系中的有限差分法。

近十年来,由于麦克斯韦方程组和流体力学中欧拉方程都同属于双曲型偏微分方程组,同样的数学特性使得计算流体力学中非定常方法逐渐被应用于电磁问题,其中有限体积法由于求解守恒形式的欧拉方程或 N– S 方程,能保持质量、动量、能量等物理守恒量在整体和局部积分空间中守恒,并在模拟复杂流场运动中取得成功而得到了广泛应用,因此 Shankar[3]和 Shang[4]等开始把计算流体力学技术中时域有限体积方法应用于电磁散射和辐射问题的计算。

本文利用时域有限体积法求解散射场形式的麦克斯韦积分方程组,空间方向使用 MUSCL 格式和基于特征值的近似黎曼解构造单元边界通量,时间推进使用四阶龙格– 库塔法,数值计算了二维完全导电体(圆柱和 NACA0012 翼型)电磁散射的表面诱导电流密度和雷达散射截面,并与精确解或文献结果作了比较。

1 控制方程

1.1 麦克斯韦方程组

电磁场满足时变麦克斯韦方程组:

法拉第电磁感应定律:

$$\frac{\partial \boldsymbol{B}}{\partial t} + \nabla \times \boldsymbol{E} = 0 \tag{1}$$

安培定理:

$$\frac{\partial \boldsymbol{D}}{\partial t} - \nabla \times \boldsymbol{H} = -\boldsymbol{J} \tag{2}$$

电场的高斯定理:

$$\nabla \cdot \boldsymbol{D} = \rho \tag{3}$$

许勇, 乐嘉陵. 基于 CFD 的电磁散射数值模拟. 空气动力学学报, 2004, 22(2): 185-189.

磁场的高斯定理：

$$\nabla \cdot \boldsymbol{B} = 0 \qquad (4)$$

式中，$\boldsymbol{B}$ 是磁感应强度矢量，$\boldsymbol{H}$ 是磁场强度，$\boldsymbol{D}$ 是电位移矢量，$\boldsymbol{E}$ 是电场强度矢量，$\boldsymbol{J}$ 是自由电流密度，ρ 是自由电荷密度。

磁感应强度和磁场强度以及电位移矢量和电场强度之间有本征关系

$$\boldsymbol{B} = \mu \boldsymbol{H} \qquad \boldsymbol{D} = \varepsilon \boldsymbol{E} \qquad (5)$$

其中 μ 是磁导率，ε 是介电常数。

电磁散射空间中电磁总场是入射电磁场和散射电磁场之和，为了避免在计算空间中传播入射波带来数值误差，使用散射场形式的麦克斯韦积分方程组。

在直角坐标系下，无源麦克斯韦方程组的两个旋度方程写为类似于欧拉方程的守恒形式：

$$\frac{\partial Q}{\partial t} + \frac{\partial F_x}{\partial x} + \frac{\partial F_y}{\partial y} + \frac{\partial F_z}{\partial z} = 0 \qquad (6)$$

其中：

$$Q = \begin{bmatrix} \boldsymbol{B} \\ \boldsymbol{D} \end{bmatrix} \quad F_x = \begin{bmatrix} \hat{x} \times \boldsymbol{E} \\ -\hat{x} \times \boldsymbol{H} \end{bmatrix}$$

$$F_y = \begin{bmatrix} \hat{y} \times \boldsymbol{E} \\ -\hat{y} \times \boldsymbol{H} \end{bmatrix} \quad F_z = \begin{bmatrix} \hat{z} \times \boldsymbol{E} \\ -\hat{z} \times \boldsymbol{H} \end{bmatrix}$$

模拟复杂外形物体，使用时不变坐标变换：$\xi = \xi(x, y, z)$，$\eta = \eta(x, y, z)$，$\zeta = \zeta(x, y, z)$ 得到曲线坐标系下麦克斯韦方程组守恒形式：

$$\frac{\partial Q}{\partial t} + \frac{\partial F_\xi}{\partial \xi} + \frac{\partial F_\eta}{\partial \eta} + \frac{\partial F_\zeta}{\partial \zeta} = 0 \qquad (7)$$

对于二维电磁散射问题，假设 $\partial/\partial z = 0$，原来六个状态变量的方程系统可分解为分别有三个状态变量的两个方程系统。即

横磁(TM)波极化：$Q = [B_x \quad B_y \quad D_z]^T$；

横电(TE)波极化：$Q = [B_z \quad D_x \quad D_y]^T$

1.2 归一化

磁感应强度矢量和电位移矢量各分量以入射平面电磁波幅度 $|B^i|$ 和 $|D^i|$ 归一化，空间坐标以入射电磁波长归一化，时间以入射电磁波周期归一化。ε 和 μ 分别以真空介电常数 ε_0 和真空磁导率 μ_0 归一化。

2 积分方法

2.1 空间离散

有限体积法能保持网格空间的通量守恒，对守恒

方程(7) 在计算空间(二维)的每个网格单元分别作时间、空间积分，得到麦克斯韦积分方程的半离散化形式：

$$\frac{\Delta \overline{Q}}{\Delta t} + \frac{\Delta \overline{F}_\xi}{\Delta \xi} + \frac{\Delta \overline{F}_\eta}{\Delta \eta} = 0 \qquad (8)$$

式中，符号"－"代表相应量的数值解，$\overline{Q}$ 为网格积分单元内变量的平均值，$\overline{F}_k(k = \xi, \eta)$ 分别代表网格积分单元分界处的通量。

对于以单元中心共置电磁场量的有限体积法，积分过程转化为平衡离散单元边界的通量，其空间精度相应体现在能否精确计算相应通量 F。数值通量 $\overline{F}$ 可定义为一个表征一维黎曼问题的基于特征值的积分，用通量差分方法求取为：

$$n \times E = n \times \frac{(\varepsilon c)_R E_R + (\varepsilon c)_L E_L + n \times (H_R - H_L)}{(\varepsilon c)_L + (\varepsilon c)_R} \qquad (9)$$

$$-n \times H = -n \times \frac{(\mu c)_R H_R + (\varepsilon c)_L H_L - n \times (E_R - E_L)}{(\mu c)_L + (\mu c)_R} \qquad (10)$$

式中 n 是波传播方向，计算中分别以 ξ，η 代替得到各个方向通量，c 是真空中的光速。

分界面处左右状态变量 $Q^L_{i+1/2}$，$Q^R_{i+1/2}$ 用 MUSCL 格式计算有：

$$Q^L_{i+1/2} = Q_i + \frac{\phi}{4}[(1-\kappa)\nabla + (1+\kappa)\Delta]Q_i \qquad (11)$$

$$Q^R_{i+1/2} = Q_{i+1} - \frac{\phi}{4}[(1+\kappa)\nabla + (1-\kappa)\Delta]Q_{i+1} \qquad (12)$$

式中 ϕ 是限制器，本文取 $\phi = 1$，κ 是格式的控制参数，∇ 和 Δ 分别是后差和前差算符。

2.2 时间推进

时间推进使用四阶 ($m = 4$) 龙格-库塔方法：

$$Q^{n+k/m} = Q^n - \lambda a_k R(Q^{n+(k-1)/m}) \qquad k = 1, m \qquad (13)$$

$$\lambda = \Delta t / V; \quad a_k = \frac{1}{m - k + 1}$$

2.3 边界条件

网格单元内无耗介质分界面处的电磁边界条件能自动满足，但有两类边界条件需要另行处理。一是完全导电体壁面的反射边界条件：

$$\hat{n} \times E^i = 0 \qquad (14)$$

$$\hat{n} \cdot \boldsymbol{B}^t = 0 \qquad (15)$$

$\hat{n}$是完全导电体表面单位外法矢, 显然以上两式是不完备的, 没有提供电场垂直于表面、磁场切向于表面分量的信息, 但贴近导电体壁面的积分单元通量计算需要磁场的切向分量值, 可补充近似边界条件:

$$(\hat{n} \times \boldsymbol{H})_B = (\hat{n} \times \boldsymbol{H})_R - \frac{\hat{n} \times \hat{n}}{\mu c} \times (E_R - E_B)$$
$$(16)$$

同时, 由于计算机存储的限制, 积分在有限的物理空间内进行, 在网格外边界使用一阶无反射边界条件, 根据信号传播的特征理论, 假设对应输入计算区域特征值的特征不变量为0。

2.4 雷达散射截面计算

入射电磁波为平面正弦波, 当雷达散射截面收敛达到相对误差小于 0.01, 时域电磁场数值解经傅立叶变换得到频域值, 并代入归一化表面诱导电流密度公式:

$$J = \frac{|\hat{n} \times \boldsymbol{H}^t|}{|\boldsymbol{H}^t|} \qquad (17)$$

和二维物体雷达散射截面计算式:

$$\sigma = \lim_{R \to \infty} 2\pi R \mid \boldsymbol{E}^s(\boldsymbol{R})/\boldsymbol{E}^i(\boldsymbol{R}) \mid^2 \qquad (18)$$

3 计算结果分析

3.1 圆柱

图1是圆柱外形的计算网格, 远场设置在3λ处, 类似于计算流体力学采用壁面加密几何级数网格, 一方面保证壁面边界通量计算的足够精度、提高表面诱导电流的计算精度, 从而得到好的雷达截面结果; 另

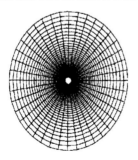

图1　圆柱计算网格

Fig. 1　Computational mesh for the circular cylinder

一方面远场稀疏网格的强耗散作用能减少计算量和远场反射杂波对内场的影响。图2是某一时刻圆柱散射电场 $\boldsymbol{B}_z^s$ 的等值线, 可见散射场离开计算区域而没有明显的反射杂波。

图2　完全导电圆柱散射电场等值线

Fig. 2　Scattered electric field contours for a perfectly conducting cylinder

图3和图4分别是TM波入射电尺寸 $ka= 10$(k是波数, a是圆柱半径)完全导电圆柱表面诱导电流

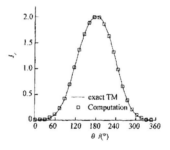

图3　归一化圆柱表面电流密度(TM波, $ka = 10$)

Fig. 3　Normalized surface current for the circular cylinder

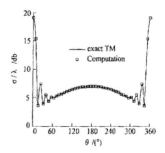

图4　圆柱双站雷达散射截面(TM波, $ka = 10$)

Fig.4　Bistatic RCS for the circular cylinder

密度和双站雷达散射截面分布，图 5 和图 6 是 TE 波
入射情况。散射角为散射方向与前向的逆时针夹角，
可见计算结果与解析级数解吻合很好。

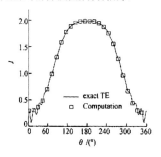

图 5 归一化圆柱表面电流密度(TE 波, $ka = 10$)
Fig. 5 Normalized surface current for
the circular cylinder

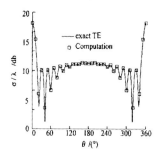

图 6 圆柱双站雷达散射截面(TE 波, $ka = 10$)
Fig. 6 Bistatic RCS for the circular cylinder

3.2 NACA0012 翼型

NACA0012 翼型是空气动力学中的实际工程外
形，本文计算 TM 波 $90°$ 垂直翼弦入射情况下，翼弦长
为 10λ，远场边界设置在 10λ 处，表面使用 200 个网
格点，辐射方向使用 100 个网格点。图 7 是时域有限
体积方法所得双站雷达截面分布与文献[5]有限差分
法结果的比较，可见在前向($\theta = 90°$)和后向($\theta = 270°$)两个最强散射峰值完全一致，其余散射波峰、波
谷的位置和大小基本吻合。

图 8 计算翼弦长为 2λ 的 NACA0012 翼型，TE 波
零度翼首入射情况下双站雷达散射截面分布。远场
设置在 3λ 处，表面使用 80 个网格，辐射方向使用 45
个网格。散射截面计算结果，与文献[6]中的紧致方
法结果比较吻合很好。

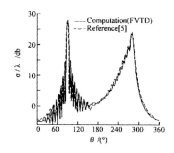

图 7 NACA0012 翼型双站 RCS(TM 波, $L = 10\lambda$)
Fig. 7 Bistatic RCS for NACA0012 airfoil
(TM wave polarization)

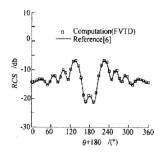

图 8 NACA0012 翼型双站 RCS(TE 波, $L = 2\lambda$)
Fig. 8 Bistatic RCS for NACA0012 airfoil
(TE wave polarization)

4 结 论

通过对二维完全导电圆柱和 NACA0012 翼型电
磁散射问题的数值计算，得到如下结论：
(1) 基于计算流体力学技术的时域有限体积法，
能够成功地推广到计算电磁学中。
(2) 直接求解麦克斯韦方程组电磁场的时域方
法，计算精度高而且能模拟电磁场时间发展变化。

参 考 文 献：

[1] YEE K S. Numerical solution of initial boundary value problems involving Maxwell's equations in isotropic media [R].
IEEE Trans. Antennas Propagation, 1966, 14: 302~ 307

[2] HOLLAND. Finite difference solution of Maxwell's equations
in generalized coordinates. IEEE Trans. Nuclear Science,
1983, 30.

[3] SHANKAR V, HILL W, MOHAMMADIAN A H. A CFD-
based finite volume procedure for computational electromagn-

ics interdisiplinary applications of CFD methods [R]. AIAA 1989-1987-CP.

[4] SHANG J S, GAITONDE D. Characteristic-based, time dependent Maxwell equations solver on a general curvilinear frame [R]. AIAA-93-3178.

[5] VINH H, DWYER H A, VAN DAM C P. Finite difference algorithms for the time domain Maxwell's equations- a numerical approach to RCS analysis [R]. AIAA-92-2989.

[6] HUH K S et al. A compact high order finite volume time domain/frequency domain method for electromagnetic scattering [R]. AIAA-92-0453.

CFD-based numerical simulation of electromagnetic scattering

XU Yong, LE Jia-ling

(China Aerodynamics Research and Development Center, Mianyang Sichuan 621000, China)

Abstract: The Maxwell equations of electromagnetism which governs electromagnetic phenomena are hypersonic sets of Partial Differential Equations(PDE) as the inviscid Euler equations of aerodynamics, then the Finite Volume Time Domain method (FVTD) of Computational Fluid Dynamics(CFD) was extended to Computational Electromagnetics(CEM), and directly numerical computed time dependent Maxwell's equations. A cell-centered, scattered field form Maxwell's equations conservative law was used in FVTD, and in which using characteristic-based approximate Riemann solution to reconstruct spatial flux and four step Runge-Kutta algorithm to time marching. The computed surface inducted current density and radar cross section of several 2-D perfectly conducting scatters in both Transverse-Magnetic(TM) and Transverse-Electric(TE) polarization demonstrated the method was effective and high accurate.

Key words: Maxwell's equations; computational fluid dynamics; finite-volume time domain method; radar cross section.

强激波阵面的非平衡特性研究

乐嘉陵[1],张若凌[1],王苏[2],崔季平[2]

(1. 中国空气动力研究与发展中心,四川 绵阳 621000;2. 中国科学院力学研究所,北京 100190)

摘　要:利用测量强激波后 N_2^+ 第一负系(0,0)带和(1,2)带的辐射,对强激波后振动温度历程的进行了测量;利用 Langmuir 探针技术,在低密度激波管中对强激波后电子数密度历程进行了测量;在弹道靶中进行了激波脱体距离的测量研究。测量和计算结果进行了对比,结果表明: $N_2^+ B^2\Sigma_u^+$ 态的激发比振动能的激发更快;实验测得的振动温度有明显的周期性振荡,在激波速度 $7.65km/s \sim 7.85km/s$、$p_1 = 1.33Pa$、实验段内径 $0.8m$ 下,实验有效时间只有约 $6.5\mu s$,实验中的电子数密度不能达到峰值,在约 10 倍波前自由程的实验有效区域内,电子密度的测量值与计算值吻合很好。激波脱体距离的测量误差约为 5%,是目前国内精度最高的结果。

关键词:强激波;振动温度;电子数密度;激波脱体距离

中图分类号:V211.7,V211.3　　　**文献标识码**:A

0 引　言

在激波波头里发生的分子的激发和弛豫过程,在很大程度上决定了高超声速再入飞行器周围的流场和辐射场。在一些飞行条件下,绕流场的非平衡效应有着强烈的影响,采用准确的计算模型有着实际的重要意义。因此,进行深入的强激波阵面非平衡结构的研究,有着重要的理论和应用价值。

1964 年 R. A. Allen[1] 在激波管中,通过测量波后气体的辐射,得到了强激波后 N_2^+ 的转动、振动温度和 N_2 的电子温度。实验气体为压强 $1mmHg$ 的纯氮气,激波速度 $V_s = 6.4km/s \sim 6.9km/s$。Park[2] 利用多温度模型详细分析了这个实验结果,发现实验测得的振动温度弛豫时间远远大于理论计算,并且实验测得的振动温度单调上升,而理论上有一个最大值。S. P. Sharma 和 W. Gillespie[3] 利用更先进的实验仪器和数据处理技术重复了这个实验,于 1990 年发表了新的不同的结果。他们利用 $N_2(2+)$ 光谱得到了波后几个点的振动温度和转动温度。根据他们得到的结果,波后的振动温度和理论计算符合要好些,但是转动温度的测量结果显示,转动和平动间达到平衡的时间较长,以至于通常认为二者处于平衡的假设

在一定区域内不成立。

上个世纪六十年代初期 Lin 等人[4,5]对空气介质里强激波阵面的电离结构进行了实验和计算研究,在激波速度小于 $7km/s$ 下实验测量了波后电子数密度峰值和电离距离。1969 年 Wison[6] 测量研究了激波速度大于 $9km/s$ 时的电离结构。Park[7] 总结了激波速度在 $2km/s \sim 13km/s$ 范围内已有的波后电子密度实验,并进行了计算比较。已经确定高温空气发生的三类电离过程,决定着强激波电离距离的行为:(1)激波速度小于 $7km/s$ 时,复合电离反应过程(如 $N + O \rightarrow NO^+ + e$)起主导作用,激波速度增大时电离距离变小;(2)激波速度在 $8km/s \sim 9.5km/s$ 之间时,电荷交换反应(如 $NO^+ + O \rightarrow NO + O^+$)和复合电离反应一起决定电离距离,激波速度增大时电离距离增大;(3)激波速度大于 $10km/s$ 时,电子冲击电离(如 $N + e \rightarrow N^+ + e + e$)起重要作用,电离距离随激波速度增大而变小。也就是说,电离距离随激波强度的变化在 $7km/s$ 和 $10km/s$ 的附近出现两个转折。

进行详细的激波脱体距离的实验和计算对比研究,对于检验物理 – 化学模型有重要的意义,Park[8] 认为是一种重要的难度适中的 CFD 验证方法。

早在上个世纪六十年代,R. K. Lobb[9] 就已经对

球模型激波体距离进行了大量的实验研究,得到了多种实验条件下的球模型激波脱体距离。1992 年,S. G. Rock、G. V. Candler 和 H. G. Hornung[10]利用在自由活塞激波风洞里得到的干涉照片,研究了 CO_2 的离解反应动力学。结果表明,Park 的反应模型比 McKenzie 的更准确。上世纪 90 年代,日本 Tohoku 大学的弹道靶实验进行得较多,曾利用弹道靶测得的激波脱体距离,研究了化学 – 振动耦合模型[11,12]。结果表明,采用 Park 模型,在有的情况下计算出的激波脱体距离比实验小 10% 左右,计算给出的激波脱体距离对于反应控制温度和振动能量源项并不敏感。他们的工作,对于理解中等高超声速区域(马赫数在 8 – 15 之间,速度 2.5km/s – 4.5km/s)的热化学过程,有着重要意义。

高超声速流动的热化学现象非常复杂,经过多年的研究已经取得了重大进展,并得到广泛应用。目前的高超声速热化学非平衡流动计算,大多采用 Park 的多温度模型[2]。从前文可以看出,关于强激波阵面的非平衡结构研究,需要进行确认和补充实验。如波后振动温度的历程需要进行确认实验;已有关于波后电离结构的研究,缺乏激波速度在 7km/s ~ 9km/s 之间的数据,这个速度区间是在电离距离随激波强度变化的第一个转折点附近。

笔者对强激波阵面的非平衡温度、电离结构和激波脱体距离进行了实验研究。首先,利用激波管测量强激波后振动温度的历程,对强激波后振动温度历程的测量过程进行探索。然后,利用 Langmuir 探针(静电探针)技术,在低密度激波管中对强激波后电子数密度历程进行测量,激波速度在 7.65km/s ~ 7.85km/s 之间,并进行测量和计算结果的对比。最后在 CARDC 弹道靶中进行了激波脱体距离的测量和计算对比研究,圆球模型直径 12mm,飞行速度 5.538km/s。

1 强激波后振动温度历程的测量与计算比较

在自由度内部平衡的情况下,强激波后气体发射光谱(化学发光除外)的强度分布和分子(或原子)总数有关,并且和某个(些)自由度的温度有关。利用发射光谱的强度及其分布,可以得到转动温度、振动温度、电子激发温度以及组分含量等物理量。

振动温度测量实验中,激波管实验段为 132mm ×

132mm 的方管。两台 0.5m 单色仪和两只光电倍增管(R456),对称地放在激波管两侧,分别测量波后 N_2^+ 第一负系(0,0)带(带头波长 3914.4Å)和(1,2)带(带头波长 4236.5Å)通过光学窗口(K9 玻璃)的辐射,并避开 CN 的紫外光(CNV)干扰。单色仪线色散率 16.6Å/mm,入缝 50μm,出缝 1mm,光谱分辨率可达 1Å,可以认为测量的是整个带的辐射。激波管实验段较细,为提高光强从而提高信号幅度,使有效光学窗口直径为 Φ5mm,这很大程度上降低了空间分辨率。Sharma 和 Gillespie 的实验利用了像增强器,因此能达到较高的空间分辨率。光学窗口距离管子尾部 128cm。高压段气体为 H_2 和 O_2 的混合气体,燃烧驱动工作气体。低压段工作(实验)气体为纯 N_2,纯度超过 99.9%,压强为 100Pa ~ 300Pa,激波速度为 5.0km/s ~ 6km/s,环境温度为 283.5K。实验中,利用 Langmuir 探针来测量实验的有效时间。

在振动模式内分子的能态遵从玻耳兹曼分布时,N_2^+ 第一负系(0,0)带和(1,2)带强度的比值为

$$ratio = \frac{I_{(0,0)}}{I_{(1,2)}} = e^{-[G_v(v_0) - G_v(v_1)]/kT_v} \cdot \frac{v_{(0,0)}^4}{v_{(1,2)}^4} \cdot \frac{q_{(0,0)}}{q_{(1,2)}} \quad (1)$$

其中:$v_0 = 0$,$v_1 = 1$,$G_v(0) = 1099.475cm^{-1}$,
$G_v(1) = 3577.605cm^{-1}$,$q_{(0,0)} = 0.6481$,
$q_{(1,2)} = 0.2889$,$v_{(0,0)} = 25566.05cm^{-1}$,
$v_{(1,2)} = 23620.5cm^{-1}$。通过实验得到(0,0)带和(1,2)带的辐射强度比值,代入(1)式就可得到波后 N_2^+ 振动温度 T_v。详细的实验原理及数据处理方法见文献[13]。

实验测量采用的是线性响应系统,标定是不需要的[13]。在实验有效时间内气体可以达到平衡态,计算得到的平衡温度值可以用来作为标定基准,避免标定过程带来的误差。振动温度的计算方法为

$$\frac{ratio}{ratio_{eq}} = \frac{e^{[G_v(v_0) - G_v(v_1)]/kT_v}}{e^{[G_v(v_0) - G_v(v_1)]/kT_{eq}}} \quad (2)$$

计算时,T_{eq} 取理论计算值,$ratio_{eq}$ 一般取在波后 2500 ~ 3000 个波前自由程处的 $ratio$。

利用 N_2^+ 第一负系(0,0)带的辐射测量结果来计算 N_2^+ 的电子温度。得到电子温度的方法为

$$\frac{I_{(0,0)}}{I_{(0,0),eq}} = \frac{\rho_{N_2^+}}{\rho_{N_2^+,eq}} \cdot \frac{e^{-36800/T_{e,B}}}{e^{-36800/T_{eq}}} \quad (3)$$

通过上式,代入计算出的 N_2^+ 密度值,可以求出

N_2^+ 的电子温度 $T_{e,B}$。Allen 在利用 N_2 第一正系 (1+) 的光谱求电子激发温度时,采用的公式形式和 (3) 式一样,并且 N_2 的密度在整个测量过程中为常数。

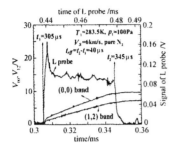

图 1　实验典型原始信号 ($p_1 = 100Pa$)
Fig. 1　Typical signals of experiment ($p_1 = 100Pa$)

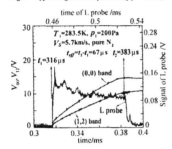

图 2　实验典型原始信号 ($p_1 = 200Pa$)
Fig. 2　Typical signals of experiment ($p_1 = 200Pa$)

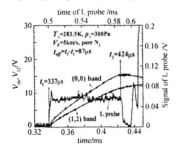

图 3　实验典型原始信号 ($p_1 = 300Pa$)
Fig. 3　Typical signals of experiment ($p1 = 300Pa$)

N_2^+ 第一负系 (0,0) 和 (1,2) 带信号及 Langmuir 探针信号如图 1~图 3 所示。由于原始信号有热噪声,并且有相对零点的漂移,在求比值 ratio 前,对原始信号作光滑和平移处理。光滑和平移处理属于纯粹的数学过程,可能会改变信号的物理性质。图 2 的原始信号经光滑和平移处理之后的信号及比值在图 4 中显示,比较图 4 和图 2 可以看出所作的光滑和平移处理没有造成信号物理性质的改变。

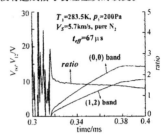

图 4　经平移和平滑处理后信号及比值
Fig. 4　Two processed signals and their ratio

温度测量的典型结果在图 5~图 7 中显示,其中理论值是利用求解一维强正激波后非平衡流动得到的,采用的是三温度优先离解模型[13]。如图所示,实验测得的振动温度单调上升,而理论上有一个最大值。实验测量的是波后 N_2^+ 的振动温度,计算中所有分子的振动温度相同。笔者认为实验结果说明 N_2^+ 分子的振动激发速率非常缓慢,不满足所有分子的振动温度相同的假设。Allen、Sharma 及笔者的测量结果比较在图 8 中显示,在非平衡区笔者得到的振动温度历程的大致趋势和文献[1]是一致的。

在图 5~图 7 中振动温度有周期性振荡。Bradley[14] 曾简单讨论过波后物理量的振荡,给出了两种可能的解释:激波沿管子运动时,由激波速度的振荡引起;或者在波后,一些横向 (和流动方向垂直) 的波叠加到气体的运动上。

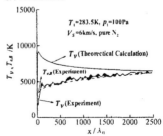

图 5　典型测量结果 ($\lambda_0 = 63.4\mu m$)
Fig. 5　Typical results of experiment ($\lambda_0 = 63.4\mu m$)

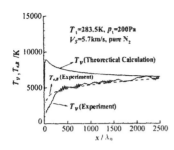

图6 典型测量结果($\lambda_0 = 31.7\mu m$)

Fig. 6 Typical results of experiment($\lambda_0 = 31.7\mu m$)

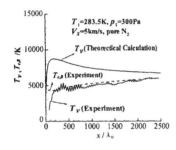

图7 典型测量结果($\lambda_0 = 21.1\mu m$)

Fig. 7 Typical results of experiment($\lambda_0 = 21.1\mu m$)

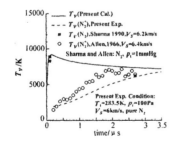

图8 测量与国外文献结果的比较

Fig. 8 Comparisons with other's results

根据国外和笔者的实验结果,可以说 N_2 与 N_2^+ 之间振动能的交换很慢,对此理论上仍然没有具体的解释。此外,通过本实验结果可以看出 N_2^+ 的 $B^2\Sigma_u^+$ 态的电子能比振动能激发得更快。

2　强激波后电子数密度研究

强激波波后气体受激波压缩加热,波后气体由于高温会发生电离,并在一定区域内经历非平衡电离过程。计算得到的电离距离,对于采用的物理 – 化学模型比较敏感。在文献[7]中,Park 的计算结果清晰地显示了不同的反应速率常数,会给出不同的波后电子数密度分布轮廓。因此强激波阵面气体电离特性的测量有着重要意义。

实验在中科院力学所 LHD 低密度激波管上进行。激波管驱动段长 1.6m、内径 22cm,低压段长 16m、内径 0.8m。为了避免杂质干扰,选用高纯氮气和氧气(纯度 >99.999%)配制的空气为实验气体,氮气和氧气的体积比为 4:1,实验压强 $p_1 = 1.33$Pa (0.01mmHg)。由于低压段初始压强很低,采用分子涡轮泵为低压段抽气,以减小激波管管壁吸附杂质的影响。可以实现极限真空 $< 10^{-5}$mmHg,关闭分子涡轮泵后,系统能在极限真空维持 10 分钟以上。

采用 Langmuir 静电探针监测波后 2 区离子数密度的变化。静电探针为圆柱型,用不锈钢丝制成,负压偏置,$V_p = 9$V,长 $l_p = 1$cm,半径 $r_p = 0.25$mm,安装在距离低压段尾盖 2m 处的管中心位置,探针轴向与气体流动方向一致。采样电阻阻值 $R_s = 100\Omega$。

在实验条件下,探针在接近自由分子流条件下工作,对气流状态没有明显扰动。根据 Langmuir 探针理论分析[15],可以得到

$$n_e = 9.42 \times 10^{15} T_e^{-1/2} j_i \quad (4)$$

其中 n_e 的单位是 cm^{-3},电流密度 j_i 的单位是 A/cm^2。利用测量得到的电流密度 j_i 和计算得到的电子温度 T_e(K),带入(4)式即可得到电子数密度 n_e。

实验测得的低压段压强 $p_1 = 1.33$Pa 的波后电子数密度以及计算结果在图9 – 图10 中显示。计算的激波速度为 7.65km/s,激波马赫数 $M_s = 22$。实验中三次运行的激波速度分别为 7.65km/s、7.85km/s 和 7.85km/s。由于球模型驻点线和正激波波后流动的物理量是近似相等的[13],因此利用球模型绕流的计算方法来计算同样的自由流条件绕球的流动,把驻点线上物理量的数值当成正激波波后物理量的数值(这样可以考虑气体的粘性效应)。在计算中,采用三温度优先离解模型,球半径取为 $R_n = 2$m,保证在脱体激波波后有适当的距离。为了减小边界条件的影响,在

计算中采用了绝热和非催化物面条件。

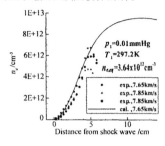

图9　强激波后电子数密度分布（空气）
Fig. 9　Electron density behind strong shock (air)

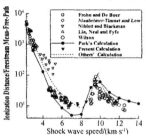

图10　电离距离随激波速度的变化（空气）
Fig. 10　Ionization distance vs. shock speed (air)

从图9可以看出，实验和计算的结果符合很好。实验的电子数密度在波后约5cm处陡然下降，是波后驱动气体的到来引起。这说明在如此低的低压段压强下，实验有效时间只有约6.5μs。在如此短的有效时间内，实验中的电子数密度不能达到峰值。在约10倍波前自由程的实验有效区域内，电子数密度的测量值与计算值吻合很好。

图10显示的是波后电离距离随激波速度的变化，其中，除了笔者给出的三个计算点外，其他数据引自文献[2,7]。笔者的计算结果和Park在文献[2]中引用的计算结果（图中的虚线）符合得较好。

3　激波脱体距离的实验与计算研究

如图11所示，弹道靶由二级轻气炮和靶室构成。二级轻气炮由火药室、压缩管、高压段和发射管组成。火药点火推动活塞压缩压缩管内的轻气体（氢气），达到极高的温度和压力。这些高温、高压的轻气体在高压段破膜后膨胀，并推动发射管内的弹丸加速运动，使

模型达到一定的发射速度。由于轻质气体的极限膨胀速度很高，所以二级轻气炮可以实现超高速发射。

利用弹道靶和阴影仪可得到模型自由飞行的阴影照片，从而获得激波脱体距离。计算出的激波脱体距离对于所采用的物理－化学模型比较敏感，因此进行详细的激波脱体距离的实验和计算对比研究，对于检验物理－化学模型有重要意义，Park[70]认为是一种重要的难度适中的CFD验证方法。

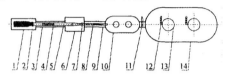

1—火药室；2—低压大膜片；3—活塞；4—压缩管；
5—轻气体；6—高压段；7—高压小膜片；
8—模型及弹托、底托；9—发射管；10—膨胀段；11—快速阀；
12—激光探测器；13—阴影仪照相窗口；14—靶室
图11　二级轻气炮弹道靶示意图
Fig. 11　Illustration of the ballistic range

由于地面飞行实验的模型尺寸不能很大，激波脱体距离的实验和计算对比，对于实验中光源的脉冲工作时间要求很高。譬如说，直径10mm的圆球在飞行速度为5km/s时，此时激波脱体距离～0.5mm。如果光源的脉冲工作时间为10ns，则在此时间内模型的飞行距离为0.05mm，激波脱体距离的测量误差达到10%。

本实验在CARDC的气动物理靶中进行。该弹道靶的二级轻气炮总长37.5m，靶室直径0.6m，靶室长度24.4m，测试段长为9m。实验中，圆球模型直径$D = 12$mm，靶室内空气的压强$p_0 = 9.9$mmHg，模型飞行速度$V = 5.538$km/s，室温$T_0 = 289.5$K。脉冲激光器为Nd－YAG，型号为SAGA PRO230/10，输出激光波长532nm，脉冲能量1320mJ。实验时脉冲工作时间5ns，激波脱体距离的测量误差约为5%。实验时脉冲工作时间和Tohoku大学[12]的一样，因而获得的激波脱体距离的测量精度不低于国外水平，在国内目前精度最高。

采用不同的振动－离解耦合模型来计算流场，结合阴影照片，来进行物理－化学模型的参数研究。具体地说，采用两温度模型，分别比较优先离解和非优先离解模型，二者差别只在振动能量源项。对优先离解模型，c分别取0.5和0.3。对于分子离解反应，

控制温度为 $T_q = T^{1-\alpha} T_v{}^\alpha, \alpha = 0.3^{[13]}$。自由流温度 T_∞ = 289.5K,压强 $p_\infty = 9.9$mmHg。计算物面条件为等温和全催化,$T_w = 800$K。

实验和计算阴影图的比较结果见图12 – 图14,图15是两种计算模型的阴影图比较。从图12 – 图15可以看出,三个计算结果的差别以及计算和实验的差别都比较微小。通过图15右侧的局部放大图可以看出,非优先离解模型给出的激波脱体距离最小。

图16是三种计算模型的驻点线密度分布的比较。可以看出,激波脱体距离约为 $0.085R_n$,非优先离解模型和优先离解模型给出的密度分布,其轮廓外缘位置差约为5%,和实验测量误差一样。要更明显地显示不同模型的差别、更清楚地判断模型的优劣,需要加大模型尺寸、降低靶室密度,并缩短光源的脉冲工作时间。在弹道靶中可以进行激波脱体距离、激波形状、密度和辐射等物理量的测量,对于检验计算模型的准确性有重要意义,文献提到NASA曾考虑过建设大尺寸的弹道靶[8]。

图12 实验与计算阴影图比较(右侧为局部放大)
非优先离解模型 上:实验;下:计算
Fig. 12 Shadowgraphs of experiment and calculation, using nPD model

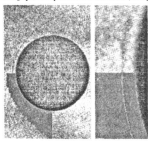

图13 实验与计算阴影图比较(右侧为局部放大)
优先离解模型 $c = 0.5$ 上:实验;下:计算
Fig. 13 Shadowgraphs of experiment and calculation, using PD model with $c = 0.5$

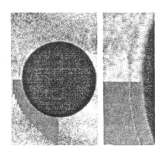

图14 实验与计算阴影图比较(右侧为局部放大)
优先离解模型 $c = 0.3$ 上:实验;下:计算
Fig. 14 Shadowgraphs of experiment and calculation, using PD model with $c = 0.3$

图15 计算阴影图比较(右侧为局部放大)
上:非优先离解模型 下:优先离解模型,$c = 0.5$
Fig. 15 Shadowgraphs of calculations, using nPD model and PD model with $c = 0.5$

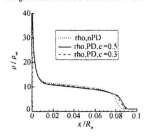

图16 驻点线密度分布比较
Fig. 16 Comparison of density along stagnation line

4 结束语

对强激波阵面的非平衡温度和电离结构进行了

激波管实验研究。在激波管中测量了强激波波后N_2^+第一负系(0,0)带和(1,2)带通过光学窗口的辐射,对强激波后振动温度历程的测量过程进行了探索。利用Langmuir探针(静电探针)技术,在低密度激波管中对强激波后电子数密度历程进行了测量,并进行了测量和计算结果的对比。最后在CARDC弹道靶中进行了激波脱体距离的测量和计算对比研究,圆球模型直径12mm,飞行速度5.538km/s。

结果表明:N_2^+ $B^2\Sigma_u^+$态的激发比振动能的激发更快;实验测得的振动温度有明显的周期性振荡;在激波速度7.65km/s ~ 7.85km/s、$p_1 = 1.33$Pa的低压段压强、实验段内径0.8m下,实验有效时间只有约6.5μs,实验中的电子数密度不能达到峰值。在约10倍波前自由程的实验有效区域内,电子数密度的测量值与计算值吻合很好。激波脱体距离的测量误差约为5%,是目前国内精度最高的结果。

参 考 文 献:

[1] ALLEN R A. Nonequilibrium shock front rotational, vibrational and electronic temperature measurements [R]. NASA CR - 205(also AVCO RR - 186), 1964.

[2] PARK C. Nonequilibrium hypersonic aerothermodynamics [M]. New York:John Wiley & Sons, 1990.

[3] SHARMA S P, GILLESPIE W. Nonequilibrium and equilibrium shock front radiation measurements [R]. AIAA 90 - 0139, 1990.

[4] LIN S C, NEAL R A and FYFE W I. Rate of ionization behind shock waves in air, I experiment results[J]. *Physice of Fluids*, 1962, 5(12):1633 - 1648.

[5] LIN S C, Teare J D. Rate of ionization behind shock waves in air, II theoretical interpretation [J]. *Physice of Fluids*, 1963, 6(10): 355 - 375.

[6] WILSON J F. Ionization rate of air behind high speed shock waves[J]. *Physice of Fluids*, 1966, 9(10):1913 - 1921.

[7] PARK C. Review of chemical - kinetic problems of future NASA missions, I: earth entries[J]. *J. Thermophysics and Heat Transfer*, 1993, 7(3):385 - 398.

[8] PARK CHUL. Validation of CFD codes for real - gas regime [R]. AIAA 97 -2530, 1997.

[9] LOBB R K. Experimental measurement of shock detachment distance on spheres fired in air at hyper velocities. The high temperature aspects of hypersonic flow[A]. Proc AGARD - NATO Specialist's Meeting [C], Belgium, pp. 519 - 527, 1962.

[10] ROCK S G, CANDLER G V, HORNUNG H G. Analysis of thermo - chemical nonequilibrium models for carbon dioxide flows[R]. AIAA 92 - 2852, 1992.

[11] FURUDATE, MICHIKO, SUZUKI, TOSHIYUKI. Vibration - dissociation coupling effects on shock standoff distances [R]. AIAA 2000 -2499, June 2000.

[12] NONAKA S, TAKAYAMA K. Overview of ballistic range program at tohoku university[R]. AIAA 98 - 2604, 1998.

[13] 张若凌. 高超声速非平衡流动的计算和实验研究[D]. [博士学位论文]. 中国空气动力研究与发展中心研究生部,2004年3月.

[14] BRADLEY, JOHN N. Shock waves in chemistry and physics [M]. Methuen & Co. LTD., London, John Wiley & Sons Inc., New York, 1962.

[15] SONIN AIN A. Free - molecule Langmuir probe and its use in flowfield studies[J]. *AIAA J.*, 1966, 4(9):1588 - 1596.

The research of nonequilibrium characteristics of strong shock front

LE Jia – ling[1], ZHANG Ruo – ling[1], WANG Su[2], CUI Jing – ping[2]

(1. *China Aerodynamics Research and Development Center*, *Mianyang* 621000, *China*;
2. *Institute of Mechanics*, *Chinese Academy of Sciences*, *Beijing* 100190, *China*)

Abstracts: The vibrational temperature and electron number density behind strong shock waves were measured in shock tubes, and the shock stand – off distance was measured in a ballistic range. The results were compared with those of theoretical calculations. The vibrational temperature was derived by measuring the radiation of (0,0) and (1, 2) bands of N_2^+ first negative system. According to the experimental results, the electronic energy of N_2^+ can be excited faster than its vibrational energy, and there are periodic fluctuations in the measured vibrational temperature. In the measurement of electron number densitiy behind strong shock waves ($p_1 = 0.01$mmHg, $V_s = 7.65 \sim 7.85$km/s) in a low density shock tube ($\Phi 0.8$m), the effective test time was only about 6.5μs, so the electron number densitiy could not reach the peak in such a short time. The agreement between measurement and calculation are good during the effective test region, which length is about 10 times of freestream mean – free – molecular path. The resolution of shock stand – off distance measurement is 5%, which is the best in China at present.

Key words: strong shock wave; vibrational temperature; electron number densitiy; shock stand – off distance

激波与流动显示研究

Mach-reflection flow at hypersonic speeds

J. Le, H. Ni, G. Hu
China Aerodynamics Research and Development Center
P.O. Box 211, Mianyang, Sichuan, China

Abstract: Experiment and computation on the reflection of a shock wave of $M_s = 7.6$ from a wedge of $\theta_w = 40°$ are investigated. Experiments are carried out in the CARDC shock tube. Computation of both equilibrium and nonequilibrium Mach-reflection flow is based on the laminar full Navier-Stokes (N-S) equation in which the inviscid and source terms are calculated by a second-order upwind TVD scheme and a point-implicit method, respectively. Computational results are in good agreement with those of experiment in all flow fields.

Key words: Mach reflection, Hypersonic flow, TVD scheme

1. Introduction

As judged by comparison with a great number of shock-tube experiments, numerical methods based on the Euler equation with second-order upwind difference schemes such as the Godunov or TVD scheme are sufficiently accurate in the analysis of the perfect inviscid, complex-wave structure of Mach-reflection flow (Glaz et al. 1986). However, for a very strong incident shock wave, the temperature behind the shock is high enough to bring about vibrational excitation and dissociation of the air, resulting in strong effects of the viscosity and nonequilibrium on the experimental flow field. Therefore, further development of the numerical modeling of the viscous nonequilibrium flow field is necessary for the analysis of experimental data and for the validation of computer codes.

To date, there have been a few typical quantitative measurements of density fields widely used for the validation of computer codes for viscous nonequilibrium usteady flow. The purpose of the experiments carried out in CARDC's shock tube at $M_s = 7.6$, $\theta_w = 40°$ is to study the detailed flow structure of the region about the two triple points, including the reflected shock wave, weak slip layer, and Mach stem.

With regard to the numerical simulation, there have been several results only based on the Euler equation (Glaz 1987, Needham 1991, Le and Hu 1992, Clinton and Groth 1993). In the present paper, the laminar full N-S equation is used for numerically simulating both thermal and chemical nonequilibrium Mach-reflection flow. The high-temperature air is assumed to consist of 11 species: O_2, N_2, NO, O, N, O_2^+, N_2^+, NO^+, O^+, N^- and the reaction-rate

Le Jialing, Ni Hongli, Hu Guangchu. Mach reflection flow at hypersonic speeds. 20th international symposium on shock waves. 1995.

coefficients in 32 chemical reactions are taken from the data set compiled by Dunn and Kang (1973). The forward reaction-rate constants K in the chemical source terms are based on Park's two-temperature model of T and Tv: $K = K(\overline{T})$, $\overline{T} = \sqrt{TTv}$. Source terms of three vibrational energy equations with vibration temperature Tv_{O2}, Tv_{N2}, Tv_{ON}. are based on the Landau-Teller method for harmonic oscillators. The relaxation times are obtained from the semi-empirical expressions from Millikan and White. The high-temperature gas-transport coefficients are taken from accurate curve fit (Roop and Gupta 1990).

2. Experiments and results

The experiments were performed in CARDC's shock tube with a test section of 132 mm × 132 mm, which has two pairs of observation windows of 80-mm diameter. The distance between the windows is 400 mm. The model located in the first window is a 2D wedge with a 40° compression corner and height of 88 mm. Double-exposure holographic interferometry was used for quantitatively measuring the density fields and differential interferometry was used for studying the flow structure. Figure 1 is a differential interferogram in which both the first and second slip layers can be distinguished easily. A strong jump in interferomentric fringes against the wall can also be seen. This jump means that values of density begin to increase sharply due to the effect of the cold wall at 298 K. Figure 2 is a holographic interferogram from which the density of the flow field can be evaluated in a very simple way by directly counting the fringes without relying on analytical considerations of three-shock theory near the triple points. The fringe-counting data-reduction technique is obviously inaccurate in chemically reacting flow. as the Gladston-Dale constant K is dependent on the mass concentration of the species. However, in the present

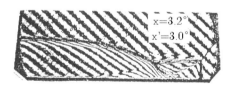

Figure 1. Differential interferogram from CARDC shock-tube experiment. $M_s = 7.6$, $\theta_w = 40°$, $P_0 = 2.63$ KPa, $T_0 = 289$ K.

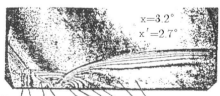

Figure 2. Holographic interferogram from CARDC shock tube. $M_s = 7.6$, $\theta_w = 40°$, $P_0 = 2.63$ KPa, $\rho_0 = 3.21 \times 10^{-5}$ g/cm^3; $\Delta\rho_0 = \lambda/KL$, $\lambda = 6.943 \times 10^{-3}$ cm, $K = 0.225$ cm^3/g. $L = 13.2$ cm; $\rho_N/\rho_0 = 1 + N(\Delta\rho_0/\rho_0) = 1 + N \times 0.729$.

test case, the value of the density some distance behind the incident shock obtained by counting fringes with $K = 0.225 \cdot 10^{-3}$ m³/kg is very close to the value calculated from the Rankine-Hugoniot conditions for the air equilibrium state. Therefore, the approximate wall density distribution shown in Fig. 3 is evaluated with the same value of K. Strong relaxation effects both behind the incident shock and reflected shock are also indicated in this figure.

3. Numerical calculation and results

Governing equations and difference schemes. The 2D laminar conservative full N-S equations with a gas of N reacting species and M vibrational excited species can be written:

$$\frac{\partial U}{\partial t} + \frac{\partial F}{\partial x} + \frac{\partial G}{\partial y} = \frac{1}{Re}\left(\frac{\partial Q}{\partial x} + \frac{\partial R}{\partial y}\right) + S, \qquad (1)$$

where $U = (\rho_s, \rho u, \rho v, \rho_m e_{vm}, E)^T$, $F = (\rho_s u, \rho u^2 + p, \rho uv, \rho_m e_{vm} u, \rho u h_0)^T$, $G = (\rho_s v, \rho uv, \rho v^2 + p, \rho_m e_{vm} v, \rho v h_0)^T$, and

$$Q = \left(\rho D_s \frac{\partial x_s}{\partial x}, \tau_{xx}, \tau_{xy}, \rho D_m \frac{\partial x_m}{\partial x} e_{vm} + k_{vm}\frac{\partial T_{vm}}{\partial x}\right.,$$
$$\left.\sum_{S=1}^{N} \rho D_s \frac{\partial x_s}{\partial x} h_s + k\frac{\partial T}{\partial x} + \sum_{m=1}^{M} k_{vm}\frac{\partial T_{vm}}{\partial x} + u\tau_{xx} + v\tau_{xy}\right)^T$$

$$R = \left(\rho D_s \frac{\partial x_s}{\partial y}, \tau_{xy}, \tau_{yy}, \rho D_m \frac{\partial x_m}{\partial y} e_{vm} + k_{vm}\frac{\partial T_{vm}}{\partial y}\right.,$$
$$\left.\sum_{S=1}^{N} \rho D_s \frac{\partial x_s}{\partial y} h_s + k\frac{\partial T}{\partial y} + \sum_{m=1}^{M} k_{vm}\frac{\partial T_{vm}}{\partial y} + u\tau_{xy} + v\tau_{yy}\right)^T$$

and $S = (\dot{\omega}_s, 0, 0, \rho_m \dot{e}_{vm} + \dot{\omega}_m e_{vm}, 0)$. The multispecies state equation is $P = \sum_{s=1}^{N} \frac{\rho_s}{M_s} RT$. Introducing the transformation of variables $\tau = t$, $\xi = \xi(x,y)$, $\eta = \eta(x,y)$, Eq. (1) becomes

$$\frac{\partial \hat{U}}{\partial \tau} + \frac{\partial \hat{F}}{\partial \xi} + \frac{\partial \hat{G}}{\partial \eta} = \frac{1}{Re}\left(\frac{\partial \hat{Q}}{\partial \xi} + \frac{\partial \hat{R}}{\partial \eta}\right) + \hat{S}. \qquad (2)$$

The discrete conservation equation (2) can be written in the following form:

$$D_{i,j}^n \Delta \hat{U}_{i,j} = -\frac{\Delta\tau}{\Delta\xi}\left(\tilde{F}_{i-\frac{1}{2},j}^n - \tilde{F}_{i-\frac{1}{2},j}^n\right) - \frac{\Delta\tau}{\Delta\eta}\left(\tilde{G}_{i,j+\frac{1}{2}}^n - \tilde{G}_{j,j-\frac{1}{2}}^n\right)$$
$$+\frac{1}{Re}[\frac{\Delta\tau}{\Delta\xi}\left(\tilde{Q}_{i+\frac{1}{2},j}^n - \tilde{Q}_{i-\frac{1}{2},j}^n\right) + \frac{\Delta\tau}{\Delta\eta}\left(\tilde{R}_{i,j+\frac{1}{2}}^n - \tilde{R}_{j,j-\frac{1}{2}}^n\right)] + \Delta\tau\hat{S}_{i,j}^n$$

$$D_{i,j} = \left(I - \Delta\tau\frac{\partial \hat{S}}{\partial \hat{U}}\right)_{i,j} \qquad (3)$$

The following two-step Range-Kutta method is used so that the conserved variables $\widehat{U}$ can be obtained with second-order accuracy in time and space

$$\overline{U}_{i,j}^{n+1} = \widehat{U}_{i,j}^n + \Delta\widehat{U}_{i,j}^n \qquad \widehat{U}_{i,j}^{n+1} = \frac{1}{2}(\widehat{U}_{i,j}^n + \overline{U}_{i,j}^{n+1}) + \frac{1}{2}\Delta\overline{U}_{i,j}^{n+1}, (4)$$

where numerical flux in ξ-direction is $\widetilde{F}_{i+\frac{1}{2},j} = \frac{1}{2}(\widetilde{F}_{i,j} + \widetilde{F}_{i-1,j} + R_{i+\frac{1}{2},j}\Phi_{i+\frac{1}{2},j})$. The vector $\Phi_{i+\frac{1}{2},j}$ in the upwind TVD scheme can be expressed by $\Phi_{i+\frac{1}{2},j}^l$: $\Phi_{i+\frac{1}{2},j}^l = \delta(a_{i+\frac{1}{2},j}^l)(g_{i-1,j}^l + g_{i,j}^l) - \Psi(a_{i+\frac{1}{2},j}^l + \gamma_{i+\frac{1}{2},j})a_{i+\frac{1}{2},j}^l; l = 1, 2, ..., N + M + 3$. The numerical flux $\widetilde{G}$ in the η direction can also be defined similarly.

3.1. Results and discussion

To judge the predictive capabilities of present computer codes, a double Mach-reflection flow of incident $M_s = 8.7$ impinging on a $27°$ wedge is numerically calculated using N-S and Euler equations. Figure 4 is the schematic diagram

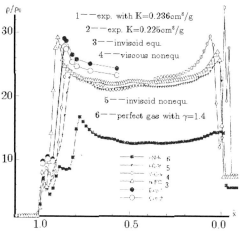

Figure 3. Comparison of density distribution on the wall. $M_s = 7.6$, $P_0 = 2.63$ KPa, $T_0 = 289$ K.

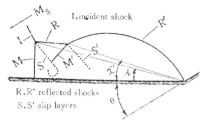

Figure 4. Schematic diagram of double Mach reflection; X, X' first and second triple-point angles.

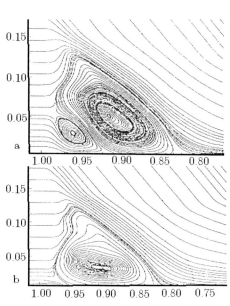

Figure 5. Computed streamlines near Mach-stem region: viscous nonequilibrium (top); inviscid nonequilibrium (bottom). $M_s = 8.7$, $\theta_w = 27°$, $P_0 = 4.1$ KPa, $T_0 = 299.2$ K.

of this flow. Figures 5a and 5b show the computed streamlines of inviscid and viscous nonequilibrium flow near the Mach-stem region. The rollup pattern with two vortices (shown in Fig. 5a) is in substantial agreement with the inter-

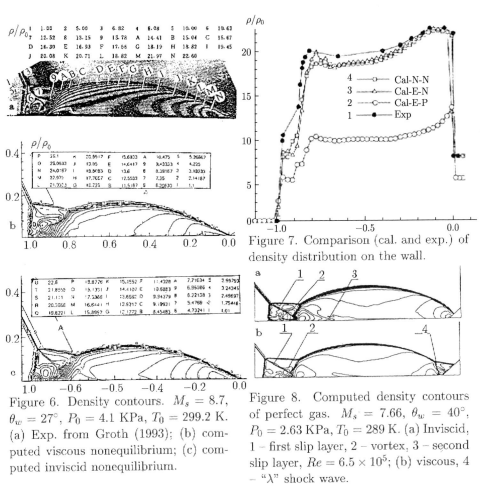

Figure 7. Comparison (cal. and exp.) of density distribution on the wall.

Figure 6. Density contours. $M_s = 8.7$, $\theta_w = 27°$, $P_0 = 4.1$ KPa, $T_0 = 299.2$ K. (a) Exp. from Groth (1993); (b) computed viscous nonequilibrium; (c) computed inviscid nonequilibrium.

Figure 8. Computed density contours of perfect gas. $M_s = 7.66$, $\theta_w = 40°$, $P_0 = 2.63$ KPa, $T_0 = 289$ K. (a) Inviscid, 1 – first slip layer, 2 – vortex, 3 – second slip layer, $Re = 6.5 \times 10^5$; (b) viscous, 4 – "λ" shock wave.

ferogram from a UTIAS experiment (shown in Fig. 6a). The computed density contours are also displayed in Figs. 6b and 6c. The computed wall density distribution shown in Fig. 7 indicates the good agreement with the measured results of UTIAS tests. A relaxation length of about $0.1L$ can be determined from Fig. 6c, which can be confirmed by the UTIAS interferogram of Fig. 6a. As a second example, the calculation of the double Mach-reflection flow in the present experimental case with $M_s = 7.66$, $\theta_w = 40°$ has also been performed. Figures 8a and 8b are the computed density contours for inviscid and viscous perfect gases, respectively; the first and second slip layer, vortex, and "λ" shock wave in the compression corner are displayed clearly. Comparison of the density distribution on the wall between experiment and computation is illustrated in Fig. 3. The computed results for $\gamma = 1.4$, equilibrium and nonequilibrium gases are reasonable and at the location of $\overline{x} = 0.4$–0.95; densities seem to tend toward equilibrium; the reason for some deviations from experimental results is now being analyzed.

4. Conclusion

The present experiment is successful in attaining quantitative density fields by directly counting fringes without relying on analytical relations of three-shock theory at the triple points. However, in chemically reacting flow, fringe-counting data-reduction techniques provide very rough estimation of the density fields, therefore the reasonable determination of Gladstone-Dale constant K is still a problem to be solved for the accurate comparison of experiment with computation. The present computer codes, validated by experiments, represent a substantial predictive capability for the detailed modeling of the complex non-stationary nonequilibrium and viscous flow with complicated shock structure.

References

Clinton PT, Groth S (1993) TVD Flux-Difference Split Methods for High-Speed Thermochemical Nonequilibrium Flows With Strong Shocks, UTIAS Report No. 350

Dunn MG, Kang SW (1973) Theoretical and Experimental Studies of Reentry Plasmas, NASA CR 2232.

Glaz HM (1987) High Resolution Calculation of Unsteady Two Dimensional Nonequilibrium Gas Dynamics With Experimental Comparisons, AIAA Paper 87-1293.

Glaz HM, Collella P, Glass II (1986) A Detailed Numerical, Graphical, and Experimental Study of Oblique Shock Wave Reflections, UTIAS Report No. 285.

Gupta R, Yos J, Thompson R, Lee K-P (1990) A Review of Reaction Rates and Thermodynamic and Transport Properties for an 11 Species Air Model for Chemical and Thermal Nonequilibrium Calculation to 30000 K, NASA RP 1232.

Le J, Hu G (1992) A high resolution calculation of reflection of oblique shock wave from a compression corner in chemical nonequilibrium gas flow. In: Proceeding of Int Conf on the Methods of Aerophysical Research, Novosibirsk, p 84.

Needham CE (1991) Chemical nonequilibrium effects of the shock reflection. In: 18th Int Symp on Shock Tube and Waves, pp 897–902.

Study of the propagation of shock waves over bodies by holographic interferometry

J. Le, X. Ye, X. Wu, H. Yang
China Aerodynamics Research and Development Center
P.O. Box 211, Mianyang, Sichuan, China

Abstract: The propagation of shock waves of $M_s = 1.8$ over various bodies such as single cylinders, double cylinders, wedges with back step, or plates with sharp-edged slits in a shock tube or in a channel bend was studied experimentally with a holographic interferometer. Quantitative density fields obtained from interferograms are in good agreement with computations for the case of the wedge with back step.

Key words: Shock wave, Holographic interferometry, Shock tube

1. Introduction

The problem of the propagation and diffraction of shock waves past solid bodies has been investigated extensively in the past both experimentally and computationally. Some excellent results of the quantitative measurement of density fields have been obtained with holographic interferograms, such as: fifteen basic cases of oblique shock-wave reflection (Glaz et al. 1986); and shock-wave diffraction through an aligned baffle system (Reichenbach 1994). However, for some problems with complicated flow structure, including shock waves reflected from a solid wall and the interaction of shock waves with vortices, the quantitative measurements of density fields from time sequences of interferogram frames are still very necessary and important for validating computer codes and for studing complicated flow phenomena. For example, only computational results of the propagation of shock waves through double cylinders in a channel bend have been completed (Fursenko and Sharov 1992), thus investigation of this kind of flow using holographic interferometry is of great value in providing insight into the physical phenomena. The purpose of this study is to quantitatively measure the unsteady flow fields of the propagation of shock waves over some typical bodies using a shock-tube technique in which the incident M_s can be precisely controlled and to compare the results of experiment and computation for one of these typical bodies.

Le Jialing, Ye Xichao, Wu Xingyan, Yang Hui. Study of the propagation of shock waves over bodies with holographic interferometry. 20th international symposium on shock waves. 1995.

2. Experiment

The experiments were carried out in CARDC's small shock tube with a test section of 32 mm × 32 mm, with two pairs of observation windows of 100 mm × 32 mm. The distance between windows is 400 mm. The driver of the shock tube has a circular section with a diameter of 70 mm. Figure 1 shows the shock tube and the optical system of the holographic interferometer. Usually, a 2D model is located in view of the first window and can be easily exchanged for a channel bend where the double cylinders are placed. Their detailed sizes are shown in Fig. 2. The incident-shock Mach number M_s can be precisely controlled by a specially made plastic diaphragm and by high-accuracy measurements of gas pressure both in the driver and in the driven section. The distances from the transducers to the model are arranged to measure the shock speed and to synchronize the ruby laser. For quantitatively measuring the density fields,

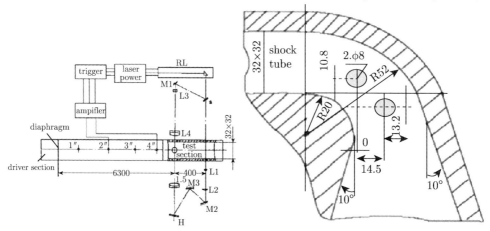

Figure 1. Shock tube and holographic interferometer system.

Figure 2. Size of channel bend with cylinders.

double-exposure holographic interferometry was used. The first exposure is performed before the incident shock wave arrives and the second exposure is triggered at the moment at which the shock waves interact with the model. By incrementing triggered delay times (with precison of 1 μs), the frame-by-frome holographic interferograms, separately obtained from a series of tests, can be considered to capture the unsteady process of shock-wave interaction with a body.

3. Results and discussion

Propagation through a slit with sharp edges. The apex angle of the sharp edge is 20° and the distance of two apices is 4 mm. Some of these experimental results are similar to those obtained by shadowgraphs (Amann and Reichenbach 1973). In Fig. 3 are holographic interferograms taken at different instants. After the incident shock wave passes over the model, two intense vortices are formed at the sharp edges but only one diffracting shock wave moves forward. The experimentally determined isopycnics are presented in Fig. 3d. The velocity of the diffracting shock wave along the slit plate can be measured by two instantaneous interferograms (Figs. 3a and b). Then its density ratio (ρ_N/ρ in Fig. 3d) can be calculated by the normal shock-wave relation.

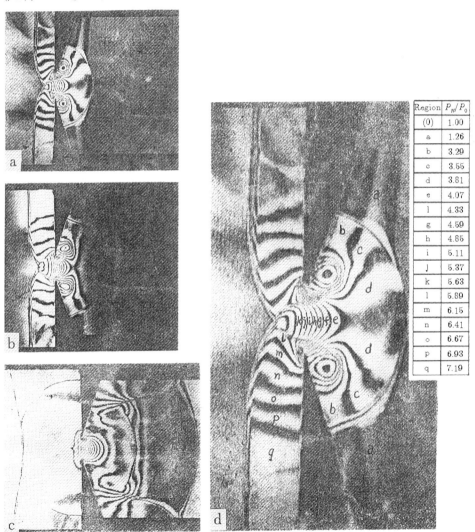

Region	P_N/P_0
(0)	1.00
a	1.26
b	3.29
c	3.55
d	3.81
e	4.07
f	4.33
g	4.59
h	4.85
i	5.11
j	5.37
k	5.63
l	5.89
m	6.15
n	6.41
o	6.67
p	6.93
q	7.19

Figure 3. Holographic interferograms, $M_s = 1.81$, $P_0 = 32$ kPa, $T_0 = 300.5$ K: (a) $t = 0$ μs; (b) $t = 2$ μs; (c) $t = 30$ μs; (d) experimentally determined density contours of case a.

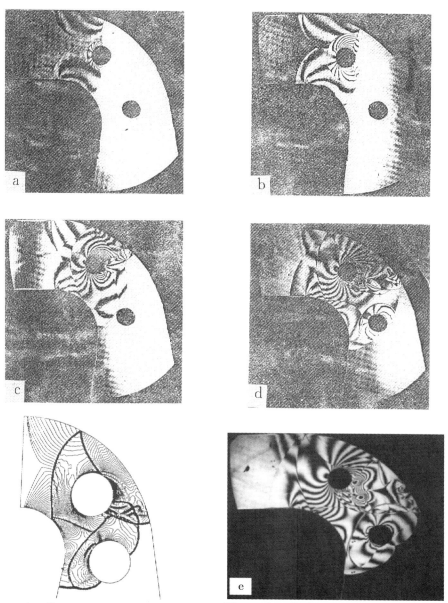

Figure 4. Computed density contours from Fursenko and Sharov (1992); $M_s = 2.1$.

Figure 5. Holographic interferograms from CARDC. $M_s = 1.82$, $P_0 = 32$ kPa, $T_0 = 288$ K: (a) $t = 10$ μs; (c) $t = 24$ μs; (d) $t = 38$ μs; (e) $t = 52$ μs.

Propagation through a channel bend with double cylinders. There is only one paper that computationally treats this unsteady problem based on

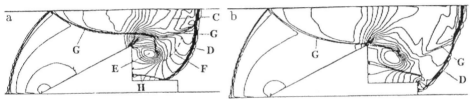

Figure 6. Computed density by ENO scheme. $M_s = 1.80$, $P_0 = 32$ kPa, $T_0 = 285$ K, $Re = 5.3 \times 10^5$, 351×121 grid points: (a) $t = 36.9$ μs; (b) $t = 44.88$ μs.

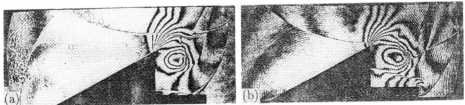

Figure 7. Holographic interferograms. $M_s = 1.80$, $P_0 = 32$ kPa, $T_0 = 285$ K, $\Delta t = 0$ is set at the moment when the Mach stem arrives at the apex of the wedge: (a) $t = 36.9$ μs; (b) $t = 44.88$ μs.

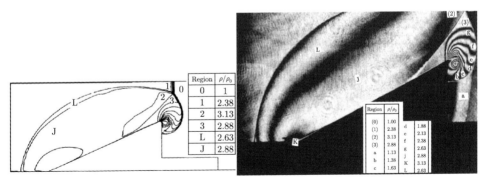

Figure 8. Comparison of density contours from computation and experiement. $M_s = 1.80$, $P_0 = 32$ kPa, $T_0 = 285$ K, $\Delta t = 12$ μs: (a) from computation; (b) from experiment.

the Euler equation with an unstructured finite-volume method (Fursenko and Sharov 1992). Figure 4 shows the result. In contrast with the computation, the experimental results shown in Fig. 5 represent a more complex wave system; in particular, the two vortices formed behind the first cylinder are more extensive. Viscous effects behind so called lambda shock wave are also stronger, thus the flow structure may depend on the Reynolds number.

Propagation past a wedge with back step. Flow fields were computed using the thin-layer approximated N-S equations with second-order ENO schemes (351×121 grid points). The model size and a detailed description of both experiment and computation are given by Le and Li (1994). As shown in Figs. 6 and 7, the vortex (E) has a significant effect on the reflected shock wave (G). The shock wave (G) is distorted a little as it approaches the vortex (shown in Fig. 7a), and then is split into three parts by the vortex (shown in Fig. 7b), its right half is accelerated and its center is completely merged by the vortex. Compared with the experiment, the interaction of the shock wave (G) with the vortex can be more clearly captured by the present computations (shown in Figs. 6a and b). In addition, as shown in Figs. 8a and b, the density contours from experiment and computation are in good agreement.

4. Conclusion

By using the present shock-tube technique, the incident M_s can be precisely controlled and time sequences of holographic interferograms can be taken, making it very ecomomical for us to study quantitatively the problems of unsteady compressible flow. However the accuracy of time measurements is very important and will be further improved.

References

Abe A, Takayama K (1989) Shock wave diffraction from the open end of a shock tube. In: 17th Int Symp on Shock Waves and Shock Tubes, pp 270–275.

Amann HO, Reichenbach H (1973) Unsteady flow phenomena in a shock tube nozzles. In: Proc 9th Int Shock Tube Symp, pp 96–112.

Fursenko AA, Sharov DM (1992) Numerical simulation of shock wave interactions, channel bends and gas nonuniformities. Computers Fluids 21(3):377–396.

Glaz HM, Collella P, Glass II (1986) A Detailed Numerical, Graphical, and Experimental Study of Oblique Shock Wave Reflections, UTIAS Report No. 285.

Le J, Li C (1994) Propagation of shock wave over a wedge. In: Proc 2nd Int Conf of Experimental Fluid Mechanics, pp 786–793.

Reichenbach H (1994) Color interferogram of shock propagation through an aligned baffle system. Shock Waves 4(2).

Numerical simulation of shock (blast) wave interaction with bodies

LE Jia-ling, NI Hong-li

(China Aerodynamics Research and Development Center, Mianyang 621000, China)

Abstract: Some typical results of computation on the shock(blast) wave interaction (2-D and 3-D) with bodies and its experimental validation in shock tube are summarized, the suggestion for improving the numerical method(difference scheme and grid systems), developing 3-D optical quantitative visualization technology and further studying the unsteady turbulent flow are put forward.

Key words: shock wave; blast wave; numerical simulation

激波(爆炸波) 与物体相互作用的数值模拟

乐嘉陵, 倪鸿礼

(中国空气动力研究与发展中心, 四川 绵阳 621000)

摘要: 给出了在二维和三维条件下激波(爆炸波) 与物体相互作用的一些典型计算结果, 概括总结了激波管中实验的有效性, 提出了改进数值方法(包括差分格式, 网格系统), 发展三维光学定量可视化技术和进一步研究非定常湍流的建议。

关 键 词: 激波; 爆炸波; 数值模拟

中图分类号: V211.3; O383   **文献标识码**: A

文章编号: 1007-3124(1999)03-0001-09

0 Introduction

shock wave(or blast wave) interaction with bodies is a basic gas dynamic problem related to some fundamental researches (such as vortex formation resulted from shocks passing over sharp corner, shock-vortex interaction) and applications (such as protective construction and the design of flight vehicle). Usually the shock wave interaction data from detailed field experiments (or flight testing) are both rather costly and complex to obtain, sometimes only the data of a few selected measurement points on the model can be effectively analyzed. The other approach to study the shock wave interaction phenomena is the combination of the numerical simulation with the shock tube experiments,

Le Jia-ling, Ni Hong-li. Numerical simulation of shock (blast) wave interaction with bodies. Communications in nonlinear science and numerical simulation, 1999, 13(3): 1-9.

such an approach can not only give a complete flow field description, but offers the possibility providing the data valuable for engineering design at a relatively low cost. In addition, the shock wave over bodies in a shock tube is also a benchmark problem both for the CFD investigation and fundamental understanding of gas dynamics due to its typical complex of this sufficient unsteady flow field.

As we know, for the shock wave interaction with a typical simple body such as a wedge, cylinder or sphere at $M_s < 5$, the numerical simulation based on the Euler equation with second order upwind difference schemes can be used in reasonable accuracy to analyze the flowfield and application. However, it is still necessary to further study by both computation and experimental validation for the following conditions:

(1) complex shock wave passing over the bodies in the shock tube due to complicated shock wave interaction.

(2) shock wave interaction with bodies in three dimensions.

(3) Shock wave interaction with moving body.

For these reasons, a series of studies of both computation and experiments on above mentioned topics are now being carried out in CARDC research group. All experimental validations were completed in shock tube with holographic interferomentry. The follows are an brief introduction of its some results in this respect.

1 Some typical results

1.1 Shock wave of $Ms= 1.8$ over 2-D wedge with back step in shock tube[1]

The experiments were conducted in a shock tube with test section of 32mm by 32mm using double exposure holographic interferometry (Fig. 1). The experimental da-

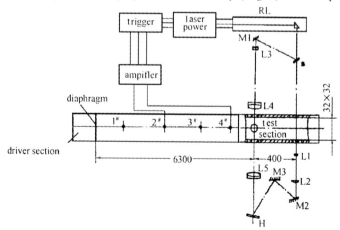

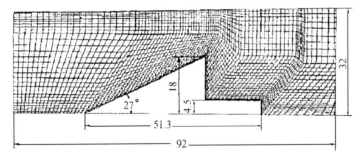

Fig. 2　Model size and grids $Ms=$ 1. 8 $Re=$ 5. 3×10⁵

图 2　模型尺寸和网格

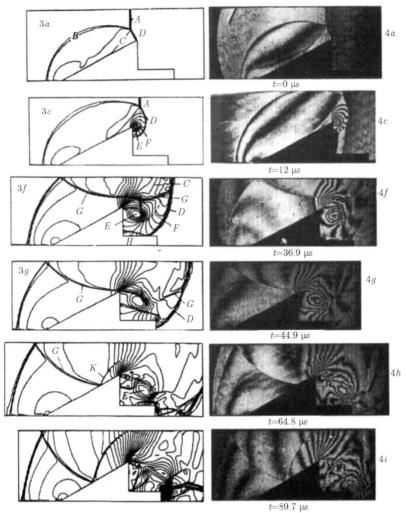

Fig. 3　Computational results with ENO scheme图 3　ENO 格式的计算结果

Fig. 4　Experimental holographic interferograms

图 4　实验的全息干涉图

ta include holographic interferograms frame by frame and one typical frame with isopycnics (Fig. 4, Fig. 5). The evolution of the diffraction of shock wave passing through the
wedge corner can be clearly visualized and quantitatively measured by present double exposure holographic interferometry (Fig. 5). Numerical simulations were carried out to

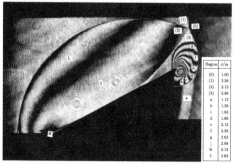

图 5 实验测定的等密度线

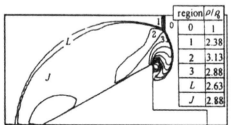

Fig. 6 **Computed density contours**
$Ms = 1.8$ $P_0 = 32kPa$ $T_0 = 285K$ $t = 12\mu s$
图 6 计算的等密度线

solve the thin-layer approximated laminar N-S equations by using second order ENO
scheme for the purpose of further confirming its capabilities of high-resolution nonosilla-
tory shock capture of unsteady complex flowfields. (model size and grid points shown in
Fig. 2). The computed density contours have good agreement with the experiments

(a) Experiment $t = 2\mu s$ (b) Experiment $t = 30\mu s$

(c) Computed interferogram   (d) Computed interferogram

Fig. 7 **Shock wave over sharp edged split plate**
$Ms = 1.807$ $P_0 = 32kPa$ $T_0 = 300K$
图 7 绕锐缘对开板的激波

(shown in Fig. 5 and Fig. 6). The comparisons between experiment and computation (Fig. 3 and Fig. 4) for the shock
diffraction process at a time series demonstrate that the complicated unsteady flow
fields, including shock, slip layers, vortices and their interactions, can be numerically simulated adequately by the present
method.

**1. 2   Study on the propagation of the
shock wave over sharp-edged split
plate**[2]

For smoothly solving the flowfields
near the sharp-edge region, the numerical
simulation using N-S equation with a finite-volume high resolution TVD scheme

and simple grid was carried out. The comparisons of computational interferograms with
measurement holograghical interferograms are satisfied with all flow fields, especially
near the sharp-edged region with complicated wave systems (fig. 7a, b, c, d for $t = 2\mu s$,
$30\mu s$), as for such region, the local densities usually can not be quantitatively evaluated

by experiment even in two dimensions condition. For this reason, the numerical simulation is needed for predicting the patterns of shock waves near the sharp-edged region in
different time steps (Fig. 8) and then the quantitative measured density fields can be obtained. (Fig. 9, $t = 0\mu s$)

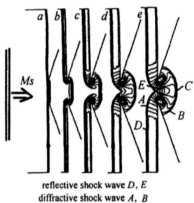

reflective shock wave D, E
diffractive shock wave A, B
incident shock wave C

Fig. 8   Flow patterns at different time steps
图 8   不同时间步的流动模型

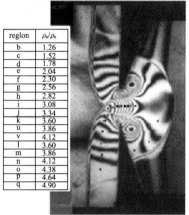

reglon	ρ_b/ρ_e
b	1.26
c	1.52
d	1.78
e	2.04
f	2.30
g	2.56
h	2.82
i	3.08
j	3.34
k	3.60
u	3.86
v	4.12
l	3.60
m	3.86
n	4.12
o	4.38
p	4.64
q	4.90

Fig. 9   Quantitative measurement density fields
图 9   等量测量密度场

1. 3   Study on the propagation of blast wave over two cylinders in a channel bend[3]

There was only one paper involved in computation of this sufficient unsteady and
complicated flow based on Euler equations with unstructured finite volume method[4]
without experimental comparison. It was the purpose for validation of this kind of flow
computation that the experiment was performed in shock tube with a channel bend
(32mm×32mm) specially designed and two cylinders of 8mm diameter were placed in
it. Experiments shown in the Fig. 10a, b, c, d represent the more complex wave systems,
especially two vortices formed behind the first cylinder are more extensive and viscous
effects behind so called "lamda" shock wave are more strong, thus the flow structure
may depend on the Reynods number. It is for this reason that the computation based on
the full NS equation is necessary. The computational results were obtained by using a
second order accurate TVD scheme with multiblock and overlap techniques. Fig. 10c, d
are the computational inteferograms corresponding experiments at different instants.
Two vortices appeared on the Fig. 10c, d are very clear and the "lamda" shock wave normal to the wall formed along with the diffraction shock wave over the second cylinder is
also reasonable. Density contours and pressure distribution on bodies are shown in Fig.
11. Computational results also give evidence of an obvious influence of the blast wave (
$Ms = 1.8$ with pulse duration of 3.8ms) duration and distance between cylinders on the
overpressure and pressure histories. Of course, due to this complicated flowfields, the
validation of the computation by detailed pressure measurement both in the shock wave
and blast wave conditions is still necessary. In addition, similar computations were also
made for the shock wave interaction with two rectangular bodies in the shock tube, the

vortices have greater effects on reflected shock waves.

(a) Experimental interogram $t=38\mu s$

(b) Experimental interogram $t=52\mu s$

(c) Computed interferogram

(d) Computed interferogram

Fig. 10   Comparison of computation with experiment for shock wave over two bodies in a channel bend $Ms = 1.8$ $P_0 = 32\text{kPa}$ $T_0 = 300\text{K}$

图 10   管道弯头处绕两体激波的实验和计算值比较

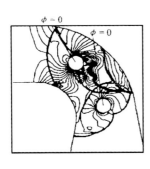

Computed density contours

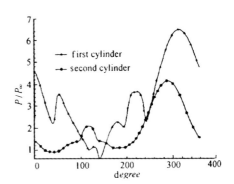

Computed pressure distribution on bodies

Fig. 11   Shock wave interaction with two cylinders $Ms = 1.8$ $P_0 = 32\text{kPa}$

图 11   双圆柱与激波的干扰

1.4   Computation and experiment on the blast wave over 3-D two bodies in channel bend[5]

For further developing our computational technology, the calculation of the more complicated flowfields for the blast wave of $Ms = 1.8$ passing over two different bodies (one is a cone, and the other, a small cylinder) in the channel bend was completed. Fig. 12 are the grids of two sections for $X-Y$ and $X-Z$ plane with multiblocks and overlaps.

The rather complex wave systems, quite different from 2-D case can be predicted (shown in the Fig. 13 and Fig. 14). Fig. 13 and Fig. 14 are the two time steps, one is the time of

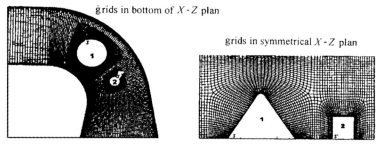

grids in bottom of X-Z plan

grids in symmetrical X-Z plan

Fig. 12   Computational grids for 3-D blast wave interaction with bodies in a channel bend
图 12   在弯管中三维爆炸波与物体相互干扰的计算网格

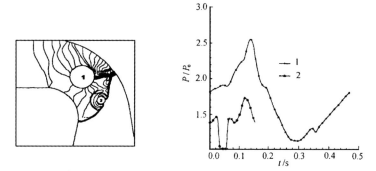

Pressure contours in bottom of X-Y plane   Pressure distribution on bodies

Fig. 13   3-D blast wave interaction with bodies in channel bend $Ms = 1.8$ $\Delta t = 3.8$ ms
图13   弯管中三维爆炸波与物体的干扰

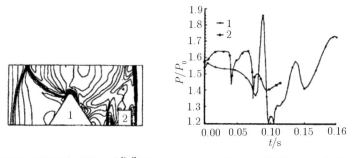

Pressure contours in bottom of X-Z plane   Pressure distribution on bodies

Fig. 14   3-D blast wave interaction with bodies in channel bend $Ms = 1.8$ $\Delta t = 3.8$ms
图 14   弯管中三维爆炸波与物体的干扰

blast wave arriving at the small cylinder, another is the blast wave passing over the cylinder. The peak value of the pressure in Fig. 14 results from reflected wave from the upper wall, the minimal values are due to the wave passing over corners. The experimental validations of the numerical simulation for similar 3-D shock wave over two bodies located in the channel bend were also conducted[5].

1. 5 Shock wave interaction with moving body[6]

Three-dimensional shock-on-shock interactions will produce a series of complicated structure of shock wave when the body moving at supersonic speed strikes a planar oblique shock wave or blast wave. This causes the body surface a high transient pressure. Although there had been numerous theoretical attempts and numerical calculation to solve this problem, however, due to the restriction of the computer condition, the wave structure and detailed overpressure distribution were not yet demonstrated in published papers. In our research, a finite difference technique of TVD scheme (with specially forth order dissipation) with multiblock and overlap technique were used for designing the grids and calculating the whole flow field. The outer regions of the 3-D flow fields of interaction can be determined by using the 1-D unsteady flow equations, which only depend on the incident blast wave. The calculation of the inner flow region are coupled with the outer one, so that results of a good resolution could be obtained throughout the whole flow region. The flow pattern for a sphere of 30mm diameter, flying with 700m/s, interacted with blast wave of $Ms = 1.8$ is shown in experimental shadowgraph (Fig. 15). The new bow shock wave, refracted wave, their weak contact surface and incident

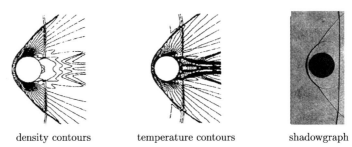

density contours   temperature contours   shadowgraph

Fig. 15   Blast wave interaction with moving body (axisymmetric condition)

Sphere $D = 30mm$ $V = 700m/s$ $Ms = 1.2$

图 15   爆炸波与运动物体的干扰(轴对称情况)

blast wave can be shown clearly from the figure. Numerical results were obtained under the same condition as the experiment and they have a good agreement each other. The wave system, especially the contact surface are shown clearly in Fig. 15, means that there exists a temperature and density jump. The results of numerical simulation of the moving body interaction with blast wave in 3D condition (interaction in impingement angle of -5 degree) are shown in Fig. 16. Although there are a few grid points ($127 \times 51 \times 19$, $22 \times 61 \times 19$) used in present 3-D calculation, the density with a jump across the contact surface in axis line means that the results are still reasonable.

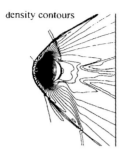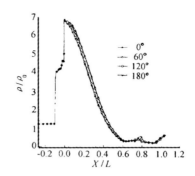

$D = 30$mm $\quad V = 700$m/s $\quad Ms = 1.2$ **Density distribution in axis line, upper and lower surface**

Fig. 16 3D Blast wave interaction with moving body, impingement angle of –5 degree

图 16 三维爆炸波与运动物体的干扰,–5°的冲击角

2 Conclusions

(1) For 2-D flow field of single body, the agreement between experiment and computation is quite good, however for shock wave interaction with moving body and static bodies (3-D), the experiment validation, including pressure measurement and flow visualization would be necessary. The 3-D optical quantitative visualization technology must be developed for this purpose.

(2) For the computational codes suitable for applications in both large scale and roughness model, the unsteady turbulent flow and its effects must be considered.

(3) To capture the contact surface in the complicated wave systems of flow fields for some fundamental research, the schemes and grid systems are needed to be improved.

References:

1) LE Jia-ling and LI Chao. Proceedings of the 2nd international conference on experimental fluid mechnics. p780.

2) LE Jia-ling and Cheng Yang-sheng. Experiments and measurements in fluid mechanics. 1998, 12(1): 29.

3) LE Jia-ling and NI Hong-li. Proceedings of the 21st international symposium on shock waves. paper 3440.

4) FURSENKO A A and SHAROV D M. Computers fluids 21 (3), 377.

5) NI Hong-li, LE jia-ling. Proceedings of the 21st international symposium on shock waves, (1997) paper 3441.

6) LE Jia-ling and NI Hong-li. Proceedings of the 20th international symposium on shock waves 1995, 387.

Research on blast wave interaction with body traveling supersonically

NI Hong-li, LE Jia-ling

(China Aerodynamics research and Development Center, Mianyang 621000, China)

Abstract: In present paper, the unsteady flow field of a blast wave interaction with a sphere traveling in supersonic speed is determined using a shock-capturing, finite difference approach and hybrid scheme. The presented numerical results compare favorably with the available experimental data, and the weak contact surface of bow shock interaction with the reflected blast wave obtained is shown to agree better with that of experiment. The flow of a blast wave encountered a sphere traveling supersonically at -5° angel of attack is studied further.

Key words: shock-capture; finite difference approach; shock-on-shock interaction; blast wave; contact surface

爆炸波与超声速飞行物体相互作用研究

倪鸿礼，乐嘉陵

(中国空气动力研究与发展中心，四川 绵阳 621000)

摘要: 本文数值模拟了超声速飞行物体与爆炸波相遇后产生的两波干扰流动，采用了激波捕捉、有限差分法和 LU-TVD 格式。同得到的实验结果对比，计算结果反映出了两波作用后的复杂流动，并且可清楚地看到飞行体头部产生的弱接触面，与实验符合较好，同时进一步模拟了飞行体以 5° 攻角飞行时与爆炸波相遇后的流动。

关　键　词: 激波捕捉; 有限差分法; 激波激波相互作用; 爆炸波; 接触面

中图分类号: O383　　文献标识码: A

0　Introduction

A tremendous transient impulsive force or some high frequency disturbances were exert-

Ni Hong-li, Le Jia-ling. Research on blast wave interaction with body traveling supersonically. Experiments and Measurements in Fluid Mechanics. 2001, 15(2): 10-15.

ed on the body traveling supersonically when it encountered a blast wave. On the one hand, the steady shock wave caused by body flying supersonically will produce effect on the body, another hand, the unsteady blast wave will also generate influence on the traveling body due to a series of complex wave systems appeared after that. This phenomenon is referred as 'double wave' or 'shock-on-shock' interaction, Fig. 1 shows the typical shock-on-shock interaction. This is an important problem both

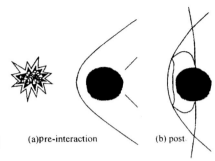

(a)pre-interaction   (b) post

Fig. 1   Typical shock on shock interaction
图 1   典型的激波与激波相互作用示意图

for fundamental research and engineering application, and is paid considerable attention by many researchers. Usually, to carry out 'shock-on-shock' interaction in experiment, a group of complicated experimental apparatus are needed since both steady and unsteady shock wave must be simulated simultaneously in the whole flow fields, and the cost is very expensive. On the other side, the numerical simulation of this kind of flow is also difficult. Firstly, the blast wave causes an additional moving boundary condition; Secondly, the overlap of two kind of shock wave will produce so large impacting force that the stability of numerical calculation can not be ensured easily. UP to now only few works were reported, and almost opened reports are related to the planar shock wave interaction with the traveling body. At present, the whole of 'shock-on-shock' interaction is simulated using numerical method, the blast wave is simulated in numerical calculation according to the course of how to generate it in experiments. By means of solving one- and three-dimensional fluid mechanics equations synchronously, steady and unsteady flows are acquired, and compared with experiments, satisfied results are obtained.

1   Experiment

The experiments of interaction of blast wave with the traveling body were carried out by using ballistic range in a quiescent atmosphere shown in Fig. 2. The sphere body is launched by a powder gun, and the blast wave is generated by conical shock tube special designed. The shock tube consists of two parts, one is a cylinder section, the other is a conical nozzle. After the detonator has exploded, a plane shock wave is formed inside a cylinder section of the shock tube, then it enters into a conic alnozzle of 30° full angle and the spherical blast wave is produced. The triggered time of a projectile can be accuratelyc ontrolled in the experimental arrangement, so shadow pictures for moving body / blast wave interaction can be captured. In this experiment, the diameter of the sphere is 30mm and flying with 700m/s, the Mach number of blast wave is 1.2 as it meets the moving body. Fig. 3 shows the exper-

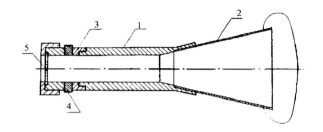

1. tube for generating a plane shock wave   2. conical nozzle
3. explosive chamber   4. electric detonator   5. steel membane

Fig. 2   Experiment for shock or shock interaction
图 2   激波—激波相互作用实验图

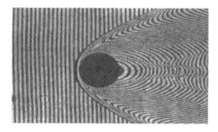

**Fig. 3   The experimental interfeyogram
of sphere traveling supersonically**
图 3   超声速球体定常飞行时的实验干涉图

imental interferogram of sphere traveling supersonically in air. Fig. 4 shows the experimental result of blast wave/moving sphere interaction for two different time steps a and b. The complicated shock wave interaction systems can be seen in the picture 4a, such as undisturbed bow shock 3-4, incident blast wave 3-5, transmitted shock wave 2-3, reflected shock wave 2-1 (regular reflection on the sphere surface point 2) and refracted shock wave 2-1. At the second time step in Fig. 4b, there are new bow shock wave (1-0), reflected wave moving off from the sphere surface as a weak disturbance (1-2) and weak contact surface are (1-6) shown clearly. The weak contact layer is generated by the reflected wave (1-2) interaction with the formed new bow shock wave (1-0.)

2   Numerical simulation

2.1   Governing equations

There are two kinds of movement in ' shock- or shock ' interaction flow field, so two groups of fluid mechanics equations are adopted, first is one-dimensional equations, second is two- or three-dimensional. The one-dimensional equations are used to calculate the movement of unsteady blast wave and to deal with the moving boundary condition; the two- or three-dimensional equations are used to

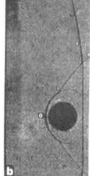

**Fig. 4   Shadows of blast wave/traveling
sphere interactions**
图 4   爆炸波与运动球体相互作用的阴影图

solve the steady and unsteady flow field of traveling body.

Three-dimensional Navier-Stokes equations can be written as following:

$$\frac{\partial Q}{\partial t} + \frac{\partial E}{\partial x} + \frac{\partial F}{\partial y} + \frac{\partial G}{\partial z} = \frac{\partial E_v}{\partial x} + \frac{\partial F_v}{\partial y} + \frac{\partial G_v}{\partial z} \tag{1}$$

where,

$$Q = \begin{pmatrix} \rho \\ \rho u \\ \rho v \\ \rho w \\ \rho e \end{pmatrix} \quad E = \begin{pmatrix} \rho u \\ \rho u^2 + p \\ \rho u v \\ \rho u w \\ h_0 u \end{pmatrix} \quad F = \begin{pmatrix} \rho v \\ \rho u v \\ \rho v^2 + p \\ \rho v w \\ h_0 v \end{pmatrix} \quad G = \begin{pmatrix} \rho w \\ \rho u w \\ \rho v w \\ \rho w^2 + p \\ h_0 w \end{pmatrix}$$

$$E_v = \begin{pmatrix} 0 \\ \tau_{xx} \\ \tau_{xy} \\ \tau_{xz} \\ u\tau_{xx} + v\tau_{xy} + w\tau_{xz} \\ + \frac{c_p}{RePr}k\frac{\partial T}{\partial x} \end{pmatrix} \quad F_v = \begin{pmatrix} 0 \\ \tau_{yx} \\ \tau_{yy} \\ \tau_{yz} \\ u\tau_{yx} + v\tau_{yy} + w\tau_{yz} \\ + \frac{c_p}{RePr}k\frac{\partial T}{\partial y} \end{pmatrix} \quad G_v = \begin{pmatrix} 0 \\ \tau_{zx} \\ \tau_{zy} \\ \tau_{zz} \\ u\tau_{zx} + v\tau_{zy} + w\tau_{zz} \\ + \frac{c_p}{RePr}k\frac{\partial T}{\partial z} \end{pmatrix}$$

and h_0 is total enthalpy $h_0 = e + \frac{p}{\rho}$; the equation of state can be expressed as

$$p = (r - 1)\left[\rho e - \frac{1}{2}\rho(u^2 + v^2 + w^2) \right] \tag{2}$$

E_v, F_v, G_v are related to the viscous terms, reduced above equations (1) some related terms or variables, one- and two-dimensional governing equations can be gotten.

2.2 Numerical method

Three independent variables (τ, ξ, η, ζ) are used, the equation (1) can be transformed in general coordinates in conservation form,

$$\frac{\partial \overline{Q}}{\partial \tau} + \frac{\partial \widetilde{E}}{\partial \xi} + \frac{\partial \widetilde{F}}{\partial \eta} + \frac{\partial \widetilde{G}}{\partial \zeta} = \frac{1}{Re}\left(\frac{\partial \widetilde{E}_v}{\partial \xi} + \frac{\partial \widetilde{F}_v}{\partial \eta} + \frac{\partial \widetilde{G}_v}{\partial \zeta} \right) \tag{3}$$

The Lower-Upper implicit algorithm is used in the equation (3), the discrete formation can be expressed as following:

$$LD^{-1}U\delta Q = \Delta\tau(RHS)$$
$$L = I + \beta\Delta\tau(D_\xi^- A^+ + D_\eta^- B^+ + D_\zeta^- C^+ - A^- - B^- - C^-)$$
$$D = [I + \beta\Delta\tau(\tilde{r}^A_{i,j,k} + \tilde{r}^B_{i,j,k} + \tilde{r}^C_{i,j,k})]^n I \tag{4}$$
$$U = I + \beta\Delta\tau(D_\xi^+ A^- + D_\eta^+ B^- + D_\zeta^+ C^- + A^+ + B^+ + C^+)$$

In right-hand terms of equation (3), the convective terms are separated using TVD scheme, the viscous terms using central difference,

$$(RHS)_{i,j,k}^{n} = - (\widetilde{E}_{i+1/2,j,k}^{n} - \widetilde{E}_{i-1/2,j,k}^{n}) - (\widetilde{F}_{i,j+1/2,k}^{n} - \widetilde{F}_{i,j-1/2,k}^{n})$$
$$- (\widetilde{G}_{i,j,k+1/2}^{n} - \widetilde{G}_{i,j,k-1/2}^{n}) + [(\widetilde{E}_{vi+1/2,j,k}^{n} - \widetilde{E}_{vi-1/2,j,k}^{n})$$
$$+ (\widetilde{F}_{vi,j+1/2,k}^{n} - \widetilde{F}_{vi,j-1/2,k}^{n}) + [(\widetilde{G}_{vi,j,k+1/2}^{n} - \widetilde{G}_{vi,j,k-1/2}^{n})$$
$$+ (d_{i,j,k+1/2}^{n} - d_{i,j,k-1/2}^{n})] / Re \tag{5}$$

The numerical flux in the ξ direction $\widetilde{E}_{i+1/2,j,k}^{n}$ is given

$$\widetilde{E}_{i+1/2,j,k} = \frac{1}{2} [\widetilde{E}_{i,j,k} + \widetilde{E}_{i+1,j,k} + R_{i+1/2} \Phi_{i+1/2} \tag{6}$$

The numerical flux in other directions can also be defined in the same way.

The following predictor-corrector method is used to keep a second-order resolution in time when the unsteady incident blast wave enters the 3-D flow field:

$$\overline{Q}_{i,j,k}^{n+1} = \overline{Q}_{i,j,k}^{n} + \frac{1}{2} \left(\delta \overline{Q}_{i,j,k(1)}^{n+1} + \delta \overline{Q}_{i,j,k(2)}^{n+1} \right) \tag{7}$$

An effective method of calculating 1-D unsteady flows is used to simulate the blast wave, meanwhile the moving boundary conditions of axissymetric flow and 3-D flow fields are also determined in this method. When the blast wave meets the traveling body, an unstable flow phenomenon often appears, so adding to specially fourth order dissipation in peripheral direction to keep the computational stability.

$$d_{k+1/2} = - \varepsilon_{k+1/2}^{(4)} \left(\overline{Q}_{k+2}^{n} - 3 \overline{Q}_{k+1}^{n} + 3 \overline{Q}_{k}^{n} - \overline{Q}_{k-1}^{n} \right) \tag{8}$$

for boundary conditions, non-slip conditions are applied for the traveling body surface, and constant temperature condition is assumed due to blast wave pass by the traveling body rapidly, almost don't cause the increase of temperature of the body surface.

3   Numerical results and discussion

Numerical simulation was done to compare with the experiment. Both the axissymetric and 3-dimensional flow fields of shock-on-shock were simulated. Numerical results were obtained under the same condition of experiment in axissymetric situation. The blast wave is simulated in numerical calculation according to the course of how to generate it in experiments.

Fig. 5 shows the density contours of interaction at different time (from blast wave just meeting the bow shock wave to passing over the sphere). The new bow shock wave, transmitted blast wave, reflected blast wave are shown obviously, the weak contact surface is also shown clearly from fig. 5d. It is noticed when the blast wave passes by the high-density zone, the surface speed of transmitted blast wave is decreased (from fig. 5b, c, d). Computational result shows a good agreement with that of experiment. The stagnation pressures are also calculated, before the blast wave meets the bow shock wave of traveling body, $p_0 / p_\infty = 6.6$, p_∞ is the far upstream pressure. When the transmitted blast wave arrives at the sphere, p_0 / p_∞ increases rapidly, at the position fig5c, is about 13. If the effect of impulsion of blast wave is considered, p_0 / p_∞ will be great more. For further, the traveling body interaction with the blast wave at some attack angle was studied. Fig. 6, fig. 7 show the results of traveling sphere at $-5°$ angle of attack interaction with the blast wave, fig. 6 shows the density contours. Not only are the refracted bow blast waves and reflected blast wave seen clearly, but also the weak contact surface in the region of stagnation is shown clearly. The jump of density along

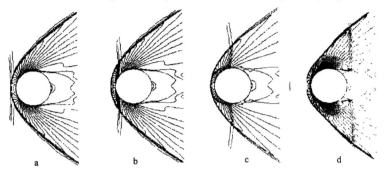

Fig. 5　Density Contours of blast wave/moving sphere interaction
图 5　爆炸波与飞行球体相互作用的密度等值线图

the contact surface in axis line can be found in fig. 7. Although there are a few grids ($127 \times 51 \times 19$) used in 3-D calculation, comparing with that of the axisymmetric, the results are still reasonable.

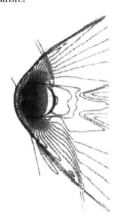

Fig. 6　Density contours
图 6　密度等值线图

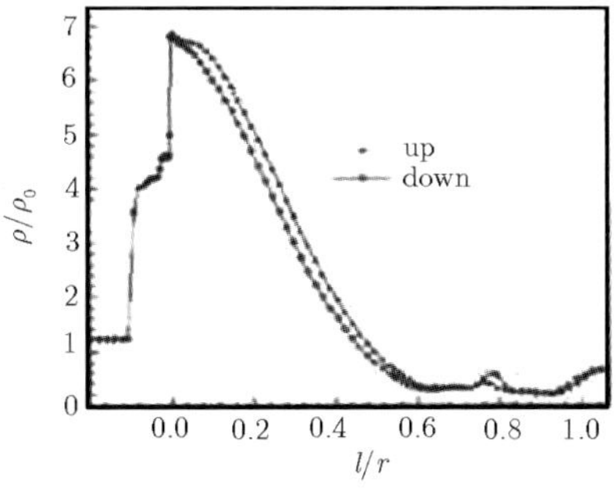

Fig. 7　The distribution of density in axis line, up and down surface
图 7　轴线和物体表面的密度分布图

4　Conclusion

(1) The complex blast waves interaction was successfully gotten both in experiment and calculation. The numerical result shows a good agreement with that of the experiment, so the effectiveness of numerical calculation was proved.

(2) Furthermore, the quantitative analysis about the effect of spherical blast wave interaction with traveling body is still studied.

References:

[1]　PAUL K, LEONIDAS S. Three dimensional shock on shock interaction problem. AIAA Paper 75~ 49.

[2]　Kovalev P I (1998) . Private communication.

彩色计算干涉技术及应用

吴颖川[1]，乐嘉陵[1]，贺安之[2]

(1. 中国空气动力研究与发展中心，四川 绵阳 621000; 2. 南京理工大学应用物理系，江苏 南京 210094)

摘要：在对高速气流进行干涉法测量时，由于激波的作用或者光的散射、斑点等现象，干涉图的干涉条纹可能会出现不连续。笔者通过仿真 M-Z 干涉仪或全息干涉仪 光学模型，并且采用对多块不规则网格的计算流体力学结果进行重构的直接体视化图形算法，得到了高分辨率的彩色数值干涉图，克服了干涉条纹的不连续性并给出了一个弹道靶中超高声速钝锥流场的例子。

关键词：干涉术; 计算流动显示; 体视化; 数值图像处理

中图分类号：O353.5　　　　**文献标识码**：A

Colored CFI-interferometry in experimental interferogram processing for hypersonic blunted cone flow in ballistic range

WU Ying-chuan[1], LE Jia-ling[1], HE An-zhi[2]

(1. China Aerodynamics Research & Development Center, Mianyang 621000, China; 2. Applied Physics Department of Nanjing University of Science & Technology, Nanjing 210094, China)

Abstract: As the result of the presence of shockwaves, or because of the spots of light, scattering, diffraction and other phenomena hiding the information searched in the experiment, discontinuities of the fringes are found in the interferogram of high velocity gas flows. A colored CFI- Interferometry that assists processing the experimental interferogram to overcome discontinuities is described. The methods include the theoretical modeling of colored CFI Interferometry simulating M-Z or holographic Interferometer and the direct volume rendering algorithms based on regularization of multi-grids CFD solutions to obtain high- resolution digital images. An example for hypersonic blunted cone flow in ballistic range is illustrated.

Key words: interferometry; computational flow imaging; volume rendering; digital image processing

0　引　言

吴颖川，乐嘉陵，贺安之. 彩色计算干涉技术及应用. 流体力学实验与测量. 2002, 16(1): 80-86,93.

计算流动显示(Computational flow imaging- CFI) 是由 George Havener[1,2] 提出。作为流动显示的一个新分支,它完全在计算机上实现。CFI 技术能够把数值计算(CFD) 获得的流场中的温度、密度、压力等各种物理场,经过与实验相同的计算光学(全息干涉、纹影、阴影、平面激光诱导荧光)的过程转换为所需的各个方向流动显示图像。与通常的实验流场显示相对应,目前最实用的 CFI 技术是计算纹影、计算阴影、计算干涉和计算平面激光诱导荧光技术。CFI 最大的特点是它的流场图像能够与试验的流场图像进行直接比较,从而验证 CFD 的有效性,使 CFD 与气动试验有机地结合,并能对复杂流动现象做出更加深入的物理分析。

对于二维流动,可以直接比较 CFD 的密度场和干涉图。但是对于复杂的三维流场,由于干涉图是沿光线的积分结果,这种直接比较是不可能的。通过 CFI 的方法,可以将 CFD 的计算结果转化为数值干涉图像,从而能够与试验干涉图进行直接的对比。

国外对 CFI 已进行了大量研究[1~8],但是许多研究仅仅涉及到二维问题,而且 Maclr Zehnder 或全息干涉法的理论光学模型也描述得不是很清楚。尽管 Tamura[4]、Yates[5] and Lanen[7] 的方法是三维的,但是在进行沿光线积分时,由于没有考虑 CFD 为多块交叉或重叠网格的情况,他们的方法有可能会产生积分错误。

CFI: 分为三个部分: 首先是准备好 CFD 的计算结果,其次是建立仿真流场光学测量系统的物理模型,最主要的是图像生成技术。直接体视化方法用来从三维体数据生成二维图像。体视化算法主要有 ray casting[9]、projection[10]、splatting[11] and shear warp[12] 等。我们在 wilhelms 和 Challinger[20] 的工作基础上,发展了一种适合于计算干涉法的直接体视化技术。通过将不规则的多块 CFD 网格数据重构为规则的密度场数据,可以得到准确的光线积分结果。

在对高速气流进行干涉法测量时,由于激波的作用或者光的散射、斑点等现象,干涉图的干涉条纹可能会出现不连续,从而造成干涉图的处理困难。我们采用彩色计算干涉法可以解决这个问题。

1 彩色计算干涉法的理论模型

对于 Maclr Zehnder 干涉仪, 亮条纹满足以下方程[21]:

$$2\varepsilon x + \int_{s_i} [\, n(x, y, z) - n_0]\, dz = i\lambda \quad (i = 1, 2, 3, \quad \quad) \quad (1)$$

其中 $[\, n(x, y, z) - n_0]$ 是相对于参考场 n_0 的密度, s_i 是光路, ε 是 Maclr Zehnder 干涉仪两块分光板的角度差, i 是条纹级数, λ 是光波长。

如果 $\varepsilon = 0$, 干涉图是无限条纹模式; 如果 $\varepsilon > 0$ 或 $\varepsilon < 0$, 干涉图是有限条纹模式。

用连续的条纹位移 F 代替 i 作为 $n(x, y, z)$ 的函数,得到

$$F(x, y) = \frac{2\varepsilon x}{\lambda} + \frac{1}{\lambda}\int_{z_1}^{z_2} [\, n(x, y, z) - n_0]\, dz, \quad (2)$$

光路沿 z 方向,在 z_1 和 z_2 之间。

代入 Gladstone- Dale 关系式

$$n - 1 = K\rho, \quad (3)$$

我们得到

$$F(x, y) = \frac{2\varepsilon x}{\lambda} + \frac{K}{\lambda}\rho_0 \int_{\xi_1}^{\xi_2}\left[\frac{-}{0} - 1\right]dz \tag{4}$$

在三维离散流场中求积分时, 一般根据梯形法则用离散点的积分和得到。如果光路被分为 n 段, 公式(5) 变为

$$F(x, y) = = \frac{2\varepsilon x}{\lambda} + \frac{K}{\lambda}\rho_0 \sum_{i=1}^{n}\left[\frac{-}{0} - 1\right]\Delta z_i, \tag{5}$$

干涉图上的光强度与条纹位移 F 的关系为

$$I \sim \cos^2(\pi, F), \tag{6}$$

将干涉图的光强度按比例转换为计算机显示器的像素亮度就可以得到最终的图像。把条纹位移的整数部分作为颜色索引以区分不同的条纹。计算机的调色板可以调整颜色与索引值的对应关系。

2 图像生成原理

对于三维流场, 沿光线路径的气体密度分布是不均匀的, 所以如何得到公式(5) 的线积分是主要问题。通常 CFD 的计算网格是多块、弯曲或不规则的, 而光线可以近似为垂直于图像平面的直线[6], 所以流场需要重新构造为规则的网格结构以便于线积分的计算。对于三维流场来说, 这个过程非常类似于关于弯曲体数据的直接体视化方法[20]。

2.1 三维流场的直接体视化方法

一般的直接体视化方法把体数据作为彩色、半透明的介质, 最终的图像是光线穿过该介质的映射结果。我们的方法假设整个区域充满了具有光学干涉性质的气体介质, 体视化的过程就是光线按照干涉条纹位移公式(5) 转化为显示屏上的光强分布。

CFD 的计算网格可能有各种类型: 正交网格、多块贴体网格、非正交网格 和自适应网格等。显示这些弯曲、多块和其它不规则网格比较困难的原因在于: 网格点定义的单元区域有非常复杂的形状, 多块网格之间有重叠和交叉部分等。

在显示三维数据场时很自然可以把网格单元作为体素(voxel)[19]。通过体素化(voxelization)将 CFD 的几何网格转化为体素的集合。为了解决多块网格的问题, 我们发展了一种预处理方法将整个流场区域划分为规则的子区域。体素被按照其几何位置分配到这些子区域中。这样公式(5) 的线积分计算只需要在各自的子区域内单独进行。另外由于每一条线积分的计算都需要搜索子区域内的所有体素, 我们可以通过划分更多子区域的方法, 使每个子区域内的体素数量减少来加快搜索速度。这是一种用空间换时间的算法。这种划分子区域的方法也便于并行计算的实现。

Wilhelms and Challinger[20]讨论了两种显示 CFD 计算结果的体视化方法: 光线投射法(ray-casting volumes) 和弯曲体数据插值法(interpolation of curvilinear volumes)。这两种方法都可以移植过来解决计算干涉法的问题。

在他们的光线投射法中, 每个六面体单元被认为由 12 个三角形包围(保证平面性)。如果单元是凸的, 光线与单元有两个交点。这非常相似于 Tamura[4], Yates[5] 和 Lanen[7] 的

方法。但是实际上不可能保证所有的体素是凸的, 因为对于两个相邻的体素来说, 它们共用一个面, 这个面对一个体素是凸的意味着它对另一个体素一定是凹的。为了解决因为交叉、重叠和非凸性带来的问题, 在我们的算法中, 沿光线积分路径进行跟踪是必要的。光线与所有体素的交点存放在一个由前至后排序的列表中, 公式(2.5)的线积分按照梯形法则由这个列表得到。这个过程如图 1 所示, 光线 S 穿过两个重叠的体素。积分的路径通过部分组成两个体素的面。一旦光线与这些面的交点求得 (图 1 的点 a, b, c 和 d), 而交点处的物理量(如密度)可由这些面的端点物理量的线性插值得到, 由梯形法则可计算出线积分的结果。

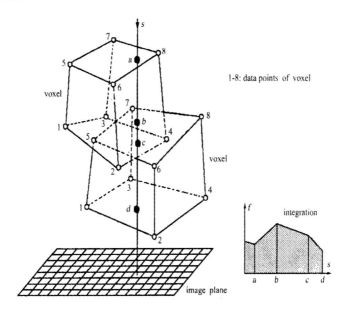

图 1　光线投射法的积分过程

Fig. 1　Schematic view of integration process of the ray casting volumes method

对于弯曲体数据插值法, 弯曲的网格必须重新插值为正交的规则网格。每个重构网格点 P 的插值过程分为两步: 找到包含点 P 的体素和由体素的角点的值算出 P 的值。Wilhelms and Challinger 完全可以用于计算干涉法, 但是第一步所用的贪婪法(greedy algorithm)[20]有时会产生错误, 所以我们发展了如下的一种新的搜索算法。

三维空间的一个平面可表示为公式,

$$ax + by + cz + d = 0 \tag{7}$$

一个有 n 个平面组成的多面体可由下面的矩阵表示,

$$[V] = \begin{bmatrix} a_1 & a_2 & \cdots & a_n \\ b_1 & b_2 & \cdots & b_n \\ c_1 & c_2 & \cdots & c_n \\ d_1 & d_2 & \cdots & d_n \end{bmatrix} \tag{8}$$

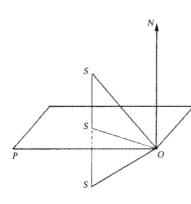

图 2 判断点 S 相对于平面 P 的位置

**Fig. 2 Schematic view for judging the position
of point S about plane P**

在图 2 中, 对于齐次坐标系中的点 S, 如果 N 是平面 P 的法向量, 设点积 $D= [S]•[P]$, $D= 0$ 时 S 在平面内; $D> 0$ 时 S 在平面上方; $D< 0$ 时 S 在平面下方。在体素化时, 使体素所有面的法向量指向体素内部, 我们可以通过所有面的法向量与某点的点积的值来判断该点是否在体素内。

2.2 生成图像

在图像平面上生成图像时, 没有必要以高于 CFD 网格的分辨率来对流场重新采样, 一般来说, 采样点之间的像素的亮度是由采样点的亮度的双线性插值得到, 但是要精确确定重新采样的粒度并不容易。对于计算干涉法, 在有限条纹模式或者有强折射区域(如激波)时, 两个相邻的采样点之间有可能会产生多个干涉条纹位移。直接由亮度插值会造成条纹不连续。如果首先用采样点的条纹位移量双线性插值得到每个像素点的条纹位移量, 再由条纹位移量根据公式(6)算出每个像素的亮度值, 就可以得到高分辨率的连续的条纹图像。

3 例 子

这个例子的目的是检验弹道靶中钝锥形状物体的高超声速气动特性。锥形的半锥角为 8°, 钝度比 0.125, 底部直径 24mm。试验的雷诺数接近 10^5, $M= 10$, 壁面温度 294K。试验干涉图由全息干涉系统得到。数值模拟采用全 Navier Stokes 方程。

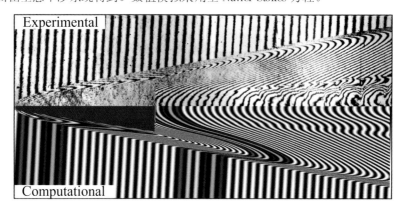

图 3 弹道靶中钝锥流场的计算干涉图与试验干涉图的比较(轴对称)

**Fig. 3 Comparison of experimental and computational interferograms for blunted
cone hypersonic flow in a ballistic range(axis symmetric flow)**

我们列举了两张试验干涉图: 图 3 是轴对称流场的; 图 4 是三维流场的, 钝锥攻角为 10°。干涉图的图像处理方法包括: 条纹跟踪(fringes tracing)、相位解调(phase unwrapping)

和光学层析(optical computerized tomography) 等, 干涉条纹的连续性是这些方法的基础。但是由于激波强折射现象, 在图 3 和图 4 中都可以发现干涉条纹的不连续性。彩色计算干涉技术是解决这些问题的有力工具。通过图 3 和图 4 中计算干涉图与试验干涉图的对比, 可以得到: 图 3 的第 23 条黄色条纹与钝锥底部后面的第一条黄色条纹相连, 图 4 的第 12 条黄色条纹与钝锥底部后面的第一条黄色条纹相连。通过条纹计数就可以把所有的参考条纹和钝锥后流场的条纹连结起来。

图4　弹道靶中钝锥流场的计算干涉图与试验干涉图的比较(三维)

Fig. 4　Comparison of experimental and computational interferograms for blunted
cone hypersonic flow in a ballistic range(three-dimensional flow)

从这个例子中可以看出, 彩色计算干涉技术可以用于辅助试验干涉图的处理, 但是它依赖于 CFD 的计算结果的准确性, CFD 的正确性需要预先进行验证。

4　结　论

事实证明, 彩色计算干涉技术可以用于辅助试验干涉图的处理。在我们的例子中, 计算干涉图与试验干涉图非常相似, 高分辨率的彩色条纹清楚地标示出通过激波的条纹的连续性。通过计算干涉图和试验干涉图的条纹一一对比, 所有的参考条纹可以连接起来, 从而使试验干涉图的处理能够顺利进行。这种方法不仅适合于空气动力学, 也可以用于其它研究领域。

体视化的方法通常是用来将三维数据场转化为二维平面图像, 所以这些方法可以移植到计算干涉法中解决图像显示问题。本文研究了生成高质量彩色计算干涉图的体视化方法, 针对计算干涉法的特殊性, 我们给出了一个物理光学模型、一个加速搜索速度的预处理方法、解决交叉重叠体素的方法、一个发现包含设定点体素的方法以及一种在图像平面生成高分辨率图像的方法。

这种直接体视化方法也可以用于其它 CFI 技术, 比如计算纹影法(CFI-Schilieren) 、计算阴影法(CFI-Shadowgraph) 和计算平面激光诱导荧光法(CFI-PLIF, Planar Laser Induced Fluorescence) 。随着 CFD 计算复杂性的增加, CFI 的并行算法也需要进行研究。

参考文献:

[1] GEORGE Havener. Computational flow imaging: fundamentals and history, AIAA-94-2615.

[2] HAVENER G, LESLIE A Yates. Visualizing flow with CFI. Aerospace America, June 1994.

[3] HAVENER A G, OBERGEFELL L A. Computational interferometric description of nested flow fields. Optical Engineering, 1985, 24(3): 441~ 445.

[4] TAMURA Y, FUji K. Visualization for computational fluid dynamics and the comparison with experiments. AIAA paper 90-3031, August 1990.

[5] YATES L A. Images constructed from computed flowfields. AIAA paper 92-4030, AIAA 17th Aerospace Ground Testing Conference, Nashville, TN, July 1992.

[6] STACEY G Rock. CFI interferometry analysis of hypervelocity ballistic flow fields. AIAA-94-2617.

[7] LANEN T A W M, Houtman E M, BAKKER P G. Comparison of interferometric measurements with 3-D Euler computations for circular cones in supersonic flow. AIAA-92-2691.

[8] HOUTMAN E M, BANNINK W J, TIMMERMAN B H. Experimental and computational study of a blunt cylinder flare model in high supersonic flow. Report LR-796. Faculty of Aerospace Engineering, Delft, The Netherlands, October, 1995, 38pp. , ISBN 90-5623-029-8.

[9] LEVOY M. Display surfaces from volumedata. IEEE Computer Graphics and Applications, 25(1988) , pp. 29-37.

[10] DREBIN R A, CARPENTER L, HANRAHAN P. Volume rendering. Proceedings of SIGGRAPH' 88, August 1988, pp. 65-74.

[11] WESTOVER L. Footprint evaluation for volume rendering. Proceedings of SIGGRAPH' 90, August 1990, pp. 367-376.

[12] LACROUTE P and LEVOY M. Fast volume rendering using a shear warp factorization of the viewing transformation. Proceedings of SIGGRAPH' 94, July 1994, pp. 451-458.

[13] LEVOY M. Efficient ray tracing of volume data. ACM Transactions on Graphics, 9(1990) , pp. 245-261.

[14] YAGEL R, SHI Z, Accelerating volume animation by space leaping. Proceedings of Visualization' 93, pp. 63-69.

[15] PFISTER H, KAUFMAN A. Cube 4-A scalable architecture for real-time volume rendering. Proceedings of 1996 Symposium on Volume Visualization, 1996, pp. 47-54.

[16] CABRAL B, CAM N, FORAN J. Accelerated volume rendering and tomographic reconstruction using texture mapping hardware. Proceedings of 1994 Symposium on Volume Visualization, October 1994, pp. 91-98.

[17] MA K L, PAINTER J S, HANSEN C, KROGH M. Parallel volume rendering using binary swap compositing. IEEE Computer Graphics & Applications, 14(1994) , pp. 59-67.

[18] LACROUTE P. Real-time volume rendering on shared memory multiprocessors using the shear-warp factorization. Proceedings of the 1995 Parallel Rendering Symposium, 1995, pp. 15-22.

[19] ELVINS T T. A survey of algorithms for volume visualization. Computer Graphics, 1992, 26(3) : 194~ 201.

[20] WILHELMS J, CHALLINGER J. Direct volume rendering of curviliner volumes. San Diego Workshop on Visualiztion, 1990.

[21] WOLFGANG Merzkirch. Flow visualiztion. Academic Press, 1974.

Some Research in Application of Holographic Interferometry and Computational Flow Imaging

Jialing Le, Yingchuan Wu, Hongli Ni, Hui Yang, Huiling Wu,
China Aerodynamics Research & Development Center
P.O. Box 211, Mianyang, Sichuan, China PR lejl@my-public.sc.cninfo.net

Abstract: Two-dimensional flow visualization by laser holograph interferometry in shock tube and some applications in Computational Flow Imaging （including color interferogram and three dimensional flow）are presented.
Keyword: holograph interferometry, Computational Flow Imaging

1. Introduction

Flow visualization is an important tool for compressible aerodynamic flow research. Though there are many advanced methods for flow visualization and quantitative analysis, such as three-dimensional particle imaging velocimeter (DPIV) and planar laser induce fluorescence （PLIF）, optical interferometry, schlieren and shadowgraph technique are still widely applied in the field of supersonic and hypersonic flow research. Therefore, further developing and fully applying these techniques combined with flow phenomena and CFD research, especially fully developing computational flow imaging technology (CFI) is still one of the important research aspects of flow visualization technology.

2. 2-D flow visualization in shock tube by laser holograph interferometry technology

2.1 Interaction of shock wave and vortex

In this field there is much theoretical research，but less detailed experimental flow observation. Figure 1 shows the interaction between the vortex generated by the wedge corner[1] （Fig.1a） and the shock wave G reflected on the upper wall （Fig.1f）(left is computational result, right experimental result). In Fig.1f,1g, the shock wave G is distorted a little as it approaches the vortex (E), then it is split into three parts by the vortex （E）. Its right half receives acceleration, its centre part seems to be merged by the vortex. The vortex finally disappears, it takes about 85 μs from the vortex formation to its complete break （Fig.1i）. In order to further observe the detailed interaction between shock wave and vortex, we specially designed an experimental section with right angle in the shock tube and a wedge model located in the vertical part of the section shown in Fig2a. The first vortex E1 （Fig.2b）, formed at the corner, gradually evolutes and moves up on the wedge surface，then second vortex E2 appears on the tip corner of the wedge （Fig.2c）,and also the incident shock wave is reflected from the end wall. Due to the more clearer picture of the flow visualization, it is interestingly observed for the first time that the reflected shock wave RW（from left to right in the Fig.2c）nearly permeates into the center of the vortex E1 (Fig. 2d). However, it is unexpected that the shock wave RW （especially the lower part） is not merged by the vortex E1 （Fig.2d）. Besides, due to the clockwise rotation of the vortex, it still does not change the characteristic that the shock wave at the upper half of the vortex decelerates while the one at the lower half of the vortex accelerates. In principle, from high accurate CFD computation, we can clearly reconstruct this kind of flow pattern structure and give it detailed description.

JiaLing Le, Yingchuan Wu, Hongli Ni, Hui Yang, Huiling Wu. Some research in application of holographic interferometry and computational flow imaging. Proceedings of spie - the international society for optical engineering, 2003.

2.2 Interaction of shock wave with two bodies in tandem

Fig.3a is the experimental shadow picture by high speed camera which shows the interaction between an incident shock wave (Ms=1.8) and two bodies (the height of the bodies is 16mm, width 10mm, the distance between the two bodies is 38.5mm). The picture of Fig.3b is the CFI-Shadowgraph obtained from CFD results. At a certain time, two shock waves , one（R1） reflected from lower wall, another （R2） diffracted from top of the second body focus onto the corner of the second body (point 4 in the Fig.3c). Fig.3c, 3d are CFD results. The CFD value P/P1 at point 4 equals to 13.99, which is higher than the value P5/P1 of 10.48, estimated from reflected shock wave of one-dimensional shock tube theory. This greatly exceeds usual expectation, but the high speed camera picture shown in fig. 3a indicates that this situation does exist.

2.3 Interaction between blast wave and body

Blast wave is generated by a conic shock tube (Fig. 4). The shape of the blast wave at the exit is measured and determined by pressure sensors, the maximum positive pulse's over-pressure $\triangle P=$ 0.1-0.13Mpa, its time duration of 400-500 μ s, the maximum negative over-pressure $\triangle P= 0.03-0.04$ Mpa, its time duration more than 1000 μ s. The diameter of 120mm uniform flow at the exit can be decided by quantitatively holographic interferogram (shown in Fig 5). This kind of device can be used to research the interaction between blast wave and body (static or moving). The shadow graph[2] （Fig.6） of the complex interaction of blast wave and moving body (the diameter of the sphere is 30mm, its velocity of 700m/s, the blast wave Ms is 1.2) can be effectively used to validate the CFD results. The interaction (Fig.6b)of the reflected wave 1-2 and the transmitted wave 1-3 generates a weak discontinuity （line 6 in the Fig.6b） of temperature and density. The computational temperature and density flow fields [3], and its value along stagnation line with a weak jump shown in Fig.7 indicate the good agreement with the picture of Fig.6b. This means our numerical method could be expanded to calculate quite complex 3-D condition. But for 3-D complicated flow such as the blast wave interaction with moving body shown in Fig.8, it's very difficulty to quantitatively measure the whole flow fields, so that CFI perhaps can provide an effective and economic way to validate this type of the computation codes.

3. Computational Flow Imagine (CFI)

CFI is the combination of Computational Fluid Dynamics, Computational Optics and High- resolution digital image processing technologies. CFD can get different kinds of physical flowfield, such as temperature, density, pressure, etc. If the CFD results are correct then the computed flowfield are thought agreeing with actual physical field exceedingly well. CFI uses theoretical predictions of the interaction and transmission of optical waves through CFD flowfield to generate digital pictures that simulate real observations. CFI is mainly used to construct flow visualization corresponding to shadowgraph, schlieren and interferometry. It is well known that the CFD technology should be experimentally validated before it is put into use, but usually the surface and integrated measurements of the object (e.g. pressure, heat flex, forces, moments) are not sufficient for such purpose. Extensive measurements in the 3-D flow field away from the surface are required to validate a code's ability to simulate the important flow phenomena. Flow visualisation techniques measuring density variations (e.g., schlieren photography, shadowgraph, interferometry) provide extensive information about flowfield parameters and can be used to evaluate effects of chemistry in CFD codes. In addition, schlieren and shadowgraph from one direction are very popular in the wind-tunnel test now, and through making full use of these flow visualization images (gray scale of the pixels), CFI has become a very effective method for CFD validation.

Gray scale of a pixel is not linearly proportional to one physical parameter (density) of a point in the field, but it is the result of integral of refractive index along the ray path. This process is similar with that the aerodynamic forces (lift, drag) are integrals of pressures along object surface. Form this viewpoint, CFI in fact represents the average effects of the physical properties along the ray path.

In CFI-Shadowgraph, CFI-Schlieren and CFI-Interferometry, if the light intensity is I, n is the gas refractive index ($n-1=K\rho$), and the ray path is *along* z-direction from z_1 to z_2, then the light intensity of shadowgraph systems is proportional to the second derivative of flow density ($I \propto \int_{z_1}^{z_2}\left(\frac{\partial^2 n}{\partial x^2}+\frac{\partial^2 n}{\partial y^2}\right)dz$) where as that of schlieren systems is proportional to the first derivative ($I \propto \int_{z_1}^{z_2}\frac{\partial n}{\partial x}dz$). The relation between the light intensity and fringe shift of Mach-Zehnder or holographic interferometry is $I \sim \cos^2(\pi F)$, and the fringe shift F is function of the wavelength of light source λ, the density of flowfield ρ and the position (x, y) on the image plane. This integral equation is $F(x,y)=\frac{2\varepsilon x}{\lambda}+\frac{1}{\lambda}\int_{z_1}^{z_2}\left[n(x,y,z)-n_0\right]dz$. Therefore from the known density field, 2-D and 3-D CFI images can be obtained and compared with experimental pictures.

For many years a research group of CARDC has developed an effective parallel adaptive ray-casting algorithm [4] for CFI application in some complicated flows. This algorithm, which can deal with any kind of complicated computational grids of CFD (intersected, multi-block, irregular grids), is different from reference 5,6 and 7.

Fig. 9 is a comparison of the experiment (Fig.9a) picture with CFI- interferogram (Fig.9b) for the shock wave over two bodies in channel bend [8] The first body is the cylinder-cone with the base diameter of 22mm and height of 16mm, the second one is the square pillar with base size of 16mm×16mm and 12mm height. Two fringe patterns of figures (computation and experiment) agree with each other very well.

In Fig.10,Fig.11, the experimental picture（Fig.11a）of only one view direction is given along with the computational interferogram[9]（Fig11b,c）of two view direction for comparison. The computational interferogram in fact can be obtained from any light direction. In addition, the CFI images especially color-marked interferograms greatly improve quantitative analyzing the flow patterns with complicated wave system

4. Conclusion

CFI can be constructed from a variety of ideal and real gas, 2-D axisymmetric and 3-D flow field solution. The CFI pictures we have shown in the examples, by high effective parallel adaptive ray-casting algorithm, are very similar to the experimental ones, and one by one comparisons can be made between them. This powerful tool actually provides a bridge between CFD and experiment.

The experiments of more details and multi-ray direction and its comparisons with CFI still need more efforts to evaluate the error of the CFI technology, to further validate CFI's data processing algorithms and CFD computation codes.

References

[1] Jialing Le, "Propagation of Shock Wave over a Wedge," *I.C.E.F.M 94 Proceedings*, pp786-793

[2] P.R. Kovalev, N.P. Mende, "An album of supersonic visualization" *National defence industrial press*, pp121-123.

[3] Hong-li Ni, Jialing Le "Research on blast wave interaction with body traveling supersonically," *EXPERIMENTS AND MEASUREMENTS IN FLUID MECHANICS*, No.2, 2001, pp10-16

[4] Yingchuan Wu, Jialing Le, "Parallel volume rendering algorithms for computational flow imaging, " *ICMAR2002 proceedings*, Part two pp168-172

[5] Tamura, Y., K. Fuji, "Visualization for Computational Fluid Dynamics and the Comparison with Experiments ", *AIAA paper 90-3031*, August 1990.

[6] Yates, L.A. "Images Constructed from Computed Flowfields, " *AIAA paper 92-4030*, AIAA 17th Aerospace Ground Testing Conference, Nashville, TN, July 1992

[7] T. A. W. M. Lanen, E. M. Houtman and P. G. Bakker, "Comparison of interferometric measurements with 3-D Euler computations for circular cones in supersonic flow", *AIAA-92-2691*.

[8] Jialing Le "Study of the propagation of shock wave over bodies by holographic interferometry, " Proceeding of the 20t[h] Inter. Sym on SHOCK WAVE. Pp525-530

[9] Jialing Le, Hong-li Ni "Research on the motion of shock wave over bodies in a channel bend, " Proceedinds of the Eight Asian Congress of Fluid Mechanics, PP392-

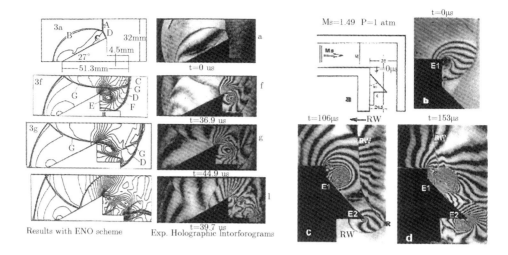

Fig. 1 2D Shock wave interaction with wedge with back step
Ms=1.8, P=32Kpa, θ=27°

Fig. 2 2D shock wave over wedge near corner

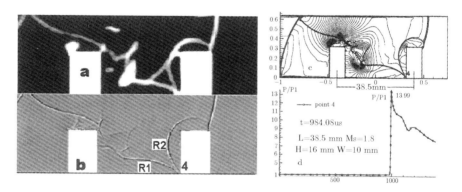

Fig. 3 Shock wave over two bodies

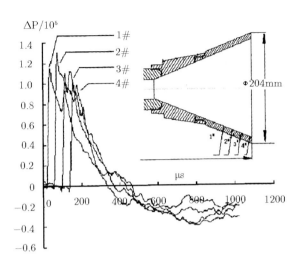

Fig. 4 Pressure trace and blast wave device

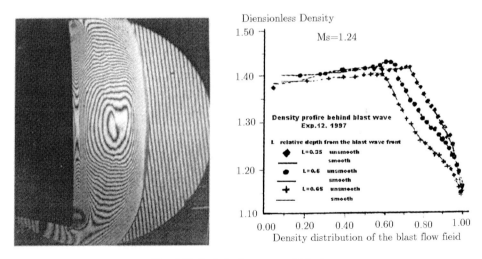

Fig. 5 Finite interferogram of blast wave

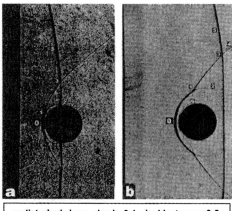

undisturbed bow shock 3-4, incident wave3-5, transmitted wave 1-3, reflected wave 2-1(regular reflection on the sphere surface point 2) and refracted wave 3-1. 0-1 new bow shock

Fig.6 2D interaction of blast wave with moving body

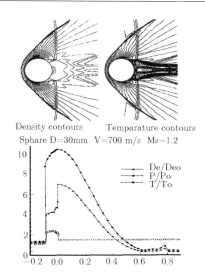

Density contours   Temparature contours
Sphere D=30mm   V=700 m/s   Ms=1.2

Fig. 7 CFD results

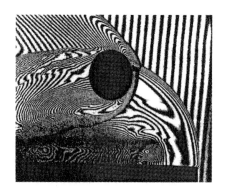

Fig. 8 3D interaction of blast wave with moving body

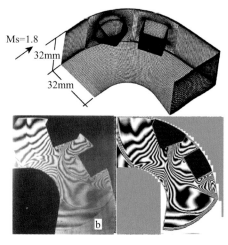

Ms=1.8
32mm
32mm

Fig. 9 3D shock wave over two bodies in bent channel

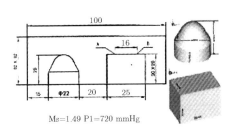

Ms=1.49 P1=720 mmHg

Fig. 10 Shock wave over two bodies in tandem

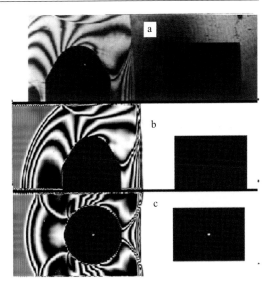

Fig. 11 Comparison of experiment with CFI
Ms=1.49, P1=720mmHg, T1=285K

激波绕楔形物体的流场分析

王惠玲, 乐嘉陵, 吴颖川

(中国空气动力研究与发展中心, 四川 绵阳　621000)

摘要: 使用数值计算方法和数值干涉技术并结合实验干涉图片, 分析了激波过拐角绕楔形物体运动的流场。并采用数值彩色干涉技术, 将数值计算结果转换为彩色条纹编码的数值干涉图, 并与实验干涉图比较, 可以辨识过激波发生跳跃后的干涉条纹的位移数, 从而确定实验干涉照片的密度场。数值计算采用显式二阶迎风TVD格式。采用数值干涉技术将计算密度场转化成数值干涉图与实验干涉图进行了比较, 检验了数值结果的可靠性。数值计算结果显示了包括激波、接触间断和旋涡在内的详细的流场结构, 与实验相比, 更为细致地刻画了流场演变过程的细节。通过数值计算结果观察到激波传播到旋涡中心附近并没有发生明显耗散, 旋涡上半部的激波得到减速, 下半部分的激波得到加速。

关键词: TVD 格式; 数值模拟; 激波; 数值干涉技术; 实验干涉图; 数值彩色干涉技术

中图分类号: O241; O242　　　**文献标识码**: A

Analysis of flow field of shock wave passing by wedge

WANG Hui-ling, LE Jia-ling, WU Ying-chuan

(China Aerodynamics Research & Development Center, Mianyang Sichuan　621000, China)

Abstract: Flow field of shock wave passing by right angle and wedge is analyzed using numerical method and numerical interference technique combining with experimental interferograms. The numerical results are converted into numerical colored interferograms through numerical colored interference technique. The displacement number of interference stripe after shock wave is confirmed by comparing numerical interferogram with experimental interferogram. Accordingly the density distribution of the experimental interferogram can be obtained. The second order explicit up-wind TVD scheme is applied in numerical calculation. The numerical density field is converted into numerical interferogram through numerical interference technique. Interferograms of numerical and experimental are compared. The accuracy of numerical result is verified. The numerical figures demonstrate a clear flow field includes shock wave, contact discontinuity and vortex. Compared with the experimental results the numerical results show the details of evolvement progress of flow field more carefully. That is observed that the shock wave is not dissipated evidently when propagating near the center of vortex through numerical results. The part of shock wave above vortex is decelerated and the part of shock wave below vortex is accelerated.

Key words: TVD scheme; numerical simulation; shock wave; numerical interference technique; experimental interferogram; numerical colored interference technique

王惠玲, 乐嘉陵, 吴颖川. 激波绕楔形物体的流场分析. 实验流体力学, 2006, 20(1): 31-35.

0 引 言

数值计算了激波经过直角拐角后绕楔形物体运动的二维流场,采用数值干涉技术将计算得到的密度转化为数值干涉图并与实验干涉图进行比较,验证了数值计算的有效性。通过数值计算结果充分研究了流场特性,采用数值彩色条纹编码技术将数值干涉图与实验干涉图比较可以辨识过激波发生跳跃后干涉条纹的位移数,从而帮助确定实验干涉照片的密度场。

数值计算用二阶显式迎风 TVD 格式求解二维层流 N-S 方程,较好地捕捉到流场中的激波、旋涡等结构。图 1 为激波绕楔形物体运动的管道及物体尺寸,管道高 32mm,拐角成直角,楔形物体顶部距离拐角 6mm,高 20.5mm,楔角为 40°,楔形物体后台阶高 6mm,长 12mm。

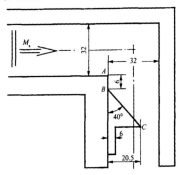

图 1　管道及模型尺寸, $M_s = 1.49$
Fig. 1 Size of model and tunnel

激波绕流实验研究是在 CARDC 设备设计与测试技术研究所的激波管上进行的。实验段内方截面尺寸为 32mm × 32mm,实验得到了激波绕流流场的无限条纹干涉照片,从而可以实现数值计算结果和实验结果的比较。

1 控制方程

控制方程为二维无量纲化层流 N-S 方程

以 L、U_0、T_0、ρ_0 为无量纲化的特征物理量,引入下列无量纲量

$$(\bar{x}, \bar{y}, \bar{z}) = (x, y, z)/L, \quad \bar{T} = T/T_0,$$
$$(\bar{u}, \bar{v}, \bar{w} = (u, v, w)/U_0, \quad \bar{e} = e/(\rho_0 U_0^2),$$
$$(\bar{H_0}, \bar{E}) = (H_0, E)/U_0^2, \quad \bar{p} = p/(\rho_0 U_0^2),$$

$$\bar{\rho} = \rho/\rho_0, \quad \bar{t} = t/(L_0/U_0)$$

为方便起见,无量纲化控制方程中的无量纲量用没有横杠的符号表示

$$\frac{\partial Q}{\partial t} + \frac{\partial E_i}{\partial x} + \frac{\partial F_i}{\partial y} = \frac{\partial E_v}{\partial x} + \frac{\partial F_v}{\partial y}$$

$$Q = \begin{Bmatrix} \rho \\ \rho u \\ \rho v \\ e \end{Bmatrix} \quad E_i = \begin{Bmatrix} \rho u \\ \rho u^2 + p \\ \rho uv \\ u(e+p) \end{Bmatrix}$$

$$F_i = \begin{Bmatrix} \rho v \\ \rho uv \\ \rho v^2 + p \\ v(e+p) \end{Bmatrix}$$

$$E_v = \begin{Bmatrix} 0 \\ \sigma_{11} \\ \sigma_{21} \\ u\sigma_{11} + v\sigma_{12} + \dfrac{c_p}{RePr}k\dfrac{\partial T}{\partial x} \end{Bmatrix}$$

$$F_v = \begin{Bmatrix} 0 \\ \sigma_{12} \\ \sigma_{22} \\ u\sigma_{21} + v\sigma_{22} + \dfrac{c_p}{RePr}k\dfrac{\partial T}{\partial y} \end{Bmatrix}$$

其中 σ_{ij} 为粘性应力张量,其表达式为

$$\sigma_{11} = \frac{\bar{\mu}}{Re}\left(\frac{4}{3}\frac{\partial u}{\partial x} - \frac{2}{3}\frac{\partial v}{\partial y}\right)$$

$$\sigma_{12} = \sigma_{21} = \frac{\bar{\mu}}{Re}\left(\frac{\partial u}{\partial y} + \frac{\partial v}{\partial x}\right)$$

$$\sigma_{22} = \frac{\bar{\mu}}{Re}\left(\frac{4}{3}\frac{\partial v}{\partial y} - \frac{2}{3}\frac{\partial u}{\partial x}\right)$$

粘性系数 μ 根据 Southland 公式给出

$$\frac{\bar{\mu}}{\mu_0} = \left(\frac{T}{T_0}\right)^{1.5}\frac{T_0 + C}{T + C}$$

$T_0 = 288.15K$, μ_0 为 μ 在 $T = T_0$ 的值,$\mu = 1.7894 \times 10^{-5} kg/(m \cdot s)$, $C = Const = 110.4K$。

2 计算方法

采用显式 TVD 格式和两步预估矫正法

$$\Delta U_{j,k}^{(1)} = -\frac{\Delta t}{\Delta x}\left(\hat{E}_{j+\frac{1}{2},k}^n - \hat{E}_{j-\frac{1}{2},k}^n\right) -$$

$$\frac{\Delta t}{\Delta y}\left(\hat{F}_{j,k+\frac{1}{2}}^n - \hat{F}_{j,k-\frac{1}{2}}^n\right) +$$

$$\frac{\Delta t}{\Delta x}\left(E_{vj+\frac{1}{2},k}^n - E_{vj-\frac{1}{2},k}^n\right) +$$

$$\frac{\Delta t}{\Delta y}\left(F^n_{v j, k+\frac{1}{2}} - F^n_{v j, k-\frac{1}{2}}\right)$$

$$U^{(1)}_{j, k} = \Delta U^{(1)}_{j, k} + U^n_{j, k}$$

$$\Delta U^{(2)}_{j, k} = \frac{1}{2}\Delta U^{(1)}_{j, k} - \frac{1}{2}\frac{\Delta t}{\Delta x}\left(\hat{E}^{(1)}_{j+\frac{1}{2}, k} - \hat{E}^{(1)}_{j-\frac{1}{2}, k}\right) -$$

$$\frac{1}{2}\frac{\Delta t}{\Delta y}\left(\hat{F}^{(1)}_{j, k+\frac{1}{2}} - \hat{F}^{(1)}_{j, k-\frac{1}{2}}\right) +$$

$$\frac{1}{2}\frac{\Delta t}{\Delta x}\left(E^{(1)}_{v j+\frac{1}{2}, k} - E^{(1)}_{v j-\frac{1}{2}, k}\right) +$$

$$\frac{1}{2}\frac{\Delta t}{\Delta y}\left(F^{(1)}_{v j, k+\frac{1}{2}} - F^{(1)}_{v j, k-\frac{1}{2}}\right)$$

$$U^{n+1}_{j, k} = \Delta U^{(2)}_{j, k} + U^{(1)}_{j, k}$$

$\hat{E}_{j+\frac{1}{2}, k}$, $\hat{F}_{j, k+\frac{1}{2}}$ 为数值通量, 采用 Yee 的对称 TVD 格式(文献[1]), 粘性项采用中心差分。

3 边界条件

在本算例的计算过程中, 相应的边界条件类型及处理方法如下:

(1) 入口边界条件: 采用激波后的值, $u = u_1$, $v = v_1$, $\rho = \rho_1$, $p = p_1$, $T = T_1$;

(2) 出口边界条件: 在激波到达出口前采用已知的波前条件, $u = 0$, $v = 0$, $\rho = \rho_0$, $p = p_0$, $T = T_0$;

(3) 固体壁面边界条件: 固壁可采用无滑移边界条件, $u = 0$, $v = 0$, 零压力梯度 $\partial p / \partial n = 0$; 等温壁 $T = T_w$;

(4) 块与块之间的边界条件: 由于采用了分块计算, 需要通过块与块之间的数据交换得到全流场的信息。这里采用重叠两层的方法, 在计算第二、第三块时计算区域向第一块内延伸一层, 使第二块、第三块的边界在计算时成为内点来计算, 将它们的边界作为第一块的次边界。计算中由于块与块之间的边界网格点相互重合, 因而保证了数据传递中通量保持一致。

4 网格生成

利用贴体结构网格在粘性计算方面的优点, 采用多块对接结构网格方法生成网格。将激波绕楔形物体运动的计算区域划分为图 2 中的 4 块。先用代数方法生成初始网格, 再利用求解椭圆型偏微分方程的方法生成光滑网格。

计算平面 (ξ, η) 与物理平面 (x, y) 的坐标变换满足 Laplace 方程。图 2 为激波绕楔形物体运动的计算网格。各区域网格点数为: 区域 1: 201×101, 区域 2:

771×201, 区域 3: 301×201, 区域 4: 201×51。

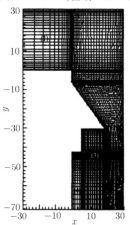

图 2 计算网格
Fig. 2 Numerical mesh

5 计算结果与分析

激波由管道左端进入并沿管道向右运动, 绕过直角拐角 A 经过楔形物体。激波马赫数为 $M_s = 1.49$, 波前温度以及壁面温度为 $T_0 = T_w = 300\mathrm{K}$, 波前压力为 $P_0 = 101300\mathrm{Pa}$。

图 3~7 是 5 个不同时刻实验得到的干涉图以及相应时刻的数值计算结果转化的数值干涉图。每组图左面一幅是实验干涉图, 右面是数值干涉图。图 8 和 9 是数值干涉图, 这两个时刻没有相应实验结果。在图 3 中激波绕过直角拐角, 在拐角 A 处区域出现旋涡和一系列膨胀波。在图 4 中入射激波一部分形成绕射激波遇到楔, 并在楔面上发生弱反射。在图 5 中入射激波在管道右端壁面反射后向左运动。直角拐点 A 处生成的涡沿楔面运动。在图 6 中入射激波经过楔面后缘楔顶点 C 处再次绕射并产生旋涡。右端壁面的反射激波 R 向左运动, 与直角拐点处生成的涡相互作用。直角拐点 C 处生成的涡由于楔面的影响靠近楔面的部分发生扭曲。在图 8 中反射激波 R 进入到了直角拐点产生的旋涡中心附近, 但没有因旋涡作用发生明显耗散。由于旋涡的顺时针运动作用, 旋涡上半分的激波减速, 下半部分的激波得到加速。此外, 反射激波 R 还传播到绕射激波在楔面后缘 C 产生的旋涡。直角拐点 A 处生成的涡靠近楔面的部分扭曲剧烈。在图 9 中 ($t = 138\mathrm{\mu s}$) 反射激

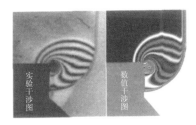

图3　干涉图比较t=0μs, M_s=1.49
Fig.3　The experimental and numerical interferograms

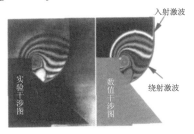

入射激波

绕射激波

图4　干涉图比较t=19μs, M_s=1.49
Fig.4　The experimental and numerical interferograms

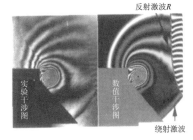

反射激波R

绕射激波

图5　干涉图比较t=73μs, M_s=1.49
Fig.5　The experimental and numerical interferograms

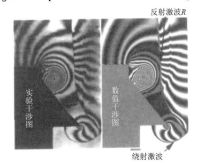

反射激波R

绕射激波

图6　干涉图比较t=106μs, M_s=1.49
Fig.6　The experimental and numerical interferograms

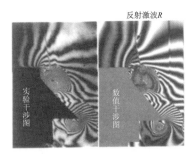

反射激波R

图7　干涉图比较t=153μs, M_s=1.49
Fig.7　The experimental and numerical interferograms

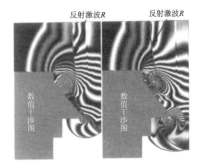

反射激波R　　　反射激波R

图8　干涉图比较　t=　　图9　干涉图比较　t=
112μs, M_s=1.49　　　138μs, M_s=1.49

Fig.8　The numerical　　**Fig.9　The numerical**
interferograms　　　**interferograms**

波R经过(楔面后缘处产生的)旋涡后发生断裂, 旋涡下半部分的激波在楔面上再次发生反射。在图7(t=153μs)中反射激波R在楔面上反射后以右向运动再次经过旋涡, 并使旋涡扭曲。

图10是实验干涉图和彩色条纹编码数值干涉图, 图中入射激波经壁面反射, 条纹过反射激波的跳跃很大, 仅凭黑白干涉条纹位移, 实验照片无法判断, 从而使得反射激波后面区域的密度值的判度很困难。采用彩色数值干涉技术, 可以解决这个问题。图10中分别用红、绿、蓝、黄、青和紫6种颜色给相邻的6个条纹做了标记, 条纹的密度值依次上升, 一个条纹位移的密度变化 $\Delta\rho = \lambda/KL$, $\lambda = 6.943 \times 10^{-7}$ 为光源波长, $K = 0.00025 \text{m}^2/\text{kg}$ 为格-代常数, $L = 0.032\text{m}$ 为实验段厚度, 所以这里 $\Delta\rho = 0.09643 \text{kg/m}^3$。由计算条件可知, 入射激波波前密度 $\rho_0 = 1.178 \text{kg/m}^3$, 波后密度与波前密度比为 1.85, 波后密度为 $\rho_1 = 2.179 \text{kg/m}^3$, 可知图10中红色条纹代表的密度值 ρ_{red}

$= \rho_1 - \Delta \rho / 2 = 2.13 \text{kg/m}^3$,所以不同颜色的条纹代表的密度值是已知的。通过计算和实验图像的比较,根据彩色条纹代表的密度值和条纹计数,可以得到整个实验流场的密度分布。

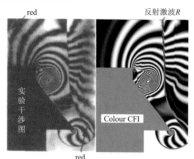

图10　实验干涉图和数值彩色条纹编码干涉图比较
$t=106\mu s$, $M_s=1.49$
Fig. 10　The experimental and numerical interferograms

6　结　论

(1) 通过上述实验和数值计算干涉图的比较,证明在采用分块对接网格技术处理复杂外形和复杂流场计算方面是成功的,不仅有效解决了分块边界区域上的流动参数的传递问题,也为从事更复杂环境下的数值计算打下了基础;

(2) 计算结果反映出了复杂的波系和涡系结构,尽管采用了分块技术,但计算结果在分界区域并没有出现任何的不连续情况,在分界区域上含有激波、旋涡、接触间断时,流场也是连续的;

(3) 通过数值计算激波绕楔形物体流动,观察到激波传播到旋涡中心附近并没有发生明显耗散。旋涡上半部的激波减速,下半部分的激波得到加速;

(4) 通过将数值计算结果转换为彩色条纹编码的数值干涉图,并与实验干涉图对照,可以辨识过激波发生跳跃后的干涉条纹的位移数,从而确定实验干涉照片的密度场。

参考文献:

[1]　YEE H C. Construction of explicit and implicit symmetric TVD schemes and their applications [J]. Journal of comp. Phys. 1987, 68: 151~ 179.

[2]　YEE H C, WARMING R F, HARTEN A. Implicit total variation diminishing (TVD) schemes for steady state calculations [R]. AIAA 83-1902.

[3]　HARTEN A. High resolution schemes for hyperbolic conservation laws [J]. Journal of computational physics 1983, (49): 357~ 393.

[4]　HARTEN A. One high order accurate interpolation for non oscillatory shock capturing schemes [C]. The IMA volumes in mathematics and ints applications, Springer Verlog 1986, (2): 71~ 106.

[5]　YEE H C, KUTLER P. Application of second order accurate total variation diminishing (TVD) schemes to the Euler equations in gerneral geometric [R]. NASA Technical Memorandum 85845.

[6]　YEE H C, SHINN J L. Semi implicit and fully implicit shock capturing methods for none equilibrium flows [R]. AIAA Journal, 1989.

[7]　COOKE C H. Application of an explicit TVD scheme for unsteady, axisymmetric, muzzle brake flow [J]. International Journal For Numerical Methods In Fluids, 1987, (7): 621~ 633.

[8]　THAKUR S S, SHYY W. Unsteady, one dimensional gas dynamics computations using a TVD type sequential solver [R]. AIAA 92-3640.

[9]　MIYAJI K, KOZOFUJII. Unsteady behavior of the shock wave passing over two and three dimensional obstacles [R]. AIAA 96-2445-CP.

双时间步方法的应用分析

赵慧勇，　　乐嘉陵

(中国空气动力研究与发展中心，四川 绵阳　621000)

[摘　要]　　对双时间步方法隐式迭代的稳定性、启动问题、子迭代初值和收敛性准则进行分析．一般认为，双时间步的真实时间步长是基于精度，而不是稳定性的基础上给定，因此真实时间步长可以取得很大，但对 Von Neumann 稳定性分析表明，对于子迭代采用隐式的算法，稳定性的要求对真实时间步有限制．并通过对 Sod 激波管问题的计算，验证该分析．

[关键词]　　非定常计算；双时间步；稳定性分析

[中图分类号]　　V211.3　　　　　[文献标识码]　　A

0　引言

计算流体力学中普遍应用的高效非定常计算方法分为两大类：双时间步方法[1](dual-time stepping) 和物理时间迭代方法[2] (physical time subiteration)．双时间步方法利用两个时间 ——伪时间和物理时间，物理时间迭代方法只利用一个时间，即物理时间．

Rumsey[3] 比较了这两种方法指出：对于层流圆弧翼型的计算，给定 CPU 时间，物理时间迭代方法得到了更准确的结果．然而，物理时间迭代方法有稳定性限制，对于某些问题不实用．而双时间步方法允许更大的时间步，所以更实用一些；对于湍流圆弧翼型的计算，加入 Spalart-Allmaras(S-A) 湍流模型后，物理时间迭代方法需要过多的子迭代次数，而双时间步方法要好得多．

双时间步方法有许多好处[1]：原来用于定常计算的预处理、当地时间步、对角化、多重网格都可以应用到计算中；一般认为真实时间步长是基于精度，而不是稳定性的基础上给定，因此真实时间步长可以取得很大；双时间法可以降低通量线性化误差和近似因子分解误差，放宽了稳定性限制．此外，利用双时间步方法可以方便地把一个定常计算的程序转化为非定常计算的程序．

本文对采用双时间步方法的隐式迭代的稳定性、启动问题、子迭代的初值和子迭代的收敛性判据进行了分析．Von Neumann 稳定性分析表明：对于子迭代采用隐式的算法，稳定性的要求对真实时间步有限制．给出一个真实时间步计算的必要条件．并通过 Sod 激波管问题的计算进行验证．

1　计算方法

流动控制方程为三维 N-S 方程，采用有限体积法离散．

$$\frac{\mathrm{d}}{\mathrm{d}t}(\int_v Q \mathrm{d}v) + \int_s (F - F_v)\mathrm{d}s = 0, \tag{1}$$

其中 Q 为依赖变量，F 和 F_v 分别为无粘通量和粘性通量．

令 Δt 为真实时间步，$\Delta\tau$ 为虚拟时间步，V 为网格体积，采用双时间步方法和 LU-SGS 隐式迭代可得

$$\begin{cases} (D - L)\Delta\overline{Q}^n = -\frac{\Delta\tau}{V}\left(\frac{3\overline{Q}^n - 4\overline{Q}^n + \overline{Q}^{n-1}}{2\Delta t}\right)V + \frac{\Delta\tau}{V}R(\overline{Q}^n), \\ (D + U)\Delta\overline{Q}^n = D\Delta\overline{Q}^*. \end{cases} \tag{2}$$

赵慧勇，乐嘉陵. 双时间步方法的应用分析. 计算物理，2008，25(3)：253-258.

其中

$$D = \left\{1 + \frac{3}{2}\frac{\Delta\tau}{\Delta t}\right\}I + \frac{\Delta\tau}{V}(\lambda_{max}^A + \lambda_{max}^B + \lambda_{max}^C)I,$$

$$L = \frac{\Delta\tau}{V}(A_{i+1,j,k}^- + B_{i,j+1,k}^- + C_{i,j,k+1}^-),$$

$$U = \frac{\Delta\tau}{V}(A_{i-1,j,k}^+ + B_{i,j-1,k}^+ + C_{i,j,k-1}^+).$$

这里 $\bar{Q}$ 为网格平均的依赖变量,矩阵 A, B, C 分别是三个方向无粘通量的 Jacobian 矩阵, $\lambda_{max}^A, \lambda_{max}^B, \lambda_{max}^C$ 分别是矩阵 A, B, C 的最大特征值的模. $A^\pm = A \pm \lambda_{max}^A I$, $B^\pm = B \pm \lambda_{max}^B I$, $C^\pm = C \pm \lambda_{max}^C I$.

方程(2)的计算需要考虑稳定性分析、启动问题、子迭代的初值和子迭代的收敛性.

1.1 双时间步的稳定性分析

Jameson[1] 指出,当子迭代的当地时间步大于真实时间步时,格式将变得不稳定.他们建议子迭代中的当地时间步取为 $\Delta\tau = \min(\frac{CFL_{sub}\cdot V_{i,j,k}}{\lambda^A + \lambda^B + \lambda^C}, \frac{2\Delta t}{3})$. 但是 Jameson[1] 的子迭代采用显式多步 Runge-Kutta. 对于子迭代采用隐式 LU 的稳定性分析,目前还比较少见.本文利用 Von Neumann 稳定性分析,得出了以下结论: 对于子迭代采用隐式的算法,对子迭代中的当地时间步没有限制,但是对真实时间步有限制.同时,给出一个真实时间步的必要条件,当子迭代的 CFL 数大于 1 时,真实时间步 $\Delta t \leqslant \frac{0.75 V_{i,j,k}}{\max(\lambda^A, \lambda^B, \lambda^C)}$, 此时对应的 $CFL = 0.75$.

为了推导方便,以下针对一维标量方程进行 Von Neumann 稳定性分析.

$$\frac{\partial u}{\partial t} + a\frac{\partial u}{\partial x} = 0, \tag{3}$$

不失一般性,假定 $a > 0$, 引入虚拟时间微分,

$$\frac{\partial u}{\partial \tau} + \frac{\partial u}{\partial t} + a\frac{\partial u}{\partial x} = 0. \tag{4}$$

令 $\Delta u_i^m = u_i^{m+1} - u_i^m$, 利用有限体积法对(4)式进行离散得

$$\frac{\Delta u_i^m}{\Delta\tau} = -\frac{3u_i^{m+1} - 4u_i^n + u_i^{n-1}}{2\Delta t} - a\frac{u_{i+1/2}^{m+1} - u_{i-1/2}^{m+1}}{\Delta x}. \tag{5}$$

线性化(5)式,且令 Δ_x 为空间后差,可得

$$\left(\frac{1}{\Delta\tau} + \frac{3}{2}\frac{1}{\Delta t} + \frac{a}{\Delta x}\Delta_x\right)\Delta u = -\frac{3u_i^m - 4u_i^n + u_i^{n-1}}{2\Delta t} - a\frac{u_{i+1/2}^m - u_{i-1/2}^m}{\Delta x}. \tag{6}$$

假设式(5)的解可表示为 $u = d + \varepsilon$, 其中 d 是真解, ε 是误差函数.由于 u 和 d 都满足式(5),两者相减得

$$\left(\frac{1}{\Delta\tau} + \frac{3}{2}\frac{1}{\Delta t} + \frac{a}{\Delta x}\Delta_x\right)\Delta\varepsilon = -\frac{3\varepsilon_i^m}{2\Delta t} - a\frac{\varepsilon_{i+1/2}^m - \varepsilon_{i-1/2}^m}{\Delta x}. \tag{7}$$

令

$$\varepsilon_i^m = G^m e^{-ikx(i)}, \qquad \varepsilon_i^{m+1} = \varepsilon_i^m G, \qquad \varepsilon_{i-1}^m = \varepsilon_i^m e^{ik\Delta x}, \tag{8}$$

将(8)式代入(7)式,经过简单的推导可得

$$G = 1 + \frac{-\frac{3}{2\Delta t} + \frac{2a}{\Delta x}\sin(\frac{k\Delta x}{2})i}{\frac{1}{\Delta\tau} + \frac{3}{2\Delta t} + \frac{a}{\Delta x}(1 - e^{ik\Delta x})} = \frac{\frac{1}{\Delta\tau} + \frac{a}{\Delta x}(1 - e^{ik\Delta x}) + \frac{2a}{\Delta x}\sin(\frac{k\Delta x}{2})i}{\frac{1}{\Delta\tau} + \frac{3}{2\Delta t} + \frac{a}{\Delta x}(1 - e^{ik\Delta x})} = \frac{G_1}{G_2}, \tag{9}$$

$$\|G_2\|^2 - \|G_1\|^2 = \frac{2a}{\Delta x}\sin^2\left(\frac{k\Delta x}{2}\right)\left[\frac{9\Delta x}{8a(\Delta t)^2} + \frac{3\Delta x}{2a\Delta t\Delta\tau} + \frac{3}{\Delta t} + \frac{4a}{\Delta x}\cos\left(\frac{k\Delta x}{2}\right) - \frac{2a}{\Delta x}\right] + \frac{2a}{\Delta x}\cos^2\left(\frac{k\Delta x}{2}\right)\left(\frac{9\Delta x}{8a(\Delta t)^2} + \frac{3\Delta x}{2a\Delta t\Delta\tau}\right), \tag{10}$$

显然,

$$\frac{2a}{\Delta x}\cos^2\left(\frac{k\Delta x}{2}\right)\left(\frac{9\Delta x}{8a(\Delta t)^2} + \frac{3\Delta x}{2a\Delta t\Delta\tau}\right) \geqslant 0, \qquad \frac{2a}{\Delta x}\sin^2\left(\frac{k\Delta x}{2}\right) \geqslant 0,$$

因此,当 $\dfrac{9\Delta x}{8a(\Delta t)^2}+\dfrac{3\Delta x}{2a\Delta t\,\Delta\tau}+\dfrac{3}{\Delta t}+\dfrac{4a}{\Delta x}\cos\left(\dfrac{k\,\Delta x}{2}\right)-\dfrac{2a}{\Delta x}\geqslant 0$ 时,

$$\|G_2\|^2-\|G_1\|^2\geqslant 0\Rightarrow\|G\|\leqslant 1.$$

令

$$\overline{CFL}=\dfrac{a\,\Delta\tau}{\Delta x},$$

即子迭代的 CFL 数,再令

$$f(\Delta t)=\dfrac{9\Delta x}{8a}+\Delta t\left(\dfrac{3}{2\overline{CFL}}+3\right)-\dfrac{6a(\Delta t)^2}{\Delta x},$$

得

$$\dfrac{9\Delta x}{8a(\Delta t)^2}+\dfrac{3\Delta x}{2a\Delta t\,\Delta\tau}+\dfrac{3}{\Delta t}+\dfrac{4a}{\Delta x}\cos\left(\dfrac{k\,\Delta x}{2}\right)-\dfrac{2a}{\Delta x}\geqslant\dfrac{1}{(\Delta t)^2}f(\Delta t). \tag{11}$$

经过简单的推导可得:由于隐式子迭代的 CFL 数大于 1,如果

$$\Delta t\leqslant\dfrac{3\Delta x}{4a}, \tag{12}$$

就有 $f(\Delta t)\geqslant 0$. 从而 $\|G_2\|^2-\|G_1\|^2\geqslant 0$, $\|G\|\leqslant 1$,格式稳定.

这表明稳定性对双时间步子迭代的虚拟时间步没有限制,但是对真实时间步有限制. 对于二维、三维和向量方程,可以做形式上的推广,

$$\Delta t\leqslant 0.75\min\left(\dfrac{\Delta x}{\chi^a},\dfrac{\Delta y}{\chi^b},\dfrac{\Delta z}{\chi^c}\right). \tag{13}$$

需要注意的是:

1) (12)式只是一个充分条件,不是充要条件. 尽管这样,(12)式仍然具有一定的指导意义.

2) (12)式针对线性标量方程,要推广到非线性向量方程,还需要实践的检验. 可能这个标准还可以放宽. 如 Buelow[4]就指出:根据稳定性分析,LU 的最大 CFL 数= 10,然而在对无粘直管的计算表明,即使 CFL 数 = 1 000 也没有出现稳定性问题. 但也有相反的例子存在.

如果对(5)式右端的对流项直接采用一阶迎风或者二阶迎风差分离散,通过类似推导得到其无条件稳定,但不是守恒型格式.

1.2 启动问题

由于双时间步一开始需要前两个时间层的 $\overline{Q}$ 值,即 $\overline{Q}^{n-1}$ 和 $\overline{Q}^{n}$,而 $\overline{Q}^{n-1}$ 和 $\overline{Q}^{n}$ 是未知的. 目前有两种解决办法:一种是使用另外的方法计算 $\overline{Q}^{n-1}$ 和 $\overline{Q}^{n}$,如显式多步 Runge-Kutta,这种方法要复杂一些,需要两套算子,但可以保证时间精度;另一种是简单地让 $\overline{Q}^{n-1}$ 和 $\overline{Q}^{n}$ 等于初场. 这种方法要简单一些,但时间精度稍有损失. 本文采用第二种方法. 数值试验表明,时间精度还是可以接受的.

1.3 子迭代的初值

子迭代的初值选取也有两种方法:①使用二阶外差,$\overline{Q}^{n,0}=\overline{Q}^{n}+\dfrac{3\overline{Q}^{n}-4\overline{Q}^{n-1}+\overline{Q}^{n-2}}{2}$. 这种方法有二阶精度,但是计算较复杂一些;②使用一阶外差,$\overline{Q}^{n,0}=\overline{Q}^{n}$. 这种方法只有一阶精度. 由于子迭代主要要求收敛,对精度的要求可以放宽,因此第二种方法更加有效. 本文采用第二种方法.

1.4 子迭代的收敛性

为保证时间精度,要求子迭代收敛. 本文的残差定义为 $\max(\Delta\rho^n/\Delta\tau)$. Christopher[5]和 Buelow[6]指出,要求残差下降到机器精度是不现实的,一般残差下降三个量级就可以满足计算的要求. 本文的收敛判据为残差下降三个量级. 对于比较复杂的流场,发现子迭代的残差不能下降三个量级,陷入死循环. 为此,在给定残差下降三个量级的收敛判据外,还限制子迭代的最大迭代步数. 实际计算表明最后的结果是可以接受的.

2 Sod 激波管问题

Sod 激波管是一个经典算例,该问题计算简单,且有理论解,可以很好地检验非定常算法的性能. 本来是

一个一维问题, 本文把它作为二维问题来计算.

计算方法: 控制方程采用二维无粘 Euler 方程, 三阶 MUSCL 外插, 无粘通量采用 AUSM PW+ 格式, 时间推进采用双时间步 LU-SGS.

初始条件: $(\rho, u, p) = \begin{cases} (1.0, 0.0, 1.0), & 0 < x < 0.5, \\ (0.125, 0.0, 0.1), & 0.5 < x < 1.0. \end{cases}$

采用 100×50 个等距网格计算, 边界条件采用镜像条件. 激波马赫数为 1.656, 激波在 0.277 s 时到达激波管的顶端. 最大子迭代数目为 4, 真实时间步的 CFL 数为 1, 虚拟时间步的 CFL 数为 2.

图 1 给出了四个时刻的密度分布, 计算与理论解吻合较好. 说明本文的方法具有比较好的时间精度.

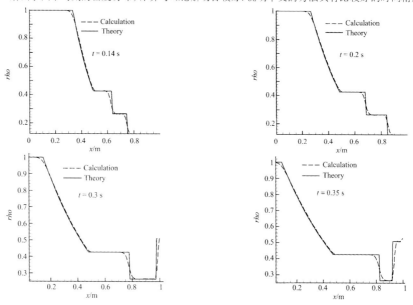

图 1 $t = 0.14, 0.2, 0.3, 0.35$ s 的密度分布
Fig. 1   Density distribution at $t = 0.14, 0.2, 0.3, 0.35$ s

为了对子迭代的时间步和迭代次数进行研究, 分别采用三种子迭代 CFL 数和三种子迭代次数. 图 2 是子迭代 CFL 数的影响. 当 $CFL = 1.0$, 五次子迭代时, 子迭代 CFL 数分别取 2.0、5.0 和 10.0, 残差均可下降三个量级. 计算得到的结果比较一致, 与理论解吻合较好. 说明只要迭代收敛, 子迭代的时间步可以取得很大. 这是因为子迭代使用隐式 LU-SGS, 稳定性比较好, 时间步长可以取得很大. 图 3 比较了子迭代数目的影响. 没有子迭代时, 计算结果很差. 与理论解相比, 稀疏波超前, 接触间断和激波滞后. 加入子迭代后, 计算结果有明显改进. 子迭代取两次、五次和十次的结果基本相同. 两次子迭代的残差下降了两个量级, 五次下降三个量级, 十次下降五个量级. 与理论解相比, 子迭代次数由五次增加到十次并没有提高多少时间精度, 而计算时间却增加了一倍. 对于空间精度, 为了比较, 采用五阶 MP5 格式[7]进行计算. 计算结果表明, 两次、五次和十次子迭代次数的空间精度几乎相同, 都不如五阶 MP5 格式的空间精度高. 说明子迭代次数对时间精度有显著影响, 五次子迭代次数可以得到比较理想的结果, 过多的子迭代数对计算结果没有改善. 子迭代对空间精度的影响很小, 需要采用高阶精度的格式才能提高空间精度.

为了验证前面的稳定性分析, 取子迭代 $CFL = 10.0$, 使用五次子迭代, 对真实时间步的 CFL 数分别取 0.5、0.75 和 1.0, 图 4、5 给出了计算的密度和温度分布, 图 6、7 给出了局部的放大图. 选取三种计算状态: 一种不满足稳定性条件, $CFL = 1.0$; 两种满足稳定性条件, $CFL = 0.75$ 和 0.5. 从图中可以看出: $CFL = 1.0$ 的结果在激波和接触间断附近有显著的波动, $CFL = 0.75$ 和 0.5 的波动减弱了许多. 与理论解相比, $CFL = 0.75$

和0.5的结果更加接近. $CFL = 0.75$ 的密度分布与 $CFL = 0.5$ 的很接近, 温度稍差一些. 因此, 本文的稳定性条件具有一定的指导意义.

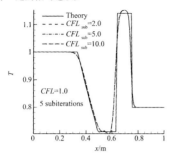

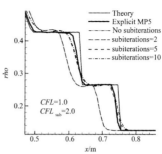

图2　$t = 0.14$ s 子迭代 CFL 数对温度分布的影响

Fig 2　Effect of subiteration CFL on temperature
distribution at $t = 0.14$ s

图3　子迭代的迭代步数对密度分布的影响

Fig 3　Effect of subiteration number on density
distribution at $t = 0.14$ s

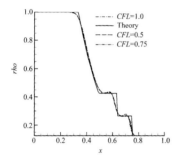

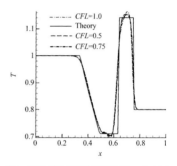

图4　不同 CFL 数下的密度分布($t = 0.14$ s)

Fig. 4　Density distribution at different
CFL numbers ($t = 0.14$ s)

图5　不同 CFL 数下的温度分布($t = 0.14$ s)

Fig. 5　Temperature distribution at different
CFL numbers ($t = 0.14$ s)

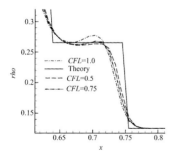

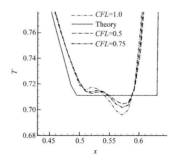

图6　不同 CFL 数下的密度分布(放大图)

Fig 6　Density distribution at different CFL
numbers (magnified figure)

图7　不同 CFL 数下的温度分布(放大图)

Fig. 7　Temperature distribution at different CFL
numbers (magnified figure)

3　结论

对双时间步方法隐式迭代的稳定性、启动问题、子迭代的初值和子迭代的收敛性进行分析,得出以下结论:

1) 子迭代的次数对时间精度有显著影响,但对空间精度的影响不大. 要提高空间精度,需要采用高阶格式. 在给定残差下降三个量级的收敛判据外,还限制了子迭代的最大迭代步数. 实际计算也表明这种收敛性判据是可以接受的.

2) 利用 Von Neumann 稳定性分析,对子迭代采用隐式的算法进行稳定性分析. 结果表明:在没有分裂以前,对子迭代中的当地时间步没有限制,但是对真实时间步有限制. 同时,还给出一个真实时间步计算的必要条件. 对 Sod 激波管的计算验证了分析结果,说明本文的稳定性条件具有一定的指导意义.

[参 考 文 献]

[1] Jameson A. Time dependent calculation using multigrid with application to unsteady flows past airfoils and wings[R]. AIAA Paper 91 − 1596, 1991.

[2] Thomas H Pulliam. Time accuracy and the use of implicit methods[R]. AIAA Paper No. 93− 3360, 1993.

[3] Rumsey Christopher L, Sanetrik Mark D, Biedron Robert T, Melson N Duane, Parlette Edward B. Efficiency and accuracy of time accurate turbulent Navier Stokes compuations[R]. AIAA Paper, No. 95− 1835, 1995.

[4] Buelow P E O, Venkateswaran S, Merkle C L. Stability and convergence analysis of implicit upwind scheme[J]. Computers & Fluids, 2001, **30**: 961− 998.

[5] Atwood Christopher A. An upwind approach to unsteady simulation[R]. AIAA Paper, No. 90− 3100, 1990.

[6] Buelow P E O, Schwer Douglas A, Feng Jinzhang, Merkle Charles L, Choi Dochul. A preconditioned dual time, diagonalized ADI scheme for unsteady computations[R]. AIAA paper 97− 2101, 1997.

[7] Suresh A, Huynh H T, Accurate monotonicity− preserving schemes with RUNGE-KUTTA time stepping[J]. J Comput Phys, 1997, **136**.

Application Analysis on Dual-time Stepping

ZHAO Huiyong,　　LE Jialing

(China Aerodynamics Research and Development Center, Mianyang　621000, China)

Abstract:　The stability, starting problem, initial value of sub iteration and convergence criterion of implicit iteration of the dual time stepping method are studied. Analysis of von Neumann stability indicates that in the implicit method employed in sub iteration real time step is limited by the requirment of stability. Computation on Sod's shock tube validates the conclusion.

Key words:　unsteady computation; dual time stepping; stability analysis

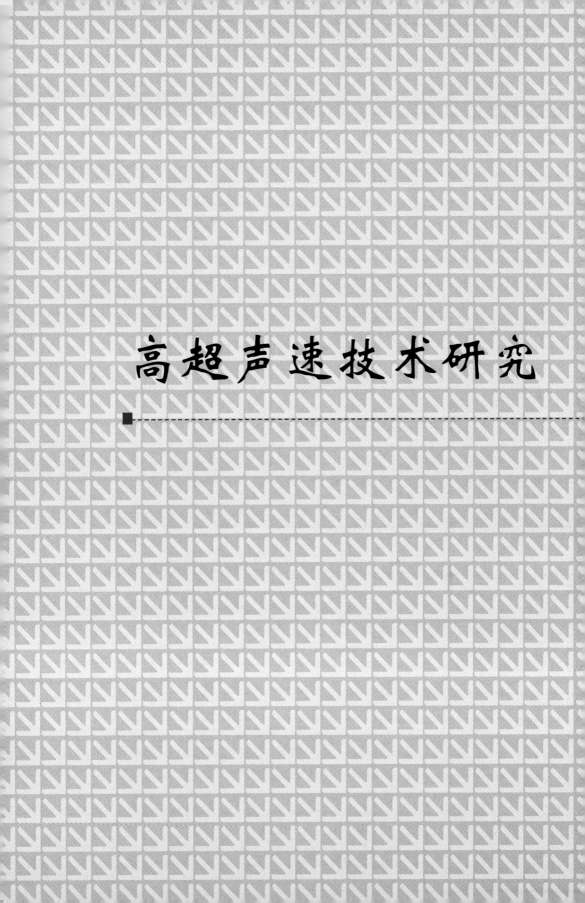

高超声速技术研究

高超声速飞行器的碳氢燃料
双模态超燃冲压方案研究

乐嘉陵

中国空气动力研究与发展中心, 绵阳　621000

刘　陵

西北工业大学, 西安　710072

摘　要　以超燃冲压作动力的高超声速巡航飞行器与火箭动力相比, 在 $M = 6$ 时, 比冲增加二倍以上; 与亚燃冲压相比, 发动机内静温、静压低, 从而减轻了结构强度负荷, 简化了结构设计。这种巡航飞行器飞行速度快, 突防与生存能力强, 具有更大的作战能力。

根据我国国情, 本文提出了一种以碳氢燃料双模态超燃冲压作动力的高超声速巡航飞行器的方案, 并针对航程 1500km, 重 1500kg, 直径 0.6m, 长 4.5m 的飞行器参数, 估算了轨道、飞行时间、燃料消耗, 确定了超燃冲压前体进气道及燃烧室的形状、尺寸, 并作了超燃冲压性能计算。

关键词　高超声速飞行器; 超声速燃烧冲压发动机; 超声速燃烧
中图号　V11, V23, V23.2

0　引　言

飞行器飞行 $M > 6$ 时, 使用超燃冲压作为动力装置具有很大的优势, 国外早已相继开展了超燃冲压的研制[1]。虽然碳氢燃料的冲压比冲只有氢燃料的 $1/3 \sim 1/2$, 但使用维护方便, 单位体积的热值高, 燃料箱体积小, 可满足飞行器体积受严格限制的要求。双模态超燃冲压的飞行 $M < 6$ 时, 超声速燃烧室内以亚声速燃烧状态工作, 称亚燃模态; 当飞行 $M > 6$ 时, 燃烧室内, 燃料在超声速气流中燃烧, 称超燃模态。双模态燃烧过程可把超燃冲压的工作下限延伸到马赫数 3, 仍保持良好的工作性能。由于无需另外安装 $M = 3 \sim 6$ 的加速动力装置, 大大减少了重量和迎风面积, 从而使飞行器具有更优良的性能。

以碳氢燃料双模态超燃冲压作动力的飞行器飞行高度可达 40km 左右, 巡航飞行 $M = 6 \sim 8$, 射程 $500 \sim 6000km$, 超燃冲压比冲高。它与火箭相比, 比冲增加 2 倍以上; 与亚燃冲压导弹相比, 超燃冲压发动机内部压力约低 $80\% \sim 83\%$, 燃气温度也大幅度下降, 从而减少了结构强度负荷与重量, 结构简化, 造价低。

根据我国国情, 本文提出了用于高超声速飞行器的超燃冲压方案的初步设计参数, 包括超燃冲压的主要尺寸、性能、飞行轨道、飞行时间及发动机耗油量等。

乐嘉陵, 刘陵. 高超声速飞行器的碳氢燃料双模态超燃冲压方案研究. 流体力学实验与测量, 1997, 11(2): 1-13.

1　飞行器总体设想

设计方案的指标是：飞行器初始状态为高度 16km 时，飞行 $M = 1.5$，由火箭发动机加速

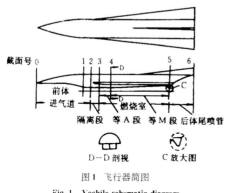

至 $M = 3.5$ 后火箭脱离，超燃冲压以亚燃模态开始工作，沿等动压轨道爬升并加速至高度 25km，$M = 6$ 左右，再转换为超燃模态工作（沿等高度，$M = 6$），总航程 1500km。飞行器直径 0.6m，长度 4.5m，重量 1500kg，飞行攻角 $\alpha = 8°$，性能数据确定如下：

图 1　飞行器简图

Fig.1　Vechile schematic diagram

升阻比	$K = 3.5$
升力系数	$C_y = 0.052$
阻力系数	$C_x = 0.01486$
等动压	$q_n = \rho M^2 a^2 / 2 = 64241\,\mathrm{Pa}$
飞行器参考面积	$S = 4.278\,\mathrm{m}^2$
飞行阻力	$X = C_x S q_n = 4084\,\mathrm{N}$

飞行器外形采用一体化设计，前体下表面构成发动机进气道的一部分，压缩来流空气，后体下表面作为发动机尾喷管的一部分，使燃气流继续充分膨胀、加速，如图 1 所示。

2　碳氢燃料双模态超声速燃烧冲压发动机性能估算

现已发表的论文没有涉及碳氢燃料双模态超燃冲压具体设计方法及非设计点的性能估算[2,3,4]。本文首先拟定 $M = 6$，$H = 25\mathrm{km}$ 飞行状态为超燃冲压设计点，进行优化设计；然后估算 $M = 3.5 \sim 6$ 非设计点亚燃模态的性能，并提供结构的主要尺寸及气流参数数据。

2.1　设计状态

2.1.1　前体进气道

前体进气道的设计原则，在文献 [1,5,6] 中已经阐明。文献 [7] 还提供了具有四个斜激波压缩的最佳前体的模型，这种模型设计方法的依据是斜激波前马赫数与激波角 ε 正弦的乘积为常数。文献 [8] 发表了 $\alpha = 9°$，飞行 $M_n = 5$ 另一种按 $\varepsilon_{in} = \mathrm{const}$ 方法设计的前体进气道参数。本文作了两种设计方法的全面比较，认为后一种方法设计更为合理。

设计时，取飞行器中心线为 x 轴，前体壁面的第一个折转角 $\theta_1 = 0$，则气流偏转角 $\delta_{,1}$ 等于飞行攻角 α。

由气体动力学斜激波公式可得到 $\delta_{,2}$、$M_{,2}$、$\delta_{,3}$、$M_{,3}$、$\delta_{,4}$、$M_{,4}$。在设计状态下，四个斜激波均交于进气道唇口，所以气流偏转角 $\delta_{,5} = \delta_{,1} + \delta_{,2} + \delta_{,3} + \delta_{,4}$。气流进入发动机通道后，在进气道内再作进一步压缩，如图 2 所示。选择 $\delta_{,6} = 12.5°$ 及选择 $\delta_{,7} = \delta_{,6}$，使气流平行进入燃烧室内。最后由计算得到 $\delta_{,6}$、$\delta_{,7}$、$M_{,6}$、$M_{,7}$。

由图 2 可知，规定其中任一尺寸，如进气道外壳唇口距飞行器前线的距离 $l_{,4}$，则其他尺寸 l_{in}、h_{in}（$n = 1 \sim 3, 5 \sim 7$）皆可确定。$l_{,4}$ 的选择原则应考虑：

（1）满足设计要求，发动机进口尺寸不超过高 0.26m、宽 0.48m 的迎风面范围；

（2）地面试验设备限制了发动机模块尺寸不能过大；

（3）飞行器重心、发动机长度和重量的综合要求。

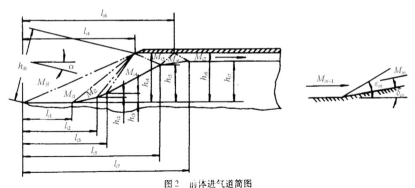

图2 前体进气道简图

Fig.2 Schematic diagram of forebody and inlet

本文选取 $l_{i4} = 1.5\text{m}$，并计算出所有 l_{in}、h_{in}（$n = 1\sim3,5\sim7$）值以及通道面积 A。

假设为理想气体，并且忽略附面层影响，可由气体动力学斜激波关系式依次计算 p_{in}、ρ_{in}、t_{in}、U_{in} 及波后总压 P_{in}（$n = 1\sim7$）。由于 γ_{in} 值与温度有关，为此，必须进行迭代运算，计算结果见表1。

表1 进气道参数

	i1	i2	i3	i4	i5	i6	i7
M	4.94	4.39	4.09	3.92	3.15	2.53	2.06
$\varepsilon(°)$	15.68	15.68	15.68	15.68	23.79	28.48	33.47
$\delta(°)$	8.00	5.58	3.64	2.25	11.47	12.5	12.5
$h(\text{m})$	0.00	0.026	0.044	0.202	0.156	0.202	0.1845
$l(\text{m})$	0.780	1.047	1.16	1.5	1.711	1.791	1.842

2.1.2 燃烧室

文献[4]估算了碳氢燃料超声速燃烧冲压发动机性能，按等马赫数加热规律设计燃烧室。文献[9]曾分析了各种加热规律对氢燃料超燃冲压发动机性能的影响并得出结论：当飞行马赫数 $M_n \leqslant 6.6$ 时，按照等 A-M，即等面积——等马赫数加热规律设计，发动机比冲最大。这种设计方法是把燃烧室分成两部分，第一段为等截面段，超声速气流燃烧后，M 数降低至接近声速；然后，进入第二段等马赫数段，即气流在马赫数不变条件下被加热。上述设计方法，由于气流是在较低马赫数下被加热，总压损失较小，总压恢复系数较高，发动机比冲大，为此本文采用等 A-M 规律设计燃烧室。

（1）等 A 段设计

由气体动力学[10]可知，气流在等截面段被加热时，若忽略壁面摩擦，则燃料质量添加流率 $\text{d}\dot{m}$，总温变化 $\text{d}T$ 和马赫数的变化 $\text{d}M$ 具有下列关系，即

$$\frac{\text{d}M}{M} = \frac{(1 + \gamma M^2)\Psi}{2(1 - M^2)}\left[\frac{\text{d}T}{T} + 2\frac{\text{d}\dot{m}}{\dot{m}}\right] \tag{1}$$

式中： $\Psi = 1 + \dfrac{\gamma - 1}{2}M^2$

又由能量守恒可知，加入的燃料流率 $\text{d}\dot{m}$ 与总温变化存在以下关系

$$\frac{\mathrm{d}T}{T} = \frac{1}{(\dot{m} + \mathrm{d}\dot{m})C_p}(\dot{m}C_p + \frac{\mathrm{d}\dot{m}H_u\eta}{T}) - 1 \tag{2}$$

选用国产 21 号煤油[11], 燃料热值 $H_u = 42378\mathrm{kJ/kg}$, 密度 $\rho_j = 0.9026\mathrm{kg/L}$, 化学恰当油气比 $f_{st} = 0.0676372$。空气的主要成分见表2。

表2 空气中的主要成分

	分子量 M_i	质量分量 f_i	莫尔分量 Y_i
N_2	28.016	0.75529	0.78088
O_2	32.0	0.23145	0.2095
A	39.944	0.01326	0.00962

注:空气分子量 28.965

由 f_{st} 可推算得知, 21 号煤油分子式近似为 C_2H_{16}, 选燃烧效率 $\eta = 0.85$。化学反应中混气成分计算如下:

若燃油空气比为 f_i, 燃烧效率为 η, 则

$$\varphi C_nH_m + (n + \frac{m}{4})O_2 + (n + \frac{m}{4})3.7274N_2 +$$

$$(n + \frac{m}{4})0.04592A \rightarrow \eta\varphi n CO_2 +$$

$$(\eta m/2)\varphi H_2O + (1 - \eta) \times (n/2)\varphi C_2H_4 + (n \div \frac{m}{4})3.7274N_2 +$$

$$(n + \frac{m}{4})0.04592A + (1 - \eta\varphi)(n + \frac{m}{4})O_2 \tag{3}$$

化学反应式中 A 为氩气, φ 为当量油气比, 即 $\varphi = \dfrac{f_i}{f_{st}}$

上式中假设未燃油为 C_2H_4, 由此可得到燃气重量组分。

加热过程中总压变化为

$$\frac{\mathrm{d}P}{P} = -\frac{\gamma M^2}{2}(\frac{\mathrm{d}T}{T} + 2\frac{\mathrm{d}\dot{m}}{\dot{m}}) \tag{4}$$

考虑了加入燃料流率 $\mathrm{d}\dot{m}$、混气组分及 C_p、γ 的变化, 数值求解式(1)、(2)、(4)直到在等截面段出口, 超声速气流被加热减速到规定的 M 值, 同时得到等截面段出口处 T、当量油气比 f_i。利用气体动力学关系式可得到相应的静压 p、静温 t、气流速度 u 及总压恢复系数 σ。

(2) 等 M 段设计

忽略壁面摩擦, 有燃料加入和总温变化时, 管道面积 A 变化为

$$\frac{\mathrm{d}A}{A} = \frac{(1 + \gamma M^2)}{2}\left[\frac{\mathrm{d}T}{T} + 2\frac{\mathrm{d}\dot{m}}{\dot{m}}\right] \tag{5}$$

用数值求解式(2)、(4)、(5), 可得到给定注入的燃料, 等 M 段的出口面积 A, 出口总温 T, 总压 P 及所有其他气流参数。

2.1.3 尾喷管后体

假设(1)气流在尾喷管后体膨胀过程中, 总温不变;

(2)总压恢复系数 $\sigma = 0.98$;

(3)不完全膨胀, 出口静压 p 大于大气压 p_H, 即 $p = p_H/\zeta$, 选取 $\zeta = 0.9$。气流 M 数与面积变化关系可由下式数值解获得, 即

$$\frac{\mathrm{d}M}{M} = -\frac{1 + \dfrac{\gamma - 1}{2}M^2}{1 - M^2}\frac{\mathrm{d}A}{A} \tag{6}$$

计算时应考虑膨胀过程中, 由于静温变化引起的比热比 γ 变化。

图3所示为 $\varphi = 0.4$ 设计状态下, 发动机内总压 P、静压 p、总温 T、静温 t、马赫数 M 等

气流参数沿各截面的变化状况。图3上截面编号的位置同图1。

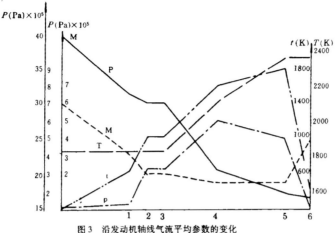

图3 沿发动机轴线气流平均参数的变化

Fig.3 Axial variation of flow paramenters with the combustion systems

2.1.4 发动机性能

发动机推力 T，比冲 I_{sp}，推力系数 C_T 和总压恢复系数 σ 计算式分别为

$$T = (1 + f_i)u_6\dot{m}_a - u_0\dot{m}_a + p_6A_6 - p_0A_0 \quad (7)$$

$$I_{sp} = \frac{T/g}{f_i\dot{m}_a} \quad (8)$$

式中：f_i 为总油气比；$\dot{m}_a$ 为发动机进口截面空气流量；g 为重力加速度。

$$C_T = \frac{T}{q_0A_0} \quad (9)$$

$$\sigma = \prod_1^n \sigma_i \quad (10)$$

图1 中已依次表示出了各截面的位置。

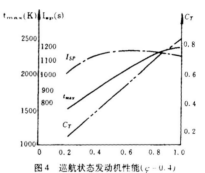

图4 巡航状态发动机性能($\varphi = 0.4$)

Fig.4 Engine performance for cruise ($\varphi = 0.4$)

所有计算数据结果如图4所示,图中同时表示出燃烧室出口截面最高静温值 t_{max}。

2.2 非设计状态

计算非设计状态的目的是研究发动机达到 $M = 6$,高度 $H = 25km$ 设计状态前,在 $M = 3.5 \sim 6$, $H = 16 \sim 25km$ 区间沿等动压 q_0 轨道爬升过程中发动机性能的变化。

2.2.1 实现亚燃模态的方案选择

根据发表的文献在双模态超声速燃烧室内实现亚燃模态,有两种方案:

(1)文献[12]提出,在燃烧室进口形成一正激波,气流速度降为亚声速,在亚声速气流中燃烧,实现亚燃模态,气流加热加速至 $M = 1$,进入燃烧室的扩张段再加速到高 M 数流出燃烧室。这一方案的最大优点是超声速燃烧室的几何形状不需要作任何变化,结构简单。但是油气混合、发生化学反应、加热气流是极其复杂的过程,利用在燃烧室不同截面的喷油,控制放热规律与气流 M 数变化,使之与燃烧室几何形状设计相互间严格、准确地配合相当困难。因此,此方案必须要作较多的数值计算及实验调试工作。

（2）文献[13]提出另一方案,即在超燃冲压发动机尾喷管处,用一阀板形成一个最小截面,即发动机的喉道,如图1所示。亚燃模态时,气流由亚声速经过喉道转为超声速流进入尾喷管后体。当飞行M数增大,在发动机内需要以超燃模态工作时,阀板打开或脱落。这一方案将增加结构的复杂性,但便于控制使用,故本文暂选用第二方案进行设计。

2.2.2　计算方法

已知前体进气道尺寸,沿等动态q_n线,在不同飞行M_n数下,可用气体动力学的斜激波关系式计算进气道出口的气流参数。在非设计状态下,前体形成的斜激波不与外壳唇口相交,计算发动机流量时,要计及溢流。气流在燃烧室等截面段无化学反应加热,进入通常为直壁扩张段的等M数段,在前半段形成正激波,气流通过正激波,降为亚声速,在后半段控制供油量形成稳定亚声速燃烧,燃烧后流出燃烧室,进入尾喷管。尾喷管系拉瓦尔喷管,气流在尾喷管内不断加速,以超声速流出尾喷管后体,产生较大的推力。

在燃烧室扩张段,M数的变化关系式为

$$\frac{dM}{M} = -\frac{\phi}{1-M^2}\left[\frac{dA}{A} - \frac{i+\gamma M^2}{2}\left(\frac{dT}{T} + \frac{2d\dot{m}}{\dot{m}}\right)\right] \tag{11}$$

已知燃烧室几何形状变化,并利用式(2)数值求解(4)、(11)及使用有关气体动力学关系式得到出口截面所有的气流参数及燃气流率$\dot{m}$。尾喷管后体无质量添加,假设不计摩擦、无化学反应和散热,总温保持不变,则流动方程式为

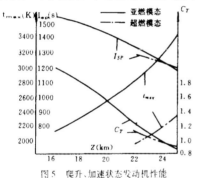

图5　爬升、加速状态发动机性能
Fig.5　Engine performance for climb and acceleration

$$\frac{dM}{M} = \frac{\Lambda}{1-M^2} \tag{12}$$

式中：

$$\Lambda = \left(1 + \frac{\gamma-1}{2}M^2\right)\left(-\frac{dA}{A}\right) \tag{13}$$

考虑了静温变化引起的比热比γ变化,数值解上式得到M,再由M数计算出所有的气流参数。

非设计状态的推力T,比冲I_{sp},推力系数C_T,总压恢复系数σ等发动机性能,可由式(7)、(8)、(9)计算获得,计算结果如图5所示。

2.3　发动机耗油率估算

2.3.1　巡航状态

飞行器如图1所示,参照图4,选取$\varphi = 0.4$、$C_T = 0.336$、$I_{sp} = 1125s$,发动机宽度

$$b_n = \frac{T}{C_T q_n h_n} = \frac{4084}{0.336 \times 64241 \times 0.421} = 0.45m$$

考虑到估算时未计及附面层厚度,选取$b_n = 0.48m$。据此超燃冲压发动机分成3个模块,每块宽0.16m,高0.202m。发动机耗油率$\dot{m}_j$与空气流率$\dot{m}_a$在巡航状态开始时分别为

$$\dot{m}_j = \frac{T/g}{I_{sp}} = \frac{4080/9.8}{1125} = 0.37kg/s$$

$$\dot{m}_a = \frac{\dot{m}_j}{\varphi \times f_{sf}} = \frac{0.37}{0.4 \times 0.06764} = 13.68kg/s$$

上述计算中仍按飞行器总重1500kg计算,所以计算结果数字略高。

2.3.2　爬升加速状态

在$M = 3.5$,$H = 16km$状态下沿等动压q_n线开始爬升,加速至$M = 6$,$H = 25km$,如图6

所示。由图6查出 $M = 3.5\sim6$ 的飞行高度 H，再由[14]查出在此高度的大气参数。

为了缩短爬升时间，设发动机以当量油气比 $\varphi = 1.0$ 工作，发动机推力最大。图5中分别表示出当发动机进口宽 $b_0 = 0.48m$，高 $b_{i4} = 0.202m$，$M = 3.5\sim6$ 时的亚燃和超燃模态的推力 T，推力系数 C_T、比冲 I_{sp} 和发动机内气流最高温度 t_{max} 变化曲线。由图5可见，在 $M = 5.8\sim6$，$H = 24.2\sim25km$ 时亚燃稳态发动机内气流最高温度高达 3232K，为了提高发动机的结构强度、减少重量，必须考虑降低 t_{max}，亦即实际飞行中要降低模态转换的 M 数。本文为简化计算，可假设 $M = 3.5\sim6$ 爬升过程中均以亚燃模态工作。

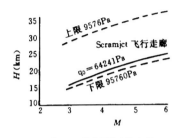

图6 超燃冲压的飞行走廊

Fig.6 Flight corridor for scramjet

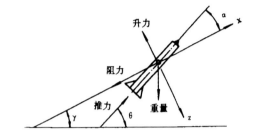

图7 飞行状态飞行器垂直平面受力简图

Fig.7 Forces actingon a vehicle in flight

图7为作用于飞行器垂直平面上各种力的示意图。以飞行器重心为原点的运动方程为

$$\begin{cases} m\dot{V}_z = L + T\sin\alpha - W\cos\gamma & (14) \\ m\dot{V}_x = T\cos\alpha - X - W\sin\gamma & (15) \end{cases}$$

飞行时仍保持攻角 $\alpha = 8°$，并忽略在 z 方向的加速度。

因沿等动压 q_0 线轨迹飞行，攻角 α 不变，飞行阻力 X 不变，已知 m、T、W、γ 从上式可计算得到平均加速度 V_x，并可计算从 $M = 3.5$ 加速到 $M = 6.0$ 所需时间 t。爬升高度 ΔH 与飞行距离 S 存在如下关系

$$\Delta H = S\sin\gamma \qquad (16)$$

因为推力不是常数，现按近拟方法计算，得到

$$t = 174s, S = 246km, \gamma = 1.7°$$

上述计算结果表明，爬升所需时间很长，如需缩短爬升时间，提高加速度，则应加大发动机进口尺寸，增加空气流量，即增加推力，但巡航时当量油气比下降，性能也将下降。这样处理的后果是，巡航时发动机 I_{sp} 降低，总耗油量增加，且发动机重量增加。

2.3.3 耗油量

(1)爬升加速度

耗油量 $\qquad\qquad m_{j1} = \sum_{1}^{n} \dot{m}_{jn}\Delta t = 137kg$

航程 $\qquad\qquad R_1 \approx S = 248kg$

(2)巡航段

航程 $\qquad\qquad R_2 = 1500 - 248 = 1252km$

由航程公式 $R_2 = I_{sp}V\dfrac{L}{X}\ln\dfrac{m_{\mathrm{I}}}{m_{\mathrm{II}}}$，得到 $m_{\mathrm{II}} = m_{\mathrm{I}}/e^{R_2/(\frac{L}{X}I_{sp})}$

式中：m_I 为巡航开始时飞行器质量；$m_Ⅱ$ 为到达目标时飞行器质量。

计算得到 $m_I = 1363\text{kg}, m_Ⅱ = 1141\text{kg}$

巡航阶段耗油量 m_{j2} $m_{j2} = m_I - m_Ⅱ = 222\text{kg}$

(3)总耗油量 $m_j = m_{j1} + m_{j2} = 359\text{kg}$

(4)巡航飞行时间 $t_2 = R_2 / V_2 = 698\text{s}$

总的飞行时间 $t = t_1 + t_2 = 174 + 698 = 872\text{s}$

3 碳氢燃料双模态超燃冲压燃烧室形状与尺寸设计

等 A-M 燃烧室由两部分组成，第一部分是等截面直管，第二部分通常为直壁扩张管。燃烧室的长度及扩张角，以及喷咀设计及喷咀位置确定是超燃燃烧室设计的难点，需要经过二维 N-S 方程数值解，配合实验最终确定。初步设计时，可按一般规律确定。

3.1 形状设计

为排除进气道与燃烧室相互干扰[14]，在等面积段前面增设一隔离段，形状与等面积段相同，长 200mm。为满足非设计状态下在扩张管产生正激波[15]，波后形成亚燃模态的要求，在扩张管的前段长 300mm 部分作为非设计状态的正激波区。据此整个模块由四部分组成(见图 8)：

(1)隔离段；

(2)等面积燃烧段；

(3)激波区段；

(4)等 M 数燃烧段。

有关超燃冲压燃烧室内煤油-空气燃烧效率的经验公式，目前未见公开发表的资料，超燃冲压燃烧室进口温度相当高，煤油喷入气流后将被迅速雾化、蒸发。假设仍然是混合速率控制了燃烧效率，可近似参照氢-空气混气的燃烧效率公式[16]估算，暂定直管段长度为 1200mm(包括隔离段 200mm，等面积燃烧段 1000mm)，扩张段长 1500mm(包括激波区段 300mm，等 M 数燃烧段 1200mm)。

超声速燃烧室扩张角 θ 范围一般为 $0.5° \sim 2°$，现选取 $\theta = 0.8°$，则燃烧室出口截面高度 ΔY_v 为

$$\Delta Y_v = \Delta Y_Ⅱ + (X_v - X_Ⅱ)\text{tg}\theta = 17.5 + 1500\text{tg}0.8° = 38\text{mm}$$

若 θ 的选择符合等 M 数燃烧规律，增加高度 ΔY_v，则增加长度，将有利于提高燃烧效率 η。

因为燃烧室通道高度已很小，只有 17.5mm，在上、下壁面设置喷咀向气流供油，不需要再在通道内设置支板。

3.2 自动着火与点火

最理想的情况是，燃烧室内煤油-空气混气能自动着火，并能保持稳定的燃烧，获得较高的燃烧效率。本文设计状态下即 $M = 6$, $H = 25\text{km}$，燃烧室进口静压与静温处于煤油-空气混气自动着火范围的下边界附近；在超燃冲压工作起始点，即飞行 $M_H = 3.5$，飞行高度 $H = 16\text{km}$，燃烧室进口总温仅为 800K，在此状态下，即使是氢-空气混气也不可能自动着火。为此，参照文献[18~21]，在亚燃模态工作时用电子点火器点燃氢-空气混气，再由氢-空气火焰点燃煤油-空气混气。超燃模态工作时，直接喷射氢气自动着火，由氢-空气火焰点燃煤油-空气混气。

使用氢-空气火焰点燃煤油-空气混气主要有两种方式,喷氢点位于煤油喷射点的上游[20],或位于煤油喷射点的下游[17,18,19],这两种方式皆能可靠地点燃煤油-空气混气。本报告采用第二种方案,优点是喷氢位置即为燃烧放热开始的截面,便于控制放热规律,优化设计。喷氢截面为一个宽15mm,深5mm的凹槽,内设电子点火器和氢气喷咀。启动时,同时打开电子点火器和氢气喷咀,形成氢-空气火炬,点燃流经凹槽的煤油-空气混气,并向主流传播,然后关闭电子点火器和氢气喷咀。凹槽中的热燃气继续点燃来流煤油-空气混气,起到火焰稳定器的作用。在直管段出口截面煤油几乎燃尽,燃气温度大大上升,其后从扩张段截面Ⅲ喷入高温燃气流的煤油将被迅速雾化、蒸发、混合和点燃,继续燃烧,在燃烧室出口截面处获得较高的燃烧效率。

理论上氢喷咀应设置在煤油蒸气和空气混气浓度场当量油气比接近1.0、静压与静温最高处。根据计算与实验经验,初步确定氢喷咀设置在煤油喷咀后,与煤油喷咀在同一轴线上,或介于两个煤油喷咀之间中心轴线上。如图8所示,截面Ⅱ距截面Ⅰ约200mm,截面Ⅳ距截面Ⅲ约200mm。

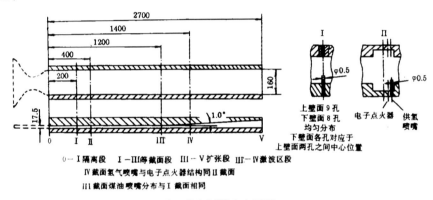

图8 超声速燃烧室通道简图
Fig.8 Detailed sizes of the scramjet burner

3.3 喷咀设计

需要确定的喷咀参数为喷咀孔径 d_j、孔数 n、喷射压力 p_j 和喷射角度 α_j。喷咀的设计应保证在喷咀后方形成较均匀的浓度场,为此,煤油喷咀沿横向间距不能太大,煤油液柱的穿透深度要适当。文献[21]提供了超声速气流中,喷射液体燃料穿透深度 Z 的经验关系式,即

$$\frac{Z}{d_j} = \frac{9.05}{1 + \cos(\alpha_j)} \sqrt{\frac{(\rho u)_j}{(\rho u)_a}} \frac{1 + \dfrac{\gamma_\infty - 1}{2} M_a^2}{\left(1 + \dfrac{\gamma_i - 1}{2} M_i^2\right) M_a} \tag{17}$$

式中: d_j 为喷咀孔径; ρ 为密度; u 为速度; M 为马赫数。下标: a 表示燃烧室内空气流; j 表示液体射流。式(17)中, α_j 为液体射流喷射角度,当垂直于超声速流喷射 $\alpha_j = 90°$ 时,液柱与超声速气流相对速度大,液柱能更快地破碎雾化、蒸发、与空气混合和发生化学反应放热,燃烧效率高,然而气流 M 数将迅速下降,容易发生热阻塞。本文中由于气流 $M \geqslant 1.5$,初次设计可以用 $\alpha_j = 90°$,试验时若发生热阻塞,可适当减少 α_j 角。

选取 $Z = 15mm$,可使得既有足够的穿透速度,又不撞溅相对壁面。

喷咀出口流速 u_j 与喷射压差 Δp_j 存在如下关系

$$\Delta p_j = p_j - p_a = (\rho u^2)_j / 2 \tag{18}$$

喷射流率 $\dot{m}_j$ 为

$$\dot{m}_j = (\rho u \cdot A)_j \times \mu_j \tag{19}$$

式中:喷咀流量系数 $\mu_j \approx 0.85$

喷射孔直径 d_j 与喷射孔总面积 A_j 及孔数 n 有关,即

$$A_j = n \, \pi d_j^2 / 4 \tag{20}$$

由前计算已知燃料室内燃油流率 $\dot{m}_j$,气流密度 ρ_a,马赫数 M_a 及速度 u_a,气流静压 p_a,空气比热比 $\gamma \approx 1.4$,燃料比热比 $\gamma \approx 1.0$。在直管段截面 I 及在扩张段截面 III 分别喷入 $\varphi = 0.2$ 的煤油,煤油密度 $\rho_j \approx 0.9 \times 10^3 \text{kg/m}^3$,则由式(18)~(20)计算得到

$$u_j = 26\text{m/s},$$
$$\Delta p_j = 3.05 \times 10^5 \text{Pa},$$
$$p_j = 9.05 \times 10^5 \text{Pa},$$
$$A_j = 9.3\text{mm}^2,$$

在直管段截面 I 处,总共设置喷咀 $n = 51$ 个。

燃烧室分三个模块,每个模块在截面 I 处初步可设 17 个喷咀,在截面 I 处沿横向上壁面设置 9 个喷咀,下壁面设置 8 个喷咀,上、下壁面喷咀相互均匀错开,即各个喷咀位于相对壁面喷咀之间的中心线上,以有利于获得均匀的浓度场。

截面 III 的喷咀设置方式与截面 I 相同。

如前所述,氢射流的作用是在凹槽内与空气混合形成可燃混气,用电子点火器点燃,或与高温空气混合自动着火形成火炬,点燃煤油-空气混气气流,因此,氢射流不要求有太大的穿透深度,选择穿透深度 $Z = 2 \sim 5\text{mm}$ 可认为是合适的。

参考文献[17],氢从平板上小孔垂直喷射至超声速空气流的穿透深度 Z 的经验关系式为

$$\frac{Z}{d_H} = 2.87 \left[\frac{(\rho u^2)_H}{(\rho u^2)_a} \right]^{0.3} \left(\frac{x}{d_H} \right)^{0.143} \tag{21}$$

式中:d_H 为氢气喷咀孔径;$(\rho u^2)_a$ 为空气的动压;$(\rho u^2)_H$ 为氢气的动压。x 为表示下游不同截面的座标值,设喷咀孔中心线为 0 截面。同理,可类似于式(18)、(19)、(20),选择氢气喷咀 $d_H = 0.5\text{mm}$,孔数 $n = 17$。又氢-空气恰当油气比 $f_0 = 0.0293$,选取氢的静温 $t_H = 300\text{K}$。通常氢喷咀压差处于超临界状态,即喷射速度 u_H 与声速 a_H 相等。选用喷氢的当量油气比 $\varphi_H = 0.05$,则可得到每一模块的

$$\dot{m}_H = 6.8\text{g/s},$$
$$\rho_H = 1.5\text{kg/m}^3,$$
$$\rho_H = 12.7 \times 10^5 \text{Pa},$$
$$\Delta p_H = 12.7 \times 10^5 \text{Pa},$$
$$Z = 2.32\text{mm}$$

4 关于实验的建议

建议首先作超声速燃烧室的实验,设发动机由三个模块组成,每块宽 0.16m,高 0.202m。

直连式超声速燃烧室实验设备中燃烧室前无激波,这就降低了对设备的总压要求,下面列出由设计状态计算得到的实验参数:

飞行马赫数 M_H	6
燃烧室进口 M 数	2.06
燃烧室进口总压 P	29.9×10^5 P.
燃烧室进口总温 T	1830K
燃烧室进口静压 p	3.66×10^5 Pa
燃烧室进口静温 t	1072K
空气流率 $\dot{m}$	4.56kg/s
燃料流率 $\dot{m}_j$	0.123kg/s
试验时间 τ	10～16s

第二步作自由射流模型超燃冲压实验,表3列出了模拟高空飞行时,飞行 M_H、大气静压 P_H 及发动机总空气流率 m_a、燃油流率 m_j 等数据。若用一个模块试验,则燃油及空气流率为表3所列数据的1/3。

表3 发动机总空气流率,燃油流率

飞行 M 数 M_H	6	5.56	5.14	4.76	4.42	4.07	3.77
高度 H(km)	25	24	23	22	21	20	19
空气流率 m_a(kg/s)	13.68	12.68	12.25	11.7	11.28	10.79	10.39
燃油流率 m_j(kg/s)	0.37	0.89	0.86	0.82	0.79	0.72	0.73
环境静压 P_H(Pa)	2549	2971	3467	4048	4729	5529	6468
进口总压 P(Pa)	40.25×10^5	29.5×10^5	21.6×10^5	16.1×10^5	12.4×10^5	9.2×10^5	7.2×10^5
进口总温 T(K)	1816	1587	1382	1211	1070	934	832

5 结束语

本文对于小型高超声速飞行器的碳氢燃料双模态超燃发动机方案进行了研究。估算表明这一方案可以获得更大的飞行速度和航程,有很大实用性。由于飞行马赫数 $M = 6$、飞行高度 $H = 25$km,因而与各类空天飞机相比,发动机的地面试验要求更低更容易实现。更重要的是我国需要有一个技术目标更先进的国家高超声速飞行器计划来推动航天高技术的发展,这一方案可以作为这个计划的现实的起点。

参 考 文 献

1 刘陵,刘敬华,张榛,唐明.超声速燃烧与超声速燃烧冲压发动机.西北工业大学出版社,1993.1

2 Charles H Carlson.Preliminary Scramjet Design for Hypersonic Airbreathing Missile Application. NASA CP 3742,1983.

3 Kay I.W, Peschke W T and Guile R N. Hydrocarbon-Fueled Scramjet Combustor Investigation, AIAA 90-2337

4 Helgeson J H and Chinitz W. A Performance Assessment of Hydrocarbon Scramjet Engines Using a Generalized Cycle Analysis Code. AIAA 95-2768

5　Edwards C L W, Small W J, Weidenr J P and Johnston P J. Studies of Scramjet/Airframe Integration Techniques for Hypersonic Aircraft. AIAA Paper 75-58

6　Manuei Martinez-Sanchez. Fundamentals of Hypersonic Airbrathing Propulsion. AIAA Professional Study Series, 1986. 6

7　约翰. 霍甫金斯大学应用物理实验所编, 李存杰等译, 王树声校. 冲压发动机技术. 国防工业出版社, 1987.

8　Robert E Coltrin. High-Speed Inlet Research Program and Supporting Analyse. NASA-CP-10003, 1988.

9　刘 陵, 张 榛. 超音速燃烧室最佳设计参数. 推进技术, 1988. (1)

10　左克罗 M J, 霍夫曼 J D. 气体动力学. 国防工业出版社.

11　刘兴洲, 于守志, 李存杰等编. 飞航导弹动力装置. 宇航出版社, 1992.

12　Leon H. Schindel. Design Model of High-Performance Ramjet or Scramjet-Powered Vehicles. J. Spacecraft, 1990. 27(6) Nov-Dec.

13　Hunt J L, Lawing P L, Marcum D C and Cubbage J M. Conceptual Study of Hypersonic Airbreathing Missiles. AIAA 16th Aerospace Sciences Meeting 1978.

14　Richard D Stockbridge, Joseph A Schetz, Paul J Waltrup and Frederich S Billig. Combustor/Inlet Interactions and Modeling of Hypersonic Dual Combustor Ramjet Engines. AFOSR-TR-84

15　Sullins G A. Demonstration of Mode Transition in a Scramjet Combustion. Journal of Propulsion and Power, 1993, 9(4)

16　刘 陵, 张 榛, 牛海发, 刘敬华. 超音速燃烧室燃烧效率数学模型及气流状态的计算. 推进技术, 1989. (2)

17　Vyacheslav A Vinogradov, Sergey A Kobigsky and Michael D Petror. Experimental Investigation of Kerosene Fuel Combustion in Supersonic Flow. Journal of Propulsion and Power 1995. 11(1)

18　Roudakov A S, Schickhman Y, Semenov V. Flight Testing an Axisymmetric Scramjet Russian Recent Advances. IAF-93-S. 4. 485

19　Sosounov V. Study of Propulsion for High Velocity Flight. ISOABE September 1~6, 1991.

20　Andrews E H, Trexler C A and Emami S. Tests of a Fixed-Geometry Inlet-Combustor Configuration for a Hydrocarbon-Fueled Dual-Mode Scramjet. AIAA-94-2817

21　Gregory A Molvik, Jeffrey V Bowles and Loc C. Huynh. Analysis of a Hydrocarbon Scramjet with Augmented Preburning. AIAA-92-3425

THE PRELIMINARY DESIGN OF A HYDROCARBON-FUELED DUAL MODE SCRAMJET ENGINE FOR THE HYPERSONIC CRUISE VEHICLE

Le Jialing

China Aerodynamics Research & Development Center, Mianyang　621000

Liu Ling

Northwest Polytechnical University Xian　710072

Abstract　The scramjet has always been competitive with the rocket for the propulsion of the hypersonic cruise vehicle in the atmosphere at $M = 6$ because it has specific impulse more than two times; Meanwhile, compared with the ramjet, it has lower static temperature and pressure in the combustion chamber, thus obviously decreasing its structure loads as well as simplying its structure designing.

This kind of hypersonic cruise vehicles will be of more advanced flight performance with the higher speed and better penetration as well as high probability survival.

Based on our current conditons, a concrete scheme for a vehicle powered by a hydrocarbon-fueled, dual-mode scramjet engine integrated with the airframe has been put forward.

Most of the paramenters needed for the evaluation of the vehicle performance, such as the trajectories and the time of the flight, the fuel mass flow rate, the configurations for the forebody and the inlet, the detailed size of the combustion chamber, the engine perfomance are estimated in accordance with the preliminary desired vehicle (1500kg weight, 1500km range) size of the 0.6m diameter and 4.5m length.

Finaly, the required flow paramenters simulating the scramjet engine in feasible ground test facilities are given.

Key words   hypersonic vehicle; scramjet engine; scramjet combusion

Preliminary investigation of full model of two-mode scramjet

J.L. Le　　Z.C. Zhang　　H.C. Bai

Centre of Aerodynamics Research and Development of China, Mianyang Sichuan, 621000, China

M.A. Goldfeld　　R.V. Nestoulia　　A.V. Starov

Institute of Theoretical and Applied Mechanics, Novosibirsk, Russia

In this paper results of previous researches of two-mode full engine model for Mach number range from3 to 6 are presented. The main aims of researches are to obtain characteristics of full engine model and its components (the inlet, the combustion chamber and the nozzle) in ramjet and scramjet regimes and to compare these characteristics with the calculation results according to CFD and to approximate (engineering) methods. Tests were carried out without fuel supply (cold tests) and with fuel supply (studyof ignition and combustion).Hydrocarbon and hydrogen fuel (gaseous or liquid) were used during the tests.

Keywords: full engine, hypersonic flow, fuel, inlet, combustion, wind tunnel, experiment.

Introduction

The latest studies of the overall characteristics of hypersonic flying vehicles with airbreathing engines have shown that these vehicles are fairly promising. Some additional problems arise, however, that are related to the definition of the general shape of the aircraft, the choice of propulsion type and operation regimes, the engine size and position on the aircraft body.

The aircraft version under consideration implies the use of a combined propulsion unit with a two-mode combustor (subsonic or supersonic combustion). Such an engine should combine the advantages of subsonic combustion at supersonic speeds ($M_\infty < 5$) and supersonic combustion at hypersonic speeds ($M_\infty \geq 5 \sim 6$). The designing of this engine faces a number of problems:

- inlet configuration,
- combustor type and geometry,
- fuel injection mode
- type and size of the exit nozzle,
- possibility, necessity and limits of adjustment of the engine duct.

Another important requirement is the necessity of providing a uniform flow field upstream of the inlet and the definition of conditions for transition from subsonic to supersonic combustion. Naturally, to solve these problems, one has to formulate the requirements to the engine as a whole and to its individual elements so that to achieve the maximum efficiency.

When choosing a ramjet/scramjet for hypersonic flight, it should be clarified which level of perfection should be required from its component (inlet, combustor, nozzle). A profound answer to this question needs the knowledge of the influence of gas dynamic and geometric parameters on specific characteristics of the engine thrust and fuel consumption.

It follows from the analysis of scramjet operation and equations for gas dynamic phenomena in the engine that the engine thrust is determined by a great number of parameters. The mutual influence of propulsion elements, especially inlet and combustor, should also be taken into account. Therefore, the choice of propulsion elements with the optimum (maximum) characteristics for a hypersonic engine is nearly not possible. The most convenient method for evaluation of scramjet parameters can be the use of specific thrust parameter I_{sp} or SFC, T_{sp} or C_T. These parameters are uniquely correlated through formulas for determination of engine characteristics in terms of its operation parameters, have different physical

J.L. Le, Z.C. Zhang, H.C. Bai, M.A. Goldfeld, R.V. Nestoulia, A.V. Starov. Preliminary investigation of full model of two – mode scramjet. Journal of Thermal Science, 2001, 10(2): 97-102.

meaning, and are important characteristics of the engine efficiency. At the same time, these parameters can be used to choose the engine size when determining its thrust-aerodynamic characteristics and for trajectory calculations.

The above mentioned implies the necessity of systematic calculations of propulsion. They should include the internal engine characteristics and an effective thrust taking into account an additional inlet drag induced by a smaller flow rate at Mach numbers lower than the design one. These calculations allow one to study simultaneously the influence of some other parameters (type of fuel, fuel-to-air ratio (air-to-fuel), combustor and nozzle geometry) on efficiency of the engine and its individual elements, and to formulate the requirements for their selection.

Model and Facilities

The model is made by a module scheme. It allows one to modify easily engine elements and carry out tests of its individual elements. A model scheme and major systems are shown in Fig.1. The model consists of a 2-D three shock wave inlet, a combustion chamber and a nozzle with flat walls. The inlet has replaceable throat inserts and cowls. It allows one to change discretely throat height and/or cowl inclination and to realize external or external-internal compression. One of the problems at scramjet development is to elaborate high-effective supersonic combustion chamber. The efficient combustion chamber means application of effective system of fuel supply to provide good mixing and minimum device drag for fuel supply in the combustion chamber. An divergence combustion chamber has two separate rows of fuel supply. The first

one realizes fuel injection from a backward step. The injecting devices (pylons) allow to fulfill parallel and opposite fuel injection in a flow. Flame stabilization is implemented on the backward step or on the pylons of the second supply row. The provision of changing of combustion chamber configuration is made to equip two regime operation of the engine at sub- and supersonic flow velocity in its duct without channel choking. Exhaust nozzle design allows to change twice relative area of nozzle exit section.

In particular, for inlet test a special flowmeter was established instead of the combustion chamber. This flowmeter was used for determination of the flow coefficient and of the total pressure recovery of the inlet.

The flowmeter consists of a bottom base, a top plate and side walls. Side walls of the main engine model are used as side walls here. The flowmeter presents a divergent channel. The initial part of metering channel for 100 mm length presents a combustion chamber initial part. Channel angle of flare is 12°. Such an angle of flare allows to exclude flow separation in the channel at fourtimes magnification of channel area before metering nozzle. The length of constant area part is 100 mm. In this part there is a grating to ensure uniform flow field before sonic metering nozzles. Before and behind the grid there are four static pressure probes for determination of pressure losses in the grating. Before metering nozzles the channel has a structure transforms from rectangular to circular cross-section. There are removable sonic nozzles on the channel exit. Reduction of area of nozzle exit cross-section allows to carry out channel throttling (including strong throttling). In critical cross section of each nozzle there are four static pressure probes, to determine flow velocity. This velocity is demanded for air mass flow calculation. The inlet and

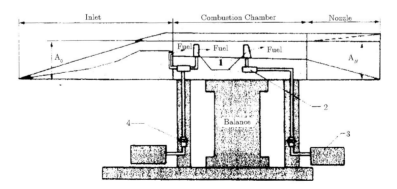

Fig.1 Scheme of full engine model and systems
1 injector   2 force separation system   3 fuel tank   4 fast acting valve

the combustion chamber has quartz side windows for flow visualization (including flame visualization at fuel combustion).

Wind tunnels

The tests without combustion (cold tests) are conducted in a blow down wind tunnel (τ =600 s) at Mach numbers of 3, 4, 5 and 6, at Re_1=10～60×10^6. The inlet start conditions were obtained in the test range. The tests with combustion are realized in hot-shot wind tunnel at Mach numbers of 5 and 6 with high enthalpy close to free flight conditions (P_f=50/90 bar, T_f=1300/2000 K, M=5/6 accordingly). The operation period of the hot-shot wind tunnel takes 60～100 ms. Tests performance in two tunnels allow one to confirm experiment adequacy and potentiality of use of short-time operation tunnels to study the inlet, the combustion chamber and the full engine.

Measurement system

1. Measurement of static pressure distribution along inlet length, combustion chamber length and nozzle length.

2. Measurement of heat fluxes distribution along inlet length, combustion chamber length and nozzle length.

3. Measurement of Pitot pressure at inlet exit by means of the Pitot pressure rakes for determination of total pressure recovery coefficient.

4. Measurement of Pitot pressure at combustion chamber and nozzle exit by means of the Pitot pressure rake for determination of exit impulse loss.

5. Measurement of Pitot and static pressure at exit of metering nozzle of flowmeter for air mass flow rate calculation through the engine.

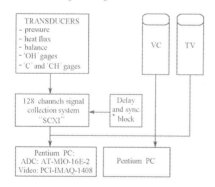

Fig.2 Functional scheme of the measurement system
VC - high speed camera for ultra-violet diagnostic
TV - TV camera for Schlieren visualization

6. Measurement of base pressure on the rear step at the inlet exit (eleven static pressure probes).

7. Measurement of forces acting on the model (three-component balance).

8. Measurement of fuel flow rate (fuel-air ratio).

9. Measurement of combustion efficiency by optical method.

10. Shadow visualization of flow.

Three-component specialized balance

Three-component specialized balance is intended for researching of full engine model in wind tunnels T-313 and IT-302M at Mach numbers of 3～6, when dynamic pressure is up to 20000 kg/m^2 and temperature range is 200～400 K in T-313 and up to 1800K in IT-302M, with regime duration of 400 s for T-313 and 200 ms for IT-302 M. Model mass up to 20 kg. Humidity 85±10%.

Balance measuring part includes elastic strain-measuring element with strain-gages pasted on it, a commutation block, a block of normalizing amplifiers and cable commutations between elastic element, separate blocks and measuring-recording complex. The elastic element of three-component balance presents a construction made as an integral part of two force-measuring blocks. The element has a common base and a common platform, which the studied model is fixed on. Every block is made as a system of force-receiving and measuring beams. Design situation of beams and their corresponding separation permits to obtain deformation response prevailing from corresponding components of total loading. Transformation of measuring deformations into electrical signal is carried out with the help of strain-gages, which are pasted on the measuring beams and collected as measuring bridges. The thermal compensatory are established on side surfaces of measuring beams and connected in correlating bridge arms.

Exit signals of measuring bridges are retranslated to entrance of measuring-recording complex directly from commutation block or in addition through the block of specialized amplifiers. The direct connection of commutation block to measuring- complex, apart from amplifier block, is possible.

The major errors of loading determination at corresponding components by the results of initial and control balance calibration were: X< 0.15%; Y<0.25%; M_z< 0.25-0.35%. An extra error of the balance by the results of temperature tests $\delta_{TX1,X2} \leq 0.004\%$ / degree; $\delta_{TY1,Y2} \leq 0.012\%$ / degree; $\delta_{TMz1,Mz2} \leq 0.008\%$ / degree.

The flame diagnostic in UV-range

For development of a technique of definition of the

completeness of fuel combustion (combustion efficiency) the methodical researches on model flames of gaseous fuel were conducted. The method is based on obtaining calibration relation of integral intensity of luminescence of OH, CH radicals in the ultra-violet range of spectrum (in the band 300~320 nm) versus the fuel flow rate under condition of complete combustion. The completeness of fuel combustion for some fuel flow rate is determined by comparison of a value of the ultra-violet radiation intensity of an investigated flame to emission intensity defined on the calibration relation.

The optical system of measurements included a quartz lens with variable aperture and two series located light filters (glass filter UvG-6 and liquid $NiSO_4$-$7H_2O$ light filter). As a detector of the optical system the photo-multiplier PEM-39A was used. It allows one to receive a high time resolution (up to 10^{-3} s) of the registration system and broad (up to 10^3) dynamic measurement ranges of emission intensity during experiment. The highest possible measurement range of flame emission intensity has been 10^5. Previously, on the basis of standard light monochromator SPM-2 the installation for definition of coefficients of light transmission of the wind tunnel windows, optical elements of the engine model and optical system in ultraviolet band was created. The data are necessary for measurement correction on attenuation of an ultra-violet radiation under conditions of the wind tunnel experiments.

The calibration desk for creation of single and double gas plumes with adjustable gas flow rate was created. Propane (C_3H_8) and ethylene (C_2H_4) were the working gases. The installation allows to conduct measurements on flames with gas mass flow rate up to 2 g/s, to study influence of distance between the optical system and plume in the range 1~3 m on output signal, to investigate influence of disposition of flames on intensity of the radicals emission. The calibration of the measured characteristics of gas pipelines and injectors of the calibration desk was conducted, the measurements on propane are carried out. In particular, the relations of intensity of ultra-violet emission of the propane flames for gas flow rate from 0.03 up to 0.8 g/s were obtained. The emission intensity measurements were conducted on the single and double plumes, on single plumes with preliminary gas-air mixing. The degree of mixing varied by change of the position of an injector in a quartz pipe of different diameter. We have obtained both steady and vibratory combustion regimes of propane. The data about influence of the disposition of two flames and distance up to the optical system on the radicals light emission intensity were obtained. It have been shown, that the influence of the shielding of one gas flame by another

takes place (5%~10%) and the changes of the integral luminosity rise with gas mass flow rate. The relation of radicals light emission I to gas mass flow rate m is nearly linear.

The fuel supply system

The fuel supply system is designed to ensure many variants of test conditions, including individual injection of different fuels along the model duct. The fuel system is designed for injection of gaseous and/or liquid fuel to the combustion chamber of the engine and includes two independent systems for fuel supply in the first (in the base region) and in the second (from the struts) supply rows. It consists of fuel tanks, fast-response valves, injectors, system for fuel supply and charging (pipelines, joining elements). Each of these two parts is carried out according to a two-channel scheme. This construction allows simultaneous injection of liquid and gaseous fuel into each row of injectors. This should provide better atomization of the hydrocarbon fuel and better conditions for its self-ignition and combustion. To conduct combustion chamber tests on hydrocarbon liquid fuel, an original fuel tank has been worked out.

The fuel tanks have a controlled volume and a displacing system of fuel supply. A special charging system allows for saturation of the hydrocarbon fuel by hydrogen or neutral gas or air to speed up the ignition and combustion. The volumes of the fuel tanks were chosen under the condition of providing a roughly constant value of the equivalence ratio during the run. This is achieved by varying the throat area of the injectors and the volume of the fuel tanks. The new fast-response valve is elaborate with two channels. It has two independent channels for fuel supply and two membranes. A common explosive charge ensures the opening of the valve (cutting of the membranes). This opens both channels for fuel supply to the injectors. The time of valve opening is no more than 1 millisecond. This time corresponds to the time required for the wind tunnel to reach the steady operation regime.

Pressure measurement in fuel tanks and additional calibration allow one to calculate fuel-air ratio. Liquid fuel can be saturate by hydrogen or another gas to maintain the ignition and the combustion. Pilot hydrogen flame will be realized for hydrocarbon fuel ignition. Also a possibility of vaporous liquid fuel supply at temperature of 500...600K is considered. It has promoted the process of mixing and ignition.

Results and Discussion

To control air inlet start, a model with various throat heights has been tested. The experiments were conducted on the model with flowmeter. It was found out that at Mach number 3, air inlet start was not realized at relative throat area of 0.15 & 0.2 A_o. Visualization of entering flow have shown that flow with intense side overflow (spilling) before a channel entry realized and it strengthened as throat area was decreased. At the same time it was established that in model channel, flows with developed separation zones realized. Boundary layer separation causes air inlet unstarted. When Mach number was 4, air inlet start realized on the model with throat height of 0.2 A_o and 0.15 A_o. In this case flow in the throat was supersonic on all its length. Pressure distribution along channel (Fig.3) and visualization of the flow in the channel

necessary to determine "cold" parameters of the engine in the tunnel of long duration and to compare it with the data in the tunnel of short duration at high temperatures of flow with and without combustion. The measurement of total pressure distribution was carried out in combustion chamber and nozzle exit, to find out exit impulse. The example of distribution of the total pressure at Mach number 6 is given in Fig.5. One may see the high nonuniformity of pressure which is typical for supersonic flow in the channel at high Mach numbers.

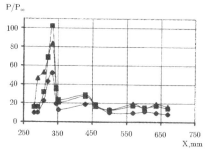

Fig.4 Static pressure distribution in the inlet throat and combustion chamber
◆ - M_∞=5, ■ - M_∞=6 blow down, ▲ - M_∞=6 hot shot

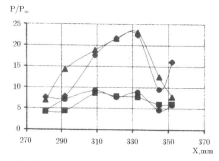

Fig.3 Static pressure distribution in the inlet throat
● - M_∞=3 h_{thr}=16mm, ■ - M_∞=3 h_{thr}=12mm,
▲ - M_∞=4 h_{thr}=16mm, ◆ - M_∞=4 h_{thr}=16mm

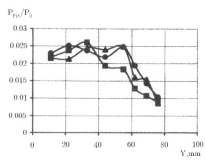

Fig. 5 Relative Pitot pressure at the nozzle exit.
●- Z=-30mm, ■- Z=0mm, ▲ - Z=40mm

Integral characteristics

According to measurement results, the dependencies of recovery coefficient of total pressure (on the exit of air inlet throat) and flow ratio for all Mach numbers on air inlet model with flowmeter have been obtained (Fig.6). One may see that high level of pressure recovery has been achieved and it corresponds to a TsAGI-CIAM standard dependence for non-regulated air inlets (or MIL-500). The air flow rate

through side windows confirms it. Increase of air inlet throat results in significant increase of air flow rate due to decrease of relative height of separation zone on channel entry. At Mach number 5 and 6 a throat of 0.15 A_o was used and the air inlet start was ensured with high compression level. It is confirmed by the data of Fig.4. At the same time it turned out that calculation regime at Mach number 6 failed to be realized because of separation of laminar boundary layer on the surface of external compression. Such a flow is typical for tests of air inlet models in tunnels with low Reynolds numbers, as a result the boundary layer remains as laminar one on most part of surfaces of external compression and its separation is carried out under the influence of shock waves. It leads to flow structure breakdown before the channel entry.

To define gas-dynamic peculiarities of supersonic flow, in combustion chamber and engine nozzle the measurements of static pressure distribution (along engine channel and on exit of the combustion chamber and the nozzle) were conducted. This information is

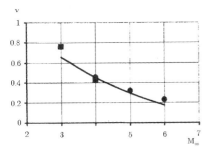

Fig.6 Total pressure recovery coefficient
■ - h=16mm, ● - h=12mm
—— - TSAGI-CIAM standard

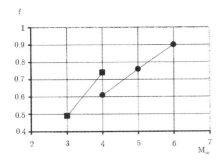

Fig.7 Air flow mass coefficient
■ - h=16mm, ● - h=12mm

obtained in the experiment is in good accordance with two-dimensional calculation within the limits of the model of non-viscous gas (Fig.7) made by Dr. V. Zudov (ITAM). Mach number on combustion chamber entry is an essential parameter for calculation and determination of flow parameters as the combustion chamber operates. Mach number is shown on the entry of the combustion chamber. It was calculated by the total and average static pressure on the exit of air inlet

throat. It is seen (Fig.8) that flow velocity was all supersonic. Supersonic velocity at Mach number of 3 is obtained through the repeat acceleration of the flow in model throat behind entry plane.

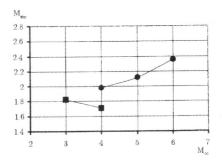

Fig. 8 Mach number in the inlet throat
■ - h=16mm, ● - h=12mm

Conclusion

The conducted experiments are the first stage of tests of the full engine model (inlet + combustion chamber + nozzles) allow one to draw the following conclusions.

1. High levels of inlet performances were obtained at test in blow down wind tunnel without adjustment.

2. The conducted measurements were shown that the inlet was started beginning with Mach number 4. Supersonic flow in channel was realized at Mach number equal to 4 and more.

3. The integral performance (mass flow rate and total pressure recovery) agree well with calculation by CFD method.

4. Balance test and calibration showed steady operation and high accuracy.

5. The conducted tests confirmed the possibility of full engine investigations in hot-short wind tunnel with free flight parameters.

脉冲燃烧风洞及其在火箭和超燃
发动机研究中的应用

乐嘉陵, 刘伟雄, 贺　伟, 谭　宇, 白菡尘

(中国空气动力研究与发展中心, 四川 绵阳　621000)

摘要: 近期美国 X－43A 的飞行试验数据表明脉冲式风洞能够预测飞行性能。中国空气动力研究与发展中心(CARDC) 20 多年来一直在发展各种脉冲燃烧风洞技术及其在火箭高空羽流、超燃发动机研究中的应用。典型的四喷管火箭底部挡板采用涡轮废气排气方案能大大减少底部热流, 这是脉冲式风洞的成功应用成果; 在 60~ 80ms 脉冲燃烧风洞中首次进行了室温煤油燃料的超燃模型发动机试验, 测量了发动机内流道中壁面压力和发动机推力, 比较了脉冲式风洞和连续式风洞的试验结果。研究表明: 在 $M = 5, 6$ 试验条件下, 煤油自发点火延滞时间约 4ms, 因而工作时间为 60~ 80ms 的脉冲燃烧风洞能够十分经济奏效地进行超燃模型发动机研究。笔者亦介绍了正在研制中的大口径脉冲燃烧风洞方案。

关键词: 脉冲风洞; 羽流; 火箭; 超燃发动机

中图分类号: V211. 74; V43　　　文献标识码: A

Impulse combustion wind tunnel and its application in
rocket and scramjet research

LE Jia-ling, LIU Wei-xiong, HE Wei, TAN Yu, BAI Han-chen

(China Aerodynamics Research & Development Centre, Mianyang 621000, China)

Abstract: The experimental data of recent X-43A flight tests indicate that impulse wind tunnel could be used to predict a vehicle's flight performance. In more than twenty years, China Aerodynamics Research and Development Center (CARDC) has been developing various impulse combustion wind tunnels and their application in the research of rocket plume at high altitude and scramjet. For a rocket with typical four-nozzles, using turbine exhaust scheme could greatly reduce the heat flux on the base plate. This is a successful example of impulse wind tunnel application. For the first time, CARDC using impulse combustion wind tunnel with duration of 60~ 80ms, carries out model scramjet tests with fuel of room-temperature kerosene, measuring the pressure on the inner wall of the engine's flow channel and its thrust, and comparing the result of impulse tunnel with that of continuous tunnel. The research shows that: under the conditions of Mach 5 and 6, the time delay of kerosene self-ignition is about 4ms, therefore, the impulse combustion wind tunnel with duration of 60~ 80ms could be effectively and economically applied in scramjet research. This paper also introduces the new scheme of the large size impulse combustion wind tunnel under construction.

Key words: impulse wind tunnel; plume flow; rocket; scramjet

乐嘉陵, 刘伟雄, 贺伟, 谭宇, 白菡尘. 脉冲燃烧风洞及其在火箭和超燃发动机研究中的应用. 实验流体力学, 2005, 19(1): 1-10.

0 引 言

脉冲燃烧风洞概念最早是由 Weatheston 和 Hertzberg 等于 1961 年提出的[1]，其运行方式基本上类同于一般的激波管与激波风洞。从 1980 年开始，中国空气动力研究与发展中心(CARDC)一直在发展这类风洞技术和开展应用研究。20 多年来，CARDC 大量的火箭发动机高空羽流试验研究和超燃冲压发动机试验研究表明，这类风洞的运行十分经济，并能十分灵活地以不同尺度方式进行基本流动和工程问题的研究。近期美国 X‒43A 的 $M = 7$、10 飞行试验结果确认了脉冲风洞中的系列试验能较好地预测飞行性能。所有这些都说明了加速建设大口径脉冲风洞对我国发展高超声速技术的重要性。

1 脉冲风洞在火箭高空羽流试验研究中的应用[2]

火箭高空羽流是指在高空时发动机喷管排出的高温燃气喷流。高空羽流特性不仅取决于发动机本身的参数，也与外界大气环境和周围物体有关。高空羽流研究是运载火箭设计中的一个十分重要的问题，研究的主要问题包括：多喷管底部的热环境、喷流对附近物体的冲击、喷流产物的光电特性等。由于火箭喷流大多是湍流流动，因而工程问题的解决往往要依赖地面和飞行试验。地面试验需要模拟火箭的高温喷管流及相应的模拟高度。大尺寸连续式地面试验需要大型真空系统、解决模型冷却等众多难题，国外一直在探索脉冲式地面试验技术。图 1 是这类脉冲

式风洞运行的原理图。模拟火箭发动机的燃料(H_2)与氧化剂(O_2)以一定比例充入一个燃烧管内并进行充分混合；用轴向热电阻丝或点火塞点火，产生火箭发动机所需的高温高压燃烧气体。燃烧管与喷管用膜片隔开，膜片打开后燃气经火箭喷管(或多喷管)流入真空箱。火箭的高空模拟高度由图 1 中真空箱中气体的真空度来确定。试验时间取决于两个方面：一个是燃烧管破膜后形成的稀疏波以当地声速向上游传播经管端返回的时间(Δt_1)，另一个是喷管排气时的扰动波从箱端返回的时间(Δt_2)。在稀疏波和扰动波返回前，火箭燃气喷流都是稳定的，试验能提供可用结果；燃烧管和真空箱尺寸越大试验时间越长。图 2(a) 是脉冲固体发动机燃烧室装置，图中固体推进剂是由厚度小于 1.5mm 的薄片组成。图 2(b) 是固体火箭燃烧室的压力时间历程。笔者研制的脉冲燃烧风洞技术，能在总压 2.6~6.5MPa，总温 2300~3500k，工作时间 6~40ms 条件下进行各类重要高空火箭羽流试验研究。

图 3 是运载火箭第三级氢氧发动机四个捆绑式喷管底部流动的示意图。在火箭上升过程中由于环境高度的变化，四喷管底部流动发生了由引射流 (30km 以下高度) 向过渡流和阻塞流(70km 以上高度) 的变化。引射流 (图 3(a)) 是火箭高温排气流引射喷管周边环境空气的情况。在高空阻塞流情况下，火箭喷流相互作用十分强烈而产生如图 3(b) 所示的

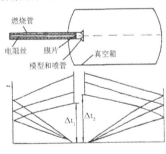

Δt_1：燃烧管等压的时间
(Time of constant pressure inside combustion tube)
Δt_2：扰动波在真空箱返回喷管的时间
(Time of disturbing wave traveling around in vacuum tank)
图 1 脉冲风洞运行原理
Fig.1 Principle of impulse wind tunnel

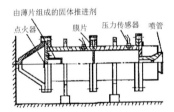

(a) 脉冲固体火箭结构

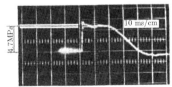

(b) 固体发动机燃烧室压力时间历程
图 2 脉冲固体火箭示意图
Fig 2 Sketch of impulse solid rocket

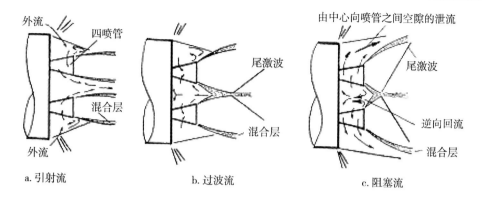

图 3　火箭底部流动结构随飞行高度变化

Fig. 3　Variation of rocket base flow patterns with flight altitude

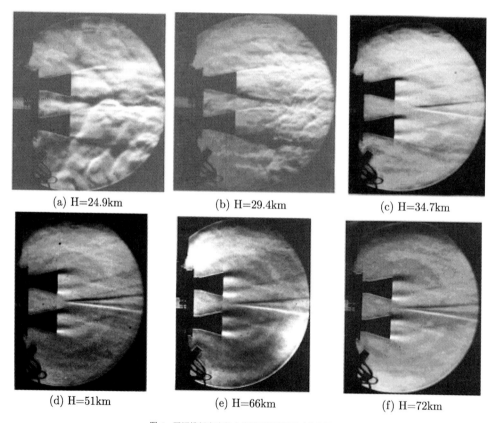

图 4　不同模拟高度的火箭四喷管底部流动纹影图

Fig. 4　Schlieren pictures of four nozzle rocket base flow at different simulation altitude

尾激波. 从而导致火箭高温喷流向底部中心的逆向回流. 随着环境高度进一步增加, 逆向回流与流向喷管周边间隙(图3(c)中喷管之间的间隙面积)的泄流达到平衡, 此时四喷管底部流动构型与特性不再随高度变化而形成了阻塞流动状态. 在阻塞流情况下, 火箭四喷管底部产生了相当严重的热环境, 这正是试验所要确定的状态. 虽然这一问题的研究是在20多年前进行的, 但试验结果充分显示了脉冲式风洞的特点.

图4是脉冲式风洞中火箭四喷管底部流动特性随高度变化的纹影图片. 火箭燃烧室是由图1的单燃烧管产生的, 总温为2906K, 总压为2.65MPa, 燃气比热比 γ 为1.24, 模拟高度从24.9Km 至72Km. 四喷管几何参数如图5、图6所示. 从图4可知, 火箭四喷管底部流动的流型是逐步建立的. 在25Km 以下, 底部是引射流型; 在66Km 以上高空, 能明显地看到喷流之间强烈撞击产生的尾激波, 并形成向底部中心的、有逆向回流的阻塞流型. 与图4(e)、(f)相比, 这种流型已不再随高度增加而变化. 流型达到这种状态的高度称之为阻塞高度. 阻塞高度的数据也可以通过测量底部中心的压力获得. 图5是挡板中心压力随不同模拟高度变化的测量结果. 图中显示在超过60Km 高度时中心压力 P_b 值已逐渐趋于稳定. 该文条件下阻塞高度约为65Km. 这一测量结果与纹影流场显示结果是一致的. 图6为阻塞高度下测量的底部挡板的热流和压力分布值(燃烧室总压为2.65MPa, 总温为2800K, 燃气比热比 γ 的计算值为1.25).

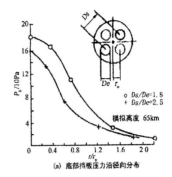

(a) 底部挡板压力沿径向分布

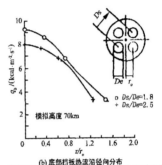

(b) 底部挡板热流沿径向分布

图6 四喷管底部挡板压力和热流沿径向分布(在阻塞流型条件下)

Fig. 6 Radius distribution of pressure and heat flux on the base plate with four nozzles (under choking flow condition)

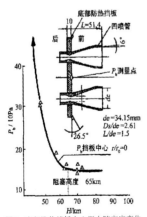

图5 底部防热挡板中心压力随高度变化

Fig. 5 Variation of pressure with altitude at base plate center

脉冲风洞的另一个重要应用是研究底部中心有涡轮废气排气流对四喷管底部热流的影响. 由于四喷管底部有较强的燃气回流, 因而在底部中心产生了较大的热流. 如果在发动机底部中心处按装一个如图7(a)中所示的小尺寸涡轮废气排气喷管, 就能大大减少逆向高温燃气的回流量, 从而减少底部热流. 要在脉冲风洞中进行实验以确认这一设想, 必须解决火箭喷流和涡轮废气排气流的同步模拟问题. 为此, 在燃烧管末端安置了一个小减压室作为涡轮废气排气喷管的前室, 如图7(b)所示, 前室的压力可由小减压室上的小孔面积和涡轮废气排气喷管喉道面积两者的比值进行调节. 涡轮废气排气喷管出口马赫数 M＝3.7, 喷管出口直径 D＝17.5mm. 图7(c)为排气喷管出口处皮托压力的时间历程. 从图7(c)可知在燃烧管点火后约7ms 左右, 皮托压力达到了要求值, 稳定的压力平台的持续时间约3ms. 对于测量热流和压力,这一 3ms 的平台时间已经足够了。图 7(d) 是底部挡板有无排气管情况下热流分布的比较。图中清晰显示: 由于底部中心的排气大大减弱了火箭喷管燃气流对底部逆向回流,挡板的热流大大减少了 (试验条件下热流减少为 40%)。

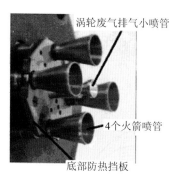

(a) 底部挡板中心的涡轮废气排气喷管

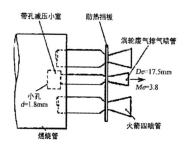

(b) 涡轮废气排气喷管与带孔减压小室

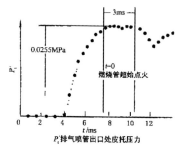

(c) 涡轮废气排气喷管出口处皮托压力历程

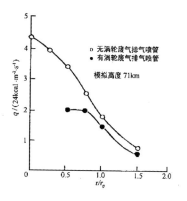

(d) 有无涡轮废气排气管情况下底部挡板热流分布比较

图 7　四喷管底部挡板与涡轮废气排气喷管

Fig. 7　Base plate with turbine exhaust nozzle

2　脉冲燃烧风洞技术(双管形式)及其在超声速燃烧模型发动机试验中的应用[3,4]

2.1　风洞工作原理与运行

基于上述脉冲风洞技术在高空火箭羽流试验中的成功应用经验,CARDC 于 1997 年起着手一类新型的、能进行超燃模型发动机研究的脉冲燃烧风洞的研制。图 8 是双管式脉冲风洞示意图。双管均以路德维希管方式运行,用喉道控制其流量。试验前分别充入氢氮和氧氮混合气体。双管的膜片打开后,管中气体就进入燃烧加热器。燃烧加热器结构类似氢氧火箭发动机,用点火器启动加热器工作。加热器后接喷管,模型发动机安装在真空箱内。燃烧加热器的温度、压力由双管初始充气压力(初始压力为 8~11MPa)和两管之间的氢和氧比例来控制。为进行超燃发动机试验,两管之间充气比例还要保证加热器燃烧后燃烧气体中 N_2、H_2O 和 O_2 两者之间的莫尔比为 79/21,以模拟大气中氧的莫尔比。一旦点火器工作、加热器中燃烧气体达到所要求的总温总压(总温可达 2400K,总压可达 6MPa)风洞就启动运行。为保证风洞快速启动,双管的两个膜片的破膜要有较高的同步精度。图 9 中的 Δt 值是氢、氧两个管中喉道处压力阶跃信号,反映了膜片破膜的同步时序。从图 9 中的 Δt 值可知膜片破膜同步精度在 2~3ms 之内,加热器的氢氮和氧氮气源压力稳定持续时间可达 100ms 左

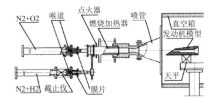

图8 双管式脉冲燃烧风洞
Fig. 8 Duat tube impulse combustion wind tunnel

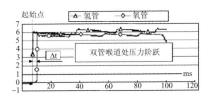

图9 双管喉道外压力时间历程
Fig. 9 Time history of pressure at dual-tube throats

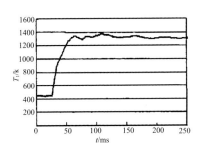

图10 脉冲式风洞喷管出口气流总温时间历程($M=5$ $Tt=1400K$ $Pt=4MPa$)

Fig. 10 Total temperature history at impulse wind tunnel nozzle exit

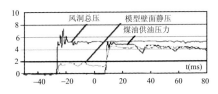

图11 脉冲风洞运行与燃料注射同步时序
Fig. 11 Time sequence of impulse wind tunnel operation and fuel input

右。图10是$M5$喷管出口气流总温的时间历程，图中约有30ms响应时间，这是由特制的铂铑热电偶决定的而不是风洞建立稳定流动时间。风洞启动与建立稳定总压的时间要短得多，约小于5ms(见图11)。从图9、11可知风洞可提供的工作时间约为60~80ms。从图8中的脉冲风洞运行原理可知，只要双管和真空箱容积足够大，工作时间还可进一步增加。实际上由于发动机模型没有冷却系统，为了避免模型受到热损伤，在氧氮管中安装了关断时间约10ms的快速截止伐(见图8)以控制高温气流的截止时间。

在脉冲燃烧风洞中进行超燃模型发动机试验还需解决燃料的注射与风洞同步启动问题。图11是风洞启动后流场建立、模型壁面压力、燃料注射、稳定燃烧流动建立的时序图。燃料注射由燃料箱通过挤压方式完成，但燃料加注与截止由CARDC自行研制的快速阀的开启时间(小于3ms)控制。

2.2 超燃模型发动机性能研究需要的试验时间

脉冲风洞试验时间多长才能满足发动机性能研究的要求这一问题至今并没有得到确切的回答。氢燃料的点火延滞时间短，尤其是在飞行马赫数大于7以上、气流总温大于2100K的情况。美国、日本、澳大利亚等国都采用工作时间小于10ms、试段出口直径大于1m的激波风洞进行模型长度大于2m氢燃料的发动机试验[5]。NASA在成功地进行了X-43A的马赫7、10飞行试验后指出:"飞行数据的初步分析表明在5ms的反射激波风洞中进行的系列试验能够预测飞行性能"[6]，从而肯定了脉冲风洞数据的可靠性。但是对于煤油燃料、飞行马赫数小于7条件下的发动机性能试验(气流总温小于2100K)，由于煤油自发点火的延滞时间比氢燃料要长得多，风洞要提供多长试验时间才能满足要求，至今国内外已发表的文献中都没有这方面的报道。

对脉冲风洞提供的试验时间的要求与液体煤油的自发点火延滞特性有关。室温液体煤油的自发点火的延滞实际上经历了十分复杂的过程。图12中显示了室温液体煤油注入发动机后在燃气温度达到平衡之前，大致经历了液滴破碎、蒸发、化学反应延滞等三个复杂的物理化学过程，而且这三个过程不是串接而是相互有重叠。此外，在燃烧过程中有些参数如:火焰稳定器形状、激波强度、湍流度、液滴尺寸与分布、来流和燃料喷射之间相对速度、燃料的反应动力学特性等众多参数都会对自发点火延滞时间产生影响。分析如此众多的影响参数来确定煤油自发点火延滞时间是一个十分困难的问题，只能通过试验确定煤油燃料超燃发动机性能研究所要求的脉冲风洞工作时间。

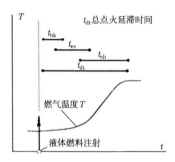

t_{bk}液滴破碎, t_{ev}蒸发, t_{ch}化学反应延滞

图12　液体碳氢燃料点火延滞特性

Fig. 12　Ignition time delay of liquid hydrocarbon fuel

作者在小尺寸直联式煤油自点火试验方面的结果见文献[2, 4]。自由射流试验的典型超燃模型发动机见图13(a)。模型发动机长 1m，三波系进气道宽高为120mm×80mm。外直径为 8mm、内直径为 1mm 的压阻式压力传感器安装在发动机内流道上、下壁面。脉冲风洞型面喷管出口直径为 300mm。风洞启动、煤油注入、燃烧流动建立时序可见图 11。图13(b) 是图13(a) 中燃烧室中凹腔附近压力测量点位置的局部放大。图 14(a) 是 $M = 5$，$P_t = 4$MPa，$T_t = 1400$K 风洞试验条件下燃烧室下壁面几个典型的测压点的时间历程。从 1、2、4、5、6 这些典型点的压力曲线可知在煤油注入后燃烧很快达到稳定。自发点火首先出现在凹腔内部(点 1、2)，然后逐渐向上游传播，因为试验观察到点 5 压力阶跃比点 1 晚约 2ms，但两者距离为25mm。点 1、2、4 处的燃烧十分强烈，持续时间约60~ 80ms。从凹腔附近 1、2、4、5 几个点压力历程判断，自发点火延滞时间约 4ms 如图 14(b) 所示，这一延滞时间还包括了压力测量孔(孔直径 1.5mm，孔长100mm)、传感器及测量系统的响应时间。图 15 为 $M = 6$ 条件下凹腔内壁面点 2 压力的时间历程，图中指出自发点火延滞时间与 $M = 5$ 基本相同。因此通过 $M = 5$、6 的试验可以获得结论: 煤油注入发动机后液滴破碎、蒸发、化学反应延滞三个物理化学过程的总时间约为 4ms。

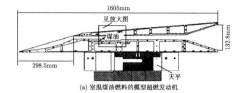

(a) 室温煤油燃料的模型超燃发动机

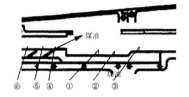

(b)　燃烧室的局部放大

图13　模型超燃发动机

Fig. 13　Sketch of model scramjet

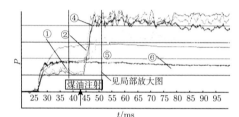

$M=5$, $T_t=1500$K, $P_t=4$MPa, $\Phi=1.0$

(a) 典型点的压力时间历程

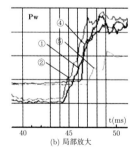

(b) 局部放大

图14　燃烧室下壁面典型点的压力

Fig. 14　Pressure history at typical points on lower wall of combustor

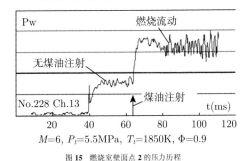

$M=6$, $P_t=5.5MPa$, $T_t=1850K$, $\Phi=0.9$

图15 燃烧室壁面点2的压力历程

Fig.15 Pressure history at point 2 on combustor wall

2.3 模型发动机典型试验结果

脉冲风洞中发动机试验及壁面压力测量结果表明试验的重复性很好[2]。此外对不同的油气比有很灵敏的反应。图16是4个不同油气比条件下模型内流道壁面的压力分布。当油气比从0.206增大时,壁面压力也相应增加,并逐步向上游发展。在油气比为0.756时,由燃烧产生的激波串仍然处于隔离段下游,表明进气道工作仍然是正常的。

图17是应变天平测量的燃烧推力增益曲线,这是由燃烧和无燃烧两者的差值组成。图中显示了天平的响应时间(而不是燃料注入后燃烧响应时间)约为15ms,天平信号稳定的持续时间与风洞工作时间相同约为60~80ms。说明了由于应变天平响应比压力测量慢得多,因此测量发动机推力要求风洞提供更长的工作时间。图18是由天平测量数据整理的燃烧推力增益随不同油气比变化的结果。显然,随着燃料的增加,推力增益基本上是线性变化的,这一结果与图16测力测量的结果完全一致。

与连续风洞一样,在脉冲风洞中亦可进行不同位置燃料喷射系统及各类凹腔、支板对燃烧影响的研

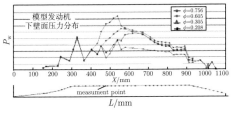

图16 不同油气比(煤油)情况下模型发动机下壁面压力分布($M=5$, $T_t=1500k$, $P_t=4MPa$)

Fig.16 Pressure distribution on model scramjet lower wall with different equivalent ratio(kerosene)

究。图19为$M=5$条件下,燃料喷射为多点(①②③三个位置,油气比 $\Phi=1$)和单点(②处一个位置,油气比$\Phi=0.6$)时对燃烧影响的压力分布比较。试验结果表明,合理布置燃料喷射点将会对燃烧室中加热量分布、燃烧效益产生明显影响,因而是一项重要的研究工作。

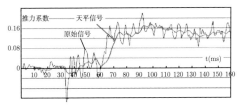

图17 燃烧推力增益的天平响应曲线($M=5$、$\Phi=0.83$、$T_t=1500K$、$P_t=4$ MPa)

Fig.17 Balance response of thrust increase of combustion

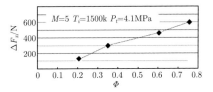

图18 不同油气比条件下模型发动机燃烧推力增益(进气道进口几何参数 120mm×80mm)

Fig.18 Thrust increase of combustion with different equivalent ratio(inlet intake size 120mm×80mm)

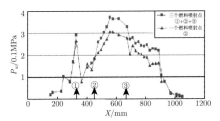

图19 单点和三点燃料喷射对模型发动机壁面压力分布影响

Fig.19 Effect of fuel input(single and three points) on pressure distribution on model scramjet lower wall

图20　连续式风洞与模型发动机(喷管出口直径 $D =$ 320mm,加热器与脉冲式相同为 H_2 Air O_2)

Fig. 20　Continuous wind tunnel and model scramjet(nozzle exit diameter $D = 320mm$, H_2 Air O_2 heater as impulse tunnel)

2.4 脉冲风洞和连续风洞中自由射流模型发动机试验结果比较

2003 年 5 月,CARDC 连续式燃烧风洞投入使用。该风洞具有与脉冲式相同的 H_2 空气-O_2 燃烧加热器, 图 20 为连续风洞与模型的照片。为了进一步确认在脉冲风洞中获得的结果,用同一个发动机在几乎相同的来流条件下进行发动机点火特性、内流道压力和推力测量比较(迄今为止国内外还没有这种比较报道)。下表是两座风洞的参数比较。

表 1　两类风洞运行参数

Table 1　Operational parameters of two wind tunnels

风洞类型	M	P_t(MPa)	T_t(K)	喷管出口直径(m)	油气比
脉冲	5.0	4	1460	0.3	0.878
连续	5.0	4	1400	0.32	0.874

两类风洞比较结果表明经过脉冲风洞试验的模型发动机在连续式风洞中能获得自点火与稳定燃烧;由于燃料注射、伐门控制特性不同,因而两者的压力和天平响应特性也不相同,但流道内两者的壁面压力峰值十分一致。图 21 是两座风洞中模型下壁面沿整个内流道压力分布测量结果的比较(比较试验中,连续式风洞运行时间为 5s),除了燃烧室前段部分两者压力有一点差别外,其它部位都很一致;两者压力有一些差别可能是由于来流参数不同(包括不同总温)引起的。脉冲式和连续式两者的比较结果对如何进一步发展高超声速地面设备和超燃发动机试验技术都是十分重要的,因而还要进一步开展这种比较研究(包括更大尺度的发动机、脉冲式和运行时间更长的连续式之间的比较)。

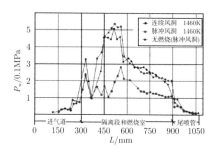

图 21　脉冲风洞和连续风洞比较

Fig. 21　Comparison of impulse and continuous results

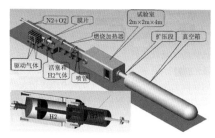

图 22　脉冲燃烧风洞进一步发展示意图

Fig. 22　Further development of impulse combustion wind tunnel

2.5 脉冲燃烧风洞的进一步发展

综上所述,脉冲式风洞具有造价低,运行经济,模型不需冷却,热流和压力测量、流动显示容易进行等独特优点,在超燃发动机基本性能和流动机理研究中已经发挥了很大作用。与连续式风洞相比,更重要的是这类设备能更经济地扩展到大尺寸风洞以进行工程性试验。CARDC 从 1997 年以来研制了 1kg/s 流量、10kg/s 流量两个脉冲燃烧风洞,积累了有关设计和运行经验,目前正在进行 25~ 40kg/s 流量、喷管出口直径大于 450mm 的脉冲式风洞研制与调试。图 22 是这一新型风洞的原理图,该图中氢燃料采用了活塞直接压缩氢气的方法以代替图 8 中的氢氧混合管,这种方式能使氢气用量大大减少,因而更加安全,也毋需安置图 8 中的截止阀。

3 结 论

(1) 脉冲燃烧风洞是进行火箭发动机高空羽流研究(包括火箭底部流动特性与光电效应)的一类十分重要的研究设备;

(2) 用双管式脉冲燃烧风洞进行了煤油燃料的

超燃模型风洞试验, 通过壁面压力测量确定了总的自发点火延滞时间; 在本文模型和 $M = 5$、6 试验条件下, 从煤油注入到获得燃烧的时间约为 4ms, 这一试验数据为研究复杂的煤油(包括其它碳氢燃料)点火延滞机理提供了重要依据;

(3) 通过连续式和脉冲式风洞中煤油燃料超燃模型发动机性能试验和压力测量比较表明: 笔者介绍的工作时间为 60~80ms 的脉冲燃烧风洞, 由于模型不用冷却、热流测量和光学流动显示容易进行等特点, 能够十分经济凑效地进行碳氢燃料超燃发动机的各类性能和机理研究;

(4) CARDC 已经积累了脉冲燃烧风洞研制与运行经验, 可以研制更大尺度的风洞, 用以研究带前体和后体的大尺寸发动机性能和带动力的高超声速飞行器性能。

致谢: 该研究工作得到了国家" 863-702" 的支持; 得到了 CARDC 超高速研究所的大力支持; CARDC 吸气式高超声速技术实验室的曾来荣, 李向东, 焦伟, 黄为民参加了本项目研究工作, 特此向他们表示衷心感谢。

参考文献:

[1] HERTZBERG A. et al. Investigation of rocket flow problem by means of short duration flow devices, ARSJ, No. 8, 1149, 1961.

[2] 乐嘉陵, 刘伟雄等. CARDC' s New short duration facility and its application in scramjet research [C]. International Conference on Hypersonic Propulsion, Beijing, China ICHP 2003-004.

[3] 乐嘉陵, 刘伟雄等. Performance study of model scramjet with fuel of kerosene in pulse facility [R]. AIAA 2003-6936.

[4] 乐嘉陵, 刘伟雄等. Study on model scramjet performance with fuel of kerosene [R]. ICMAR 2004 Proceedings Part 2. 120~124.

[5] HOLDEN M S. Studies of scramjet performance in the LENS facilities [R]. AIAA 2000-3604.

[6] MICHAEL A D. Mach 10, But Now What? Aviation week & Space Technology. Nov. 22, 2004, 24~26.

超燃冲压发动机再生冷却热结构设计的计算工具

蒋　劲[1], 张若凌[2], 乐嘉陵[2]

(1. 西北工业大学动力与能源学院, 陕西 西安　710072; 2. 中国空气动力研究与发展中心, 四川 绵阳　621000)

摘要: 为分析在电弧加热器上进行的超燃冲压发动机再生冷却热结构试验的热交换, 用了准三维的热分析工具和三维内流场 CFD 计算平台来作为分析工具。在换热计算及试验中用了水和煤油作为冷却剂, 而且用煤油作为燃料。计算和试验结果吻合较好, 表明换热分析工具和国内航空煤油物性表达式可以在深入的热结构试验和设计中应用。

关键词: 超燃冲压发动机; 再生冷却; 试验; 热分析计算工具

中图分类号: V211.74　　　**文献标识码:** A

The calculational tool of thermal structure design for regeneratively cooled scramjet

JIANG Jin[1], ZHANG Ruo ling[2], LE Jia ling[2]

(1. School of Power and Energy, Northwestern Polytechnical University, Xi'an '/10072, China; 2. China Aerodynamics Research and Development Center, Mianyang 621000, China)

Abstract: In order to analyze the heat exchange in the thermal structure tests of regeneratively cooled scramjet conducted in an arc heater, one quasi three dimensional thermal evaluation tool and one three dimensional internal CFD platform are used together as calculational tools. Water and kerosene can be chosen as coolant in heat exchange calculations as well as in tests, while kerosene is used as fuel. Good agreements between calculations and tests are derived, which indicats that the thermal evaluation tool and the analytical expressions of domestic aviation kerosene properties can be used extensively in thermal structure designs and tests.

Key words: scramjet; regenerative cooling; test; thermal evaluation tool

0　引　言

长时间工作的超燃冲压发动机需要采用再生冷却(regenerative cooling)。这种冷却途径的优点在于可以减小冷却系统的重量并利用燃料的吸热性质。燃料的吸热使得传向壁面的热能得到了利用(能量的再生), 并且, 碳氢燃料的大分子经吸热会分解(裂解)为一些较轻分子, 如甲烷、乙烯、氢气和一些从 C_2 到 C_{10} 的烃, 这既是燃料附加的吸热能力(化学热沉能力), 又使燃料在冷却发动机的同时分解为更容易燃烧的组分(利于燃烧)。燃料吸收的热量最后被带入燃烧室, 因此再生冷却不降低飞行器的性能[1]。

超燃冲压发动机燃烧室的再生冷却结构设计中, 已经建立了结合试验研究的若干热分析计算研究程序。M. H. N. Naraghi 建立了再生冷却火箭发动机推力室和喷管的三维热分析理论模型, 发展了计算程序 RTE[2]。该程序经过适当修改可应用于超燃冲压发动机再生冷却的热分析。应用于 NASP 的飞行器综合热管理分析程序(VITMAC)[3], 已用于以吸热碳氢燃料为冷却剂的超燃冲压发动机冷却系统的设计

蒋劲, 张岩凌, 乐嘉陵. 超燃冲压发动机再生冷却热结构设计的计算工具, 实验流体力学, 2006, 20(3): 1-7.

分析,可以模拟传热/流动的耦合系统,包括相互影响的流体回路、多个热交换部件以及承受不同热载荷的表面。经过试验验证的法国MBDA的一维NANCY程序,已经深入应用于主动冷却面板的设计中[4]。其它还有法国ONERA发展的MOSAR程序[5],它和MARC程序结合用于再生冷却的设计研究。

以上程序的主要缺点是采用一维发动机内流场的计算结果作为换热计算的输入条件。实际上,超燃冲压发动机不象火箭发动机那样有简单的轴对称结构,其流场具有显著的多维特征。将发动机内部流动作均匀的一维简化,达不到热分析研究的目的和要求。

为了克服上述缺点,笔者紧密结合中国空气动力研究与发展中心吸气式高超声速技术实验室(CARDC-AHL)发展的超燃冲压发动机三维内流场计算软件平台AHL3D和二维换热分析计算工具,建立了一套热结构设计的计算工具,来进行发动机再生冷却热分析。在二维换热分析计算里采用了基于国内航空煤油组分测量结果而得到的煤油物性解析表达式。首先对二维换热分析计算工具进行了验证,然后利用这套热结构设计计算工具来分析CARDC电弧风洞上的超燃发动机热结构试验,其中冷却剂有水和煤油。

1 超燃冲压发动机三维内流场计算平台

CARDC吸气式高超声速技术实验室的的计算组,从1997年开始进行超燃发动机流场的CFD研究。为适应复杂外形和不同燃料(主要是氢气和碳氢燃料)化学反应的超燃冲压发动机部件和整机内部流场的高效、实用的三维大规模并行数值模拟,计算组从2001年开始研制流场的并行计算软件平台,已经对超燃发动机流场进行了大量的并行计算研究。

目前的软件平台AHL3D基于MPI并行环境,运用时间相关法,采用强耦合的全隐式格式有限体积离散,数值求解带化学非平衡流动的全N-S方程。该平台支持完全气体和燃烧化学非平衡流计算、两相流计算、定常和非定常计算、层流和湍流计算、有喷流和无喷流计算,支持多块网格、错位拼接网格和重叠网格,可以求解二维、轴对称或三维全N-S方程、欧拉方程和PNS方程。现在正进行非结构网格的计算方法研究。

大量的试验验证和应用表明,软件平台具有较高的计算精度和可靠性,不仅可以用于基础理论的研究,而且还可以应用于重大的工程实际问题,在发动机的优化设计、快速有效的性能分析和对试验的指导中,发挥了重要的作用。

笔者利用AHL3D对试验中的发动机内流场进行数值模拟,采用了10组分、12步化学反应和k-ωTNT湍流模型。

2 二维换热分析计算工具

AHL3D的计算流场作为换热分析工具的输入条件。发动机沿燃气流动方向(x向)被细分成许多的站,如图1所示。图1还示意了发动机燃烧室横截面。把冷却通道局部和相应的发动机燃气内流边界层外缘局部——一对应,从中抽取相应位置的静压、静温、总温、速度、燃气各组分的质量百分比等流场参数,作为换热分析工具的输入条件。这里假定从燃气到燃烧室壁的传热(壁面热流)只对燃气流动有非常小的影响。

对于一条冷却通道回路,冷却液沿x或负x方向来回流动。先计算不同x坐标位置的燃气向室壁、室壁到冷却剂的一维传热,其结果作为室壁内二维热交换的边界条件。然后通过迭代求出这个x位置上的二维温度场截面[2]。

2.1 燃气向室壁的传热

超燃冲压发动机燃烧室内的流动是高速可压缩流动,传热计算采用的是Eckert参考焓方法(The Eckert reference enthalpy method)[2,3]。

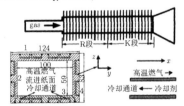

图1 超燃冲压发动机模型、燃烧室横截面和冷却通道示意图

Fig 1 The sketch of the scramjet model, cross section of its combustor and cooling channel

这个方法假设:整体上,变物性的边界层可以用一个常物性的参考边界层来代替,这个参考边界层的物性参数和一个参考的焓(或温度)对应。

用下标G表示燃气,X表示参考状态,n表示站

点的序数, W 表示壁面, A 表示绝热, S 表示静止状态, 第 n 个站点上燃气的参考焓可表示为

$$h_{GXn} = 0.5(h_{GSn} + h_{GWn}) + 0.22(h_{GAWn} - h_{GSn}) \tag{1}$$

上式 h_{GXn} 表示燃气的参考焓, h_{GSn} 是燃气的静焓, h_{GWn} 是燃气恢复焓, h_{GWn} 是燃气在壁温下的焓。h_{GAWn} 由恢复因子、燃气总焓和燃气静焓决定, 恢复因子由参考温度决定。先假设第 n 个站点上的气壁温度 T_{GWn}, 由 (1) 式可以确定参考温度, 从而确定燃气边界层参考状态。燃气热力与输运性质由多组分公式计算[7], 第 n 个站点上的气壁温度 T_{GWn} 结合下面的热平衡条件来确定。

燃气传到气壁的对流热流[6]和斯坦顿数 St 有关:

$$q_c = St_{GXn} \cdot \theta_{Xn} V_{GSn}(h_{GAWn} - h_{GWn}) \tag{2}$$

燃气对壁面的辐射热流

$$q_r = \sigma \varepsilon_{W,ef}(\varepsilon T_{GSn}^4 - a_W T_{GWn}^4) \tag{3}$$

其中 q_r 为辐射热流, T_{GSn} 为燃气静温, ε_c 为燃气黑度, a_W 为壁面吸收率, $\varepsilon_{W,ef}$ 为壁面有效黑度, σ 为斯忒藩·波尔兹曼常数。

综合上两式得到第 n 个站点上由燃气传到气壁的总热流

$$q_n = q_c + q_r \tag{4}$$

2.2 通过衬层的导热

燃烧室衬层把燃烧系统和冷却系统分隔开来。由傅立叶定律, 第 n 个站点上通过衬层从气壁到液壁的热传导热流:

$$q_n = \frac{k_W}{\delta_W}(T_{GWn} - T_{LWn}) \tag{5}$$

其中 δ_W 是燃烧室壁衬层的厚度, T_{LWn} 是冷却剂侧壁面温度, k_W 是燃烧室壁的导热系数。

2.3 由液壁至冷却液的传热[3]

冷却剂在冷却通道内流动是充分发展湍流流动, 液壁至冷却剂的换热是管内强制对流换热, 可以采用和努塞尔数相关的准则关系式。通常为得到更为准确的对流换热系数, 需要再加上几个修正: 二维几何修正 ϕ_{2D}, 液体物性膜温度变化 ϕ_{prop}, 通道的进口效应 ϕ_{in}, 通道弯曲效应 ϕ_{curv}, 通道的粗糙度 ϕ_R 影响,

$$Nu = \frac{h_c D_h}{k_{ref}} = C Re_m^{0.8} Pr_m^{0.4} \phi_{2D} \phi_{prop} \phi_{in} \phi_R \phi_{curv} \tag{6}$$

上式用来确定对流换热系数 h_c, 其略去了下标 n。这里 D_h 是水力直径, k_{ref} 是以液体参考温度 T_{ref} 为定性温度计算的液体导热系数, 其它以液体平均温度 T_{mn} 来定性。参考温度 T_{ref} 由液壁温度 T_{LWn} 和液体平均温度 T_{mn} 确定。(6) 式表明液体与液体之间的对流换热系数 h_c 和液壁温度 T_{LWn} 及液体平均温度 T_{mn} 有关。T_{mn} 和小单元冷却通道的进口温度 T_n、出口温度 T_{n+1} 有关 (图2)。

从液壁到液体的对流换热热流为:

$$q_n = h_c(T_{LWn} - T_{mn}) \tag{7}$$

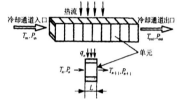

图2　冷却通道和单元段

Fig. 2　Cooling channel flow and its segment

和燃烧室三维 CFD 计算网格对应, 冷却通道也被分割成许多的站, 如图3标示。两站之间的部分可看成一小单元通道, 对于这一小单元通道有

$$q_n \times d \times L = c_{pm} \dot{m}_c(T_{n+1} - T_n) \tag{8}$$

这里 L 是这一小段冷却通道的长度, c_{pm} 是以平均温度 T_{mn} 定性的冷却剂的比热, $\dot{m}_c$ 是冷却通道中冷却剂的质量流率。这里假定从燃气传来的热被冷却剂全部吸收。这是一个保守的做法, 实际上会有部分热传向环境。

由 (4~8) 式建立热平衡关系, 联立迭代求解就可以求得气壁温度 T_{GWn}。

2.4 冷却剂的物性模型

考虑了水和煤油作为冷却剂。对于水, 选用文献[4]提供水的物性模型来确定已知温度下的物性参数。

对于煤油采用 Peng-Robinson 状态方程[8]。单组分的 Peng Robinson 状态方程可以写为

$$p = \frac{RT}{V - B} - \frac{A}{V^2 + 2VB - B^2} \tag{9}$$

其中, p 为压强, R 为气体普适常数, V 是 1mol 物质的体积, A、B 为和物质性质相关的常数。

将国内航空煤油组分的测量结果输入到成熟软

件进行计算, 可以得到煤油的物理性质, 如密度随压强、温度的变化, 以及比热, 粘性系数等等。对计算结果进行拟合, 从而得到换热计算需要的煤油物理性质的解析表达式。

2.5 冷却通道室壁内二维温度分布的求解方法

用热平衡法离散二维导热的控制微分方程:

$$\frac{\partial}{\partial x}\left(k\frac{\partial T}{\partial x}\right) + \frac{\partial}{\partial y}\left(k\frac{\partial T}{\partial y}\right) = 0 \qquad (10)$$

整理成二维有限差分方程用来模拟室壁内的热传导, 室壁内的二维温度分布用迭代的方式求解, 迭代中用逐次超松弛方法来加快收敛[2]。输入这个模块的参数有: 燃烧室壁的几何尺寸, 冷却通道的几何尺寸, 所用材料的类型, 由前面计算获得的冷却剂与液壁之间的对流换热系数、冷却剂的温度以及燃气到气壁的热流。冷却剂与液壁之间的对流换热系数假定在通道的上面、侧面、下面均是相等的。由于结构的对称性, 冷却通道的每个截面只需计算一半。假定截面的两个侧边之间是绝热的, 外壁与环境也假定是绝热的。如图3所示, 把有限差分网格先添加在该截面上, 然后对每个节点把有限差分方程表达出来。对于非边界节点, 有限差分方程可写为

$$T_{i,j} = \frac{\dfrac{T_{i+1,j}}{R_1} + \dfrac{T_{i,j-1}}{R_2} + \dfrac{T_{i-1,j}}{R_3} + \dfrac{T_{i,j+1}}{R_4}}{\dfrac{1}{R_1} + \dfrac{1}{R_2} + \dfrac{1}{R_3} + \dfrac{1}{R_4}} \qquad (11)$$

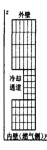

图3 半个冷却通道上的网格
Fig. 3 The grid inside the wall of cooling channel

用类似的办法可以写出边界节点的方程。总之, 有限差分方程每个节点的温度表示为相邻节点温度和(或)对流换热系数、热流、导热系数、冷却剂温度的形式。和 n 对应于 x 方向不同, 这里的 i、j 分别对应于 y、z 方向。

3 计算结果和讨论

3.1 二维换热分析计算工具的验证

为验证二维换热分析计算工具的可靠性, 选用 MBDA 的试验作为算例[4]。试验中, 横截面是矩形的冷却通道由不锈钢制成, 通道壁厚 1.5mm, 通道内截面 7mm × 17mm, 通道总长 2.5m, 焊接在通过热电阻来加热的铜块上。测温度的热电偶的位置如图4所示。试验用了煤油和水作为冷却剂, 各有两次试验结果, 即 Water2、Kero2、Water10、Kero10。试验的目的是用来验证在法国 MBDA 广泛用于主动冷却研究的一维 NANCY 程序。选用 Water2 和 Kero2 两次试验来验证笔者发展的二维换热分析工具。

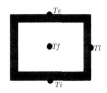

图4 冷却通道里温度热电偶测量点
Fig. 4 Locations of the thermocouple

Water2 的计算结果如图5~8, 图中有 NANCY 程序的结果。由图可知, 对水冷的情况, 计算结果与 NANCY 程序、试验测量吻合很好。Kero2 的计算结果如图9~10, 图中也有 NANCY 程序的结果。由图可知, 对于煤油冷却的情况, 笔者的计算在大部分位置比 NANCY 程序更接近于试验测量值。

通过水冷和油冷的计算验证, 可以看出所发展的换热分析工具是可靠的, 可以用来分析超燃冲压发动机燃烧室冷却结构试验。

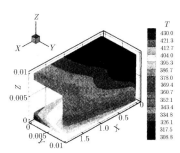

图5 Water2 冷却通道的准三维温度分布
Fig. 5 The quasi 3 D distribution of wall temperature for Water2

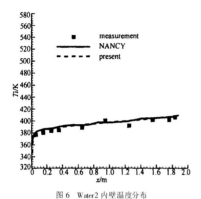

图 6   Water2 内壁温度分布

Fig. 6   The distribution of inner wall temperature for Water2

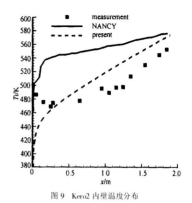

图 9   Kero2 内壁温度分布

Fig. 9   The distribution of inner wall temperature for Kero2

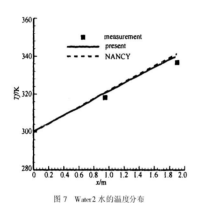

图 7   Water2 水的温度分布

Fig. 7   The distribution of water temperature for Water2

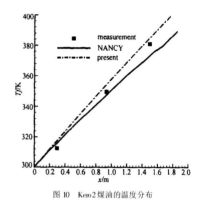

图 10   Kero2 煤油的温度分布

Fig. 10   The distribution of kerosene temperature for Kero2

3.2   超燃发动机燃烧室冷却结构试验和计算对比

超燃发动机燃烧室冷却结构试验是在 CARDC 的电弧风洞中进行的。

试验采用了 DJ-21 高压管式电弧加热器,该设备的直流整流电源功率可以运行到 6MW 以上,弧室压力可达 10MPa,流场温度可达 3000K。喷管出口的气流达到超燃模型的入口来流条件,超燃发动机冷却结构模型直接连接在喷管的出口上进行直联式试验研究。燃烧室分成 R 段和 K 段,组成两段的四个板的标号 1、2、3、4 在如图 1 中标明。

在 100# 水冷试验中发动机不点火,运行了 30s。各板的冷却水质量流量和水的温升如表 1 所示。计算值除底面 R3 外均比试验值大。计算值偏大是合理的,因为在计算里假定热量全部被水所吸收。R 段底面 R3 有一段凹槽(起稳定火焰作用),其冷却水温

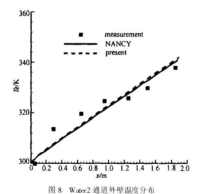

图 8   Water2 通道外壁温度分布

Fig. 8   The distribution of outer wall temperature for Water2

升的测量值比计算值大较多,可能和模型加工方式有关。图11是燃烧室R段和K段的左壁面的计算和试验的温度分布比较(图中注明了试验的发动机入口条件,也是超燃发动机三维内流场计算的入口条件,下同),计算结果和试验的测量值吻合较好。

表1 100#试验各壁面的冷却水流量和温升
Table 1 The mass flow rates and temperature rise of coolant for test 100#

100#	壁面	冷却水流量 (kg/s)	冷却水温升(K) 试验	冷却水温升(K) 计算
燃烧室	上面 R1	0.087	23	25.3
	左面 R2	0.276	3	4.1
	底面 R3	0.11	17	10.9
	右面 R4	0.276	3	4.1
扩张段	上面 K1	0.138	13	13.0
	左面 K2	0.407	2	3.8
	底面 K3	0.355	4	4.4
	右面 K4	0.407	3	3.8

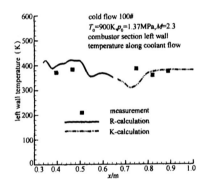

图11 100#试验R段和K段左壁面中心线温度
Fig.11 Wall temperature along the middle line of R2 & K2(test 100#)

14#试验采用煤油冷却,发动机不点火,运行30s。图12是R段左壁面的计算和试验的温度分布比较。测量值与计算值非常接近,符合了试验快到稳态的事实。21#试验采用煤油冷却,发动机点火,运行10s。图13是R段右壁面的计算和试验的温度分布比较。计算值高于测量值,符合试验未达热平衡的事实。

图14是对14#试验的R段左壁面中心线温度采用不同煤油物性模型的计算比较。将文献[4]和笔者采用的煤油物性模型用于相同的侧壁换热计算。试验基本达到稳态,图示是壁面中心线温度的比较,可以看出笔者采用的煤油物性模型要稍好一些。

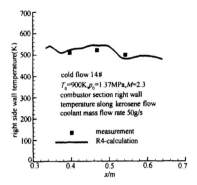

图12 14#试验R段右壁面中心线温度
Fig.12 Wall temperature along the middle line of R4 (test 14#)

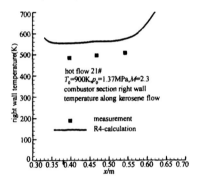

图13 21#试验R段右壁面中心线温度
Fig.13 Wall temperature along the middle line of R4 (test 21#)

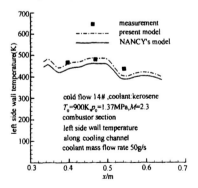

图14 采用不同煤油物性模型的计算结果
Fig.14 Wall temperature along the middle line of R2 using different thermophysical models(test 14#)

4 结 论

建立了一套热结构设计的计算工具,包括二维换热分析工具和超燃发动机三维内流场 CFD 计算平台 AHL3D,后者输出的流场结果作为前者的输入条件。煤油物性的解析表达式是成熟软件的计算结果经拟合得到的,其中利用了国内航空煤油组分的测量结果。利用所发展的计算工具,分析了电弧风洞上的超燃发动机热结构试验,其中冷却剂有水和煤油。在发动机热结构试验达到稳态热平衡后,可以得到与试验非常吻合的结果;对于非稳态的试验情况,计算结果也有一定的参考价值。

对比结果表明热结构设计计算工具和煤油物性表达式是可靠的,可以在深入的热结构试验和设计中应用。

参考文献:

[1] KHODABANDEH J W, FREDERICK R A. Experimentation and modeling of Jet A thermal stability in a heated tube[R].
AIAA 2005 3769, 2005.

[2] NARAGHI M H N. A computer code for three dimensional rocket thermal evaluation, user manual for RTE2002 version 1 [R]. Tara Technologies ,LLC, 2002.

[3] RICHARD M, TRACI John L, FARR Jr. and LAGANELLI Tony. A thermal management systems model for the NASA GTX RBCC concept[R]. NASA/CR 2002 211587.

[4] DUFOUR E, BOUCHEZ M. Semi empirical and CFD analysis of actively cooled dual mode ramjets[R]. AIAA 2002 5126, 2002.

[5] BOUQUET C,HAUBER B,THEBAULT J. Validation of a leak free C/SiC heat exchanger technology [R]. AIAA 2003 6918.

[6] HEISER William H, PRATT David T. Hypersonic airbreathing propulsion[M]. American Institute of Aeronautics and Astronautics Inc. , 1994.

[7] 郑忠华. 双模态超燃冲压发动机燃烧室流场的大规模并行计算及试验验证[D]. 长沙: 国防科技大学博士学位论文, 2003.

[8] WANG Hai, FRENKLACH M. Transport properties of poly cyclic aromatic hydrocarbons for flame modeling[J]. Combustion and Flame, 1994, 96: 163~170.

甲烷点火燃烧的简化化学反应动力学模型*

肖保国，钱炜祺，杨顺华，乐嘉陵

(中国空气动力研究与发展中心，四川 绵阳 621000)

摘　要：从甲烷点火燃烧的 GRI-Mech 1.2 详细化学反应动力学模型出发，采用"准稳态"假设方法来对其进行简化，得到了包含 18 组分 14 步总包反应的简化反应模型。通过典型试验结果对比和基于均匀试验设计的校核计算结果比较可以看出，采用"准稳态"假设方法得到的简化反应动力学模型能有效地再现详细基元反应模型的反应机理，具有较高精度，是可靠而有效的。

关键词：甲烷；燃烧；点火延迟；简化化学反应动力学模型*

中图分类号：V235.213　　**文献标识码**：A　　**文章编号**：1001-4055（2006）02-0101-05

Reduced chemical reaction kinetic model for combustion of methane

XIAO Bao-guo，QIAN Wei-qi，YANG Shun-hua，LE Jia-ling

(China Aerodynamics Research and Development Center, Mianyang 621000, China)

Abstract　A reduced chemical kinetic model was obtained by reducing the detailed chemical kinetic model GRI-Mech 1.2 for the combustion of methane based on quasi-stationary state approximation (QSSA). This reduced model is composed of 18 species with 14 global reactions. The comparisons between the reduced model and the typical experimental results were made. Computational results of the reduced model were also compared with that of the detailed model by adopting the uniform experimental design technique. The little difference between these results showed that the reduced model using QSSA has the ability to represent efficiently the reaction mechanisms of detailed element reaction model. Therefore, the reduced model is of high precision and the can be used reliably and efficiently.

Key words　Methane; Combustion; Ignition delay; Reduced chemical kinetic model*

1 引　言

超燃发动机的燃烧室是整个超燃发动机研究的关键所在，有效缩短碳氢燃料的点火延迟时间，提高点火性能，对于缩短发动机长度、减轻发动机重量、提高其性能具有极其重要的意义[1]。碳氢化合物的燃烧是一个很复杂的化学过程，以甲烷为例[2]，描述其详细燃烧反应机理的 GRI－Mech 化学动力学模型由 32 种组分，175 个基元反应组成；Seiser 等人[3]提出的庚烷燃烧的详细化学动力学模型则包含了 160 种组分，1540 个基元反应。如此庞大的化学反应模型给理论分析和数值模拟带来了巨大困难，耦合到流场计算中时会出现两个问题：一是"刚性问题"，即实际化学动力学模型中一般都既包括反应速率快、特征时间尺度小的基元反应，也包括反应速率慢、特征时间尺度大的基元反应，反映到微分方程中就是方程的特征值差别明显，特征矩阵的条件数很大，形成了计算的刚性问题；二是计算效率的问题，当化学反应动力学模型包含较多的组分和基元反应时，会占用大量的计算机内存和 CPU 时间，从而影响到计算效率。因此，近年来，许多学者提出了灵敏度分析[4]、计算奇异值摄动[5]、"准稳态"假设 (Quasi Stationary State Approximation)[6]等方法来对类似这些庞大的化学反应动力学模型进行简化分析。文献 [7] 对上述简化

肖保国，钱炜祺，杨顺华，乐嘉陵. 甲烷点火燃烧的简化化学反应动力模型. 推进技术, 2006, 27(2): 101-105.

方法分析后指出:"准稳态"假设方法较适用于工程实际应用。比如美国加州大学 Berkeley分校的 Chen J Y 教授开发的一套基于准稳态假设方法的化学反应动力学模型简化软件 CARM (Computer Assisted Reduction Method); 在国内,中国科学技术大学的陈义良教授等人对部分碳氢燃料的简化化学反应机理也进行了相关研究。本文则在"准稳态"假设方法的基层上自主研发了一套简化化学反应动力学的分析软件 SPARCK (Software Package for Reduction of Chemical Kinetics), 并针对甲烷的点火燃烧的详细基元反应动力学模型,采用该软件来对其进行简化,得到了比较好的简化结果。

2 "准稳态"假设方法

对由 N 种组分和 K 个基元反应组成的化学反应系统,各组分的反应速率可写为常微分方程形式

$$\mathrm{d}\vec{y}/\mathrm{d}t = \vec{g}(\vec{y}) = \sum_{r=1}^{K} \vec{s}_r R_r (k_r, \vec{y}) \quad (1)$$

式中 $\vec{y}$ 为组分浓度矢量, y_i ($i = 1, N$) 为第 i 种组分的浓度, $\vec{s}_r$, R_r 分别是第 r 个基元反应的化学当量系数矢量和化学反应速率。所谓"准稳态"假设,是指假设反应系统中某些组分是最活跃、最容易发生反应的物质,其生成率近似等于消耗率,这些组分(设有 M 种)处于准稳定状态,即

$$\mathrm{d}y_i/\mathrm{d}t = \sum_{r=1}^{K} s_{ir} R_r \approx 0 \quad (i = 1, M) \quad (2)$$

这样就得到表示反应速率的 M 代数方程。利用这些稳态条件,同时可以去掉代数方程中消耗反应速率相对较快的 M 个基元反应。进一步对化学反应的当量系数矩阵进行分析,将稳态条件以特定形式加到非稳态组分的反应速率表达式中,由于稳态组分的反应速率为零,因此这样相加后不会给非稳态组分本身的反应速率值带来影响,但会使其反应速率对应的基元反应表达形式发生变化。此后,对非稳态组分反应速率中的基元反应速率项进行合并处理,最终可以得到具有总包反应形式的简化化学反应动力学模型。

这一简化方法的关键是准稳态组分的选取和代数方程中消耗反应速率相对较快的 M 个基元反应的识别。准稳态组分的判别方法是考察组分 i 的产生率 g_i^p 与消耗率 g_i^c, 若产生率 g_i^p 与消耗率 g_i^c 满足

$$|g_i^p - g_i^c| / \max(g_i^p, g_i^c) \quad (3)$$

式中 δ 为一较小的阈值,则可认为组分 i 为准稳态组分,该组分满足关系式 (2)。代数方程中消耗反应速

率相对较快的基元反应识别则主要采用的是针对稳态组分计算各基元反应消耗率贡献的方法。比如对于准稳态组分 i 的消耗而言,如果基元反应 r 在这一过程中占主导地位,则可以认为基元反应 r 为消耗反应速率相对较快的基元反应。

本文研发的分析软件 SPARCK 对于碳氢燃料点火燃烧反应的简化,主要包括以下步骤:

(1)将碳氢燃料点火燃烧的详细基元反应模型代入良搅拌器 (PSR Perfect Stirred Reactor)的反应环境下,目的是使碳氢燃料与氧化剂充分反应,通过对反应过程进行详细计算分析,利用式 (3) 识别出反应系统中的准稳态组分以及对应于各稳态组分消耗最快的基元反应号; (2)识别出稳态组分和可消去的基元反应号以后,利用该软件建立具有总包反应形式的简化化学反应动力学模型。

3 甲烷点火燃烧反应动力学模型的简化

采用"准稳态"假设方法对甲烷燃烧的化学反应动力学模型进行简化。采用本文介绍的软件包 SPARCK, 将甲烷燃烧反应的详细基元反应模型 GRI Mech 1. 2代入良搅拌器 (PSR)的反应环境下,可分析出如下 13 种组分为准稳态组分: C_2H_3, HO_2, HCO O, C_2H_5, C, H_2O_2, CH_2OH, CH, C_2H, CH_2, CH_3 O HCO, CH_2 (S), 同时识别出对各稳态组分消耗最快的基元反应号分别为: 70, 45, 78, 73, 121, 88, 55, 124, 169, 125, 56, 165, 141。通过矩阵运算,可得到包含 18 个组分 14步总包反应形式的简化化学反应动力学模型

$$\begin{array}{ll}
\mathrm{I} & 2O \Rightarrow O_2 \\
\mathrm{II} & O + H_2O \Rightarrow 2OH \\
\mathrm{III} & O + CH_2O \Rightarrow H + OH + CO \\
\mathrm{IV} & O_2 + C_2H_2 \Rightarrow H_2 + 2CO \\
\mathrm{V} & O_2 + CO \Rightarrow O + CO_2 \\
\mathrm{VI} & O_2 + CH_2O \Rightarrow 2OH + CO \\
\mathrm{VII} & CH_2O \Rightarrow H_2 + CO \\
\mathrm{VIII} & 2H + CH_2O \Rightarrow OH + CH_3 \\
\mathrm{IX} & C_2H_4 \Rightarrow H_2 + C_2H_2 \\
\mathrm{X} & OH + C_2H_2 \Rightarrow H + HCCOH \\
\mathrm{XI} & O + CO + CH_2O \Rightarrow O_2 + CH_2CO \\
\mathrm{XII} & O + H_2 + CO + H_2O \Rightarrow O_2 + CH_3OH \\
\mathrm{XIII} & CH_3 + H_2O \Rightarrow OH + CH_4 \\
\mathrm{XIV} & 2CH_3 \Rightarrow C_2H_6
\end{array}$$

各总包反应的反应速率具体表达式为 (式中 w

后的罗马数字代表总包反应号,阿拉伯数字代表详细　基元反应模型 GRI-M ech 1. 2中的基元反应号):

w Ⅰ $= w\,1+w\,2+w\,6+w\,8+w\,10+w\,12+w\,13+w\,14+w\,16+w\,17-w\,25+\;2w\,27+w\,29+w\,30+w\,33+w\,34+w\,35+$
$w\,36+w\,38+w\,39+w\,40+\;w\,41+w\,42+w\,48+w\,51+w\,53+w\,54+w\,58+w\,59+w\,61+w\,62+w\,64+w\,66+$
$w\,69+w\,71+w\,72-w\,75+w\,76+2w\,77+w\,79-w\,81+w\,84+w\,94+w\,95+w\,96+w\,99+w\,101+w\,102-106-$
$w\,109+w\,110+\;2w\,112+w\,113+w\,123+w\,127+w\,128+w\,129+w\,130+w\,133-w\,135+w\,136+w\,137-w\,138+$
$w\,143-w\,144+w\,147-w\,148+w\,152+w\,154+w\,157+\;w\,158+\;2w\,163+w\,166+w\,167+w\,168+w\,173$

w Ⅱ $= w\,5+w\,11+w\,18+w\,19-w\,42-w\,43+w\,46-w\,51+w\,52-w\,58-w\,61+w\,62-w\,66+w\,67+w\,68-w\,83-w\,84-$
$w\,85-w\,86-w\,92-w\,94-w\,95-w\,96-w\,99-w\,100-w\,101-w\,102-w\,108-w\,110-w\,111-w\,112-w\,113-$
$w\,114-w\,115-w\,117-w\,120+w\,126+w\,129+w\,138-w\,143+w\,148-w\,158-w\,159-w\,162-w\,163$

w Ⅲ $= -w\,2+w\,3+w\,4+\;2w\,7+2w\,9-w\,10-w\,13+w\,15-w\,16-w\,17+w\,18+w\,19+w\,22+\;3w\,25+w\,26-\;3w\,27+$
$2w\,28-w\,29-\;2w\,33-\;2w\,34-w\,35-\;2w\,36-w\,37-\;2w\,38-\;2w\,39-\;2w\,40-\;2w\,41-\;w\,42-w\,46-\;2w\,48+$
$w\,49-w\,51-w\,52-\;2w\,53-\;2w\,54-\;2w\,58-\;2w\,59-\;2w\,61-\;2w\,62-\;2w\,64-\;w\,66-\;2w\,69-\;2w\,71-\;2w\,72+$
$2w\,75-\;2w\,76-\;4w\,77-\;2w\,79+w\,80+w\,81+w\,83-w\,84+w\,86+\;w\,89+w\,90+w\,91+w\,92+w\,93-\;2w\,94-$
$2w\,95-\;2w\,96+w\,98-w\,99+\;w\,100-w\,101-w\,102+w\,103+w\,104-w\,105+w\,106+w\,108+\;2w\,109-w\,110-$
$w\,111-\;3w\,112-w\,113+w\,114+w\,115+w\,116+w\,117+w\,119-w\,120+\;2w\,122-w\,123-w\,128-\;2w\,129-$
$2w\,130+w\,134+\;3w\,135-\;w\,137+\;2w\,138+\;2w\,140+w\,142+\;3w\,144-w\,147+\;2w\,148-w\,152-\;w\,153-$
$2w\,154-\;2w\,157-\;w\,158+w\,159+w\,160+w\,161+w\,162-\;3w\,163-\;2w\,166-\;2w\,167-\;2w\,168-\;2w\,173+w\,174+$
$2w\,175$

w Ⅳ $= w\,21+w\,22+w\,23+w\,24+w\,25+\;w\,26-w\,27-w\,69+w\,75-w\,77+w\,81+w\,106+w\,108+w\,109-w\,112-$
$w\,123-w\,127-w\,128-w\,129-\;w\,133-w\,136-w\,137-w\,140-w\,147-w\,152-w\,157-w\,163-w\,170+$
$w\,171-w\,175$

w Ⅴ $= w\,12+w\,14+w\,30+w\,31+w\,98+w\,119-w\,131-w\,151$

w Ⅵ $= w\,2-w\,4-w\,7-w\,9+w\,10+w\,13+w\,16+w\,17-w\,18-w\,19-\;2w\,25-\;w\,26+\;3w\,27-w\,28+w\,29+w\,32+$
$2w\,33+\;2w\,34+2w\,35+\;2w\,36+w\,37+w\,38+w\,39+w\,40+w\,41+w\,42-w\,44+w\,48-w\,49+\;w\,51+w\,53+$
$w\,54+\;2w\,58+w\,59+\;2w\,61+2w\,62+w\,64+\;2w\,66-\;w\,67-w\,68+w\,69+w\,71+w\,72-\;2w\,75+w\,76+\;3w\,77+$
$w\,79-w\,80-\;w\,81+w\,84-w\,86-w\,89-w\,90-w\,91-w\,93+\;2w\,94+\;2w\,95+\;2w\,96-w\,98+w\,99+w\,101+$
$w\,102-w\,103-w\,104-w\,106-\;2w\,109+w\,110+\;3w\,112+w\,113-w\,114-w\,115-w\,116-w\,117-w\,119-$
$w\,122+w\,123-\;w\,126+w\,128+w\,129+w\,130-\;2w\,135+w\,137-\;2w\,138-w\,140+w\,143-\;w\,144+w\,147-$
$2w\,148+w\,152+\;w\,153+\;2w\,154+\;2w\,157+w\,158-w\,160-w\,161+\;3w\,163+\;2w\,166+\;2w\,167+\;2w\,168+$
$2w\,173-w\,175$

w Ⅶ $= -w\,3-w\,7-w\,9-w\,18-w\,19-w\,22-w\,24-\;2w\,25-\;2w\,26+\;2w\,27-w\,28+w\,29+w\,30+w\,38+w\,39+w\,40+$
$w\,41+w\,44+w\,46+\;w\,48-w\,49+w\,52+w\,54+w\,57+w\,58+w\,59+w\,61+w\,62+w\,64+w\,66+\;w\,69+w\,71+$
$w\,72-\;2w\,75+w\,76+\;3w\,77+\;2w\,79-w\,81-w\,82-w\,83-\;w\,91-w\,92-w\,93+w\,94+w\,95+w\,96-w\,103-$
$w\,104+w\,105-w\,106-\;w\,108-w\,109-w\,111+\;2w\,112+w\,113-w\,116-w\,122+w\,123+w\,128+\;2w\,129+$
$w\,130-w\,134-\;2w\,135+w\,137-w\,138-w\,139-w\,140-w\,142-w\,143-\;2w\,144+w\,147-w\,148-w\,151+$
$w\,152+\;2w\,157-w\,160-w\,161-w\,162+\;2w\,163-w\,171-\;w\,174-w\,175$

w Ⅷ $= -w\,10+w\,25+w\,26-\;2w\,27+w\,49+w\,60+w\,65+\;2w\,75-\;2w\,77+w\,80-w\,94-w\,95-w\,96+w\,109-$
$2w\,112-w\,118-w\,123-w\,128-w\,129+w\,135-w\,137+w\,138+w\,144-w\,147+w\,148-w\,152-w\,153-$
$w\,154-\;2w\,157-\;2w\,163$

w Ⅸ $= w\,25+w\,26-w\,27-w\,71+w\,74+w\,75-w\,77+w\,111-w\,112-\;w\,129-w\,137-w\,147-w\,152-w\,157+$
$w\,162-w\,163+w\,172$

w Ⅸ $= -w\,81+w\,107$

w Ⅺ $= w\,24-w\,29-w\,30-w\,79-w\,80+w\,81+w\,106-w\,113+w\,132+\;w\,139$

$w\ XII = -w\ 18 - w\ 19 + w\ 58 + w\ 62 - w\ 67 - w\ 68 + w\ 94 - w\ 103 - w\ 104 + w\ 145 - w\ 160 - w\ 161$

$w\ XIII = -w\ 11 + w\ 51 - w\ 52 - w\ 97 + w\ 117 - w\ 129 - w\ 138 - w\ 148 + w\ 155 + w\ 158 + w\ 159 + w\ 160 + w\ 161 + w\ 162 + w\ 163$

$w\ XIV = -w\ 27 + w\ 75 - w\ 77 - w\ 112 - w\ 152 + w\ 156 - w\ 163$

对此简化的化学反应动力学模型进行分析验证, 首先分析 0. 1% CH₄ + 1% H₂ + 1% O₂ + 97. 9% N₂ 的燃料混合气体在激波管中的点火工况[8]。采用甲烷燃烧反应的详细基元反应模型 GRI-Mech 1. 2 和本文导出的简化模型来对此工况进行数值模拟。模拟过程中, 燃料在激波管中激波反射区的点火燃烧过程可简化为一等容反应过程[9], 用常微分方程组来描述

$$\begin{cases} dY_k / dt = \dot{\omega}_k W_k / \rho & k = 1, \cdots, K \\ dT / dt = -\dfrac{1}{\rho \overline{c_v}} \sum_{k=1}^{K} (h_k - R_k T) \dot{\omega}_k W_k \end{cases} \quad (4)$$

式中 K 为组分数目, T 为反应混合气体的温度; $Y_k (k = 1, K)$ 为混合气体中第 k 种组分的质量分数; $W_k (k = 1, K)$ 为第 k 种组分的分子量; $h_k (k = 1, K)$ 为第 k 种组分的焓值; ρ 为混合气体密度; R_k 为气体常数; $\overline{c_v}$ 为混合气体的平均定容比热, $\dot{\omega}_k$ 为第 k 种组分的摩尔产率。采用 Gear 算法对式 (4) 进行数值求解, 应用详细基元反应模型 GRI-Mech 1. 2 和本文导出的简化模型计算了混合燃料在不同初始温度 T_0 下的点火延时。图 1 示出了不同初始温度 T_0 下详细基元反应模型 (图中记为 " Full mechanism ")、简化模型 (图中记为 " Reduced mechanism ") 计算出的点火延时值与实测值 (图中记为 " Experiment ") 的比较, 由于此时点火延时定义为反应开始至 H₂O 组分生成率最大时刻的时间间隔, 故记为 τ_{H_2O}。从图中可以看出, 简化模型的计算结果与详细基元模型的计算结果很接近, 与实测结果也符合较好。图 2 示出了初始温度 $T_0 = 1\ 666K$ 时详细基元模型和简化模型计算出的混合气体温度的时间变化历程比较, 从中也可以看出两组结果符合较好。

图 3 示出了 20% CH₄ + 13% O₂ + 67% N₂ 的燃料混合物在激波管中反应的工况下用详细基元模型和简化模型计算出的点火延时与实测结果[2], 图 4 示出了初始温度 $T_0 = 1428K$ 时混合气体温度变化历程的比较, 从这一工况的计算结果同样也可以看出简化模型的计算结果与详细基元模型的计算结果符合较好。

由于点火延时主要与混合气体初始温度、初始压力、油气比这三个参数有关, 因此, 为进一步检验简化模型的可靠性, 可以考虑对某一范围内的初始温度、压力和油气比各取 M 个水平值, 针对不同参数组合分别用详细反应动力学模型和简化模型来进行对比计算, 如果详细模型和简化模型的计算结果都吻合, 则表示简化模型在这些参数范围内是适用的。一般而言, 这需要进行 M^3 全组合计算, 计算量较大。为此, 本文引入均匀试验设计的思想[10], 当 $M = 5$ 时, 只需验算其中的 15 组数据即可, 如表 1 所示, 表中前三列分别表示初始压力、初始温度和油气比, 其中, 温度的 5 个水平值取为 1000, 1250, 1500, 1750, 2000K; 压力的水平值取为 0. 1, 0. 5, 1, 2, 2. 5MPa; 油气比的水平值取为 0. 5, 0. 8, 1. 0, 1. 5, 2. 0。表中组分 CH₄, O₂ 和 N₂ 的值表示摩尔百分数; τ_1 表示简化反应模型所计算出的点火延时, τ_2 则是详细模型所计算出的甲烷点火延时。从表中可以看出, 简化模型计算出点火延时与用详细反应动力学模型计算出的点火延时是基本一致的。这一结果进一步表明, 采用 "准稳态" 假设方法得出的甲烷点火燃烧的简化模型具有较高精度, 并且在一个比较宽的参数范围内适用。

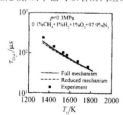

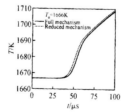

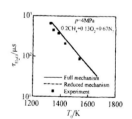

Fig 1　Comparison of experimental and calculated ignition delay of methane by full and reduced mechanism

Fig 2　Comparison of calculated mixture temperature history by full and reduced mechanism

Fig 3　Comparison of experimental and calculated ignition delay of methane by full and reduced mechanism

Table 1 Comparison of calculated ignition delay of methane by full and reduced mechanism

p /MPa	T_0 /K	ϕ	CH_4	O_2	N_2	τ1 /μs	τ2 /μs
0. 1	1000	0. 5	0. 04	0. 16	0. 8	4.6675×10^5	5.06505×10^5
0. 1	1500	2. 0	0. 1	0. 1	0. 8	2.9250×10^3	2.99180×10^3
0. 1	2000	1. 0	0. 0667	0. 1333	0. 8	47. 500	49. 0960
0. 5	1250	1. 5	0. 0857	0. 1143	0. 8	1.2250×10^4	1.25120×10^4
0. 5	1500	0. 8	0. 0571	0. 1429	0. 8	4.4000×10^2	4.63800×10^2
0. 5	1750	1. 5	0. 0857	0. 1143	0. 8	71. 000	73. 5000
1	1000	2. 0	0. 1	0. 1	0. 8	1.4200×10^5	1.48120×10^5
1	1250	0. 8	0. 0571	0. 1429	0. 8	3.7700×10^3	3.95360×10^3
1	2000	0. 5	0. 04	0. 16	0. 8	5. 0500	5. 50000
2	1250	1. 5	0. 0857	0. 1143	0. 8	3.2450×10^3	3.33610×10^3
2	1750	0. 8	0. 0571	0. 1429	0. 8	17. 000	18. 3700
2	1750	1. 0	0. 0667	0. 1333	0. 8	19. 700	20. 9490
2. 5	1000	1. 0	0. 0667	0. 1333	0. 8	3.1350×10^4	3.56300×10^4
2. 5	1500	0. 5	0. 04	0. 16	0. 8	0.9500×10^2	1.01300×10^2
2. 5	2000	2. 0	0. 1	0. 1	0. 8	4. 9500	4. 93000

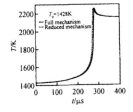

Fig 4 Comparison of calculated mixture temperature history by full and reduced mechanism

4 结 论

本文从甲烷点火燃烧的 GRI-Mech 1. 2详细化学反应动力学模型出发，采用"准稳态"假设方法来对其进行简化，得到了包含 18组分 14步总包反应的简化反应模型，通过典型试验结果对比和基于均匀试验设计的校核计算结果比较可以看出，采用"准稳态"假设方法得到的简化反应动力学模型能有效地再现详细基元模型的反应机理，具有较高精度，是可靠而有效的，能够很好地应用于实际工作中。与原来的详细反应动力学模型相比，简化模型有效地减少了反应组分，耦合到流场计算中时有望大大减少求解变量和求解微分方程数目，从而有效地提高计算效率。

参考文献:

[1] 刘 陵. 超音速燃烧与超音速燃烧冲压发动机 [M]. 西安: 西北工业大学出版社, 1993.

[2] Peterson E L, Davidson D F, Hanson R K. Kinetics modeling of shock-induced ignition in low-dilution CH_4 / O_2 mixtures at high pressures and intermediate temperatures [J]. Combustion and Flame, 1999, 117: 272 ~ 290.

[3] Seiser H, Pitsch H, Seshadri K, et al. Extinction and autoignition of n-heptane in counterflow configuration [C]. Proceedings of the Combustion Institute, 2000, 28.

[4] Turanyi T, Berces T, Vajda S. Reaction rate analysis of complex kinetic system [J]. Int. J. of Chemical Kinetics, 1989, 21: 83 ~ 99.

[5] Lam S H. Using CSP to understand complex chemical kinetics [J]. Combust Sci and Tech, 1993, 89.

[6] Warnatz J, Maas U, Dibble R W. Combustion [M]. Germany: Springer-Verlag, 1999.

[7] 钱炜祺, 乐嘉陵, 肖保国. 复杂化学反应动力学模型的简化方法 [C]. 第十一届全国激波与激波管学术会议论文集, 2004: 348 ~ 354.

[8] Hidaka Y, Sato K, Henmi Y, et al. Shock-tube and modeling study of methane pyrolysis and oxidation [J]. Combustion and Flame, 1999, 118: 340 ~ 358.

[9] Cribb P H, Dove J E, Yamazaki S. A kinetic study of methanol using shock tube and computer simulation techniques [J]. Combustion and Flame, 1992, 88: 169 ~ 185.

[10] 方开泰, 马长兴. 正交与均匀试验设计 [M]. 北京: 科学出版社, 2001.

一体化高超声速飞行器气动- 推进性能评估

贺元元, 倪鸿礼, 乐嘉陵

（中国空气动力研究与发展中心，四川 绵阳　621000）

摘要：吸气式高超声速飞行器的一个重要特点就是机体和推进系统的高度一体化设计。在这类高超声速飞行器的发展中，机体- 推进系统内外流场相互干扰的评估以及飞行器气动- 推进性能的研究是非常重要的。文中阐述了 CFD 和风洞试验结合评估一体化飞行器气动- 推进性能的近似方法，涉及一体化飞行器进气道和发动机的三个工作状态：进气道关闭、进气道打开发动机不工作以及进气道打开发动机工作。针对进气道关闭的工作状态，大量气动数据可由试验获得。但是，受模型尺寸和设备的限制，试验模拟进气道打开发动机不工作特别是进气道打开发动机工作的飞行状态是非常困难的。因此，首先根据进气道关闭和进气道打开发动机不工作两种情况下风洞试验数据与 CFD 计算结果的对比得到计算误差，在此基础上，结合内外流数值模拟，预测不同进气道和发动机工作状态下一体化飞行器的气动- 推进性能。

关键词：一体化高超声速飞行器；气动- 推进性能；CFD；试验数据；计算误差

中图分类号：V211.7；V511$^+$.1　　**文献标识码**：A

Evaluation of aero-propulsive performance for integrated hypersonic vehicle

HE Yuan-yuan, NI Hong-li, LE Jia-ling

（China Aerodynamics Research & Development Center, Mianyang Sichuan　621000, China）

Abstract：Hypersonic airbreathing vehicle configurations are characterized by highly integrated propulsion flowpath and airframe systems. The assessment of propulsion-airframe flow field interactions and the integrated aero-propulsive performance of candidate systems are very important in the development of this class of hypersonic vehicle. This paper describes the approximate methods to obtain pre-flight predictions of longitudinal performance from CFD and experimental data. Three mission points are analyzed: cowl-closed, cowl-open unpowered, and cowl-open powered. For the cowl-closed configuration, a large amount of wind tunnel data can be obtained. Because of model scale and facility limitations, it is difficult to simulate the complete internal and external flow fields when the cowl-open unpowered or the cowl-open powered. Therefore, in this paper, firstly, computational errors are determined by the comparison of numerical results with experimental data for the cowl closed model and cowl open but unpowered model. These errors are then applied to develop predictions for longitudinal performance in each of the three mission phases.

Key words：integrated hypersonic vehicle; aero-propulsive performance; CFD; experimental data; computational errors

贺元元, 倪鸿礼, 乐嘉陵. 一体化高超声速飞行器气动-推进性能评估. 实验流体力学, 2007, 21(2): 63-67.

0 引 言

发展吸气式高超声速技术是实现可持续高超声速飞行(尤其是在大气层以内)的重要途径。吸气式高超声速技术的核心是超燃冲压发动机技术和机体/推进一体化飞行器技术。吸气式高超声速飞行器各部件必须经过精确设计以确保获得最优的气动-推进性能,由于这类飞行器推力有限,为了有效减小阻力,获得尽可能高的推力(有效比冲),必须采用机体系统和推进系统的高度一体化设计。经过一体化设计的高超声速飞行器,其前体作为进气道的预压缩面,气流经前体初步压缩后由进气道流入发动机,后体是使气流进一步膨胀的"外喷管",经发动机燃烧后的高温燃气流作用在飞行器后体下表面上,进一步产生推力并对飞行器气动性能产生重大影响。

采用一体化设计的高超声速飞行器各部件紧密关联,相互影响。由于机体和发动机共用前体和后体尾喷管,前体和尾喷管设计不仅影响了飞行器所能获得的气动性能,同时也影响了推进系统的推进性能。内外流的相互干扰和强耦合也是一体化飞行器的重要特点,机体下表面的外流一部分进入发动机成为内流,从燃烧室排出的高温气流在飞行器后体进一步膨胀并与外流相互作用。因此,发展一体化高超声速飞行器的最大挑战在于准确评价机体和推进系统的相互影响,确定飞行器的气动-推进性能。

以超燃冲压发动机为动力的一体化飞行器在实际飞行过程中需经历发动机和进气道的不同工作状态。例如地面发射的吸气式飞行器,飞行器首先将由助推火箭推动达到动力试验的高度和马赫数条件,在这期间,为了保护发动机,降低热载荷,一体化飞行器的进气道是关闭的;助推火箭与飞行器分离后,飞行器的进气道打开,外界气流流入发动机,此时,发动机还没有开始工作,需经历几秒钟流动建立过程;接着,发动机点火,飞行器将由自身动力推进进入飞行试验状态。因此,吸气式高超声速飞行器在实际飞行中经历了进气道-发动机的三种组合工作状态:(1) 进气道关闭;(2) 进气道打开,发动机不工作;(3) 进气道打开,发动机工作。为了全面考核一体化飞行器在实际飞行过程中的气动-推进性能,必须对上述三种飞行状态进行深入细致研究。针对进气道关闭的工作状态,大量气动数据可由试验获得。但是,受模型尺寸和设备的限制,试验模拟进气道打开发动机不工作特别是进气道打开发动机工作的飞行状态是非常困

难的。因此,首先根据进气道关闭和进气道打开发动机不工作两种情况下风洞试验数据与CFD计算结果的对比得到计算误差,在此基础上,结合内外流数值模拟,预测不同进气道和发动机工作状态下一体化飞行器的气动-推进性能,描述了CFD和风洞试验相结合评估一体化飞行器气动-推进性能的近似方法和过程。

1 数值研究方法

在研究过程中,涉及到飞行器处在发动机工作和不工作两种状态,数值模拟将分别求解两类流动控制方程:(1) 发动机不工作时,采用理想气体的可压缩 Navier Stokes 方程组;(2) 发动机工作时,采用可压缩化学反应 N-S 方程组。湍流模型采用 k-ω 双方程湍流模型。另外,为了提高计算效率,采用了基于 MPI 的大规模并行算法。网格采用多区域结构网格结合重叠网格。具体数值方法参见文献[1]。

2 一体化飞行器气动-推进性能评估

将一体化飞行器按照机体和推进系统进行部件划分,如图1,其中机体系统包括:前体、进气道外压缩部分、机翼、尾翼、飞行器上表面和侧面、发动机外罩,推进系统包括发动机内流道和尾喷管。参考美国 X-43A 高超声速飞行器性能评价方法[2-4],形成了根据试验和CFD评估一体化飞行器升阻、推阻特性的分析方法,包括通过对无动力一体化飞行器(包括进气道关闭和进气道打开发动机不工作两种情况)的三维数值模拟结合风洞试验评估一体化飞行器在无动力条件下的气动性能,以及将发动机工作条件下的一体化飞行器二维数值计算结果应用到三维一体化飞行器气动-推进性能评估的分析方法。

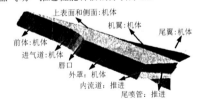

上表面和侧面:机体
机翼:机体
尾翼:机体
前体:机体
进气道:机体
唇口
外罩:机体
内流道:推进
尾喷管:推进

图1 气动力及推进力的划分体系

Fig 1 Aerodynamic and propulsion force accounting system

步骤(1)-(6)说明了通过对无动力一体化飞行器(包括进气道关闭和进气道打开发动机不工作两种情况)的三维数值模拟结合风洞试验评估一体化飞行器

在无动力条件下气动性能的分析方法:

(1) 进气道关闭模型计算、试验,确定计算误差

$$\Delta X_{\text{CFD Aero}}^{\text{Close}} = X_{\text{Tunnel}}^{\text{Close}} - X_{\text{CFD}}^{\text{Close}}$$

(2) 通气模型计算,得到机体系统气动性能 $X_{\text{CFD Aero}}^{\text{Open}}$

(3) (1)、(2)项综合,预测试验条件下通气模型机体系统气动性能

$$X_{\text{Aero}}^{\text{Open}} = X_{\text{CFD Aero}}^{\text{Open}} + \Delta X_{\text{CFD Aero}}^{\text{Close}}$$

(4) 通气模型试验结果与(3)项比较,得到推进系统气动性能

$$X_{\text{Prop}}^{\text{Open}} = X_{\text{Tunnel}}^{\text{Open}} - X_{\text{Aero}}^{\text{Open}}$$

(5) 通气模型计算中得到的推进系统气动性能与(4)项比较,得到推进系统计算误差

$$\Delta X_{\text{CFD Prop}}^{\text{Open}} = X_{\text{Prop}}^{\text{Open}} - X_{\text{CFD Prop}}^{\text{Open}}$$

(6) 一体化飞行器在进气道打开发动机不工作条件下的气动性能由数值计算结果与机体系统计算误差以及推进系统计算误差叠加得到

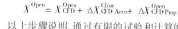

$$X^{\text{Open}} = X_{\text{CFD}}^{\text{Open}} + \Delta X_{\text{CFD Aero}}^{\text{Close}} + \Delta X_{\text{CFD Prop}}^{\text{Open}}$$

以上步骤说明,通过有限的试验和计算的对比得到的机体系统和推进系统的计算误差,使得今后大量采用 CFD 预测一体化飞行器气动性能成为可能。

图 2~6 是根据上述步骤得到的针对 CARDC 一体化高超声速飞行器在无动力条件下的气动性能分析结果。试验在 CARDC Φ1m 高超声速风洞中进行,模拟来流马赫数 $M_\infty = 6$。图 2 给出了计算得到的进气道关闭模型升阻力系数与试验结果的比较以及计算与试验的误差,可以看出,数值模拟得到的升力系数与试验结果非常吻合,阻力系数基本吻合,计算得到的阻力系数略低于试验结果,其误差约为 0.03。图 3 是根据通气模型计算结果,积分机体系统表面压力和摩擦力得到的通气模型机体系统计算结果以及修正结果,进气道关闭模型计算和试验表明,数值模拟能够较为准确地给出升力特性,因此对升力系数的修正量很小,修正主要表现在阻力特性上。图 4 是通

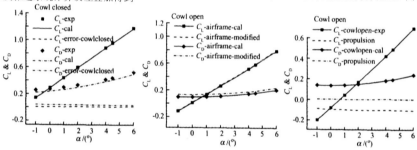

图 2 进气道关闭模型计算和试验的误差
Fig. 2 Error between computation and experiment of cowl closed model

图 3 通气模型机体系统气动性能
Fig.3 Aerodynamic performance of airframe system for cowl open model

图 4 通气模型推进系统气动性能
Fig. 4 Aerodynamic performance of propulsion system for cowl open model

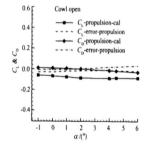

图 5 通气模型推进系统计算和试验的误差
Fig.5 Error between computation and experiment of propulsion system for cowl open model

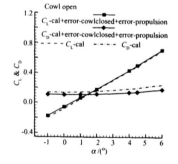

图 6 通气模型气动性能
Fig. 6 Aerodynamic performance of cowl open model

气模型风洞试验结果减去图3的通气模型机体系统气动性能修正结果, 得到的推进系统气动性能。由此通过对比通气模型推进系统计算结果和试验结果, 可得到无动力一体化飞行器通气模型推进系统计算误差, 如图5。将通气模型三维计算结果与机体系统计算误差以及推进系统计算误差叠加, 可得到一体化飞行器在无动力(进气道打开)条件下的气动性能, 如图6。

一体化飞行器在发动机工作条件下的气动–推进性能研究是一体化研究的重点和难点, 受客观条件的限制, 发动机工作的一体化飞行器性能研究无论是试验还是数值模拟都存在很大的挑战。以下在对无动力一体化飞行器进行大量数值计算和试验模拟的基础上, 结合所研究的一体化飞行器具有较为明显的二元特性, 从相对比较容易得到的二维一体化飞行器热流计算结果入手, 通过一系列比较、综合, 得到了一体化飞行器在发动机工作条件下的气动–推进性能, 主要概括为以下四步:

(1) 无动力(进气道打开)一体化飞行器计算和试验, 得到计算误差

$$\Delta X_{CFD}^{Power\ off} = X_{Tunnel}^{Power\ off} - X_{CFD}^{Power\ off}$$

(2) 无动力(进气道打开)一体化飞行器二维和三维流场数值模拟, 确定二维计算误差

$$\Delta X_{2DCFD}^{Power\ off} = X_{3DCFD}^{Power\ off} - X_{2DCFD}^{power\ off}$$

(3) 发动机工作条件下一体化飞行器二维内外流场模拟, 得到二维计算结果 $X_{2DCFD}^{Power\ on}$

(4) 将(3)得到的计算结果与(1)、(2)的误差叠加, 评估带动力一体化飞行器的气动–推进性能。

图7给出了针对无动力(通气)一体化飞行器, 分别采用二维数值模拟、三维数值模拟以及试验研究得到的升阻力系数及力矩系数, 可以看出, 升力和力矩

特性, 三维计算结果非常接近于试验测量值, 阻力性能, 二维计算结果与试验更为吻合, 在进气道打开, 无动力条件下, 飞行器的升阻比可达2.5。图8给出了超燃发动机试验条件下($M\infty= 6$, $\alpha= 4.5^\circ$), 通过二维流场数值模拟得到的流场马赫数和O_2质量分数分布, 可以看出飞行器前缘激波基本打在进气道唇口附近, 流量捕获较好; 从O_2质量分数分布云图发现主流区域氧消耗量较壁面附近少, 在近壁面附近燃烧充分。根据前面得到的计算和试验误差以及二维和三维计算误差, 可预测一体化飞行器在发动机工作时的气动–推进性能。图9是一体化飞行器在进气道和发动机不同工作状态下的升力、阻力(推力)以及力矩性能, 由于封堵面的存在, 进气道关闭模型的升阻力较大, 进气道打开, 封堵面上的压力消失, 飞行器的升阻力下降, 发动机工作, 燃气流作用在后体尾喷管上, 使得飞行器升力提高, 同时发动机和尾喷管产生的推力不仅克服了飞行器阻力, 还有一部分富余使得飞行器在巡航状态下获得加速。另外, 图9还给出了在发动机工作条件下, 通过二维发动机内流计算结合三维飞行器外流计算(包括尾喷管喷流模拟)得到的飞行器升阻力系数和力矩系数, 可以看出, 采用上述方法获得的升力和力矩特性与三维计算结果较为吻合。图10是一体化飞行器流场压力分布, 进气道唇口和尾喷管压力较高, 图中可以看到清晰的喷流边界。

3 结 论

内外流的相互干扰和强耦合是一体化飞行器的重要特点, 发展一体化高超声速飞行器的最大挑战在于准确评价机体和推进系统的相互影响, 确定飞行器的气动–推进性能。通过大量的数值模拟和试验, 阐述了在有限的试验资源条件下, 采用CFD预测机体/

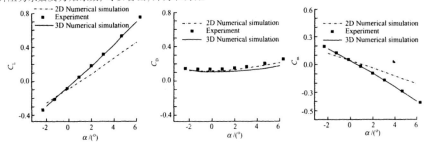

图7 无动力一体化高超声速飞行器二维、三维计算结果及试验结果

Fig.7 Results of 2D (or 3D) numerical simulation and experiments for unpowered integrative vehicle

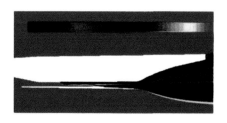 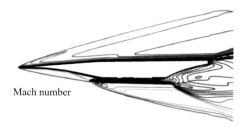

Mach number

图 8　带动力一体化高超声速飞行器数值模拟流场

Fig. 8　Flow fields of powered integrative hypersonic vehicle

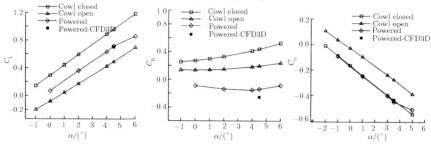

图 9　机体/推进一体化高超声速飞行器纵向气动性能

Fig. 9　Longitudinal performance of airframe/ propulsion integrated hypersonic vehicle

图 10　带动力一体化飞行器壁面压力分布和尾部喷流
流场

**Fig. 10　Wall pressure distribution and exhaust flow fields
of powered integrated vehicle**

推进一体化飞行器在进气道和发动机不同工作状态
下气动- 推进性能的方法，并进行了实际运用，介绍
的方法和获取的计算误差为一体化飞行器的性能评
估奠定了基础。

参考文献:

[1] 贺元元. 机体/推进一体化高超声速飞行器数值研究
[D]. 中国空气动力研究发展中心博士学位论文, 2004.

[2] ABDELKADER Frendi. On the CFD support for the Hyper X
aerodynamic database[R]. AIAA 99 0885.

[3] CHARLES E Cockrell, et al. Integrated aero propulsive CFD
methodology for the Hyper X flight experiment [R]. AIAA
2000 4010.

[4] WALTER C Engelund, et al. Propulsion system airframe integration issues and aerodynamic database development for the
Hyper X flight research vehicle[R]. ISOABE 99 7215.

超声速气流中液体燃料雾化数值模拟*

杨顺华，乐嘉陵

(中国空气动力研究与发展中心，四川 绵阳 621000)

摘　要： 为了模拟超声速气流中液体射流的雾化，发展了一种液滴破碎混合模型。该模型将 Kelvin-Helmhotz (K-H) 模型和 Rayleigh-Taylor(R-T) 波动模型耦合在一起，首先计算 R-T 波增长模型，当 R-T 模型不能导致液滴破碎时，然后计算 K-H 波增长模型。利用该模型对来流 $Ma = 1.94$ 的超声速气流中的水射流的雾化进行了数值模拟。模拟结果再现了超声速气流中射流的雾化结构，计算得到的射流穿透深度、颗粒直径分布和速度分布与实验结果吻合较好。

关键词： 超燃冲压发动机；雾化；超声速；数值仿真

中图分类号： O359.1; V235.213　　**文献标识码：** A　　**文章编号：** 1001-4055 (2008) 05-0519-04

Numerical simulation of liquid fuel atomization in supersonic crossflow

YANG Shun-hua，LE Jia-ling

(China Aerodynamics Research and Development Center, Mianyang 621000, China)

Abstract　In order to simulate the liquid fuel atomization in supersonic crossflow, a hybrid spray atomization model was developed. The model combines two different Kelvin-Helmhotz (K-H) and Rayleigh-Taylor (R-T) wave models. The calculations of the R-T model are performed firstly. If the R-T model does not result in a breakup, the droplet breakup is then simulated using the K-H model. Using the theory model, a water jet atomized in a supersonic crossflow with Mach number 1.94 was simulated. Numerical results regenerate the spray structure of jet atomized in supersonic crossflow, and give good agreement with experimental data for spray penetration, droplet size and velocity.

Key words　Scramjet; Atomization; Supersonic flow; Numerical simulation

1 引言

超燃冲压发动机的燃烧室是整个超燃冲压发动机研究的关键所在，有效地缩短燃料点火延迟时间，提高燃烧性能，对于发动机设计具有及其重要的意义。在脉冲试验设备上，对于 1m 量级的超燃冲压发动机，测量得到的燃料总的点火延迟时间 t_{dt} 大约为 4 ~ 5ms，其包括三个部分

$$t_{dt} = t_{bk} + t_{ev} + t_{ch} \qquad (1)$$

式中 t_{bk} 为液滴破碎时间，t_{ev} 为液滴蒸发时间，t_{ch} 为燃料化学延迟时间。t_{ch} 主要取决于燃料自身的化学特性，一旦燃料选定，t_{ch} 也就确定了。t_{bk} 和 t_{ev} 主要取决于燃料的雾化效果，因此燃料的雾化对发动机的性能有重大影响。目前对 t_{bk} 和 t_{ev} 两个时间特性我们仍然

不够清楚。由于试验测量手段的限制，对这两个时间的测量比较困难，因此，有必要发展一套考虑燃料雾化、蒸发和燃烧的数值模拟体系，对这两个时间特性进行分析。

Wu PK 等人对亚声速横向气流中液体射流的雾化做了大量研究[1,2]，对比了不同射流 /来流动量比、喷射角度、来流马赫数、不同液体物性 (粘性系数、表面张力、密度等) 等条件对射流破碎过程的影响。Lin KC 等人对超声速横向气流中射流雾化做了系统的实验研究[3-5]，研究了不同液体物性、喷嘴尺寸、气液动压比、气液流量比、喷射角度等因素对雾化效果的影响。采用阴影照相对喷雾结构进行了显示，实验同时测量了喷雾的穿透深度，总结给出了穿透深度的经验表达式。目前对超声速横向气流中液体射流雾化

杨顺华, 乐嘉陵. 超声速气流中液体燃料雾化数值模拟. 推进技术, 2008, 29(5): 519-522.

的数值模拟比较少，In Kyoung Su, Lin K C 和 Lai Ming Chia[6]等人提出了一种混合破碎模型，他们的模型比经典的 TAB 模型给出的颗粒直径和速度分布更加接近实验测量值。但他们的模型没有给出与实验一致的颗粒直径分布和速度分布规律。

目前，数值模拟采用的喷雾模型主要有 TAB 模型[7]、Reitz 的波增长模型[8]以及在这些模型基础上发展起来的一些修正模型。其中 TAB 模型主要应用于内燃机中的喷雾模拟，著名的内燃机模拟软件 KIVA 就采用该模型。Reitz 的波增长模型考虑气动力引起的液滴表面波的增长，其是目前广泛应用的喷雾模型，已经成功应用于内燃机、火箭发动机燃烧室的喷雾模拟。上述这些模型大都是在低速条件下发展起来的理论模型，在将这些模型应用到高速气流中射流的雾化过程时不能取得满意的结果[6]。为了模拟超燃冲压发动机燃烧室内的燃料雾化过程，评估燃料雾化特征时间，必须建立可靠的喷雾模型，本文正是在这样的背景需求下，在总结现有理论模型的基础上，发展了一种用于模拟超声速气流中液体燃料射流雾化的混合模型，数值结果表明，理论模型能够很好的模拟超声速气流中液体射流的雾化特征。

2 雾化模型

在超燃冲压发动机的应用中，燃料往往是通过壁面或者支板注入燃烧室，并在超声速亚声速横向气流中经历雾化。对于横向气流中射流的雾化，液体射流首先经历变形，然后发生表面破碎，小的"子液滴"从"父液滴"表面脱落。这些"子液滴"随后经历液滴之间的碰撞和聚合，以及二次破碎。当这些液滴随气流一起运动时，还会与周围气体进行质量、动量和热量交换。因此，一个合理的雾化模型应当能够用来预测液滴的形成过程、喷雾结构以及液滴尺寸和速度。

近年来，许多学者对横向气流中射流的雾化理论进行了广泛的研究。基于线性稳定性分析，Reitz 提出了波动模型（1987）[8]，该模型已经成功应用于内燃机的数值模拟。Patterson 和 Reitz（1998）[9]提出了一种混合破碎模型，提高了喷雾模拟精度。在近喷嘴的稠密喷雾区域，Reitz 提出射流可以通过注入"块状液滴"来模拟，这些"块状液滴"的尺寸与喷嘴出口直径相当。

目前，我们对 Patterson 和 Reitz 的混合模型进行了改进，并成功应用于超声速气流中的液体射流雾化的模拟。

我们的雾化模型将 Kelvin-Helmholtz 和 Rayleigh-Taylor 波增长模型耦合在一起。假设液滴在超声速气流中的破碎过程同时受 Kelvin-Helmholtz 和 Rayleigh-Taylor 两种不稳定波的共同作用[6]。破碎开始发生通过破碎长度 l_{bu} 来决定，当液滴离开喷嘴的距离小于这个破碎长度时，认为液滴不发生破碎。l_{bu} 是一个经验关系式，可以通过实验给出。如果液滴离开喷嘴的距离大于这个破碎长度，目前的模型就用来计算液滴的破碎。首先采用 R-T 模型来判断液滴是否破碎，当液滴破碎的特征时间大于 R-T 模型的破碎时间尺度时，液滴发生 R-T 类型的破碎；如果 R-T 模型不能导致液滴破碎，然后采用 K-H 模型来决定液滴的破碎，当液滴的 Weber 数大于 K-H 模型破碎所需的临界 Weber 数时，液滴发生破碎。

在 K-H 模型中，仅允许当液滴的 Weber 数大于临界 Weber 数 We_c 时才发生破碎，液滴 Weber 数定义为 $We_1 = \rho_l U_r^2 a / \sigma$，式中 ρ_l 为液体密度，U_r 为液滴与气体间的相当运动速度，a 为液滴半径，σ 为液滴表面张力系数。由半径为 a 的"父液滴"破碎后产生的"子液滴"半径 r 可以通过下式得到

$$r = B_0 \Lambda_{KH} \quad (2)$$

式中 B_0 是一个经验常数，Λ_{KH} 是最不稳定 K-H 波的波长，可以通过耗散关系式得到

$$\frac{\Lambda_{KH}}{a} = 9.02 \frac{(1 + 0.45 Z^{0.5})(1 + 0.4 T^{0.7})}{(1 + 0.87 We_g^{1.67})^{0.6}} \quad (3)$$

式中 $Z = \sqrt{We_l} / Re_1$ 为 Ohnesorge 数，$We_g = \rho_g U_r^2 a / \sigma$ 为气体 Weber 数，$T = Z \sqrt{We_g}$ 为 Taybr 数。

在破碎过程中，由于"子液滴"从"父液滴"表面发生剥落，"父液滴"的直径会减小，这种"父液滴"半径的变化可以通过半径衰减方程给出

$$\frac{da}{dt} = -(a - r) / \tau_{KH} \quad (r \leq a) \quad (4)$$

式中 $\tau_{KH} = 3.726 B_1 a / \Lambda_{KH} \Omega_{KH}$ 是破碎的特征时间，$B_1 = 1.732 \sim 40$，最不稳定 K-H 波的增长速率 Ω_{KH} 为

$$\Omega_{KH} \left[\frac{\rho_l a^3}{\sigma} \right]^{0.5} = \frac{0.34 + 0.38 We_g^{1.5}}{(1 + Z)(1 + 1.4 T^{0.6})} \quad (5)$$

R-T 破碎模型同样通过最不稳定波的增长来确定液滴什么时候以及怎样破碎。增长最快的 R-T 波的波长 Λ_{RT} 为

$$\Lambda_{RT} = 2\pi C_1 \sqrt{\frac{3\sigma}{a_p(\rho_l - \rho_g)}} \quad (6)$$

式中 a_p 为液滴加速度，C_1 是依赖于喷嘴条件的可调参数。对应的破碎时间尺度 τ_{RT} 可以通过增长最快的

R-T波的频率来计算

$$\tau_{RT} = C_2 \sqrt{\frac{\delta^{0.5}(\rho_l + \rho_g)}{2}} \left[\frac{3}{a_p(\rho_l - \rho_g)}\right]^{1.5} \quad (7)$$

由 R-T 破碎新产生的"子液滴"半径为

$$r = 0.5\Lambda_{RT} \quad (8)$$

在计算过程中,跟踪"父液滴"质量的减少。当累计脱落质量达到初始液滴质量的 3% 时,我们就允许产生新的"子液滴"。

与 Patterson 和 Reitz 的混合模型 (简称 P-R 模型,下同)相比,我们的模型在以下几个方面进行了改进:(1) P-R 模型中液滴破碎长度是基于 K-H 不稳定波得到的,我们模型的破碎长度是基于实验数据的经验关系式;(2) P-R 模型在破碎长度内采用 K-H 模型来决定射流的破碎,我们认为在破碎长度内射流不发生 K-H 失稳破碎;(3)对 P-R 模型中的有关参数进行了修正。对超声速射流破碎,式(7)中 $C_2 = 1.5$ 区别于 P-R 模型中的 1.0。

3 数值方法

上述雾化模型已经被耦合到并行 CFD 流场软件平台—AHL3D 中[10,11]。AHL3D 中控制方程采用格心有限体积法离散,无粘项离散采用多种数值格式,包括 Steger-Warming 矢通量分裂、AUSM、AUSMPW、LDFSS。网格界面左右两侧的守恒变量通过 MUSCL 插值得到,为减小截断误差,MUSCL 参数取 1/3。粘性项计算采用散度定律在偏置网格上得到。时间推进采用隐式 LU-SGS 方法。非定常计算采用双时间步方法。

对于喷雾计算,采用离散液滴模型。每一个计算液滴代表具有相同尺寸、位置和速度的大量液滴。颗粒相控制方程采用龙格-库塔方法求解。

为了确定液滴位于哪个计算网格,我们采用了一种新的液滴跟踪技术,这种技术用于确定液滴位于哪个网格内是非常有效和鲁棒的。这种方法判断液滴轨迹与网格面的是否相交,如果液滴轨迹穿过该网格面,那与该面相邻的网格就为搜索的下一网格。

4 模型应用

采用上述模型对来流马赫数 1.94 的超声速气流中水射流的雾化进行了数值模拟。来流气流总压和总温分别为 206 kPa 和 533 K。一共计算了 4 组不同的射流/空气动压比 $q_0 = \rho_l V_j^2 / \rho_\infty V_\infty^2$,分别对应 3、7、10、15。

图 1 给出了数值仿真得到的喷雾结构,其中 ■ 代表了计算得到的水液滴的分布;实线代表 Lin K C 等人在超声速风洞中测量得到的射流穿透深度[3-5](Experiment 下同);虚线为 Kyoung-Su In 等人采用他们提出的混合模型计算得到的射流穿透深度[6]。可以看出:Kyoung-Su In 等人的雾化模型低估了射流的穿透深度,我们的模型在较大范围内与实际的射流穿透深度吻合较好,在射流发生偏折的区域,我们的模型高估了射流的穿透深度。

图 2 对比了 $q_0 = 7$ 条件下距离喷嘴下游 50 mm 横截面上的水液滴 Sauter 平均直径 (SMD)和流向速度分布。我们的模型很好地预测了液滴的平均直径和速度分布规律,而 Kyoung-Su In 等人的模型未能正确给出液滴的分布规律。同时,我们模型预测的颗粒平均直径大小和速度大小更加接近于实验测量结果。

图 3、图 4 为不同 q_0 条件下不同横截面上的水液滴平均直径和速度分布与实验值的比较。从这些图中可以看出,我们模型给出的颗粒直径分布和速度分布与实验结果吻合较好。预测的颗粒速度分布要优于颗粒平均直径分布。

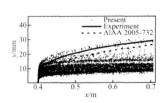

Fig. 1　Comparison of spray penetration height at the center plane ($q_0 = 7$)

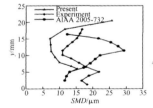

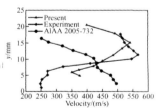

Fig. 2　Comparison of SMD (up) and velocity (down) of droplet along y-direction at $x = 50$ mm from the injector location ($q_0 = 7$)

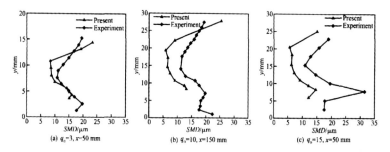

Fig. 3　Comparison of droplet SMD with experimental results along y-direction at different location

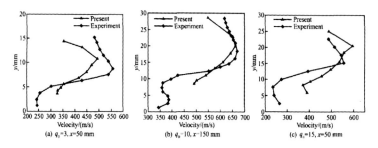

Fig. 4　Comparison of droplet velocity with experimental results along y-direction at different location

5　结　论

　　本文基于 P-R 混合模型, 建立了适用于超声速气流中射流雾化的物理模型, 并将其耦合到三维并行 CFD 程序中。将此模型应用于超声速气流中水射流的雾化数值模拟, 模拟结果给出的水射流破碎后的喷雾结构、液滴平均尺寸和速度分布与实验吻合较好, 计算结果表明所用的雾化模型是可靠的。此项工作为今后深入研究超燃冲压发动机中的燃料雾化特性奠定了基础。下一步的工作是对超燃冲压发动机燃烧室的燃料雾化、燃烧全过程进行数值仿真。

参考文献:

[1]　Wu P K, Kirkendall K A, Fuller R. Breakup processes of liquid jets in subsonic crossflows[R]. AIAA 96-3024.

[2]　Fuller R, Wu P K, Kirkendall K A. Effects of injection angle on the breakup processes of liquid jets in subsonic crossflows[R]. AIAA 97-2966.

[3]　Lin K C. Structures of water jets in a Mach 1.94 supersonic crossflow[R]. AIAA 2004-971.

[4]　Lin K C, Kennedy P J. Spray penetration heights of angled-injected aerated liquid jets in supersonic crossflows

[5]　Lin K C, Kennedy P J. Penetration heights of liquid jets in high speed crossflows[R]. AIAA 2002-0873.

[6]　Kyoung Su In, Lin K C, Ming Chia Lai. Spray atomization of liquid jet in supersonic cross flows[R]. AIAA 2005-732.

[7]　Amsden A A, O'Rourke P J, Butler T D. KIVA-II A computer program for chemically reactive flows with sprays [R]. *Los Alamos Scientific Laboratory Report LA*-11560-*MS* 1989.

[8]　Kuo K K. Recent advances in spray combustion[M]. *Progress in Astronautics and Aeronautics*, 1996.

[9]　Patterson M A, Reitz R D. Modeling the effects of fuel spray characteristics on diesel engine combustion and emission[R]. *SAE Paper* 98031, 1998.

[10]　YANG Shunhua, LE Jialing. Computational analysis of kerosene fueled scramjet [C]. *ISABE* 2005-1195 *Munich, Germany*, 2005.

[11]　LE Jialing, YANG Shunhua, LIU Weixiong et al. Massively parallel simulations of kerosene fueled scramjet [R]. AIAA 2005-3318.

[R]. *AIAA* 2000-0194.

火焰面模型在超燃冲压发动机数值模拟中的应用

邢建文, 乐嘉陵

(中国空气动力研究与发展中心吸气式高超声速技术实验室, 四川 绵阳 621000)

摘要: 采用 k-ω 两方程湍流模型、火焰面模型和质量加权平均的 Navier Stokes 方程组解耦求解的方法, 模拟了 DLR 氢燃料直连式超燃冲压发动机湍流燃烧流场, 并与化学反应源项直接采用 Arrhenius 公式模拟及实验得到的结果进行对比, 可以发现湍流脉动对化学反应的影响不能忽略。

关键词: 火焰面模型; 湍流燃烧; 超燃冲压发动机; 两方程湍流模型; 有限速率反应

中图分类号: V235.213   **文献标识码**: A

Application of flamelet model for the numerical simulation of scramjet

XING Jian-wen, LE Jia-ling

(China Aerodynamics Research and Development Center, Mianyang Sichuan   621000, China)

Abstract: A hydrogen fueled scramjet of DLR was numerically simulated by the combination of a k-ω two equation turbulence model, the flamelet model and Navier-Stokes equations. The numerical results are compared with experimental data and finite rate reacting model. Computational results show that the interaction of turbulence and combustion in scramjet can't be ignored. Applying flamelet to simulate turbulent combustion in scramjet is a good choice.

Key words: flamelet model; turbulent combustion; scramjet; two equation turbulence model; finite rate reacting

0 引 言

在亚声速湍流燃烧中湍流脉动的影响不可忽略; 在超声速流场中湍流脉动更加剧烈, 反应速率受湍流脉动控制, 湍流脉动的影响更是不可忽略。在过去的 10 多年间, 由于高超声速吸气推进研究的再一次兴起, 一些研究者将注意力转向高速湍流燃烧。实验数据表明在超声速燃烧室组分浓度和温度脉动可分别达到 40% 和 20%[1]。因此忽略这些脉动对平均反应速率的影响会导致对流场的错误描述, 进而可能错误地预测流场中的主要组分分布。甚至, 湍流燃烧相互作用可能会影响一些设计参数, 如: 燃料的加注方式、火焰稳定、点火延迟、燃烧效率等[2]。因此应该采用湍流燃烧模型来模拟超声速湍流燃烧。目前, 应用于超声速湍流燃烧比较有代表性的湍流燃烧模型有概率函数密度方法(PDF methods) 和火焰面模型(flamelet models) 。

由于在概率函数密度方法中, 与湍流输运和化学反应速率有关的项都以封闭的形式出现, 可以精确计算, 从而避免了对一些重要过程的模拟; 同时, 还可以提供比通常的湍流模型更多的信息, 应该说 PDF 方法是解决有限反应速率和污染物生成等诸类湍流燃烧问题的最合适和最理想的方法, 但联合概率密度函数求解的复杂性和计算量之大给其在工程中的广泛应用带来了很大的困难。火焰面模型将湍流火焰看作嵌入湍流流场内的局部具有一维结构的薄的层流火焰的一个系综, 湍流场中平均火焰结构由层流火焰面系综作统计平均得到。在燃烧模型中既考虑混合过程的影响, 同时也考虑有限化学反应速率的影响, 能用于详细反应机理, 而计算量大大小于 PDF 方法。目前能用于工程应用的仅有火焰面模型。

采用并行软件 AHL3D+ 火焰面模型模拟了 DLR 氢燃料直连式超燃冲压发动机, 并与目前常用的方法一直接采用有限速率反应模型(Arrhenius 公式) 模拟得到的结果及试验结果进行对比。

邢建文, 乐嘉陵. 火焰面模型在超燃冲压发动机数值模拟中的应用. 实验流体力学, 2008, 22(2): 40-45.

1　物理模型和控制方程

　　湍流燃烧的火焰面模型在湍流非预混燃烧和湍流预混燃烧中的具体形式不同,该文所有计算只涉及湍流非预混燃烧的火焰面模型。

1.1　火焰面模型

　　在低马赫数流动中,在一定的假设下[3],当地的火焰在混合分数空间中呈现一维结构,这就使得随时间变化的化学反应计算能与流动计算分开。这就是火焰面模型的实质——将复杂化学反应动力学和湍流运动之间强烈的耦合关系解耦。为了获得这个一维结构的薄的层流火焰面,需要求解层流火焰面方程:

$$\rho \frac{\partial Y_i}{\partial \tau} = \frac{\chi}{2 L e_i} \rho \frac{\partial^2 Y_i}{\partial Z^2} + \rho \dot{\omega}_i \quad (1)$$

$$\rho_p \frac{\partial T}{\partial \tau} = \rho_{cp} \frac{\chi}{2} \frac{\partial^2 T}{\partial Z^2} + \sum_{i=1}^{ns} h_i \rho \dot{\omega}_i + q_R + \frac{\partial p}{\partial t} \quad (2)$$

其中 q_R 为辐射热, χ 为标量耗散率,即

$$\chi = 2D (\nabla Z)^2$$

　　求解火焰面方程,可得到火焰面数据库。求解火焰面方程需要给定标量耗散率 χ 和混合分数 Z 之间的函数关系。Pitsch 等人[4]提出可以用公式

$$\chi = \chi_{st} f(Z) / f(Z_{st})$$

来模拟。其中,$f(Z) = Z^2 \ln Z$, χ_{st} 为当量条件下的标量耗散率。这样 χ_{st} 就作为一个独立的摄动量引入到火焰面方程中,所有的 ϕ 标量都可以表示为 $\phi = \phi (Z, \chi_{st})$。给定一系列当量标量耗散率,求解火焰面方程可得到层流火焰面数据库。

　　湍流扩散火焰中的平均热力学参数可由层流火焰面系作统计得到

$$\tilde{\Phi}(x_i, t) = \int_0^1 \int_0^\infty (Z, \chi_{st}, t) P(Z, \chi_{st}; x_i, t) dZ d\chi_{st}$$

$$(3)$$

　　上标“~ ”为密度加权平均。为此需要给出混合分数 Z 和当量耗散率的联合概率密度函数。但在工程应用中,通常假定混合分数 Z 和耗散率统计独立,联合概率密度函数可以用两个边缘概率密度函数的乘积近似

$$P(Z, \chi_{st}; x_i, t) = P(Z; x_i, t) P(\chi_{st}; x_i, t) \quad (4)$$

事实上,式(4)的二重积分很耗时间,对于非定常火焰面,二重积分导致计算时间的花费几乎是不可忍受的,为此,在实际应用中我们忽略标量耗散率 χ_{st} 的脉动,认为标量耗散率 χ_{st} 的概率服从 δ 函数分布。

假定混合分数 Z 满足 β 函数,要得到这个概率密度函数,需要两个平均值 $\tilde{Z}$ 和 $\tilde{Z}''^2$。这两个平均值需要从湍流流场计算中获得。

1.2　控制方程

　　在直角坐标系下,三维质量加权平均的 Navier Stokes 方程组的形式如下

$$\frac{\partial Q}{\partial t} + \frac{\partial F}{\partial x} + \frac{\partial G}{\partial y} + \frac{\partial E}{\partial z} = \frac{\partial F_v}{\partial x} + \frac{\partial G_v}{\partial y} + \frac{\partial E_v}{\partial z} \quad (5)$$

　　式中, 变量 $Q = (\rho, \rho u, \rho v, \rho w, \rho E_t)^T$, E, F, G 表示无粘通量。湍流模型采用 k-ω 两方程湍流模型,与 Navier Stokes 方程组解耦求解。混合分数的平均值和方差的方程也与 Navier Stokes 方程组解耦求解。混合分数的平均值和方差的方程为

$$\frac{\partial \bar{\rho} \tilde{Z}}{\partial t} + \cdot (\bar{\rho} \tilde{Z} \tilde{v}) = \cdot \left[\left(\bar{\rho} D + \frac{\mu_t}{\sigma_Z} \right) \nabla \tilde{Z} \right] \quad (6)$$

$$\frac{\partial \bar{\rho} \tilde{Z}''^2}{\partial t} + \cdot (\bar{\rho} \tilde{Z}''^2 \tilde{v}) = \cdot \left[\left(\bar{\rho} D + \frac{\mu_t}{\sigma_Z} \right) \nabla \tilde{Z}''^2 \right]$$

$$+ 2 \frac{\mu_t}{\sigma_Z} (\nabla \tilde{Z} \cdot \nabla \tilde{Z}) - \bar{\rho} \tilde{\chi} \quad (7)$$

1.3　数值方法

　　采用粘性 N-S 方程,MUSCL 外插,无粘通量采用 Steger Warming 格式,粘性项采用 Gauss 定理计算,时间推进采用 LU-SGS,湍流模型分别采用高雷诺数的 Kok 的 TNT 模型。

1.4　高速可压缩流动中湍流非预混火焰面模型的一些新特征

　　相对于低马赫数流的火焰面模型而言,超声速可压缩流的火焰面模型更加复杂。使得可压缩火焰面模型复杂化的主要问题是:

　　(1) 可压缩流动的一个特征是速度、密度、压力和温度之间的强耦合,在超声速流中动能的增加甚至能达到与化学反应释热相同的量级[5],完整的火焰面数据库应该包括所有的这些信息[6]。相反,基于低马赫数假设的一维扩散火焰面的计算导致一个常压空间,也忽略了动能的影响。火焰面数据库仅包含混合分数空间中的密度和温度信息。以往的高速流计算工作[7]中,火焰面数据库中的温度和连续方程的解被一起存放,并被用于计算压力。然而,当使用数据库中的温度时,就不能计及速度和温度的强耦合作用。更重要的是,激波捕捉算法不能再使用。

　　采取 Oevermann[8]的解决办法:在超声速燃烧流场中,只有数据库中的组分质量分数被使用。当地的

温度由能量方程隐式得到。从这个意义上讲,整个层流火焰面模型能被当作可压流状态方程的一个比较复杂的修正。值得注意的是,将不可压火焰面数据库(没考虑动能影响)应用到高速反应流中时,从理论上讲并不完备。但是,不可压火焰面数据库是采用有限速率反应得到的,湍流对平均反应速率的影响通过统计平均得到。因此,采用不可压火焰面数据库的火焰面模型模拟超声速湍流燃烧比直接采用有限速率反应(Arrhenius 公式)更合适。

(2) 在有间断的可压流中,用于计算湍流扩散火焰的火焰面模型有另一个关键特征。如果一道激波扫过火焰前峰,火焰前峰和激波引起的温度变化量可能处于可比量级。这就意味着,火焰面模型假设一温度和质量分数型面沿火焰前峰(燃烧波)切向的改变远小于沿火焰前峰法向的改变量不再成立。如果出现这种情况,一维火焰面结构不再适用,计算结果也必须谨慎解释。

尽管将火焰面模型应用于超声速燃烧有一些限制和不确定性,但它的经济适用性很可观(相比PDF),并且火焰面模型也能直接用于 LES。因此,可以说火焰面模型是目前在工程应用上模拟超声速湍流燃烧的最佳选择。

2 计算模型及边界条件

DLR氢燃料直连式发动机[8-9]实验示意图见图1。预热空气通过喷管膨胀进入燃烧室。燃烧室入口高50mm,宽40mm。支板长32mm,高6mm。氢气通过支板上的15个喷孔被注入,喷孔直径1mm,喷孔间距2.4mm。为了二维模拟计算将喷孔等效为相同面积的狭缝(slot)。

(1) 来流条件和喷流条件

来流和喷流条件见表1。

(2) 边界条件

来流由于是超声速来流,其上的流动参数全部固定;出流认为是超声速出流,采用外插边界条件。壁面为绝热无滑移壁。喷口参数固定。

(3) 计算网格

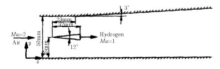

图1 DLR氢燃料直连式发动机实验示意图
Fig.1 Sketch of the DLR scramjet combustion chamber

表1 DLR氢燃料直连式发动机来流和喷流条件
Table 1 Inflow conditions of the air stream and the hydrogen jet

	Air	Hydrogen
Ma	2.0	1.0
$v[m/s]$	730	1200
$T[K]$	340	250
$p[10^5 Ps]$	1	1
$\rho[kg/m^3]$	1.002	0.097
Y_{O_2}	0.232	0
Y_{N_2}	0.736	0
Y_{H_2O}	0.032	0
Y_{H_2}	0	1

网格总数约25000。壁面附近网格间距 1×10^{-5} m,在支板后网格加密。

3 计算结果与分析

3.1 喷氢不燃烧时的计算结果

保持喷氢但不点火,此时流场中温度很低(< 700K)不燃烧。图2给出了喷氢不点火计算所得流场(Sarkar 可压缩修正)与实验所得阴影图片的对比。由图2可以看出整个流场是一个复杂的流场结构,在斜劈尖端产生斜激波,在斜劈尾部产生膨胀波,流场中有一系列的激波和反射激波。由图2中的计算和实验结果对比可以看出,计算和实验所得波系比较相似,波系位置也比较接近。由于数值模拟时采用的是二维模型,燃烧室中的激波较强,衰减较慢,激波与剪切层作用明显;而实验的模型是三维的,支板高度只有6mm,其波系衰减很快。图3显示喷氢不点火计算所得沿上下壁面压强分布和中心面压强分布($y = 25mm$)与实验结果对比,虽然计算结果和实验结果比较一致,但在 $x = 120mm$ 处下壁面压强, $x = 100mm$ 处

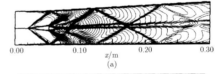

0.00   0.10   0.20   0.30
x/m
(a)

(b)

图2 喷氢不点火计算所得流场与实验所得阴影图片的对比
Fig.2 Comparison between shadow picture and hydrogen injection without combustion

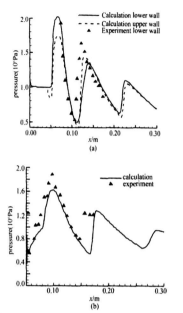

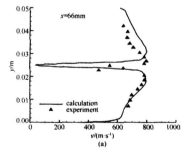

图3　喷氢不点火计算所得沿上下壁面压强分布和中心面
　　压强分布
**Fig. 3　Pressure distributions along the upper and lower
channel wall and the middle channel
(hydrogen injection without combustion y = 25mm)**

中心面压强与实验结果相比差别较大, 这可能是三维
效应导致的。在实验中, 氢被通过 15 个孔注入流场,
这是一个三维流场。而且, 压强测量点靠近流道侧
壁, 那里拐角和边界层影响比较明显。在计算过程中
发现, 采用增加混合分数方程来模拟混合过程, 比传
统方法—求解组分连续方程快捷得多, 尤其是组分比
较多时。

　　图4给出了沿流向 4 个不同位置的速度剖面。
由图可以看出, 计算所得剖面与激光多普勒速度仪
(LDV) 测得的速度剖面比较接近, 说明给定的边界条

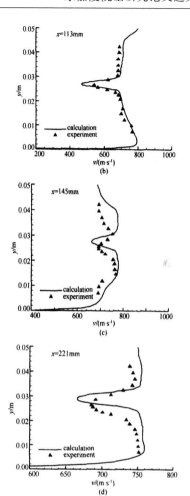

图4　喷氢不点火计算所得不同流向位置的速度剖面
**Fig. 4　Cross stream velocity profiles at different streamwise
locations x (hydrogen injection without combustion)**

件是合理的, 计算软件能比较准确地模拟该状态。

3.2　喷氢燃烧时的计算结果

　　火焰面数据库生成采用的是 9 组分 19 方程的化
学反应模型。当量标量耗散率变化范围从 $x_{st} =$
$0.001s^{-1}$ 到 $x_{st} = 130.625s^{-1}$。对应不同当量标量耗
散率间的火焰面内最高温度相差不大于 50K, 一共生
成了 40 个火焰面。在获得火焰面数据库时, 对所有
组分取单位刘易斯数($Le = 1$)。

　　为了和没有湍流燃烧模型的计算结果进行对比,
还计算了只采用有限速率反应时的燃烧结果, 采用的

仍是9组分19方程化学反应模型。两种方法都采用Sarkar的可压缩修正。由图5可知,没有湍流燃烧模型时计算得到的燃烧区比实验的小,燃烧区最窄处在 $x \approx 0.11$m 处;而采用火焰面模型得到的燃烧区比实验的大,燃烧区最窄处在 $x \approx 0.13$m 处,这与实验比较接近。

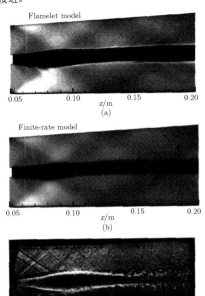

图5 喷氢燃烧时实验所得阴影图与计算所得密度云图的对比
Fig.5 Density contour plots and shadow picture (hydrogen injection with combustion)

图6对比了实验和两种计算方法得到的速度剖面。由图可知,流向速度沿 $y = 25$mm 的分布中,靠近喷油嘴附近流速被油流加速,随后流速递减然后增加。在 $x = 100$mm 以前,两种方法得到的流向速度都比实验值低,有限速率法得到的流速甚至出现比较大的负值。图6中在沿流向三个不同位置中,火焰面模型得到速度剖面在 $x = 66$ 和113mm 处和激光多普勒速度仪(LDV)得到的比较接近,在 $x = 195$mm 处差别较大。在 $x = 66$mm 处,有限速率反应模型得到的速度剖面与实验所得差别较大,明显不如火焰面模型的计算结果好。

图7给出了喷氢点火计算所得沿流向三个截面静温的分布,可以看到,在喷口附近($x = 66$mm)火焰

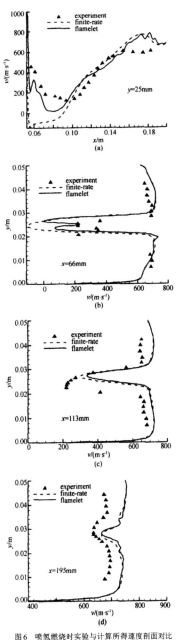

图6 喷氢燃烧时实验与计算所得速度剖面对比
Fig.6 Velocity profiles along the middle of the channel at $y = 25$ mm (upper left) and at three different streamwise cross sections x (hydrogen injection with combustion)

面模型得到的静温剖面与实验结果(CARS 方法得到)比较接近,高温区的宽度明显较有限速率反应模型大。由于低温氢的喷入,静温剖面应该像实验测量和火焰面模型的剖面那样,呈双峰结构(Oevermann[8]采用RANS+ 火焰面模型模拟 2D 流场和 Berglund[9]等人采用大涡模拟+ 火焰面模型模拟 3D 流场都得到相近的结果)。而有限速率反应模型得到的静温剖面与实验结果差别较大,预测的静温过高,而且仅有一

个峰值(Dinde[10]等人 RANS+ 有限速率模型模拟 3D流场得到相近的结果),说明反应比较剧烈。在 $x =$ 113mm 处,火焰面模型得到的静温剖面与有限速率方法得到的剖面和实验结果比较接近。而在 $x =$ 221mm 处,有限速率反应的结果其峰值与实验结果比较接近,火焰面模型得到的结果其高温区宽度与实验值比较接近。

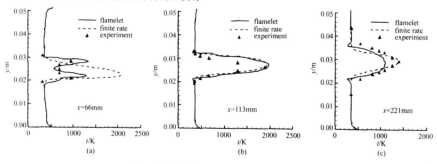

图 7 喷氢燃烧时计算所得沿流向三个截面静温的分布
Fig. 7   Cross stream temperature profiles at different streamwise locations x (hydrogen injection with combustion)

4   结   论

从两者的数值模拟结果对比,可以确定:

(1)对本算例而言,与有限速率模型的结果相比,湍流和燃烧的相互作用增大了喷孔附近的燃烧区域,减弱了喷孔出口附近燃烧的剧烈程度,有限速率模型由于没有考虑湍流和燃烧的相互作用而过高地预测了喷孔出口附近的燃烧剧烈程度。因此,湍流和燃烧的相互作用不能忽略;

(2)火焰面模型和有限速率反应模型得到的高温区分布有明显差异,尤其是在喷孔附近,这和 Baurle[11]的论述"湍流燃烧相互作用的影响主要体现在点火点附近"一致。

参考文献:

[1]   VILLASENOR R, CHEN J Y, PITZ R W. Interaction between chemical reaction and turbulence in supersonic nonpremixed H₂ Air combustion[R]. AIAA 9+ 0375.

[2]   JOHNNY R N. Effect of turbulence chemistry interactions in compressible reacting flows[R]. AIAA 94 2311.

[3]   PETERS N . Laminar diffusion flamelet models in non- premixed turbulent combustion[J]. Progress in Energy and Combustion Science. 1984. 10: 319-339.

[4]   PITSCH H, CHEN M, PETERS N. Unsteady flamelet modeling of turbulent hydrogen/air diffusion flames[R]. Proc. Combust. Inst., 27: 1057-1064, 1998.

[5]   SABEL' NIKOV V, DESHAIES B. et al. Revisited flamelet model for non premixed combustion in supersonic turbulent flows[J]. Combustion and Flame 114: 577-584, 1998.

[6]   SECUNDOV, et al. Flamelet model application for non- premixed turbulent combustion. final report under cooperative agreement NoNCCW- 75 with NASA[R]. July 1996.

[7]   KUMAR S , TAMARU T . Computation of turbulent reacting flow in a jet assisted RAM combustor[J]. Comput. Fluids 26 (2): 117-133, 1997.

[8]   MICHAEL Oevermann. Numerical investigation hydrogen combustion in a SCRAMJET using flamelet modeling[J]. Aerosp. Sci. Technol. 4, 463-480, 2000.

[9]   MAGNUS Berglund, NIKLAS Wikstrom, CHRISTER Fureby. Numerical simulation of scramjet combustion[R]. FOI R- 1650- SE, JUNE 2005.

[10]   PRASHANT D, RAJASEKARAN A, and BABU V. abu. 3D Numerical simulation of the supersonic Combustion of H₂[J]. Aeronautical Journal, 110(1114), 2006.

[11]   BAURLE R A. Modeling of high speed reacting flows: established practices and future challenges[R]. AIAA 2004 0267.

煤油超燃冲压发动机三维大规模并行数值模拟

杨顺华[1,2]，   乐嘉陵[1]，   赵慧勇[1]，   郑忠华[1]

(1. 中国空气动力研究与发展中心计算空气动力学研究所，四川 绵阳   621000;
2. 中国科学技术大学力学与机械工程系，安徽 合肥   230026)

[摘　要]　在我国巨型计算机上，采用 1024 个 CPU 对煤油燃料超燃冲压发动机燃烧流场进行三维大规模并行数值模拟. 计算软件采用自主研发的并行软件 AHL3D，控制方程采用雷诺平均的 N-S 方程，无粘项计算采用 Steger-Warming 矢通量分裂格式，湍流模型采用 k-ω 双方程模型，煤油分子式采用正癸烷代替. 计算给出的发动机壁面压强分布与试验测量结果有较好的一致性. 结果表明，凹槽是发动机主要的着火区和火焰稳定区，同时，由于燃料的喷注形成的回流区也起到一定的稳焰作用. 计算结果验证了 AHL3D 程序和采用的理论模型可以用于模拟煤油燃料超燃冲压发动机内部的复杂流动.

[关键词]　超燃冲压发动机；数值模拟；负载平衡；大规模并行；煤油；化学动力学

[中图分类号]　O359+.1　　[文献标识码]　A

0　引言

超燃冲压发动机是一种新型的高超声速吸气式动力装置，其特点是氧化剂直接取自大气中的氧气，没有复杂的转动装置. 燃料在燃烧室内组织超声速燃烧，具有较高的比冲，在军事、民用领域具有广阔的应用前景. 过去几年里，中国空气动力研究与发展中心(CARDC) 已经开展了许多针对碳氢燃料超燃冲压发动机的直连式和自由射流式试验研究[1].

燃料混合、火焰稳定、总压恢复和热结构等问题对煤油超燃冲压发动机设计至关重要. 为解决这些问题，计算流体力学(CFD) 已经成为超燃冲压发动机设计的重要手段之一，准确、快速地模拟超燃冲压发动机内部流动是一个具有挑战性的课题. 由于其流动具有很强的湍流效应，并伴随有激波与激波、激波与边界层和湍流与化学反应等多种相互作用，需要可靠的 CFD 程序及湍流、化学动力学模型. 因此，在将 CFD 程序应用于分析和设计之前，有必要对 CFD 程序进行验证，以确认其物理化学模型和数值方法的可靠性.

目前，对煤油超燃冲压发动机的数值模拟仍然存在许多困难. 一方面是由于超声速燃烧中复杂的物理和化学问题，包括液体碳氢燃料的雾化、蒸发、混合和燃烧，湍流、激波与燃烧的相互作用等，由于对这些机理的认识不够深入，缺乏可靠的理论模型，CFD 还不能完全真实地模拟这些流动现象. 另一方面就是计算效率问题. CFD 计算中所需的 CPU 时间和内存强烈依赖于网格数和所采用的化学动力学模型中的组份个数. 目前，CFD 计算中可以承受的组份个数大约为 10~20. 对于煤油，其详细的动力学模型包含上百个组份，因此，如何获得一个可靠的简化学动力学模型一直是许多研究者长期为之奋斗的目标. 为了获得可靠的数值结果，CFD 所需的网格数往往超过百万量级. 对于如此规模的 CFD 计算，靠单机完成至少需要数月的时间，对于设计周期显得过长，必须依靠大规模并行计算.

本文对 CARDC 试验用的煤油燃料超燃冲压发动机模型进行三维大规模并行数值模拟. 通过对比计算和试验的壁面压强分布数据，验证 CFD 程序和理论模型的可靠性.

1　计算几何模型和计算条件

计算的几何模型为 CARDC 设计的超燃冲压发动机试验模型，如图 1 所示，该模型在 CARDC 的脉冲燃烧

杨顺华, 乐嘉陵, 赵慧勇, 郑中华. 煤油超燃冲压发动机三维大规模并行数值模拟. 计算物理, 2009, 26(4): 534-539.

风洞上进行了多次试验, 为CFD计算提供了可靠的试验数据. 模型总长 2.15 m, 宽度 0.15 m, 隔离段入口高度 30 mm. 模型依次包括进气道、隔离段、燃烧室和尾喷管, 各段的具体几何尺寸如图1所示.

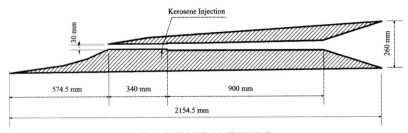

图 1   超燃冲压发动机模型示意图
Fig. 1   Schematic of a scramjet

计算采用的来流条件和煤油喷流条件见表 1. 由于试验设备采用燃烧加热器来提高试验气体的焓值, 因此来流气体中含有一定比例的 H_2O.

表 1   计算采用的来流条件和喷流条件
Table 1   Inflow and injection conditions in computation

参数	马赫数	总温/K	总压/Pa	油气比	煤油质量分数	O_2 质量分数	N_2 质量分数	H_2O 质量分数
来流条件	6.0	1 650	$60.0 \times 101\ 325$	—	0.0	0.255	0.574	0.171
喷流条件	1.0	300	—	1.1	1.0	0.0	0.0	0.0

煤油的喷注位置在燃烧室入口略微靠前处, 实际的煤油是在室温条件下以液态的形式注入燃烧室的, 煤油首先经历雾化和蒸发, 但试验结果表明, 煤油的雾化和蒸发速度非常快[2], 煤油一离开喷嘴很快就雾化和蒸发了, 因此本文的计算忽略煤油的雾化和蒸发过程.

2   计算方法

计算采用自主研发的三维并行软件平台 – AHL3D. AHL3D 采用结构化网格, 可以求解 Reynolds 平均的守恒方程(二维、轴对称和三维). 控制方程采用格心有限体积法离散, 气体模型可以为完全气体和带化学反应的任意组份的完全气体混合物. 无粘项离散格式有 Steger-Warming 分裂, AUSM, AUSMPW+, LDFSS 等, 控制体界面采用 MUSC 方法重构, 为减小截断误差, MUSCL 参数取 1/3. 限制器采用 Van Albada 限制器. 为了提高计算效率和稳定性, 时间推进采用隐式 LU-SGS 方法, 并对化学反应源项采用点隐式处理.

湍流模型是超燃冲压发动机数值模拟的主要困难之一. 本文采用 $k\text{-}\omega$ 双方程湍流模型, 然而, Menter 指出, $k\text{-}\omega$ 湍流模型强烈依赖于初值(特别是 ω)的选取, 这种初值依赖性对于自由剪切层特别强烈, 对于边界层流动也比较明显. 为了消除这种依赖性, 我们采用 J. C. Kok[3] 提出的 TNT $k\text{-}\omega$ 湍流模型.

由于煤油的成分复杂, 并且成分随产地不同而不同, 因此目前还没有一套公认的化学动力学模型, 这给 CFD 计算带来了极大的困难. CARDC 与天津大学、中国科学技术大学合作, 对国产航空煤油成分进行了分析, 通过数据拟合, 我们提出了煤油的替代分子式 – $n\text{-}C_{10}H_{22}$. 同时, 建立了一套 10 组分($n\text{-}C_{10}H_{22}$, CO_2, CO, H_2, H_2O, H, OH, O_2, O, N_2) 和 12 步反应的两阶段化学动力学模型[4, 5].

由于流动是对称的, 为了减少计算网格, 计算实际采用的宽度为模型实际宽度的一半, 对称面按对称条件处理. 总的计算网格数为 220×10^4, 第一层网格距离壁面 0.001 mm.

计算采用的边界条件如下: 来流给定入口马赫数、压强和温度; 出流采用超声速出口条件; 壁面采用无滑移、等温壁条件, 壁温取为 300 K.

本文根据燃烧后 CO_2 的生成量定义燃烧效率

$$\eta_{c} = \frac{\dot{m}_{CO_2, x} - \dot{m}_{CO_2, \infty}}{\dot{m}_{CO_2, ideal} - \dot{m}_{CO_2, \infty}}. \tag{1}$$

这里 $\dot{m}_{CO_2, x}$ 表示沿流向 x 处的 CO_2 的质量流量, $\dot{m}_{CO_2, \infty}$ 表示来流的 CO_2 的质量流量, $\dot{m}_{CO_2, ideal}$ 表示理论生成的 CO_2 的质量流量.

3 大规模并行计算技术

为了模拟燃烧室内部复杂的湍流、激波和漩涡流动, 必需要有足够密的网格, 并且网格要能够对发动机的几何细节进行完整捕捉, 这对网格的精度和数量提出了要求. 同时, 由于增加了对组分方程的求解, 计算量大增. 仅依靠目前单个计算机的能力, 已难以满足发动机三维燃烧流场的数值模拟需求, 必需发展大规模的并行计算技术.

本文采用 MPI 并行编程语言实现 CFD 程序的并行化. 并行计算技术的关键是负载平衡算法, 为了使每个 CPU 获得均匀的负载, 有时需要将原始的多块网格进一步剖分成更小的网格子块, 每个 CPU 计算一个或者多个剖分后的子块. 一个好的负载平衡算法必需满足以下两条准则: ①剖分后的最大子块的网格量尽可能小; ②为了减小通讯量, 总的通讯面的面积尽可能小. 假设原始网格块 i 的网格量为 V_i, 被剖分成 n_i 个子块, 那么负载平衡准则一就转化为求这样的 n_i, 使其满足 $\min\left[\max_i\left(\frac{V_i}{n_i}\right)\right]$.

我们可以采用"贪婪算法"来获得最优的 n_i 整数解. 为了获得尽可能小的通讯量, 在进行网格块的剖分时总是沿网格块的最长边进行. 图 2 给出了采用上述负载平衡算法获得的程序加速比随 CPU 数目的变化, 可以看出, 加速比随 CPU 数目呈线形变化.

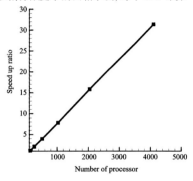

图 2 加速比随 CPU 数目的变化
Fig. 2 Speed-up ratio with CPU numbers

4 计算结果分析

本文在我国的巨型计算机上, 采用 1024 个 CPU 对图 1 所示的超燃冲压发动机试验模型进行了三维大规模并行计算. 图 3 对比了有/无燃烧条件下发动机的下壁面压强分布(– 计算, ·试验). 对于无燃烧情形, 计

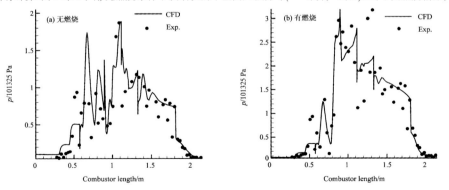

图 3 计算和试验测量的发动机下壁面压强分布比较
Fig. 3 Computed pressure distribution on bottom wall and experimental data

算值和试验值之间吻合得非常好,说明我们采用的湍流模型和计算网格是非常成功的;对于有燃烧情形,计算值和试验值在绝大部分区域是吻合的,在凹槽后台阶附近,试验值有一个明显的峰值,计算值明显低于试验值,分析其原因如下:由于凹槽后台阶的阻挡,在凹槽后台阶前会出现一道很强的脱体激波,脱体激波后温度升高,形成高温区,燃烧非常剧烈.由于数值粘性等因素造成计算得到的脱体激波要弱于实际情形,因此计算压强值没有达到实际峰值.总体上,计算值和试验值之间的吻合程度还是令人满意的,进一步表明了我们提出的化学动力学模型可以模拟超燃冲压发动机燃烧流场.

图4为计算得到的燃烧后的静温分布.从图中可以看出,进气道的三道斜激波刚好相交于发动机外罩唇口,与设计条件一致.燃烧后高温区域主要集中在凹槽以后的区域,并且越往后,燃烧越均匀.同时,一部分火焰已经向前传播到了隔离段出口,在隔离段出口下壁面附近形成了一小片高温区,这种"回火"现象会破坏隔离段的抗反压性能,特别是有可能进一步向前扰动到隔离段入口,导致进气道的不起动.从图中还可以看出,在喷孔附近,有一段明显的低温区,这是由于燃料从喷孔出来以后,由于温度低,并没有马上着火,而是经过一段距离后才开始燃烧.从计算过程看,煤油首先在凹槽后台阶的地方着火,然后火焰向下游传播,火焰高度逐渐扩大.同时凹槽内的火焰也缓慢向前传播,最后火焰逐渐充满整个发动机流道.因此,凹槽提供了煤油着火的条件,同时也起到了稳焰的作用.

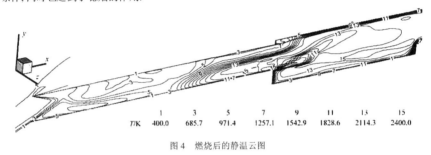

	1	3	5	7	9	11	13	15
T/K	400.0	685.7	971.4	1257.1	1542.9	1828.6	2114.3	2400.0

图 4　燃烧后的静温云图

Fig. 4　Contours of static temperature in combustion case

图5给出了燃烧后的 CO_2 分布.由于燃料的穿透深度有限,在发动机上壁面附近生成的 CO_2 较少,并且靠近壁面生成的 CO_2 要比中心面上的多, CO_2 整体呈"U"字型分布.

Mass fraction of CO_2

1	2	3	4	5	6	7	8	9	10	11	12	13	14	15
0.005	0.017	0.029	0.041	0.053	0.065	0.077	0.089	0.101	0.113	0.125	0.137	0.149	0.161	0.173

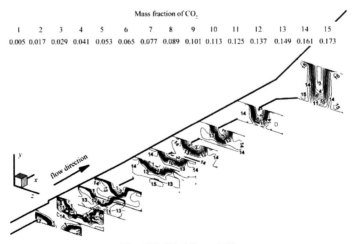

图 5　燃烧后生成的 CO_2 云图

Fig. 5　Contours of CO_2 in combustion case

图6为凹槽附近的流线分布.在隔离段出口,由于燃料的喷射,在燃料喷孔前,形成了一个非常强的漩涡.受燃料射流柱的阻挡,在喷孔前,边界层发生分离,并且产生分离激波,同时形成回流区.在回流区内,一部分燃料被输送到上游,由于回流区内温度升高,燃料着火燃烧,这就是图4中火焰前传的原因.这个着火区能够起到火焰稳定器的作用.在凹槽内,流动非常的复杂.一个非常明显的漩涡位于凹槽中间靠后的位置,燃烧主要发生在这个区域.这个漩涡不仅增加了燃料的驻留时间(流线变长),还将燃料向壁面输送,这也进一步解释了图5中为何 CO_2 分布呈"U"字型.

为了进一步分析凹槽内的流动结构,图7给出了凹槽内不同截面上的流线.从左到右,在第一个截面上,在对称面的上下壁面附近出现两个小的漩涡;在第二个截面,在凹槽底部形成了一个非常大的漩涡,并且在侧壁附近也出现了一个较大的漩涡,对称面上的漩涡明显减小;在第三个截面,凹槽侧壁和凹槽底部的漩涡合并成一个大的漩涡,而对称面上的漩涡消失.

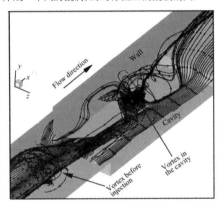

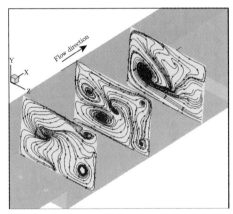

图6　凹槽附近的流线和漩涡结构(有燃烧,半宽)

Fig. 6　Streamlines and vortex structure near the cavity (combustion, half width)

Fig. 7　Flow patterns at different locations in the cavity (combustion, half width)

图8~11为一维质量加权的参数分布.

图8给出了加权后的马赫数分布,燃烧室内加权马赫数都大于1,在燃烧扩张段,马赫数变化不大,基本实现了"变截面+等马赫数"的释热规律.

图9给出了加权后的总温和静温分布,在燃烧室入口前,总温基本保持不变;在隔离段接近出口的位置,

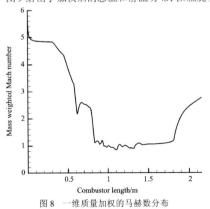

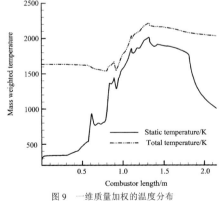

图8　一维质量加权的马赫数分布

Fig. 8　1-D mass weighted Mach numbers

图9　一维质量加权的温度分布

Fig. 9　1-D mass weighted temperature

总温略有下降, 是由于燃料的注入导致的. 燃烧室内由于煤油的燃烧, 总温迅速爬升, 在接近凹槽的出口位置, 总温爬升达到最大值, 燃料的释热基本完成, 总温随后缓慢下降, 主要是由于壁面带走一部分热量所致.

图 10 为加权的组分质量分数分布. 可以看出, 由于煤油的注入, 煤油组分在喷孔处达到峰值, 然后煤油的量迅速下降. 煤油的消耗主要集中在凹槽内, 出了凹槽, 煤油的量基本保持不变. CO 的生成主要集中在凹槽内, 其生成率与煤油的消耗相对应. 出了凹槽, 由于一部分 CO 转变成 CO_2, CO 的量缓慢下降.

图 11 给出了根据(1) 式计算得到的燃烧效率曲线, 计算评估的燃烧效率大约为 60%.

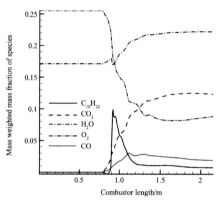
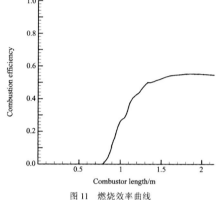

图 10　一维质量加权的组分质量分数分布
Fig. 10　1-D mass weighted mass fraction of species

图 11　燃烧效率曲线
Fig. 11　Combustion efficiency

5　结论

本文在我国巨型计算机上, 采用大规模并行计算, 对我国 CARDC 的超燃冲压发动机试验模型进行了数值模拟, 验证了 CFD 软件的可靠性和化学动力学模型的有效性. 详细的分析了发动机内部的三维流动特征, 分析指出, 凹槽是发动机主要的着火区和火焰稳定区, 同时, 由于燃料的喷注形成的回流区也起到了一定的稳焰作用. 计算结果证明了软件平台 AHL3D 可以模拟超燃冲压发动机内部的复杂流动.

致谢: CARDC 的超高速空气动力研究所的超燃课题组为本文计算提供了模型型面和试验数据, 在此表示感谢.

[参 考 文 献]

[1] LE Jialing, Liu Weixiong, Yu Gang. Recent progreee of our scramjet reserarch [R]. ISABE- 2005- 1009.

[2] Le Jialing, Yang Shunhua, Liu Weixiong, et al. Experimental and numerical investigation of ignition delay for kerosene fueled scramjet [R]. 7th Sino Russia High speed FLow Conference. July, 2008.

[3] Kok J C. Resolving the dependence on free stream values for the k omega turbulence model[R]. NLR TP-99295, July, 1999.

[4] Yang Shunhua, Le Jialing. Computational analysis of a kerosene fueled scramjet [R]. ISABE- 2005- 1195.

[5] Le Jialing, Yang Shunhua, Liu Weixiong, et al. Massively parallel simulations of kerosene fueled model scramjet and comparisons with experiments [R]. AIAA- 2005- 3318, 2005.

Three-dimensional Massively Parallel Numerical Simulation of Kerosene-fueled Scramjet

YANG Shunhua[1,2],　　LE Jialing[1],　　ZHAO Huiyong[1],　　ZHENG Zhonghua[1]

(1. CAI, China Aerodynamics Research & Development Center, Mianyang　621000, China;

2. Department of Modern Mechanics, University of Science and Technology of China, Hefei　230026, China)

Abstract:　　Massively parallel numerical simulation of a kerosene-fueled scramjet on national MPP computer with 1024 CPUs is presented. A self-developed software AHL3D is employed. Reynolds averaged N-S equations are adopted in governing equations. Inviscid flux is computed with Steger-Warming flux-splitting scheme. Turbulent effects are studied in a k-ω two-equation turbulence model. Kerosene is replaced by a simple surrogated fuel of n-decane. Computed wall pressure along combustor exhibits good agreement with measured data. Computation shows that cavity provides main ignition zone and flame holder, and recirculation zone resulting from fuel injection contributes to flame stability. It indicates that the AHL3D code and methodologies can be used to simulate complicated flow patterns in kerosene-fueled scramjets.

Key words:　　scramjet; numerical simulation; load balance; parallel computation; kerosene; chemical kinetics

Experimental and numerical investigation of air vitiation effects on

scramjet test Performance

Le Jialing[1,2], Liu Weixiong[1], Song Wenyan[2], Xing Jianwen[1],Yang Yang[1]

1. China Aerodynamics R&D Center, Mianyang, Sichuan, China, 621000

2. Northwest Polytechnical University, Xi'an, China, 710072

Abstract

Understanding the complicated issues with regard to the effects of airstream vitiation on scramjet test performance requires a lot of efforts involving experiment, modeling and CFD simulation. Based on a direct-connected resistance heated wind tunnel, a test facility has been developed to support comparative experiments for combustion in clean air and air vitiated with H_2O/CO_2. Under the same test gas parameters of nominal Mach 2.0, total temperature of 800K and total pressure of 760kPa, a set of well –controlled comparative combustion tests for a hydrogen-fueled combustor with a cavity were conducted. Then, massively parallel CFD simulations of reacting flowfield in the combustor for three cases corresponding to experimental conditions were conducted with a relatively detailed chemical kinetic model involving 13 species and 33 reactions. The simulated results are close to measured wall pressures of comparative experiments, and could "distinguish" the vitiation effects of H_2O in test gas. The simulated results show that for overall performance of the combustor, 7.4% H_2O and 18.3%H_2O contaminants of inflow decrease the thrust of the combustor by 4.3% and 15.2% respectively.

1. Introduction

In order to help provide effective means for extrapolating scramjet performance with vitiated air to flight, experimental and numerical investigations were carried out to study air vitiation effects on test scramjet. On account of the combustion-heated facilities are widely applied in propulsion tests below Mach 8.0 for scramjet development, we need to elaborately evaluate the effects of species contaminants on scramjet combustion processes. The species contaminants from combustion air preheated include major contaminants, namely H_2O or H_2O and CO_2, and minor contaminants, e.g. OH and O. The study reported herein focuses on the effects of major species contaminants on combustion in a scramjet combustor.

Differences of scramjet performance between those in flight environment and in vitiated air could be mainly due to differences of air inflow in thermodynamic properties and chemical kinetic effects. When amounts of H_2O and CO_2 in vitiated air are substantive, their larger heat capacities and molecular dissociation at high temperature could lower combustion-generated temperature rise and internal thrust generated in a scramjet engine. Furthermore, H_2O and CO_2 could exert kinetic effects by changing third body efficiency of some key third-body reactions etc. Owing to coexistence of various effects and competition from each other, integrated effects of H_2O and CO_2 on scramjet combustor performance often depend on the contaminants content, the specific operating condition and the configuration of the combustor.

Some attempts in the past have been made to deal with effects of major species contaminants on hydrogen-fueled scramjets. Most of them were numerical studies based on one-dimensional or two-dimensional flow models. And few experimental studies were conducted. Noticeably, two important facility-based studies have been carried out using comparative experiments between the clean test gas and vitiated test gas for a scramjet engine [1] and a scramjet combustor [2],

JiaLing Le, Weixiong Liu, Wenyan Song, Jianwen Xing, Yang Yang. Experimental and numerical investigation of air vitiation effects on scramjet test performance.16th AIAA/DLR/DGLR international space planes and hypersonic systems and technologies conference. AIAA 2009-7344, 2009.

respectively. In Ref. [1], an H2-fueled scramjet engine was tested at Mach 6 flight condition with the air supplied by a vitiation air heater and storage-heater. It was found that artificially ignition was easier in vitiated air than in clean air, and that thermal choking occurred at lower fueling rate in clean air. In Ref. [2], a direct-connected resistance heated facility was modified to obtain H_2O/CO_2 vitiated air, and comparative experiments for a scramjet combustor with a divergent extender were conducted with clean and H_2O/CO_2 vitiated air at total temperature of about 1200K. The addition of H_2O or H_2O plus CO_2 yielded lower wall pressure at a fixed equivalent ratio of 0.27, and the combustor operated in a dual mode with clean air, but in the supersonic mode when vitiated with 7% H_2O.

To understand complex issues related to major species contamination, we badly need systematic studies which combine well-controlled comparative experiments and serious-minded numerical simulation since simplex means often could not provide researchers sufficient confidence in conclusions and in-depth understanding of contamination effects. The purpose of this research is to obtain the influence of major species contaminations through well-controlled comparative experimental method using a modified resistance heated facility, with a combination of CFD simulation with a full 3-D reacting flow model. Attentions are paid to match test parameters of clean and vitiated air in experiments and to reduce uncertainty in numerical simulation by adopting a relatively detailed chemical kinetic model involving 13 species and 33 reactions [3].

2 Experimental investigation

2.1 Experimental method and facility

It is inevitable that not all test parameters can be matched since vitiated air is different from clean air in composition. We need to select appropriate test parameters to be matched while softly relaxing the matching of other parameters for clean air and vitiated air. Generally, the molecular weight and specific heat of test gases are not considered to be matched. Following are the preferential test parameters that could be considered to be matched:

(1) Velocity and Mach number of test gas. Using the same nozzle for comparative experiments, the difference of Mach number for the two test gases (clean air and vitiated air) is negligible. However, herein lies some difference in velocity of test gases depending on their static temperature and Mach number at the exit of nozzle. For the present study, we keep the Mach number nearly the same for the two test gases by using the same nozzle.

(2) Oxygen concentration. Maintaining 21 molar percent of oxygen in vitiated air permits proper stoichiometry with fuel on a molar basis and mass fraction of oxygen is variable relying on composition makeup of vitiated air. For the present study, we maintain 21 molar percent of oxygen in vitiated air.

(3) Total pressure, static pressure and dynamic pressure. These parameters cannot be matched at the same time due to difference of specific heat of clean and vitiated air. Total pressure reflects the capacity of air stream for doing work. Static pressure of test gas exerts direct influence on reaction with the fuel. And dynamic pressure influences aerodynamic force and thermo-structure characteristics for scramjet engine. For the present study, we keep the total pressure the same for the two test gases since total pressure of vitiated was matched with flight environment in our early experiments.

(4) Total temperature, static temperature and enthalpy. Total temperature is usually to be matched between combustion-heated wind tunnel and flight. Enthalpy is relevant to heat release of the fuel and static temperature has an influence on chemical reaction with fuel. For the present

study, we keep the same for the two test gases since total temperature of vitiated air was matched with flight environment in our early experiments.

In addition to the inflow conditions, clearly the equivalence ratio should be kept the same for clean and vitiated air. Based on the direct-connected resistance-heated facility of Northwest Polytechnical University (Fig.1), a test system was developed to comprehensively study air vitiated effects, which can supply the test gas with clean air and vitiated air. Vitiated air is formed by adding specific flow rate of H_2O and/or CO_2 to clean air exited from the resistance heater.

A tube heat exchanger for H_2O was designed to increase temperature of vitiated air, and a mixer and a Venturitube were designed to conveniently vary the amount of H_2O and CO_2 contaminants in test gas to desired level. Molar percentages of H_2O and CO_2 in vitiated test gas can be varied within range of 3.5%~30% and 3%~10%, respectively, allowing molar contents of H_2O and CO_2 in test gas to cover the range of major species contaminants in several typical vitiated air simulated for Mach 4~7. Key test parameters of vitiated air can be controlled to match those of clean air.

Fig. 1 Direct-connected resistance heated facility of Northwest Polytechnical University

2.2 Experimental results and discussion

Well–controlled comparative combustion tests with clean and vitiated air were conducted under the same operation parameters of nominal Mach 2.0, total temperature of 800K and total pressure of 760kPa by Prof. Song working with colleagues. Data processing results of pitot pressure measurements show that the average Mach number at the exit of nozzle is 1.982, and root mean square error of Mach number is 2.2%. The geometry of scramjet combustor model tested is show in Fig. 2. The combustor, with a length of 770mm and a width of 40mm, consists of the isolator, the upper wall with a step and the lower wall with a cavity. Hydrogen was injected at Mach 1.0 and with total temperature of 300K, perpendicular to the air stream upstream of the cavity through an injector with 9 holes with a diameter of 0.8 mm, and ignited by a spark plug in cavity.

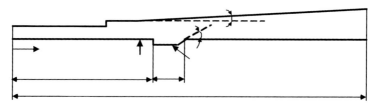

Fig. 2 H_2-fueled scramjet combustor tested in the present study

Fig. 3 shows plots of static pressure on nozzle and combustor walls, normalized by the total pressure P_t of test gas vitiated H_2O, vs. axial position. The first pressure tap and the third pressure tap are located at the nozzle entrance ($X = -145mm$) and at the combustor model entrance ($X = 5mm$), respectively. All the inflow conditions and composition contents of test gas are nearly the same for two runs. The pressure distributions are essentially identical for the two runs whether at a fuel/air equivalence ration ϕ of 0.54 or without fuel injection ($\phi = 0$). Thus, these plots illustrate that combustion tests have a good repeatability although our test system is rather complex in order to produce well-controlled vitiated air.

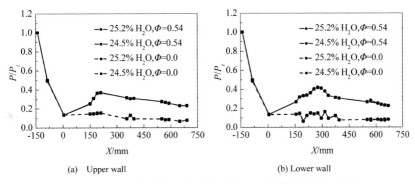

(a) Upper wall   (b) Lower wall

Fig. 3 Axial pressure distributions along the combustor wall with H2O vitiated airflow

Fig. 4 shows wall pressure distributions for the combustor model at a fixed equivalence ratio $\phi = 0.42$ for three cases, namely clean air, air vitiated with 7.4% H_2O and air vitiated with 18.3% H_2O. The fuel-off case with clean air is shown together for observation of pressure rise due to combustion. At the pressure rise segment of the upper wall ($X=150\sim280mm$), the pressure decrease relative to clean air is 7%~9% with 7.4%H_2O and 26%~28% with 18.3% H_2O. The pressure distributions with the clean air indicate a thermally choked flow, with the shock train attached to the entrance of the isolator. Using 1-D reacting flow analysis, it was found that combustion is in dual-mode operation with the clean air but in supersonic mode with the addition of 18.3% H_2O. That is, by adding 18.3% H_2O in airstream, the dual mode has transitioned to the supersonic mode.

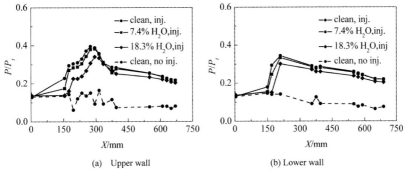

(a) Upper wall   (b) Lower wall

Fig. 4 Axial pressure distributions along the combustor wall with clean and H2O vitiated airflow($\phi= 0.42$)

Fig. 5 shows wall pressure distributions for nozzle and combustor model at a fixed equivalence ratio $\phi=0.53$ for four cases, namely clean air, air vitiated with 7.5% H_2O, air vitiated with 16.6% H_2O and air vitiated with 25.2% H_2O. It can be seen that for clean air, pressure rise due to combustion propagates into the facility nozzle. By adding 7.5%~25.2% H$_2$O, the combustion-generated pressure rise cannot propagate into the nozzle. This would suggest that H$_2$O vitiation probably transits inlet unstart with clean air to inlet startup.

At the pressure rise segment of the upper wall (X=150~280mm), the pressure decrease relative to clean air is about 10% with 7.5% H$_2$O, 28% with 16.6% H$_2$O and 33% with 25.2% H$_2$O. The molar percent increment of H$_2$O vitiation from 7.5% to 16.6% and from 16.6% to 25.2% are approximately equal, but the pressure decrease delta of the former is much larger than that of the latter. Therefore, the combustion-generated pressure decreases nonlinearly with increasing molar percentage of H$_2$O.

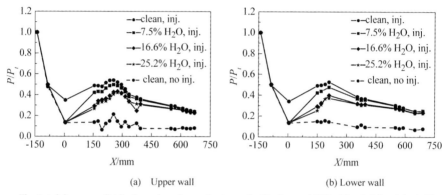

(a) Upper wall   (b) Lower wall

Fig. 5 Axial pressure distributions along the combustor wall with clean and H2O vitiated airflow(ϕ= 0.53)

Fig. 6 shows wall pressure distributions for nozzle and combustor model at a fixed equivalence ratio ϕ= 0.42 for four cases, namely clean air, air vitiated with 3.4% CO$_2$, air vitiated with 7.4% CO$_2$ and air vitiated with 7.7% CO$_2$ plus 7.4% H$_2$O. At the pressure rise segment of the upper wall (X=150~280mm), the pressure decrease relative to clean air is about 20% with 3.4% CO$_2$, 23% with 7.4% CO$_2$ and 25% with 7.7% CO$_2$ plus 7.4% H$_2$O. The follow-up pressure decreases from 3.4% CO$_2$ addition to 7.7% CO$_2$ addition is rather slight, so the vitiation effects of CO$_2$ on combustion pressure has also nonlinear characteristics.

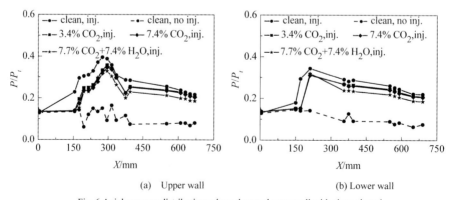

(a) Upper wall   (b) Lower wall

Fig. 6 Axial pressure distributions along the combustor wall with clean air and

air vitiated with CO$_2$ or CO$_2$ plus H$_2$O (ϕ= 0.42)

3 Numerical investigation

3.1 Numerical technique

The massively parallel CFD code AHL3D was used to evaluate the vitiation effects of H_2O. Using RANS approach, the reacting flowfields in the combustor model have been simulated. Finite volume method is used to discretize Favre-averaged Navier-Stokes equations, and k-ω turbulence model is employed to account for the turbulent effect. The conservative form of the equations is solved using a diagonal implicit finite-volume method, which solves this system using two sweeps of point Gauss-Seidel relaxation. The inviscid fluxes are computed using Steger-Warming scheme with 3rd-orer accurate MUSCL interpolation. The combustion is modeled with a chemical kinetic model involving 13 species and 33 reactions proposed by Jachimowski. The flow configuration is shown in Fig. 7, with a computational mesh number of about 2830,000.

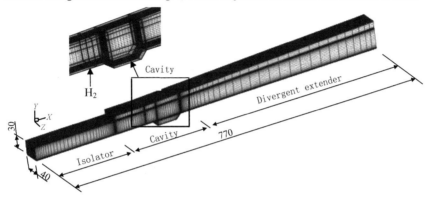

Fig. 7 Computational mesh of the combustor (unit in mm)

The computational conditions are chosen corresponding to the test conditions with inflow condition of Mach 2.0, total temperature 800K, and total pressure 760kPa. Hydrogen is injected normal to test gas at an equivalence ratio of 0.42, with injection Mach 1.0. Three cases for clean air, air vitiated with 7.4% H_2O and 18.3% H_2O, which correspond to the operation conditions shown in Fig.4, are simulated by using 2048, 1024 and 1024 CPU in a Super Computer.

3.2 Definitions for analysis of H_2O contamination effects

In order to "take out" the effects of the composition differences between clean air and vitiated air with H_2O on the reacting flowfield, we define a set of physical quantities referred to Ref. [4].

(1) Pseudo mole fraction of H_2O

The pseudo mole fraction of H_2O in the combustion products with vitiated air is the ratio of the moles of water produced from hydrogen fuel to the moles of mixture at local point of flowfield, and is expressed as

$$x_{H_2O}^* = x_{H_2O} - x_{H_2O,\infty}\left(x_{N2} / x_{N_2,\infty}\right) \tag{1}$$

Where x_{N_2}, x_{H_2O} are the mole fractions of N_2 and H_2O at that location respectively, and $x_{N_2,\infty}$, $x_{H_2O,\infty}$ are the mole fractions of N_2 and H_2O in test gas.

(2) Local total hydrogen/nitrogen ratio and local water/nitrogen ratio

The local mixing extent of H_2 fuel can be represented by the local total hydrogen/nitrogen ratio, which is expressed as

$$r_{H,N} = \frac{\sum\limits_{i=1}^{n} a_{H,i} x_i - 2x_{H_2O,\infty}\left(x_{N_2} / x_{N_2,\infty}\right)}{2x_{N_2}\left(1 - x_{O_2,\infty}\right)/ x_{N_2,\infty}} \tag{2}$$

The local combustion extent of H_2 fuel can be represented by the local water/nitrogen ratio, which is expressed as

$$r_{H_2O,N2} = \frac{\sum\limits_{i=1}^{n} x_{H_2O} - x_{H_2O,\infty}\left(x_{N_2} / x_{N_2,\infty}\right)}{x_{N_2}\left(1 - x_{O_2,\infty}\right)/ x_{N_2,\infty}} \tag{3}$$

(3) Mixing efficiency and combustion efficiency based on heat release

The mixing efficiency can be defined as

$$\eta_{mix} = \frac{\sum\limits_{l=1}^{L}\left\{\dot{n}_l \min\left[(\Omega_H - \Lambda_H), f(\Omega_O - \Lambda_O)\right]\right\}_l}{\min\left\{\sum\limits_{l=1}^{L}\left[\dot{n}_l(\Omega_H - \Lambda_H)\right]_l, \sum\limits_{l=1}^{L}\left[\dot{n}_l f(\Omega_O - \Lambda_O)\right]_l\right\}} \tag{4}$$

Where f = 0.126, and $\dot{n}_l$ is the molar flow rate in the computational cell l. Ω_H、Ω_O are the total mass of H atom and O atom in any form per mole of mixture in the cell l respectively. Λ_H and Λ_O are the total mass of H atom and O atom from the test gas per mole of mixture in the cell l respectively.

The combustion efficiency is the ratio of the practical heat release due to hydrogen combustion divided by the possible heat release if the Hydrogen fuel were converted to H_2O, and is expressed as

$$\eta_{c,q} = \frac{\sum\limits_{l=1}^{L}\left\{\dot{n}_l\left[\sum\limits_{i=1}^{n}\left(x_i \Delta H_{f,i}^0\right) - \left(x_{H2O,\infty}\Delta H_{f,H_2O}^0 + x_{CO_2,\infty}\Delta H_{f,CO_2}^0\right)\frac{x_{N_2}}{x_{N_2,\infty}}\right]_l\right\}}{\sum\limits_{l=1}^{L}\left\{\dot{n}_l\left[\sum\limits_{i=1}^{n}\left(x_i a_{H,i}\Delta H_{f,H_2O}^0 / 2\right) - x_{H_2O,\infty}\Delta H_{f,H_2O}^0 \frac{x_{N_2}}{x_{N_2,\infty}}\right]_l\right\}} \tag{5}$$

3.3 Numerical results and discussion

(1) Comparison with the experimental measurement

We compared the simulation results with experimental data, which is depicted in Fig. 8. The results show that the wall pressure distributions in the scramjet combustor predicted by the simulation are close to the experimental measurements(considering the complexity of simulating the reacting field in the scramjet combustor, it can be think so). Furthermore, the simulated results show that the addition of H_2O in the test gas decreases the combustor wall pressures, illustrating the same tendency as the experimental results show.

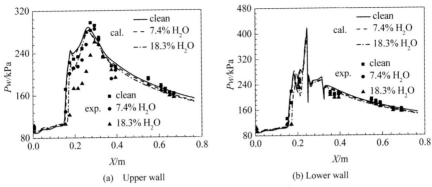

(a) Upper wall　　　　　　　(b) Lower wall

Fig. 8 Comparison of wall pressure distributions between calculated results and experimental results

(2) Local mixing and local combustion

The extents to which the hydrogen fuel mixes and combustion locally with the clean air or vitiated air are expressed by using local total hydrogen/nitrogen ratio $r_{H,N}$ and by using Pseudo mole fraction of H_2O $x^*_{H_2O}$, respectively. Fig. 9 shows the contours of $r_{H,N}$ based on equ. (2) for three cases, in which only half the combustor is illustrated for each test gas for the sake of comparison with clarity. Note that the stoichiometric value of H to N is 0.532. Clearly, the local mixing process with clean air is better than that with vitiated air.

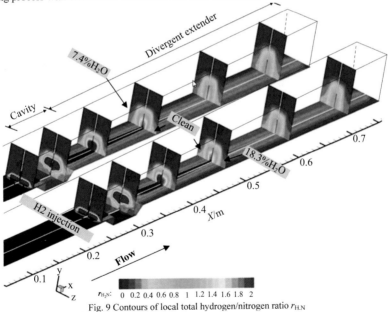

Fig. 9 Contours of local total hydrogen/nitrogen ratio $r_{H,N}$

Fig. 10 shows the contours of pseudo mole fraction of H_2O based on eqn. (1) for three cases. It is clear that the pseudo mole fraction of H_2O along flow direction for clean air is highest, and for air vitiated with 18.3% H_2O is lowest. That is, H_2O produced by combustion of hydrogen fuel with clean air are higher than that with vitiated air.

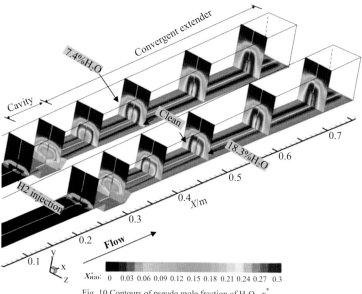

Fig. 10 Contours of pseudo mole fraction of H_2O $x^*_{H_2O}$

(3) Static temperature and Mach number

　　The 1-D mass-weighted curves of static temperature and of Mach number are shown in Fig. 11 and Fig. 12, respectively. The contours of static temperature and of Mach number are shown in Fig. 13 and Fig. 14, respectively. It is obvious that the distributing profiles of static temperature in Fig.13 are similar for three cases, but the values of static temperature in clean air are generally higher than those in two cases vitiated with H_2O as shown in Fig. 11 and Fig. 13. Similarly, the distributing profiles of Mach number in Fig.14 are similar for three cases, but the values of Mach number in clean air are generally lower than those in two cases vitiated with H_2O as shown in Fig. 12 and Fig. 14.

　　The higher static temperature for clean air can be attributed to two causes. One is that the addition of H_2O increases the heat capacity of test gas. Another is that the amount of heat release from hydrogen combustion in vitiated air is lower than that in clean air, considering H_2O produced by hydrogen in vitiated air is lower as shown in Fig. 10. Meanwhile, the smaller amount of heat release from hydrogen combustion in vitiated air can lead to lower static temperature as shown in Fig. 11 and Fig. 13.

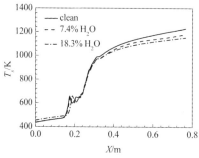

Fig. 11 1-D mass-weighted static temperature

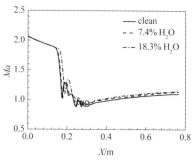

Fig. 12 1-D mass-weighted Mach number

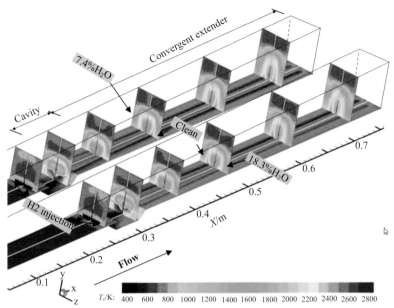

Fig. 13 Contours of static temperature T_s

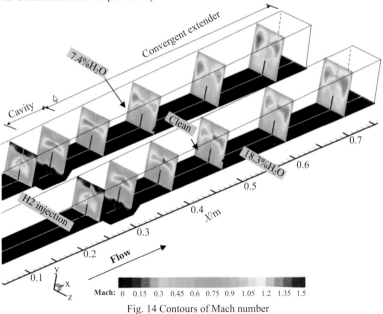

Fig. 14 Contours of Mach number

(4) Mixing efficiency and combustion efficiency

Fig.15 and Fig. 16 show the mixing efficiency η_{mix} based on eqn. (4) and the combustion efficiency $\eta_{c,q}$ based on eqn. (5) along the flow direction for three cases, respectively. The mixing efficiency η_{mix} and the combustion efficiency $\eta_{c,q}$ for clean air are highest, and for air vitiated with 18.3% H_2O is lowest. Furthermore, the differences of η_{mix} and of $\eta_{c,q}$ for clean air and vitiated air gradually become larger along the flow direction. The higher combustion efficiency in clean air is in accordance with the better local combustion in clean air (Fig. 10).

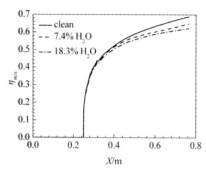 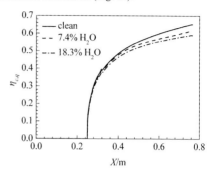

Fig. 15 mixing efficiency η_{mix} along flow direction   Fig. 16 combustion efficiency $\eta_{c,q}$ along flow direction

（5）Overall performance

To evaluate the vitiation effects of H_2O on overall performance, we consider about two parameters: internal thrust and combustion efficiency at the combustor exit. They are listed for three cases in Tab. 1. In Tab.1, we can see that the additions of 7.4% H_2O and of 18.3% H_2O to test gas decrease the thrust of the combustor by 4.3% and 15.2% respectively, and decrease the combustion efficiency at the combustor exit by 5.5% and 9.5% respectively.

Tab. 1 Overall performance of combustor for three cases

Test gas	Thrust/N	Combustion efficiency at exit
Clean air	117.0	0.652
Air vitiated with 7.4% H_2O	112.0	0.616
Air vitiated with 18.3 H2O	99.2	0.590

4 Conclusions

Understanding the complicated issues with regard to the effects of airstream vitiation on scramjet test performance requires a lot of efforts involving experiment, modeling and CFD simulation. Based on a direct-connected resistance heated wind tunnel, we developed a test facility to support comparative experiments for combustion in clean air and air vitiated with H_2O/CO_2. Under the same test gas parameters of nominal Mach 2.0, total temperature of 800K and total pressure of 760kPa, a set of well –controlled comparative combustion tests for a hydrogen-fueled combustor with a cavity were conducted. Then, massively parallel CFD simulations of reacting flowfield in the combustor for three cases corresponding to experimental conditions were conducted.

The findings and conclusions from the comparative experiments are as follows:

(1) At an equivalence ratio of 0.42, wall pressures of the combustor due to combustion decrease nonlinearly with increasing molar percentage of H_2O and/or CO_2 in test gas. Combustion in the combustor is in dual-mode operation with the clean air, but in supersonic mode with the air vitiated with 18.3% H_2O and with 7.7%CO_2+7.4%H_2O.

(2) At an equivalence ratio of 0.53, the combustion pressure rise in the combustor propagates into the facility nozzle for the clean air, but cannot propagate into the nozzle for air

vitiated with 7.5%~25.2% H_2O. That is, H_2O vitiation in test gas could transits inlet unstart with clean air to inlet startup.

(3) Applying law of feeding obtained in vitiated test gas directly to flight could result in over-fueling of the combustor, even leading to possible inlet unstart or unexpected combustion mode transition.

The findings and conclusions from the numerical investigation are as follows:

(1) The simulated results involving a relatively detailed chemical kinetic model are close to measured wall pressures of comparative experiments, and could "distinguish" the vitiation effects of H_2O in test gas.

(2) The extents to which the hydrogen fuel mixes and combustion locally with the clean air are higher than with vitiated air, leading to the higher static temperature and the lower Mach number in reacting flowfield than with vitiated air.

(3) For overall performance of the combustor, 7.4% H_2O and 18.3%H_2O contaminants of inflow decrease the thrust of the combustor by 4.3% and 15.2% respectively, and decrease the combustion efficiency at the combustor exit by 5.5% and 9.5% respectively, mainly due to the higher heat capacities and the lower combustion efficiency for two vitiated cases.

References

[1] Mitani, T., Hiraiwa, T., Sato, et al. Comparison of scramjet engine performance in Mach 6 vitiated and storage-heated-air. Journal of Propulsion and Power, 1997, 13(5): 635-642.

[2] C. P. Goyne, J. C. McDaniel Jr., R. H. Krauss, et al. Test gas vitiation effects in a dual-mode scramjet combustor. Journal of Propulsion and Power, 2007, 23(3): 559-565.

[3] C. J. Jachimowski. An analytical study of the hydrogen-air reaction mechanism with application to scramjet combustion. NASA TP-2791, 1988.

[4] S. Srinivasan, W. D. Erickson. Influence of test-gas vitiation on mixing and combustion at mach 7 flight conditions. AIAA 1994-2816.

[5] R. R. Boyce., M. Wendt., A. Puall, et al. Supersonic combustion-a shock tunnel and vitiation-heated blowdown tunnel comparison. AIAA 1998-0941, 1998.

[6] C. P. Goyne, D. Cresci. Hy-V program overview and status. AIAA 2008-2577, 2008.

[7] W. Chinitz, J. I. Erdos. Test facility chemistry effects on hydrocarbon flames and detonations. AIAA 1995-2467, 1995.

[8] G. L. Pellett, C. Bruno, W. Chinitz. Review of air vitiation effects on scramjet ignition and flamholding combustion processes. AIAA 2002-3880, 2002.

超声速化学反应流动的 LU-SGS 伪时间迭代空间推进求解

贺旭照[1,2]，乐嘉陵[1,2]，宋文艳[2]

(1. 中国空气动力研究与发展中心 计算空气动力研究所，621000 绵阳；

2. 西北工业大学 动力与能源学院，710072 西安)

摘　　要： 介绍了求解超声速多组分有限速率化学反应流动的伪时间迭代(lower upper symmetric Gauss Seidel, LU-SGS) 方法. 空间推进求解多组分抛物化(parabolized Navier Stokes, PNS) 方程时，在一个推进面上采用修正的 LU-SGS 方法迭代至收敛，把得到的结果作为初值赋给下一个推进面. 沿推进面依次迭代直至求解完整个流场. 采用伪时间迭代 LU-SGS 方法求解化学反应 PNS 方程，计算结果的准确性和时间迭代求解完全 Navier Stokes (N-S) 方程相当，求解效率提高一个数量级.

关 键 词： 超声速流动；化学反应；PNS (parabolized Navier Stokes) 方程；空间推进；

　　　　　LU-SGS (lower upper symmetric Gauss Seidel) 方法

中图分类号： V 211. 3　　　　　**文献标识码：** A

LU-SGS pseudo-time iteration space marching method for solving supersonic chemical reaction flow

HE Xu-zhao[1,2], LE Jia-ling[1,2], SONG Wen-yan[1]

(1. Computational Aerodynamics Institute,

China Aerodynamic Research and Development Center, Mianyang 621000, China;

2. School of Power and Energy,

Northwestern Polytechnical University, Xi'an 710072, China)

Abstract: A pseudo time iteration lower-upper symmetric-Gauss Seidel (LU-SGS) method for solving supersonic multi-species chemical reaction flow was introduced. When solving parabolized Navier Stokes (PNS) equations by space marching method, the modified LU-SGS method was used in the marching plane until reaching convergence. This plane's results were transferred to next plane and the entire flow field was solved plane by plane. Using pseudo time iteration LU-SGS method to solve the chemical reaction PNS equation, the result has the same level of accuracy compared with those obtained from Navier-Stokes (N-S) equations solved by time iteration method, and the computation efficiency is improved to an order or more.

Key words: supersonic flow; chemical reaction;

　　　　　　PNS (parabolized Navier Stokes) equation; space marching;

　　　　　　LU-SGS (lower upper symmetric-Gauss-Seidel) method

贺旭照，乐嘉陵，宋文艳. 超声速化学反应流动的 LU-SGS 伪时间迭代空间推进求解. 航空动力学报，2010，25(5)：1043-1048.

空间推进方法适用于求解超声速流动占主导的沿主流方向无分离的流场. 采用空间推进方法求解这类流场时, 在不失求解流场准确性的情况下, 时间效率比时间迭代方法提高一个数量级以上. 空间推进方法可应用在高超飞行器气动力快速评估, 可作为高效可靠的数值工具应用在高超飞行器气动优化设计中.

在国外, 空间推进方法已经成为各大气动实验室的重要研究工具[1]. 国外具备了采用空间推进方法求解各类超声速流动的能力[2-4], 包括超声速无化学反应流动、超声速化学反应流动等; 并把空间推进方法和时间迭代方法相结合, 完成了对超声速飞行器分区求解[1,5].

在国内, 随着吸气式高超声速推进技术的深入研究, 空间推进求解超声速流动重新受到关注. 贺旭照[6]、陈兵[7]等人分别进行了空间推进求解超声速量热完全气体流动的研究, 并开发了相应的计算流体动力学(CFD)软件, 具备了空间推进求解超声速层流流动的能力. 超声速化学反应流动在超燃冲压发动机中普遍存在, 发展超声速化学反应流动的空间推进求解, 可以为超燃冲压发动机尾喷管的快速评估及优化设计奠定技术基础.

空间推进求解超声速化学反应流动有以下困难: 在数值求解方面, 现有的时间迭代方法不能直接应用在每个推进面的化学反应流动的求解中,

需要基于现有时间迭代方法, 发展超声速化学反应流动的空间推进求解方法; 有限速率化学反应流动的控制方程中, 包含主流控制方程、湍流控制方程和组分控制方程, 对组分控制方程的扩散项及源项的抛物化处理, 需要仔细研究和实践, 一方面要保证控制方程的抛物化性质, 另一方面要保证控制方程抛物化处理后, 化学反应的扩散及反应规律不被破坏; 由于化学反应源项加入, 使求解在每个推进平面上的收敛变得困难.

本文详细分析了抛物化化学反应流动控制方程的数学性质, 给出了化学反应 PNS (parabolized Navier-Stokes) 程的伪时间 LU-SGS (lower upper symmetric Gauss-Seidel) 求解方法. 开发了超声速化学反应流动的空间推进求解 CFD 软件. 采用 H₂-Air 的 7 组元 8 步化学反应模型求解了两个超声速化学反应流动问题, 验证了计算方法和计算结果, 为进一步的快速评估和优化工作奠定了基础.

1 化学反应抛物化 Navier-Stokes (N-S) 方程的伪时间 LU-SGS 迭代方法

在计算坐标系中以 ξ 方向作为主流方向, 抛物化 Navier-Stokes(N-S) 方程是通过对完全 N-S 方程丢掉 ξ 方向的黏性耗散项以及对 ξ 方向的无黏通量作 Vigneron[9] 修正得到.

$$\frac{\partial Q}{\partial t} + \frac{\partial E}{\partial \xi} + (\omega - 1)\frac{\partial E^p}{\partial \xi} + \frac{\partial F}{\partial \eta} + \frac{\partial G}{\partial \zeta} = \frac{1}{Re}\left(\frac{\partial F^{*v}}{\partial \eta} + \frac{\partial G^{*v}}{\partial \zeta}\right) + S \tag{1}$$

$$E^p = \frac{|\Delta\xi|}{J}\begin{bmatrix} 0 & \xi_x p & \xi_y p & \xi_z p & 0 & 0 & \cdots & 0 & 0 & 0 \end{bmatrix}^T \tag{2}$$

$$\omega = \min\left[1, \frac{\sigma\gamma M_\xi^2}{1 + (\gamma - 1)M_\xi^2}\right] \qquad M_\xi^2 = \frac{U^2}{\gamma p/\rho} \tag{3}$$

$Q = \frac{1}{J}[\rho \; \rho_u \; \rho_v \; \rho_w \; \rho E \; \rho_{c_1} \; \cdots \; \rho_{c_{ns-1}} \; \rho_k \; \rho_\omega]^T$ 包含了 5 个主流控制方程变量, $ns-1$ 个组分变量和 k-ω 两方程湍流变量. ns 为反应气体组分数目.

湍流模型采用了 Kok[10] 提出的 TNT (turbulent non turbulent) k-ω 模型. S 为化学反应和湍流控制方程的源项

$$S = \begin{bmatrix} 0 & 0 & 0 & 0 & 0 & \frac{L}{\rho_\infty V}w_1 & \cdots & \frac{L}{\rho_\infty V}w_{ns-1} & \frac{P_k}{Re} - \beta_k\rho\omega k & \frac{P_\omega}{Re} - \beta\rho\omega^2 + \sigma_1\frac{\rho}{\omega}\max(\omega,0) \end{bmatrix}^T \tag{4}$$

对第 i 种组分的化学生成源项可以表示为

$$w_i = M_i \sum_{j=1}^n (\beta_{ij} - \alpha_{ij})(R_j - R_{-j}) \tag{5}$$

M_i 为第 i 种组分的克摩尔质量, R_j 和 R_{-j} 分别代表以浓度表示的第 j 个基元反应的正反应和逆反应的速率. F^{*v} 和 G^{*v} 略去了所有沿 ξ 方向偏导数

项的 η 和 ζ 方向的黏性通量.

为了解决空间推进求解超声速化学反应流动的困难, 本文采用修正的 LU-SGS 迭代方法, 配合化学反应源项的点隐式处理、当地时间步法实现了对化学反应流动的空间推进求解. 比较了多种化学反应源项的抛物化处理方法, 认为源项的流向偏导数采用向后一阶差分能较好的模拟化学反应流动, 且求解具有较好的收敛特性. 具体采用

的数值求解方法如下文所述.

对式(1)的空间推进求解, 在推进平面上采用伪时间迭代 LU-SGS 方法. 主流控制方程和组分控制方程耦合求解, 湍流控制方程解耦求解. 对于式(1)在 i 平面上进行的推进求解, 在 $i-1$ 和 $i+1$ 平面上, $\Delta Q = 0$ 消除了 $A^{\pm}_{i,j,k}$ 对迭代过程的贡献, 经过修正的 LU-SGS 方法为

$$LD^{-1}U\Delta Q_{ijk} = \frac{\Delta t}{V} R^n_{i,j,k} \tag{6}$$

$$D = \left[1 + \frac{\Delta t}{V}\left(\lambda^{A^*}_{max} + \lambda^B_{max} + \lambda^C_{max} + 2\lambda^v_B + 2\lambda^v_C \right) \right] I \tag{7}$$

$$L = D + \Delta t^* \Omega - \frac{\Delta t}{V}\left(B^{+n}_{i,j-1,k} + C^{+n}_{i,j,k-1} \right) \tag{8}$$

$$U = D + \frac{\Delta t}{V}\left(B^{-n}_{i,j+1,k} + C^{-n}_{i,j,k+1} \right) \tag{9}$$

$$\Delta Q_{ijk} = Q^{m-1}_{ijk} - Q^n_{ijk} \tag{10}$$

$$R^n_{i,j,k} = -\left[\left(\frac{\partial E^*}{\partial \xi} + \frac{\partial F}{\partial \eta} + \frac{\partial G}{\partial \zeta} + S \right)^n - \frac{1}{Re}\left(\frac{\partial F^v}{\partial \eta} + \frac{\partial G^v}{\partial \zeta} \right)^n \right] \tag{11}$$

其中 B 和 C 为守恒变量 Q 对无黏通量 F 和 G 的偏导数, $\lambda^{A^*}_{max}$ 为 A^* 矩阵的最大特征值

$$A^* = \frac{\partial E}{\partial Q} + (\omega_r - 1)\frac{\partial E^p}{\partial Q} =$$

$$\begin{bmatrix}
0 & \xi_x & \xi_y & \xi_z & 0 & 0 \\
-Uu + \xi_x\omega_r p_\rho & U + \xi_x(u + \omega_r p_{\rho u}) & \xi_x u + \xi_y\omega_r p_{\rho v} & \xi_x u + \xi_z\omega_r p_{\rho w} & \xi_x\omega_r p_{\rho E} & \xi_x\omega_r p_{\rho_{s1}} \\
-Uv + \xi_y\omega_r p_\rho & \xi_y v + \xi_x\omega_r p_{\rho u} & U + \xi_y(v + \omega_r p_{\rho v}) & \xi_y v + \xi_z\omega_r p_{\rho w} & \xi_y\omega_r p_{\rho E} & \xi_y\omega_r p_{\rho_{s1}} \\
-Uw + \xi_z\omega_r p_\rho & \xi_z w + \xi_x\omega_r p_{\rho u} & \xi_z w + \xi_y\omega_r p_{\rho v} & U + \xi_z(w + \omega_r p_{\rho w}) & \xi_z\omega_r p_{\rho E} & \xi_z\omega_r p_{\rho_{s1}} \\
-U(H + p\rho) & \xi_x H + Up_{\rho u} & \xi_y H + Up_{\rho v} & \xi_z H + Up_{\rho w} & U(1 + p_{\rho E}) & Up_{\rho_{s1}} \\
-Uc_s & c_s\xi_x & c_s\xi_y & c_s\xi_z & 0 & U\delta_{s,s1}
\end{bmatrix} \tag{12}$$

其中下标 s 和 $s1$ 代表组分在行和列的下标. 关于压力 p 的各项偏导数可以由下面的式子求到:

$$dp = p_{\rho E}\left[d(\rho E) - ud(\rho u) - vd(\rho v) - wd(\rho w) \right] + \left[p_{\rho E}\left(\frac{u^2 + v^2 + w^2}{2} - e_{ns} \right) + \frac{T}{M_{ns}} \right]d\rho - \sum_{s=1}^{ns-1}\left[p_{\rho E}(e - e_{ns}) + \frac{T}{M_s} - \frac{T}{M_{ns}} \right]d\rho_s \tag{13}$$

无量纲形式的 $p_{\rho E} = \dfrac{\sum_{s=1}^{ns}\dfrac{\rho_s}{M_s}}{\rho_{cv}}$, 其中 M_s 为对应的 s 组分的分子量, $\rho_s = \rho c_s$, c_v 为无量纲的比定容热容. 式(12)对应的特征值为

$$\lambda^*_{1,2,3} = U$$

$$\lambda^*_{4,5} = \frac{1}{2}\cdot\left[(b + 1)U \pm \sqrt{(b - 1)^2 U^2 + 4\omega_r c^2} \right]$$

$$\lambda^*_{max} = U \tag{14}$$

其中 $b = \omega_r + \gamma(1 - \omega_r)$, $\gamma = 1 + p_{\rho E}$ 为气体的冻

结比热比, $c = \sqrt{\gamma p/\rho}$ 表示冻结声速. 通过式(14)可以求出 $\lambda^{A^*}_{max}$.

对黏性通量的线性化, 采用 Tysinger[11] 的方法, 只保留黏性通量的最大特征值.

$$\lambda^v_i = \frac{s_i^2}{VRe}\max\left(\frac{4}{3}\frac{\mu + \mu_t}{\rho}, \frac{K + K_t}{\rho c_v}, D_s \right) \tag{15}$$

i 代表 B 和 C. λ^v_B, λ^v_C 和 s_B, s_C 分别是 η 和 ζ 面黏性通量的最大特征值和 η 、 ζ 方向的面积, D_s 为第 s 个组分的扩散系数, μ 为动力黏性系数, K 为传热系数.

对化学反应源项的线性化,采用 Imlay[12] 的对角隐式方法.

$$\Omega = \frac{\partial S}{\partial Q} = \mathrm{diag}\left\{ \sqrt{\sum_{N=1}^{n-1}\left[\frac{\partial w_i}{\partial \rho_i}\right]^2} \right\} \quad (16)$$

这样就得到了在一个推进平面上的伪时间迭代方法.应用上述方法,在一个推进面上时间迭代求解,直至收敛,把得到的计算结果作为初值赋给下一个推进面,计算在下一个推进面上进行至收敛,循环迭代直至求解完整个流场.在一个区域的初始推进面上,给定自由来流值.湍流方程的求解方法和主流相同.

2　超声速化学反应流的空间推进求解

2.1　二维超声速氢/空气燃烧的推进求解

这个实验主要用来验证多组分有限速率化学反应气体的数值模拟方法,计算模型来自 Burrows 的实验[13].实验模型见图 1.氢气从下壁面顺流平行喷入.计算区域布置两块网格,网格数为 $151\times 81 + 121\times 141$,网格在壁面加密大约在 10^{-5} m 左右.边界条件:给定来流值,下壁面为等温壁,壁温 298 K,上壁面为绝热壁.

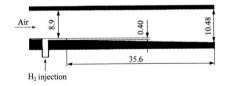

图 1　实验模型的示意图 (单位: cm)

Fig 1　Experimental model (unit: cm)

化学模型采用 Evans[14] 的氢/空气的 7 组分 8 方程模型.来流条件在表 1 中列出.

表 1　来流气体条件

Table 1　Inflow conditions

	氢气喷流	空气来流
Mach 数	1.0	2.44
压强/Pa	101325	101325
温度/K	254	1260
H_2 质量分数	1.0	0
H_2O 质量分数	0	0.256
O_2 质量分数	0	0.258
N_2 质量分数	0	0.486
湍流强度/%	10	10

图 2 给出了反应产物 H_2O 摩尔组分云图的空间推进结果,从云图中可以清楚的看到化学反应发生在来流和氢气的扩散交界面上.图 3 为出口截面摩尔组分沿法向分布的空间推进结果、时间迭代结果和实验测量值的比较.图 4 为出口截面总温分布和实验测量值的比较.从计算的结果看,空间推进方法和时间迭代方法在求解这个问题中有相当的精度,与实验测量结果在峰值区域略有差别,这可能是由于化学反应机理不能完全重现实验状态.图 5 为空间推进方法的时间收敛历程与时间迭代方法的收敛历程的比较,图中显

H_2O　0.146　0.236　0.317　0.338　0.355　0.366　0.374　0.388　0.397

图 2　流场中 H_2O 摩尔组分分布云图

Fig. 2　H_2O mole fraction in flow field

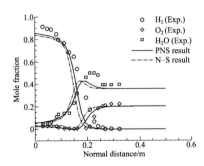

图 3　组分在出口截面浓度的比较

Fig. 3　Species mole fraction density in exit plane

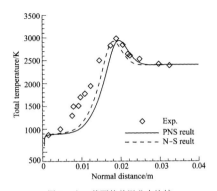

图 4　出口截面的总温分布比较

Fig. 4　Total temperature comparison at exit plane

示了空间推进方法在最后几个推进截面上的收敛
过程. 空间推进求解在每个截面上的 max(Δρ/ρ)
收敛到 10^{-5}. 时间迭代方法采用 LU-SGS 方法求
解与式(1)对应的完全 N-S 方程. 表 2 给出了空
间推进和时间迭代的耗时比较. 本文的计算是在
同一个 2.8 GHz 的 CPU 上完成的. 可以看出, 采
用空间推进方法, 大大的节约了计算时间.

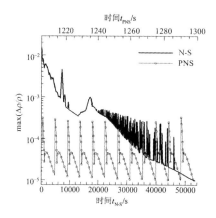

图 5　空间推进/时间迭代的收敛历程

Fig. 5　Convergence rate of space marching and
time iteration method

表 2　求解时间效率比较

Table 2　Time efficiency comparison

时间 t_{N-S}/ s	时间 t_{PNS}/ s	时间比
3 450	1 298	41. 2

2. 2　三维超声速氢/空气燃烧的推进求解

为了显示目前的空间推进方法在处理超声速
三维化学反应流动的能力, 推进求解了一个带侧
壁和上壁面的三维压缩拐角流动. 物体的几何模
型如图 6 所示, 压缩拐角的角度为 α = 15°. 整个
计算区域的网格为 121 × 61 × 62, 约 45. 8 万. 采
用 k-ω 湍流模型, 网格在壁面加密 10^{-5} m 左右.

数值模拟的边界条件为[3]: 来流马赫数
$Ma = 7.0$, 静温 $T = 1200 K$, 单位长度的雷诺数
为 $Re = (1.013 × 10^{6})$/ m, H_2 的质量分数为
0. 032 07, O_2 的质量分数为 0. 254 47, N_2 的质量
分数为 0. 713 46.

图 7 为对称面上压力等值线云图, 可以清楚
的看到前缘平板产生的激波以及由压缩拐角产生
的激波. 图 8 为对称面出口处沿法向的温度分布,

和文献[3]结果比较, 温度在下壁面一致, 在上壁
面的峰值区域, 本文的温度分布略高于文献结果.
图 9 为对称面出口处沿法向的 H_2O 摩尔分布. 本文

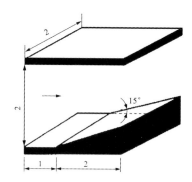

图 6　计算模型的示意图 (单位: cm)

Fig. 6　Computation model map (unit: cm)

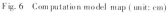

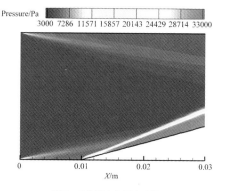

图 7　对称面上的压力云图

Fig. 7　Pressure contour in symmetry plane

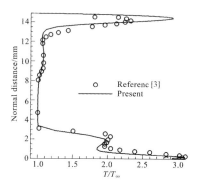

图 8　对称面出口截面温度分布

Fig. 8　Temperature distribution at symmetry exit plane

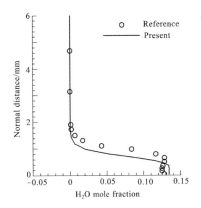

图9 对称面出口截面 H_2O 摩尔组分

Fig. 9 H_2O mole distribution at symmetry exit plane

的计算结果和文献[3]的计算结果作了比较,除了在局部峰值区域存在差别外,其他区域和文献吻合.文献[3]采用的是9组分11步反应模型.这个例子有45.8万个网格点,每个推进面 $\max(\Delta\rho/\rho)$ 收敛到 10^{-4},求解12个控制方程,采用空间推进方法在一个2.8GHz的CPU上,耗时5.72 h.

3 结 论

本文介绍了一种空间推进求解超声速化学反应流动的"伪时间LU-SGS迭代"方法.得到以下两点结论:

1)"伪时间LU-SGS迭代"方法可以成功的应用在求解超声速多组分有限速率化学反应流动当中,求解的准确性和时间迭代方法相当.

2) 采用"伪时间LU-SGS迭代"方法空间推进求解超声速化学反应流动,求解效率比采用相应的时间迭代方法提高一个数量级.开发的空间推进求解超声速化学反应流动的CFD软件可应用在超燃冲压发动机的快速评估、带动力状态下发动机和尾喷管的优化设计等方面.

参考文献:

[1] Anderson J D. hypersonic and high temperature gas dynamics[M]. 2nd. New York: AIAA Education Series, 2006: 445.

[2] John J K. An explicit upwind algorithm for solving the parabolized Navier-Stokes equations[R]. NASA Technical Paper 3050, 1991.

[3] Gerbsch R A. Solution of the parabolized Navier-Stokes equations for three-dimensional real gas flows using Osher's upwind scheme[R]. AIAA 91-0248, 1991.

[4] Wadawadigi G, Tannehill J C, Buelow P E. A three dimensional upwind PNS code for chemically reacting scramjet flowfields[R]. AIAA 92-2898, 1992.

[5] Charles E C, Jr, Wakter C E. Integrated aero-propulsive CFD methodology for the hyper-X flight experiment[R]. AIAA 2000-4010, 2000.

[6] 贺旭照,乐嘉陵.空间推进方法求解抛物化 Navier-Stokes 方程及其验证[J].空气动力学学报,2007, 26(2): 189-193.
HE Xuzhao, LE Jialing. Solving PNS equations with space marching method and it's validation[J]. Acta Aerodynamica Sinica, 2007, 26(2): 189-193. (in Chinese)

[7] 陈兵,徐旭,蔡国飙.一种求解抛物化 Navier-Stokes 方程的空间推进算法[J].力学学报,2008,40(2): 162-170.
CHEN Bing, XU Xu, CAI Guobiao. A space marching algorithm for solving the parabolized Navier-stokes equations[J]. Chinese Journal of Theoretical and Applied Mechanics, 2008, 40(2): 162-170. (in Chinese)

[8] 贺旭照.高超声速飞行器气动力气动热数值模拟和超声速流动的区域推进求解[D].绵阳:中国空气动力研究与发展中心,2007.
HE Xuzhao. Hypersonic vehicle's aerodynamic force/heat numerical simulation and supersonic flow's region marching simulation[D]. Mianyang: China Aerodynamic Research and Development Center, 2007. (in Chinese)

[9] White J A, Morrison J H. A pseudo-temporal multi-grid relaxation scheme for solving the parabolized Navier-Stokes equations[R]. AIAA 99-3360, 1999.

[10] Kok J C. Resolving the dependence on free stream values for the k-omega turbulence model[R]. NLR-TP-99295, 1999.

[11] Tysinger T L, Caughey D A. Implicit multigrid algorithm for the Navier-Stokes equations[R]. AIAA Paper 91-0242, 1991.

[12] Imlay S T, Roberts D W, Soetrisno M, et al. Nonequilibrium thermo-chemical calculations using a diagonal implicit scheme[R]. AIAA Paper 91-0468, 1991.

[13] Burrows M C, Kurkow A P. Analytical and experimental study of supersonic combustion of hydrogen in a vitiated airstream[R]. NASA TM X-2828, 1973.

[14] Evans J S, Schexnayder C J. Influence of chemical kinetics and unmixedness on burning in supersonic hydrogen flames[J]. AIAA Journal, 1979, 18(2): 805-811.

超声速冷态流场液体射流雾化实验研究

陈　亮[1]，乐嘉陵[1,2]，宋文艳[1]，杨顺华[2]，曹　娜[3]

(1. 西北工业大学动力与能源学院，西安　710072; 2. 中国空气动力研究与发展中心，四川 绵阳　621000; 3. 西北核技术研究所，西安　710024)

摘要: 采用全息诊断和高速纹影，对不同来流总压、来流马赫数、喷孔直径和喷射压力等条件下超声速冷态流场液体射流雾化进行了研究。初步了解了超声速流场中液体射流的雾化过程和机理，得到了射流的 Weber 数和 Oh 数等雾化参数，比较了不同条件下射流穿透高度的差异，得到了液滴平均直径和数量密度的空间分布。研究表明: 射流表面不稳定波的增长是超声速流场中射流破碎的主要原因; 射流与气流的动量通量比和喷孔直径影响射流穿透高度，动量通量比和喷孔直径增加都会增加穿透高度; 实验中液体射流的雾化过程非常迅速，在喷嘴下游 20mm 处，直径 0.5mm 的射流就破碎成平均直径 10μm 左右的液滴群，随着液滴向下游运动，平均直径逐渐减小，平均直径和数量密度分布逐渐均匀。

关键词: 超声速流场; 全息诊断; 雾化

中图分类号: O348.12　　**文献标识码**: A

Experimental investigation of liquid jets atomization in supersonic cold crossflow

CHEN Liang[1]，LE Jia-ling[1,2]，SONG Wen-yan[1]，YANG Shun-hua[2]，CAO Na[3]

(1. School of Power and Energy, Northwestern Polytechnical University, Xi'an　710072, China;
2. China Aerodynamics Research and Development Center, Mianyang Sichuan　621000, China;
3. Northwest Institute of Nuclear Technology, Xi'an　710024, China)

Abstract: Holographic diagnosis and high-speed schlieren photography were applied to study the atomization of liquid jets in supersonic crossflow under various experimental conditions, which contain the stream total pressure, the stream Mach number, the nozzle diameter and the injection pressure. The process and the theoretical principal of liquid jets atomization in the supersonic crossflow have been studied preliminary. Atomization parameters such as Weber number and Oh number have been acquired as well. The jet penetration heights under different experimental conditions and the spatial distributions of diameters and number density of the droplets have been measured. Results indicate that the instability wave growth on the surface is one of the primary factors to liquid jet breaking up in supersonic crossflow. The jet penetration height will increase if the jet-to-air momentums flux ratios or orifice diameters increase. Moreover, the study demonstrates the process of liquid jet atomization is very speedy in supersonic crossflow, i.e., the liquid jet is broken into droplets of mean diameter 10μm at 20mm downstream the nozzle bank. As the droplets move downstream, the mean diameter decreases and distributions of mean diameters and number density become uniform gradually.

Key words: supersonic flow; holography diagnosis; atomization

0　引　言

液体碳氢燃料，特别是煤油，在安全性和存储性等方面具有很多优点，是未来飞行马赫数小于 7 超燃冲压发动机的首选燃料。超燃冲压发动机燃烧室入口气流是超声速的，气流在燃烧室内的停留时间仅为毫秒量级，在如此短的时间内，液体燃料要完成自身的雾化蒸发、和来流空气的混合以及混合物的燃烧化学反应，才能满足发动机的要求。在液体燃料燃烧的三个阶段中，液滴的蒸发是最缓慢的环节，要加速燃烧，提高雾化效果、缩短蒸发时间是关键。因此，对于液体碳氢燃料超燃冲压发动机，研究燃料在超声速气流中的雾化机理,寻找提高燃料雾化效果的措施显得尤为重要。

目前,超声速条件下燃料雾化研究主要依靠实验测量。超声速燃烧室气流速度快、温度高,流场复杂,气流和射流的相互作用强烈,实验测量受到很大的限制。相位多普勒粒子分析仪(PDPA)是使用较广泛的一种标准液滴尺寸和速度测量的光学诊断系统,其测量精度高、技术成熟、使用方便,很多研究者采用该系统分别对静态空气、亚声速纵向气流以及超声速横向气流中液体射流雾化特性进行了研究与分析[1-3],测量了流场中液滴的尺寸、速度和数量密度。但是PDPA测量时每个测量点需要采集足够的样本,无法在实验时间很短的脉冲风洞中使用,同时在热流状态下无法进行测量,而且在近喷嘴区域,液滴的非球形特征使得运用PDPA测量很困难。为了解决上述问题,近来研究者开始采用激光全息技术对横向流场中特别是近喷嘴区域射流雾化进行研究[4-6]。笔者采用激光全息诊断和高速纹影进行超声速流场中液体射流雾化的研究,测量三维瞬态流场液滴平均直径和数量密度分布以及射流的穿透高度。实验在冷流的情况下进行,为了减少环境污染,大部分实验采用纯净水代替航空煤油作为测量对象。尽管实验与实际发动机中的雾化存在一定的差异,而且纯净水和航空煤油的物性不同,但是该实验对于评估喷嘴性能以及认识雾化机理仍然具有十分重要的意义。

1 设备实验条件与实验方法

实验在中国空气动力研究与发展中心的Φ450mm直联式脉冲燃烧风洞设备上进行,该设备提供的最大空气流量为25kg/s,加热器工作时,模拟的总温为900~2100K。实验模型分为三段:第一段为设备喷管,通过更换不同的型面,可以将气流加速到马赫数2、2.5或3;第二段是燃料喷注、实验测量段,在该段模型的左右侧壁上,有两个对称布置的观测窗,射流喷孔垂直安装在模型的底板上;第三段为连接真空箱的等直段。

实验使用压阻传感器测量压力,使用64通道最高采样率为100k的并行数据采集系统对压力信号进行放大调制和采集记录。高速纹影系统用于喷雾流场流动显示与拍摄,纹影图像的采集使用分辨率1024×768,帧频1k的高速CCD。液滴平均直径和数量密度的三维分布采用激光全息诊断系统进行测量。激光全息技术使用激光作为光源,利用干涉原理,将与物体光波有关的振幅和相位以强度变化的方式记录在感光材料上,与光波的振幅和相位对应的分别是干涉条纹的反差(灰度)和条纹间距。实验采用的激光全息系统是离轴式的,由两部分组成。一部分是实验现场记录部分,包含布置在厂房里的激光器、分光镜、反射镜、扩束准直镜、光学传输成像镜和记录介质;另一部分是再现采集分析系统,包含光学再现系统、三维空间图像采集系统和图像处理系统。激光器脉冲宽度为170ps,对于1000m/s的流场,曝光时间内的位移不大于0.17μm,可以满足测量要求。实验现场光路布置如图1所示,激光器接受控制系统的触发后,发出的脉冲激光束经过分光镜分为两路(物光路和参考光路),物光路光束经反射镜改变方向之后,通过扩束准直,得到直径为60mm的物光光斑。物光光斑垂直模型侧壁入射,依次穿过流向左侧观测窗、燃烧室内流场和右侧观测窗,之后通过用于保护记录介质的光学传输成像镜,照射在记录介质上。参考光束经过几次反射,从模型下方绕过,通过扩束准直之后形成参考光斑,从侧向照射到记录介质相同的位置上,与物光光斑形成干涉图像,成像在记录介质上。现场实验完成之后,使用相同的参考光照射经过定影处理的记录介质,将原始喷雾场的三维图像再现,然后使用由计算机控制的分层再现和图像采集系统,得到不同层面的粒子图像,最后由图像处理软件,对粒子进行判断识别,处理计算出粒子的三维坐标、粒径和相关统计信息。

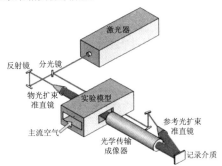

图1 实验现场全息光路示意
Fig.1 Optical layout of the experimental site

2 实验结果与分析

实验状态参数如表1所示。总共有12个实验状态,研究来流总压、来流马赫数、喷孔直径、喷射压力和液体类型对雾化的影响。实验状态1~6采用激光全息诊断系统进行实验测量,实验状态7~12采用高速纹影系统进行流动显示测量。

表1　实验状态参数
Table 1　State parameters in the experiment

实验状态	总温/K	总压/kPa	马赫数	喷孔直径/mm	液体工质类型	喷嘴喷射压力/MPa	动量通量比
1	305	182	2.0	0.5	纯净水	2.0	31
2	307	116	2.0	0.5	纯净水	2.0	48
3	303	205	2.5	0.5	纯净水	2.0	38
4	306	215	2.0	0.3	纯净水	1.5	19
5	306	212	2.0	0.3	纯净水	2.0	26
6	306	210	2.0	0.3	航空煤油	2.0	27
7	305	165	2.0	0.5	纯净水	2.0	34
8	299	107	2.0	0.5	纯净水	2.0	52
9	303	114	2.0	0.5	纯净水	3.0	74
10	305	103	2.0	0.5	纯净水	1.5	41
11	305	142	2.5	0.5	纯净水	2.0	55
12	304	192	2.0	0.3	纯净水	2.0	29

2.1　射流与主流的相互作用

图2为通过喷嘴中心平面的全息再现图像和流场纹影图像的对比,分别属于实验状态1和状态相近的实验状态7,气流方向从左到右。从图中可以看到,在超声速燃烧室中,射流与主流之间相互作用强烈。射流从喷嘴出来之后,液束逐渐变粗,射流受到气流的强烈作用,向主流方向弯曲,呈现出羽状,从喷嘴出来的液束表面存在波动现象,并且这种不稳定波在下游逐渐增长,最后在离喷嘴不远的地方液束分裂成液体段和液体团,并进一步破碎成更小的液滴。在液束的背风面,从喷嘴出口开始,不断有小的液滴脱落进入主流,并随流运动。纹影图显示,水平向右的主流流动受到射流的阻碍,在弯曲射流的前方产生一道明显的脱体激波。

2.2　Weber数和Oh数

Weber数和Oh数是和射流雾化紧密相关的两个无量纲参数。液体射流在空气中做相对运动,一方面它将受到空气动力(惯性力)的作用,这个力促使其破碎;另一方面是液体表面张力的作用,这个力力图射流不破碎。Weber数表征了空气动力和液体表面张力的比值,其表达式为[7]

$$We = \frac{\rho_g u_d^2 d_0}{\sigma} \qquad (1)$$

其中,ρ_g为空气密度,u_d为气液体的速度差,d_0为射流直径或液滴直径,σ为液体表面张力系数。

Oh数是表征粘性力与表面张力之比的无量纲参数,表示液体粘性对破碎过程的影响,其表达式为[7]

$$Oh = \frac{\mu}{\sqrt{\rho_l d_0 \sigma}} \qquad (2)$$

其中,下标l代表液体的参数。

实验中液体射流的初始Weber数和Oh数如表2所示。从表中可以看出,纯净水和航空煤油由于物性的差异,两种射流的Weber数和Oh数相差较大。喷孔直径、主流来流条件和射流喷射压力的变化都会引起Weber数的变化,喷孔直径的变化会引起Oh数的变化。实验状态的Weber数显示射流属于典型的剪切破碎。实验状态的Oh数均小于0.1,液体粘性系数对破碎状态转换的影响不大。

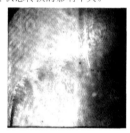

(a) 实验状态1通过喷嘴中心平面的全息再现图像

(b) 实验状态7纹影图像
图2　全息再现图像和纹影图像
Fig. 2　Image from the holographic reconstruction and schlieren image

2.3　射流的穿透高度

图3为实验状态1~6通过喷嘴中心平面的全息再现图像,从图中可以看出,在喷孔直径相同的情况下,射流和主流空气动量通量比越大,射流的穿透高度越大;在射流和主流空气动量通量比相同的情况

下,喷孔直径越大,射流的穿透高度越大。来流状态基本相同的情况下,喷射压力相同的纯净水和航空煤油喷射高度基本相同。

图4所示是实验状态7~12纹影图像的对比,从图中可以看出,随着射流穿透高度发生变化,由于射流阻碍产生的弓形脱体激波在燃烧室上壁面的反射位置也发生变化,穿透高度降低,反射点后移。

表2 液体射流初始 Weber 数和 Oh 数
Table 2 Weber number and Oh number of jets

实验状态	Weber 数	Oh 数
1	920	0.00464
2	582	0.00464
3	740	0.00464
4	651	0.00599
5	643	0.00599
6	1765	0.03077

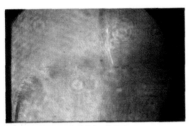

(a) 实验状态1(纯净水)

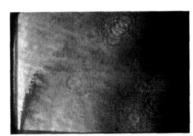

(b) 实验状态2(纯净水)

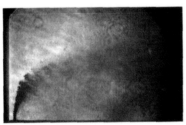

(c) 实验状态3(纯净水)

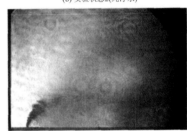

(d) 实验状态4(纯净水)

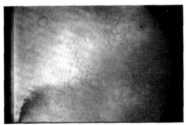

(e) 实验状态5(纯净水)

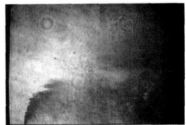

(f) 实验状态6(航空煤油)

图3 通过喷嘴中心平面的全息再现喷雾场图像
Fig. 3 Images going through the center plane of the nozzles from the holographic reconstruction

(a) 实验状态7(纯净水)

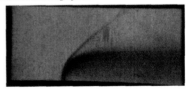

(a) 实验状态8(纯净水)

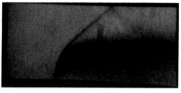

(c) 实验状态9 (纯净水)

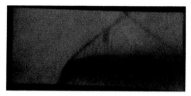

(d) 实验状态10 (纯净水)

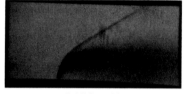

(e) 实验状态11 (纯净水)

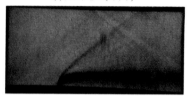

(f) 实验状态12 (航空煤油)

图4 高速纹影图像

Fig. 4 High-speed schlieren images

图 5 是根据纹影图得到的纯净水射流穿透高度与喷孔直径的比值 (h/d_0) 随动量通量比变化的关系。从图中可以看到，随着动量通量比的增加，该比值也增加，但变化趋势不是线性的，而是呈抛物线形，随着动量通量比的增大，比值增加趋缓。

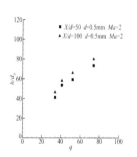

图 5 射流穿透高度与喷孔直径的比值 (h/d_0) 随动量通量比变化

Fig 5 h/d_0 value for different jet to air momentums flux ratios

2.4 液滴的空间分布

图 6 是 3 个不同流向位置 y-z 平面上液滴数量密度和 SMD 的分布云图。坐标原点位于喷嘴出口中心，x 方向为流向，y 方向为射流的出射方向，z 方向为光轴方向，W 和 H 分别是燃烧室横截面的宽度和高度。从图中可以看出，液滴在喷嘴下游 20mm 的地方平均直径已经不到 $10\mu m$。随着液滴向下游运动，平均直径逐渐减小，同时平均直径和数量密度的分布都趋向均匀。

3 结 论

实验采用激光全息诊断和高速纹影测量方法，对超声速冷态流场液体射流雾化进行了实验研究，主要结论如下：

(1) 在超声速冷态内流场中，液体射流与主流气体间的相互作用非常强烈：射流从壁面出射以后，受到主流气动力的影响，向主流方向弯折，呈现出羽状；同时，主流由于受到射流的阻碍，在弯曲射流的前方产生一道明显的脱体激波。实验得到的全息再现图片清晰地展示了超声速流场中液体射流的雾化过程：液束从喷嘴出来以后，表面存在波动现象，并且这种不稳定波逐渐增长，液束变粗，在距离喷嘴不远的地方分裂成液体段和液体团，并进一步破碎成更小的液滴；另一方面，从喷嘴出来的液束背风面不断有小的液滴脱落进入主流，并随流运动；

(2) 射流与主流的动量通量比和喷孔直径是影响射流穿透高度的主要因素：喷孔直径相同的情况下，动量通量比越大，射流的穿透高度越大；动量通量比相同的情况下，喷孔直径减小，射流的穿透高度减小。穿透高度和孔径的比值随动量通量比的增加而增加，但不是线性的，趋势呈现抛物线形，随着动量通量比的增大，比值增加趋缓；

(3) 超声速流场中液体射流的雾化过程非常迅速，在喷嘴下游 20mm 的地方，直径 0.5mm 的射流就会破碎成平均直径 $10\mu m$ 左右的液滴群。随着液

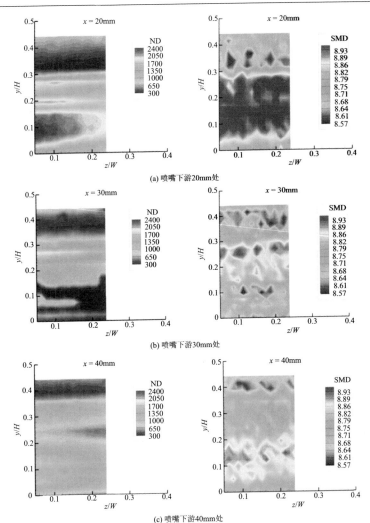

图6　y-z 平面液滴数量密度(1/mm³)和SMD(μm)分布云图
Fig. 6　y-z plane distributions of droplet number density (1/mm³) and droplet SMD(μm)

滴向下游运动,平均直径逐渐减小,平均直径和数量
密度分布逐渐均匀。

参考文献:

[1]　YU G, LI J G, YUE L J, et al. Characterization of kerosene combustion in supersonic flow using effervescent atomization[R]. AIAA 2002-5225, 2002.

[2]　LIN K C, KENNEDY P J, JACKSON T A. Structures of aerated liquid jets in high speed crossflows [R]. AIAA 2002-3178, 2002.

[3]　LIN K C, KENNEDY P J, JACKSON T A. Structures of water jets in a Mach 1.94 supersonic crossflow[R]. AIAA 2004-971, 2004.

[4]　MILLER B, SALLAM K A, BINGABR M, et al. Secondary breakup of aerated liquid jets in subsonic crossflow[R]. AIAA 2007-1342, 2007.

[5]　OLINGER D S, SALLAM K A, LIN K C, et al. Effects of GLR on the spray in the near injector region of aerated liquid jets in crossflow [R]. AIAA 2009-1373, 2009.

[6]　SALLAM K A, LIN K C, CARTER C D. Spray structure of aerated liquid jets using double view digital holography[R]. AIAA 2010-1940, 2010.

[7]　曹建明. 喷雾学[M]. 北京: 机械工业出版社, 2005.

NUMERICAL INVESTIGATIONS OF UNSTEADY SPRAY COMBUSTION IN A LIQUID KEROSENE FUELED SCRAMJET

Jialing Le, Shunhua Yang, Xiyao Wang
China Aerodynamics R&D Center, Mianyang, Sichuan, 621000, P.R.China

Abstract

Numerical simulation of spray combustion coupling the atomization and combustion in a liquid kerosene fueled scramjet is performed. Numerical results indicate the overall ignition delay is about 5ms, which is close to our experimental data. The time for fuel breakup and evaporation is about 0.2ms and 0.8ms respectively, the time for chemical delay is about 4ms.

Introduction

As we know, injection of liquid fuel into high-speed cross flows is a critical process for liquid fueled scramjet/ramjet engine applications. A great number of experimental data, including what we have obtained in our facility, indicates that the combustion performance depends strongly on fuel atomization and the mixing process of the free stream and liquid fuel spray. Understanding the whole process from liquid breakup to combustion is also very important to our pulse combustion facility.

As we have reported that the measured self-ignition delay time for our typical scramjet model of 1m long is about 4~5ms under M5, 6 test condition in our pulse combustion facility. This delay time includes three parts: time for fuel breakup, fuel droplet evaporation and chemical reaction.

In the past, numerical investigations have been carried out by us to evaluate the self-ignition delay time in a liquid kerosene fueled scramjet[1], but only separated two stages of computation were performed in our previous calculation. For the first stage, an unsteady computation with gaseous kerosene injection and combustion is performed without considering fuel atomization. For the second stage, computation is performed with only consideration of fuel atomization to evaluate the time needed for fuel breakup and evaporation. In fact, those two processes (atomization and combustion) are coupled and can not be divided separately. Combustion is strongly dependent on the size and distribution of fuel droplets after atomization; on the contrary, heat release and species change caused by combustion will accelerate the processes of breakup and evaporation.

The purpose of the present study is to numerical investigation of unsteady spray combustion in a kerosene fuelled scramjet coupling two processes of atomization and combustion together.

Numerical Method

A fully implicit and time-accuracy numerical method is developed for analysis of the unsteady combustion in a scramjet. The dual-time stepping with sub-iteration method is used to ensure the second-order time accuracy. In the method, a pseudo-time derivative term is added to the governing equations,

$$\frac{\partial Q}{\partial \tau}+\frac{\partial Q}{\partial t}+\frac{\partial F}{\partial \xi}+\frac{\partial G}{\partial \eta}+\frac{\partial H}{\partial \varsigma}=W \tag{1}$$

Where, $Q=\left(\rho, \rho u, \rho v, \rho w, \rho E_t, \rho_i\right)^T$ is the vector of conservation variables, i stands for the ith species, ns is the total number of species, W is the source vector for chemical reaction, ρ_i is the density of ith species and can be written as following:

Jialing Le, Shunhua Yang, Xiyao Wang. Numerical investigations of unsteady spray combustion in a liquid kerosene fueled scramjet. 20th International symposium on air breathing engines. Goteborg, Sweden, 2011, ISABE-2011-1524.

$$\rho_i = \rho \cdot Y_i \quad (2)$$

E_t is the total energy:

$$E_t = \frac{1}{2}\left(u^2 + v^2 + w^2\right) + \sum_{i=1,}^{ns} Y_i h_i - \frac{p}{\rho} \quad (3)$$

The physical time derivative term is discretized in a second-order time accuracy form:

$$\frac{\partial Q}{\partial t} = \frac{3Q^m - 4Q^n + Q^{n-1}}{2\Delta t} \quad (4)$$

Where m is the sub-iteration number of pseudo-time, n is the iteration number of physical time. The pseudo-time derivative is discretized with first-order backward differentiation,

$$\frac{\partial Q}{\partial \tau} = \frac{Q^{m+1} - Q^m}{\Delta \tau} \quad (5)$$

Finally, the governing equations (1) can be written in the discretized form as following:

$$\left(\frac{1}{\Delta \tau} + \frac{3}{2\Delta t}\right)\Delta Q^m = -\left(\frac{3Q^m - 4Q^n + Q^{n-1}}{2\Delta t}\right)$$
$$+ RHS\left(Q^{m+1}\right)$$
$$(6)$$

Where $\Delta Q^m = Q^{m+1} - Q^m$, the residual vector RHS can be evaluated as following:

$$RHS\left(Q^{m+1}\right) = -F_{i+\frac{1}{2}}^{m+1} + F_{i-\frac{1}{2}}^{m+1} -$$
$$G_{j+\frac{1}{2}}^{m+1} + G_{j-\frac{1}{2}}^{m+1} - H_{k+\frac{1}{2}}^{m+1} + H_{k-\frac{1}{2}}^{m+1} + W^{m+1} \quad (7)$$

The chemical source and RHS term in equation (7) can be linearized, which result in the following equation:

$$\left(\frac{1}{\Delta \tau} + \frac{3}{2\Delta t} - \frac{\partial RHS}{\partial Q} - \frac{\partial W}{\partial Q}\right)\Delta Q^m =$$
$$-\left(\frac{3Q^m - 4Q^n + Q^{n-1}}{2\Delta t}\right) + RHS\left(Q^m\right) \quad (8)$$

Equation (8) can be solved using the LU-SGS method. After linearization of the numerical fluxes by a first-order Steger-Warming vector splitting scheme, equation (8) can be rewritten in the following form:

$$\left(L + D + U\right)\Delta Q^m = \overline{RHS} \quad (9)$$

Where

$$\overline{RHS} = -\left(\frac{3Q^m - 4Q^n + Q^{n-1}}{2\Delta t}\right) + RHS\left(Q^m\right)$$

$$L = -\left(A_{i-1/2}^+\right)_{i-1,j,k} - \left(B_{j-1/2}^+\right)_{i,j-1,k} - \left(C_{k-1/2}^+\right)_{i,j,k-1}$$
$$- Z_{i,j,k}$$

$$U = \left(A_{i+1/2}^-\right)_{i+1,j,k} + \left(B_{j+1/2}^-\right)_{i,j+1,k} + \left(C_{k+1/2}^-\right)_{i,j,k+1}$$

$$D = \left(\frac{1}{\Delta \tau} + \frac{3}{2\Delta t}\right)I + \left(\begin{array}{c}A_{i+1/2}^+ - A_{i-1/2}^- + B_{j+1/2}^+ - \\ B_{j-1/2}^- + C_{k+1/2}^+ - C_{k-1/2}^-\end{array}\right)_{i,j,k}$$

$$A = \frac{\partial F}{\partial Q}, B = \frac{\partial G}{\partial Q}, C = \frac{\partial H}{\partial Q}$$

Where A, B and C are the Jacobian matrices of the flux vector, Z is the Jacobian matrix of chemical source vector W.

The LU-SGS method is obtained when the LHS of equation (9) is factorized as following:

$$\left(L + D\right)D^{-1}\left(U + D\right)\Delta Q^m = \overline{RHS} \quad (10)$$

The LU-SGS scheme is now inverted by a forward and a backward sweep:

$$\left(L + D\right)\Delta Q^* = \overline{RHS}$$
$$\left(U + D\right)\Delta Q^m = D\Delta Q^* \quad (11)$$

The above mentioned method is integrated into CFD code AHL3D, which can be used to solve the Reynolds-averaged Navier-Stokes equations (in 2-D, axisymmetric, or 3-D form) with a structured cell-centered finite volume method, which is appropriate for flows of calorically perfect gas or for flows of an arbitrary mixture of thermally perfect gases undergoing non-equilibrium chemical reactions.

Numerical Method Validation

To validate the numerical method, two unsteady cases are simulated. The first one is the unsteady shock-induced combustion, the second one is the oscillation of the cold field in the scramjet observed in our experiments.

Lehr[2] and Jeong-Yeol Choi[3] studied the unsteady shock-induced combustion experimentally and numerically respectively, we use the Ma 4.48 case to validate our numerical method. Fig.1 shows the density history along the stagnation streamline, Table. 1 gives the frequency comparison.

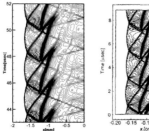

Fig.1 Temporal variation of density along stagnation streamline. Left: our result; Right: Jeong-Yeol Choi result

Talbe. 1 comparison of frequency

	Frequency, KHz
Experiment	425
Present result	426
Jeont-Yeol Choi	426

In experiments we observe the oscillation of scramjet cold field, so we numerically simulated this phenomenon using 2-dimentional U-RANS. The direct-connected combustor is shown in Fig.2, the cavity of the combustor had a L/D ratio more than 10. The stagnation temperature and pressure is 1500K and 1.5MPa respectively, the entrance Mach number is 2.5. A uniform flow field with T=298K, P=100Pa and Mach=0.0 is initialized at the start of the computation (t=0), then the unsteady numerical computation was performed until t=60ms. The time traces of pressure at two wall points (ponit1 and point2 in Fig. 2) are monitored. Fig.3 shows the numerical wall pressure oscillation. The numerical pressure frequncies are 50HZ, which agrees well with our experimental results. As our estimation, the oscillation is caused mainly by the impingement of shear layer onto the rear step of cavity.

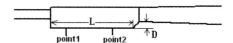

Fig.2 schematic of scramjet

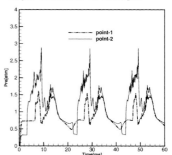

Fig.3 wall pressure oscillation

Atomization Model

Based on the hybrid model of Patterson and Reitz[4], we have developed a new droplet breakup model to simulate the atomization process of fuel jet in supersonic cross flows[5]. In supersonic cross flows, we assume that Kelvin-Helmhotz(K-H) and Rayleigh-Taylor(R-T) instabilities are simultaneous phenomena in the breakup process, which is similar to the breakup model proposed by Patterson and Reitz. Under this assumption, the K-H and R-T wave models are used together to model the two types of instability. The initiation of the breakup is determined by the breakup length l_{bu}, which is an empirical correlation and can be deduced from experiment. Thus, if the walk distance of droplets after injection is within the breakup length, no breakup occurs. If the distance is greater than the breakup length, the present model is used to simulate the droplet breakup. First, the R-T model is used. If no breakup occurs, then the K-H model is used.

In the K-H model, we only allow the breakup to occur when the Weber number of a droplet is greater than the critical Weber number, We_{cr}. The radius, r, of a newly produced parcel from a parent parcel with radius, a, is given by

$$r = B_0 \Lambda_{KH} \qquad (12)$$

where B_0 is a constant. In supersonic cross flows, we suggest that B_0 is equal to 1.0 rather than 0.61 proposed by Reitz. In equation (12), Λ_{KH} is the wavelength of most unstable K-H wave, which is given as following:

$$\frac{\Lambda_{KH}}{a} = 9.02 \frac{\left(1 + 0.45 Z^{0.5}\right)\left(1 + 0.4 T^{0.7}\right)}{\left(1 + 0.87 We_g^{1.67}\right)^{0.6}} \qquad (13)$$

where $Z = \sqrt{We_l}/\mathrm{Re}_l$ is the Ohnesorge number, $T = Z\sqrt{We_g}$ is the Taylor number.

During the breakup, the parent parcel will reduce in diameter due to the mass stripping off from the parcel surface. The change of the radius of a parent droplet is assumed to follow the rate equation:

$$\frac{da}{dt} = -(a-r)/\tau_{KH} \qquad (r \le a) \qquad (14)$$

where $\tau_{KH} = 3.726 B_1 a / \Lambda_{KH}\Omega_{KH}$ is the breakup time, and $B_1 = 1.732\sim 40$ depending on the injector characteristics. The growth rate corresponding to the most unstable wave, Ω_{KH}, is given by

$$\Omega_{KH}\left[\frac{\rho_l a^3}{\sigma}\right]^{0.5} = \frac{0.34 + 0.38 We_g^{1.5}}{(1+Z)(1+1.4 T^{0.6})} \qquad (15)$$

The R-T breakup model also determines when and how droplets break up by predicting the fastest growing disturbances. The fastest growing wavelength, Λ_{RT}, is given by

$$\Lambda_{RT} = 2\pi C_1 \sqrt{\frac{3\sigma}{a_p(\rho_l - \rho_g)}} \qquad (16)$$

where a_p is the acceleration of the droplet, C_1 is an adjustable number depending on nozzle condition. The associated breakup timescale, calculated from the frequency of the fastest growing wavelength, is given by

$$\tau_{RT} = \sqrt{\frac{\sigma^{0.5}(\rho_l + \rho_g)}{2}\left(\frac{3}{a_p(\rho_l - \rho_g)}\right)^{1.5}} \qquad (17)$$

After a certain enough time, the breakup is allowed to occur. The radius of the newly formed parcels is defined as

$$r = 0.5\Lambda_{RT} \qquad (18)$$

The liquid mass striped from the parent parcel is tracked during the computation. A new parcel is introduced once a sufficient amount of mass, 3% of the initial particle mass, has been shed from the parent particle (Reitz, 1987).

Facility, Engine Model and Experiment

As described in references [6~8], experiments have been carried out in CARDC pulse combustion facility to investigate the time characteristics of self-ignition of pure liquid kerosene, using a typical model combustor of 0.58m in length (consisting of a constant area section and an expanding section). The size of the isolator entry is 40mm×18mm, and the entry conditions are: Mach number 2.09, static pressure 0.32MPa, static temperature 1100K and velocity 1440m/s. Typical wall pressure traces indicate that all the three processes, i.e. fuel drop breaking, evaporation and chemical delay of ignition are completed within about 4 ms.

To further examine the time characteristics of kerosene self-ignition for model engine under free jet condition, we test a typical model engine using profile nozzle with exit diameter of 300mm. Fig.3 shows the principle of the pulse combustion facility. Gaseous fuel (H_2+N_2) and oxidizer (O_2+N_2) are stored in a pair of Ludwieg tubes contained by a pair of quick-acting valves at the end of each tube. The gases are discharged into an un-cooled combustion heater, where combustion takes place, and the desired test gas enthalpy and pressure can be reached.

The total temperature, pressure, and mass flow rate in the combustion heater can be controlled by changing the pressure and composition ratio of fuel (H_2+N_2) to oxidizer (O_2+N_2) in the two Ludwieg tubes, while the mole-fraction of oxygen after combustion heating remains 0.21 as the atmosphere. The synchronization operation of the two Ludwieg tubes can be controlled by the time difference of tubes diaphragms (<3ms).

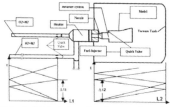

Fig. 3 Principle of pulse combustion facility

The scramjet model has a 2-D inlet of three-wave system with the size of 120mm×80mm, and the whole length of the model is about one meter (Fig4). Fig5 is an enlarged picture of the part marked in Fig. 4.

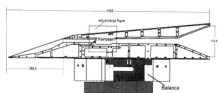

Fig. 4 Model engine for test

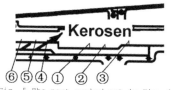

Fig. 5 The part marked out in Fig. 4

The pressure gauges placed on the lower wall of the engine are shown in Fig. 6. Fig. 7,8 gives a typical traces of pressure at M=5 condition (Tt=1500K, Pt=4MPa). They indicate excellent combustion (point 1,2,4,5, and 6), and the combustion at these

points lasts about 60-80 ms. The self-ignition delay time near the cavity region is indicated in Fig. 7, which is an enlarged part of Fig. 6. It takes about 4 ms to transit from steady flow without fuel to combustion flow at all the four points. Fig. 8 is a typical pressure trace in the cavity region of the model engine at the M=6 condition (Pt=5.5MPa, Tt=1850K), showing that the ignition delay time is the same as at M=5. This also means that fuel drop breaking, evaporation and chemical reaction delay can be completed within about 4 ms under our M=5, 6 test conditions .

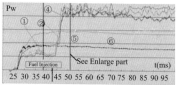

Fig. 6 Wall pressure traces (M=5)

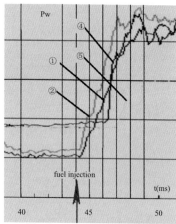

Fig. 7 Enlarqed part of Fig. 6

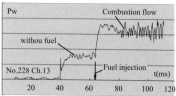

Fig. 8 Wall pressure traces in the Cavity (M=6)

Numerical results

Numerical simulation of the experimental scramjet model in Fig. 4 was carried out in the present study. Table 2 gives the conditions of the computation. A multi-block structured mesh with total cell number of 2.6M is employed. The present computation is carried on our cluster computer system with total CPU number of 240, which requires one week for computing one condition.

Table 2 Conditions for vitiated air stream and fuel injection

Parameter	Vitiated air	Fuel
Mach Number Total	6.0	1.0
Temperature(K)	1800	300
Total Pressure (atm)	54.7	—
Equivalence Ratio	—	0.877
Kerosene Mass Fraction	0.0	1.0
O_2 Mass Fraction	0.256	0.0
N_2 Mass Fraction	0.5656	0.0
H_2O Mass Fraction	0.1784	0.0

Fig. 9 gives the computed pressure traces of typical locations, the pressure of point No. 3 increases firstly, then point No. 2 follows, which indicates the combustion firstly occurs near the rear step of cavity, then the flame propagates upstream in the cavity. The computed ignition delay of about 5ms can be drawn from Fig. 9, which is close to our experimental result.

Fig. 10 shows the comparison of computed pressure along the lower wall with the experimental data under the condition of fuel-on. The computed wall pressure agrees very well with the experimental data. In

the cavity, the computed wall pressure is slightly higher than the experimental dada, which indicates the heat released for computation is larger than experiment in this region.

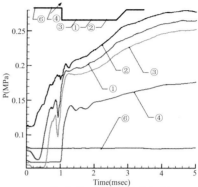

Fig. 9 Pressure trace of typical points

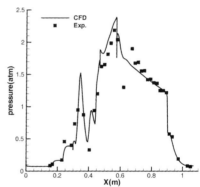

Fig. 10 Comparison of wall pressure

Fig. 11 gives the distribution of droplets in the combustor after the establishment of stable combustion. The un-evaporated droplets are mainly concentrated inside the cavity. Out of the cavity, most of the droplets are evaporated completely.

Fig. 12 shows the time evolution of droplets after injection. At t=0.01ms, there are only a small amount of droplets in the combustor. At t=0.2ms, the droplets concentrate only at the front step of the cavity, when the droplets move downstream, the heat is transferred from the hot air to the droplets, thus increasing the temperature of the droplets. After t=0.3ms, the cavity is almost filled with droplets, most of the droplets are deflected to the bottom wall of the cavity. When the droplets moved down to the rear step of the cavity, the droplets are circulated back to the main flow. At t=1.0ms, the number of the droplets are clearly reduced at the rear step of the cavity because most of the droplets are evaporated completely at this region.

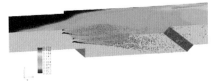

Fig. 11　Contours of CO2 and distribution of droplet in the combustor

Fig. 13 shows the time evolution of kerosene vapor flow field. At t=0.1ms, a small mount of the kerosene vapor is mainly concentrated at the front step of the cavity. At t=0.2ms, the kerosene vapor spreads more downstream. The diffused kerosene vapor is mainly in the cavity. At t=0.3ms, the entire cavity was almost filled with the kerosene vapor. At the rear step, the kerosene vapor mix with the air and self-ignition occurs. At t=0.7ms, the concentration of kerosene vapor at the rear step tends to decreases because of the kerosene vapor decomposition resulted from combustion in this region. At t=1.0ms, the kerosene vapor diffuses into the main flow, which is caused by the increasing of the size of the recirculation zone in the cavity and thus lifting the particles stream.

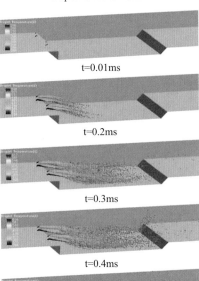

t=0.01ms

t=0.2ms

t=0.3ms

t=0.4ms

t=1.0ms

Fig. 12　Development of droplet from t=0ms to 1ms

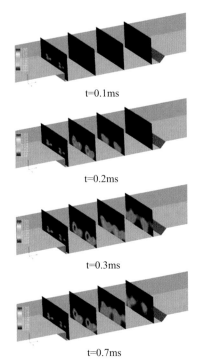

t=0.1ms

t=0.2ms

t=0.3ms

t=0.7ms

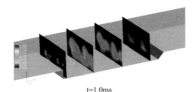

Fig. 13 Development of kerosene vapor field
from t=0ms to 1ms

Fig. 14 shows the time evolution of CO_2 flow field. At time t=0.4ms, the flame develops at the rear step of cavity and propagates both downstream and upstream. At t=1.0ms, the flame is almost full of the combustor. This characteristics of flame propagation is observed and confirmed in our experiments.

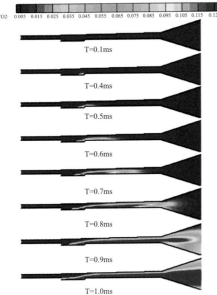

Fig. 14 Development of CO2 field from
t=0.1ms to 1ms

Fig. 15 shows the time history of Sauter Mean Diameter (SMD) and evaporation ratio. Here, the evaporation ratio is defined as the mass of the evaporated fuel divided by the mass of the total injected fuel. In this figure, for t<0.2ms, the size of droplets decreases very quickly (diameter decreased from 300μm to 30μm), but the evaporation ratio is low (less than 30%), which indicates that the change of the droplet diameter is mostly resulted from the droplet breakup. For t>0.2ms, the change of the droplet diameter is slowly, but the evaporation ratio increases quickly, it takes about 1ms for 80% of liquid droplets evaporating into gaseous phase, so the time for fuel evaporation is about 0.8ms.

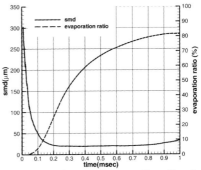

Fig. 15 Time history of Sauter Mean Diameter
(SMD) and evaporation ratio

Conclusion

With our newly developed spray atomization model and unsteady computational method, numerical evaluation of the ignition characteristics from fuel injection to combustion in a kerosene fueled scramjet was carried out. The three-dimensional computation couples the fuel atomization, droplet evaporation and combustion. Typical wall pressure traces indicate that all the three processes, i.e. fuel drop breaking, evaporation and chemical delay of ignition are completed within about 5ms, the time for fuel breaking up and evaporation is about 0.2ms and 0.8ms respectively, the chemical reaction delay is about 4ms.

REFERENCES

1. Jialing Le, Wei He, Shunhua Yang and Weixiong Liu, Investigation of ignition characteristics for

kerosene fueled scramjet, ISABE-2009-1322

2. Lehr, H. F., "Experiment on Shock-Induced Combustion," Astronautica Actu, Vol. 17, Nob. 4 and 5, 1972, pp 589-597.

3. Jeong-Yeol Choi, In-Seuck Jeung and Youngbin Yoon, "Computation Fluid Dynamics Algorithms for Unsteady Shock-Induced Combustion, Part1: Validation", AIAA Journal, Vol. 38, No. 7 .2000, pp. 1179-1187.

4. M.A.Patterson and R.D.Reitz, Modeling the effects of Fuel Spray Characteristics on Diesel Engine Combustion and Emission, SAE Paper 98031,1998

5. Shunhua Yang, Jialing Le, Fuel Atomization and Droplet Breakup Models for Parallel Spray Combustion CFD Code, AIAA 2006-8130

6. Jialing Le, et al. Pulse Combustion Facility and its Preliminary Application in Scramjet Research, ICMAR'2002, Proceedings Part II, pp.106-110

7. Jialing Le, et al. Study on Model Scramjet Performance with Fuel of Kerosene, ICMAR'2004, Proceedings, Part II, pp.120-14

8. Jialing Le, et al. Recent Progress of CARDC in Experimental and Computational Scramjet Research, EWHSFF'2005

超燃冲压发动机带凹槽的燃烧室流场振荡研究*

王西耀[1,2]，杨顺华[1,2]，乐嘉陵[1,2]

（1. 中国空气动力研究与发展中心 高超声速冲压发动机技术重点实验室，四川 绵阳 621000；

2. 中国空气动力研究与发展中心 吸气式高超声速技术研究中心，四川 绵阳 621000）

摘　要：为了分析闭式凹槽的流场振荡现象，结合试验和二维非定常数值模拟方法，开展了三种条件下的非定常研究。通过实验发现，长深比较大的凹槽作为超燃冲压发动机火焰稳定装置时，凹槽的冷流流场存在强烈的振荡现象。结合二维非定常数值模拟方法，初步分析了振荡机理：振荡的主要原因是燃烧室入口气流马赫数的大小，马赫数 2.0 条件下，流场周期振荡；马赫数 3.0 条件下，流场可以稳定存在。根据分析结果给出了流场特征与马赫数的定性依赖关系。通过增加隔离段长度的方法对分析结果进行了验证。

关键词：超燃发动机；凹槽；流场振荡；非定常计算

中图分类号：V235.21　　　文献标识码：A　　　文章编号：1001-4055（2013）05-0651-07

A Study on Flow Oscillation in Scramjet Combustor with Cavity

WANG Xi-yao[1,2]，YANG Shun-hua[1,2]，LE Jia-ling[1,2]

（1. Science and Technology on Scramjet Laboratory，China Aerodynamics Research and Development Center，Mianyang 621000，China；

2. Airbreathing Hypersonics Research Center，China Aerodynamics Research and Development Center，Mianyang 621000，China）

Abstract：To analyze the flow oscillation of closed cavity in scramjet combustor with cavity，three unsteady fields were studied by experiment and unsteady CFD method. It is discovered from experiment that intensive oscillation exists in cold field of scramjet if cavity with large L/D ratio is used as a flame holder. The mechanism of oscillation was analyzed with 2D unsteady CFD method. The entrance Mach number of scramjet combustor is the key parameter that dominates the oscillation. For Mach number 2.0 the field is oscillatory. While for Mach 3.0，the field is stabilized. From the analysis the qualitative variation of field with Mach number was given. Finally，the analysis results with longer isolator were verified.

Key words：Scramjet；Cavity；Field oscillation；Unsteady CFD

1　引　言

超燃冲压发动机燃烧室的入口气流一般为超声速，气流速度高，温度和压力低，点火和稳定燃烧难以实现。为了解决该问题，燃烧室常常采用凹槽构型[1~3]。凹槽内气流流动速度低，燃气驻留时间长，同时燃气温度和压力高，点火延迟时间短，凹槽的这些特点有利于点火和稳定燃烧。

根据长深比的不同，凹槽分为开式凹槽和闭式凹槽。一般来说，如果凹槽长深比小于 7~10，凹槽流态为开式；如果大于 10~13，则为闭式，开式流态和闭式流态的流动示意图如图 1 所示[4]。从中可以看出，开式凹槽的流动由一个大的漩涡组成，凹槽内压力分布较均匀，阻力小；闭式凹槽由前后两个漩涡流

王西耀, 杨顺华, 乐嘉陵. 超燃冲压发动机带凹槽的燃烧室流场振荡研究. 推进技术, 2013, 34(5): 651-657.

动组成,前面漩涡对应前台阶膨胀回流区,此处压力低,温度高,后面漩涡为主流和后台阶相互作用产生的激波分离区,受到激波压缩的影响,此处压力和温度比开式凹槽的高。由此可见,开式凹槽比闭式凹槽的阻力小,但开式凹槽的温度和压力低于闭式后台阶回流区的,从点火和稳定燃烧的角度出发,闭式凹槽在超燃冲压发动机中的应用具有一定的优势。

Fig. 1   Flow structures of open(left) and closed(right) cavities[4]

凹槽流场普遍存在振荡现象,类似的研究很多。Zhang 和 Edwards[5]通过实验发现开式凹槽存在横向振荡模式和纵向振荡两种振荡模式,纵向振荡现象主要是由剪切层和凹槽后台阶之间相互碰撞导致剪切层不稳定而产生的[4]。抑制振荡的措施分为主动式[6,7]和被动式[8-10],典型的主动式方法为在凹槽入口壁面上垂直来流喷射一股气体,这将抬高剪切层,抑制剪切层与后台阶的碰撞;典型的被动式方法是将凹槽后台阶带有一定角度,从而保证剪切层与后台阶碰撞前无气体进入凹槽,这两种方式如图2所示。

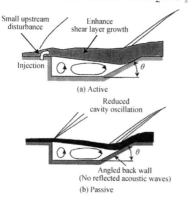

Fig. 2   Typical active and passive method suppressing cavity oscillation[11]

当超燃发动机燃烧室的凹槽流动为闭式时,由于凹槽长度增加,剪切层在到达凹槽后台阶之前就在凹槽底部再附,并产生斜激波,斜激波与上壁面相交将会导致上壁面分离并产生分离激波,进而会使燃烧

出现较强的激波串,如图3所示。

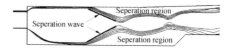

Fig. 3   Shock train structure in closed cavity

在实验中发现,闭式凹槽冷流激波串结构有时会稳定存在,有时则出现周期振荡现象。虽然关于开式凹槽的振荡研究很多,但是关于闭式凹槽振荡的研究结果很少。开式凹槽的压力振荡主要为高频(数kHz)小幅的声波振荡,而闭式凹槽的振荡则为激波导致的低频大幅振荡,开式凹槽的相关结论很难应用于闭式凹槽,本文将针对闭式凹槽的振荡现象进行研究,以期得到初步的振荡机理。

2   实验结果

实验在ϕ450毫米脉冲燃烧风洞上进行,风洞照片如图4所示。燃烧室为直连式模型,如图5所示,凹槽长深比大于13,为典型的闭式凹槽。实验过程中进行了纹影呈像和壁面压力监测,纹影呈像区域和压力监测点的位置均在图5中给出。

Fig. 4 ϕ450mm impulse combustion tunnel

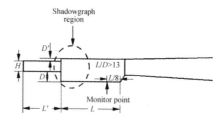

Fig. 5   Direct connected combustor model

来流条件如表1所示,实验分别对隔离段入口马赫数2.5和3.0的燃烧室冷流流场进行研究,结果表明两种情况下流场都出现了振荡现象。

Table 1   Incoming flow conditions

T_t/K	p_t/MPa	Ma	T/K	p/Pa
1500	3.0	3.0/2.5	612/757	36363/83194

入口马赫数 2.5 的实验结果如图 6 和图 7 所示，入口马赫数 3.0 的情况与之类似。图 6 给出了凹槽前台阶剪切层的纹影结果。从中可以看出，剪切层在上下摆动，且颜色在黑白之间变化，颜色的变化表明凹槽气流和主流气流之间的密度相对大小在不断变化，这两个现象都充分表明凹槽流动在开式和闭式之间周期振荡。图 7 给出了凹槽壁面压力的振荡过程，可以看出，压力振荡幅度达到 5 个大气压，因此这种振荡过程必然会对飞行器的控制和结构强度带来较大影响，对于这种现象进行研究具有十分重要的意义。

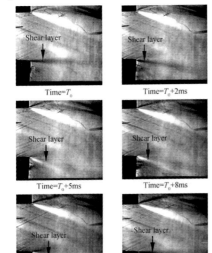

Fig. 6   Oscillation of closed cavity flow

3   计算方法

本文首先对以上两个实验结果进行了二维非定常数值模拟。计算采用我中心自行研发的 AHL3D 大规模并行软件，该软件可以模拟二维或三维、定常或非定常、完全气体或化学非平衡流动。本文采用热完全气体的二维非定常控制方程，计算区域从隔离段入口至燃烧室出口，隔离段入口的初始条件为来流条件，燃烧室内的初始压力为 $100Pa$，温度为 $300K$，马

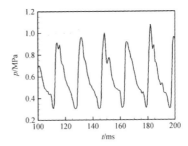

Fig. 7   Pressure oscillation of experiment

赫数为 0。计算过程中对图 5 中压力监测点的压力变化过程进行了纪录。为了能够准确模拟边界层流动，固壁附近的网格进行了加密，最小网格为 10^{-3} mm，同时保证 $y^+ < 5$ 的范围内（粘性底层）至少有 10 个网格点。

非定常算法采用二阶精度的双时间步方法，子迭代采用 LU-SGS 方法，无粘通量采用最大特征值线性化，粘性通量采用对角粘性特征值线性化。计算中真实时间步长为 $1 \times 10^{-7}s$，子迭代最大步数取 15 步。从理论上来说，当子迭代完全收敛后，所有的隐式项将为 0，此时得到的流场具有二阶时间精度，从计算过程来看，每个真实时间步的全局残差至少可以下降 2 个量级，因此时间精度可以保证。该软件模拟非定常流动的可靠性已经被各种算例验证[11~13]。

4   计算结果及分析

4.1   入口马赫数 2.5 的计算结果

本小节按照一定的流场特征将振荡过程分为四个阶段。第一阶段为正激波出现的过程，如图 8 所示。$5.20ms$ 时，凹槽流场处于典型的闭式流态，凹槽前台阶存在一个低马赫数区域，此处对应一个分离漩涡，该漩涡之后流动再附。主流气流受到凹槽下壁面的影响，在再附点产生一道分离斜激波，该斜激波与上壁面相交并导致上壁面流动分离，因此上壁面出现一个分离区和一道分离斜激波。在上下壁面分离斜激波的作用下，凹槽内出现了明显的激波串结构。

$5.20ms$ 时的激波串结构并不能满足压力和主流动量之间的匹配条件，即分离斜激波无法抵抗波前波后的逆压梯度，因此随着时间的推移，激波串逐渐被一道正激波代替，如 $8.00ms$ 所示。

第二阶段为正激波消失的过程，如图 9 所示。正激波形成后仍然无法满足压力匹配需求，激波位置继

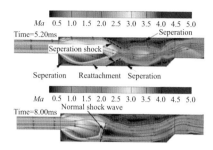

Fig. 8　Appearance of normal shock at procedure 1

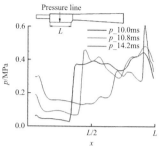

Fig. 10　pressure distribution variation along the bottom wall of cavity（pressure line）at procedure 2

续向上游移动并进入下壁面前台阶回流区。由于此处为亚声速流动，因此当正激波到达回流区之后，下壁面附近的正激波将消失，如 10.0ms 所示。正激波消失后，波后高压气体在压差作用下向前台阶低压区回流，前台阶附近压力升高，后台阶附近压力降低，图 10 给出了这一阶段凹槽下壁面压力的变化过程。

凹槽前台阶压力升高，主流在此处的膨胀效果减弱，这可以通过 10.8ms 以及 14.2ms 的流线反映出来，因此原正激波波前压力升高，逆压梯度降低，主流正激波向后移动并逐渐消失，如 14.2ms 流场所示。

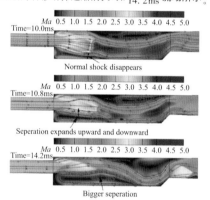

Fig. 9　Disappearance of normal shock at procedure 2

第三阶段为开式流态出现过程，如图 11 所示。上壁面分离区赖以存在的原因是主流的逆压梯度，当正激波消失后主流逆压梯度消失，分离现象自然无法稳定存在，终将完全消失，此处流动必将再附。同时前台阶回流区不断向凹槽后半部分扩展，面积增加，最终占满整个凹槽，凹槽流态变为开式。

第四阶段为凹槽闭式流态的再出现过程。由于

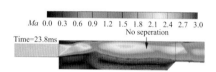

Fig. 11　Appearance of open cavity at procedure 3

主流速度大，凹槽速度小，两者之间存在动量和质量疏运，质量疏运使凹槽内气体向下游泻流，因此凹槽内的流线由凹槽内发出并流向凹槽下游，如图 11 所示。在这种作用下，凹槽内的压力逐渐降低，如图 12 所示。由于凹槽长度过大，主流动量相对较小，主流流体质点无法直接跨过凹槽而在后台阶再附，因此随着凹槽压力的逐渐减少，剪切层逐渐向凹槽底部偏转并最终在凹槽底部再附，这一过程如图 13 所示。相反，如果凹槽长度较小，主流可以跨过凹槽而在后台阶再附，那么将会呈现稳定的开式流动结构。

当主流在凹槽底部再附后，后台阶出现新的分离区，凹槽的闭式流态再现，至此流场完成了一个周期

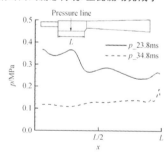

Fig. 12　Pressure distribution variation along the bottom wall of cavity（pressure line）at procedure 4

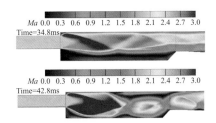

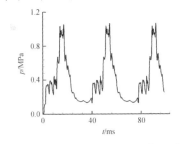

Fig. 13　Reoccurrence of closed flowfield at procedure 4

的振荡。图 14 给出了壁面监测点的压力变化过程，从中可以看出，计算得到的压力振荡幅值和实验的相同，但是频率比实验的低，实验频率约 $50\,\mathrm{Hz}$，计算频率约 $25\,\mathrm{Hz}$。这主要是因为实验是三维现象，而计算是二维的，三维和二维的影响机制将在下文给出。

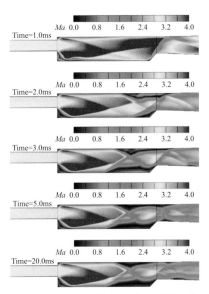

Fig. 15　Mach filed variation at the condition of entrance $Ma = 3.0$

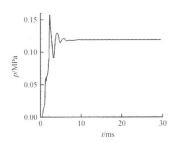

Fig. 16　Pressure history of monitor point at the condition of entrance $Ma = 3.0$

Fig. 14　Pressure history of monitor point at the condition of entrance $Ma = 2.5$

4.2　入口马赫数 3.0 的计算结果

图 15 给出了隔离段入口马赫数 3.0 的流场变化过程，图 16 给出了压力监测点的压力变化历程。从中可以看出，流场在 5 毫秒之前就可以以闭式流态稳定存在，这点和实验结果完全不同。

4.3　振荡机理分析

4.1 节和 4.2 节分别给出了入口马赫数 2.5 和 3.0 的计算结果，结果表明，前者流场出现了振荡现象，后者则无振荡现象产生。本节将对该问题进行分析。

图 17 给出了两种马赫数条件下的闭式流场结构，从中可以看出，入口马赫数 2.5 时上壁面分离区大于入口马赫数 3.0 的。图 18 和图 19 分别给出了两个算例在闭式流态下主流速度沿着燃烧室高度方向的分布以及凹槽上下壁面的压力分布。结合图 17，图 18 以及图 19 可以对流场振荡的原因进行如下分析：高马赫数条件下，主流速度大，逆压梯度小，上

壁面分离区小，主流抵抗分离的能力强，斜激波可以和逆压梯度满足匹配要求，闭式流场可以稳定存在；低马赫数条件下，斜激波无法和逆压梯度匹配，斜激波转化为正激波，当正激波也无法满足匹配要求时，正激波将逐渐向上游移动，如果在移动过程中仍然无法出现匹配点，则流场将会出现振荡现象。

根据这种解释，可以得到流场结构与隔离段入口马赫数之间定性的依赖关系，如图 20 所示。图中 Ma 为隔离段入口马赫数，且 $Ma_2 > Ma_1 > 1.0$，当隔离段入口马赫数大于 1 小于 Ma_1 时，流场是周期振荡的，如果大于 Ma_1 小于 Ma_2，流场将以定常正激波形式存

在,如果大于 Ma_2,流场将以定常斜激波串结构(闭式结构)稳定存在。Ma_1 和 Ma_2 的具体取值和燃烧室构型有关,需要进行更为深入的研究。

Fig. 17　Mach field comparison between entrance $Ma = 2.5$ and 3.0

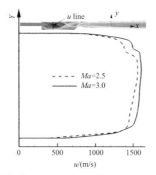

Fig. 18　Comparison of u profile across cavity (along u line) between $Ma = 2.5$ and 3.0

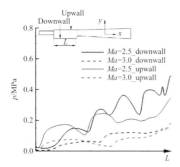

Fig. 19　Comparison of pressure distribution along the up and bottom wall of cavity between entrance $Ma = 2.5$ and 3.0

4.4　振荡机理验证

根据 4.3 节的分析可以看出,主流速度大小是影响流场振荡的主要原因,速度越大,抵抗反压能力越

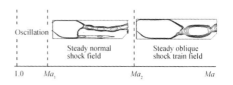

Fig. 20　Cavity field types classified by isolator entrance Mach number

强,流场越稳定,反之流场将会出现振荡。实验条件下,隔离段是带有侧板的三维模型,侧板会对主流带来摩擦损失,这种影响是二维计算无法模拟的。因此为了得到 $Ma = 3.0$ 条件下的流场,需要增加隔离段的动量损失。本文完成一个二维非定常模拟需要 72 个 CPU 并行计算两周,如果完全模拟一个 3 维非定常,其计算量巨大,无法承受。另一种可以验证本文思想的方法是将隔离段长度适当增加,以期通过增加的隔离段长度来模拟侧板的影响并得到振荡的流场,从而验证本文对振荡机理分析的正确性。本文初步将隔离段延长一倍。

图 21 给出了隔离段出口速度沿着高度方向的分布,从中可以看出,增加隔离段长度后,隔离段出口附面层大大增加,这表明主流速度损失增加。图 22 和图 23 分别给出了隔离段长度 $2L'$ 条件下马赫数流场的变化过程和监测点的压力变化过程,从中可以看出,增加隔离段长度后,流场出现了振荡现象,这表明了 4.3 节关于振荡机理的分析是正确的,同时表明侧板带来的流动损失是影响振荡的因素,这也解释了 4.2 节中振荡频率的计算与实验结果不同的原因。

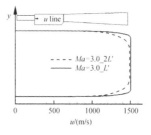

Fig. 21　Comparison of u profile across isolator height at isolator exit between isolators of length $2L'$ and L' (along u line) at the condition of $Ma = 3.0$

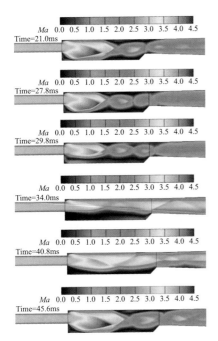

Fig. 22　Mach field oscillation procedure of cavity with isolator of length $2L'$ at the condition of entrance $Ma = 3.0$

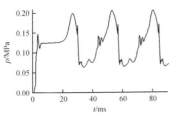

Fig. 23　Pressure history of monitor point in cavity with isolator of length $2L'$ at the condition of entrance $Ma = 3.0$

5　结　论

通过本文研究,得到以下结论:

($_1$)通过实验发现,超燃冲压发动机闭式凹槽的流场在开式和闭式之间振荡。

($_2$)数值模拟了实验状态下的振荡流场,根据数值模拟结果可以看出,高马赫数条件下($Ma = 3.0$),流场稳定存在,低马赫数条件下($Ma = 2.0$),流场则周期振荡。因此,隔离段入口马赫数大小是决定流场振荡的主要因素。

($_3$)采用加长隔离段长度的方法,在$Ma = 3.0$条件下得到了振荡流场,验证了振荡机理的分析结果。

参考文献:

[1]　Tishkoff J M, Drummond J P, Edwards T, et al. Future Direction of Supersonic Combustion Research: Air Force/NASA Workshop on Supersonic Combustion [R]. AIAA 97-1017.

[2]　Vinagradov V, Kobigsky S A, Petrov M D. Experimental Investigation of Kerosene Fuel Combustion in Supersonic Flow [J]. *Journal of Propulsion and Power*, 1995, 11(1): 130-134.

[3]　Ortweth P, Mathur A, Vinogradov V, et al. Experimental and Numerical Investigation of Hydrogen and Ethylene Combustion in a Mach 3–5 Channel with a Single Injector [R]. AIAA 96-3245.

[4]　Adela Ben-Yakar, Ronald K Hanson. Cavity Flame-Holder for Ignition and Flame Stabilization in Scramjets: An Overview [J]. *Journal of Propulsion and Power*, 2001, 17(4).

[5]　Zhang X, Edwards J A. An Investigation of Supersonic Oscillatory Cavity Flows Driven by Thick Shear Layers [J]. *Aeronautical Journal*, 1990, 94(12): 355-364.

[6]　Perng S W, Dolling D S. Passive Control of Pressure Oscillations in Hypersonic Cavity Flow [R]. AIAA 96-0444.

[7]　Zhang X, Rona A, Edwards J A. The Effect of Trailing Edge Geometry on Cavity Flow Oscillation Driven by a Supersonic Shear Layer [J]. *Aeronautical Journal*, 1998, 102(1013): 129-136.

[8]　Sarno R L, Franke M E. Suppression of Flow-Induced Pressure Oscillations in Cavities [J]. *Journal of Aircraft*, 1994, 31(1): 90-96.

[9]　Vakili A D, Gauthier C. Control of Cavity Flow by Upstream Mass-Injection [J]. *Journal of Aircraft*, 1994, 31(1): 169-174.

[10]　Lamp A M, Chokani N. Computation of Cavity Flows with Suppression Using Jet Blowing [J]. *Journal of Aircraft*, 1997, 34(4): 545-551.

[11]　杨顺华. 碳氢燃料超燃冲压发动机数值研究 [D]. 绵阳:中国空气动力研究与发展中心,2006.

[12]　赵慧勇. 超燃冲压整体发动机并行数值研究 [D]. 绵阳:中国空气动力研究与发展中心,2005.

[13]　王西耀. 超燃冲压发动机非定常流动数值研究 [D]. 绵阳:中国空气动力研究与发展中心,2012.

马赫数为 4 的超燃发动机碳氢燃料点火试验

张弯洲[1,2]，乐嘉陵[2]，杨顺华[2]，程文明[1]，邓维鑫[1,2]，周化波[2]

（1. 西南交通大学 机械工程学院，成都 610031；

2. 中国空气动力研究与发展中心 吸气式高超声速技术研究中心，四川 绵阳 621000）

摘　　要： 在直连式脉冲燃烧风洞设备上，开展了模拟马赫数为 4，总温为 935 K 的超燃发动机碳氢燃料点火试验. 试验利用了点火器加引导氢气、引导氢气自燃辅助点火、节流加引导氢气 3 种辅助点火方式成功实现了乙烯燃料的点火并维持了稳定燃烧. 试验研究发现，利用氢气自燃辅助乙烯点火，氢气质量流量范围为 0.43～12.61 g/s，氢气质量流量过大不能成功点火. 利用节流加引导氢气的辅助点火方式，节流量为 10%～30%，氢气注油压力为 5 MPa 能够可靠点火. 最后研究了乙烯从凹槽上游和从凹槽底部注油的发动机贫油点火极限和富油工作极限，研究发现两者的贫油熄火极限相近，为当量比为 0.077，而富油工作极限差别较大，当量比分别为 0.327 和 0.471.

关　键　词： 超燃发动机；点火；节流；贫油极限；富油工作极限

中图分类号： V231.3　　　　　　**文献标志码：** A

Experiments on hydrocarbon fuel ignition for scramjet at Mach 4

ZHANG Wan-zhou[1,2]，LE Jia-ling[2]，YANG Shun-hua[2]，
CHENG Wen-ming[1]，DENG Wei-xin[1,2]，ZHOU Hua-bo[2]

（1. School of Mechanical Engineering，

Southwest Jiaotong University，Chengdu 610031，China；

2. Airbreathing Hypersonics Research Center，

China Aerodynamics Research and Development Center，Mianyang Sichuan 621000，China）

Abstract： Hydrocarbon fuel ignition experiments were performed on directly connected pulse combustion facility. The test inflow parameters included stagnation temperature of 935 K and the Mach number of 4 at the entrance of isolator. Gaseous ethylene was ignited successfully and combustion was maintained stable by employing three combination assistant ignition methods：torch igniter combined with pilot hydrogen，self-combustion of pilot hydrogen and air throttle combined with pilot hydrogen. The experimental results indicate that when employing self-combustion of pilot hydrogen，the mass flow rate of hydrogen should be ranged from 0.43～12.61 g/s，and higher mass flow rate of hydrogen isn't helpful to ignition；for air throttle combined with pilot hydrogen ignition method，the injection pressure of hydrogen should be set at 5 MPa，and the throttle mass flow rate should be 10%～30% of inflow mass flow rate. At last the lean blow-out limit and rich operation limit of scramjet were researched with ethylene injected from upstream of cavity or cavity floor. The results show that two injection schemes have similar lean blow-out limits at equivalence ratio of

张弯洲，乐嘉陵，杨顺华，程文明，邓维鑫，周化波. 马赫数为 4 的超燃发动机碳氢燃料点火试验. 航空动力学报，2013，28(4)：800-806.

0.077，and the rich operation limits varies a lot at equivalence ratios of 0.327 and 0.471 respectively.

Key words：scramjet；ignition；throttle；lean limit；rich operation limit

以涡轮发动机为基础、与超燃发动机组合的循环推进系统（TBCC）中的涡轮发动机能够提供的最大速度为 $Ma=3\sim4$，超燃发动机必须在 $Ma=4$ 附近成功点火起动，才能保证 TBCC 的正常工作。但飞行速度为 $Ma=4$ 时，根据弹道参数计算，其气流的总温较低，一般为 $800\sim1000\,K$。来流总温低，燃烧室内的超声速气流静温一般仅为 $500\sim600\,K$，低于燃料的自点火温度。因此必须开展超燃发动机在低总温条件下的起动方式和过程的研究工作，为超燃发动机的低马赫数下的成功起动进行基础研究和技术储备。

总结相关的超燃发动机点火方面的文献，低总温条件下成功点火的流场具有以下特点：高温、低速、混合充分和富含自由基。低总温条件下点火必须借助辅助的点火方式。文献中的辅助装置和措施有：① 强迫点火器。电火花塞[1]、等离子炬[2]、火炬点火器、引导火焰点火、高压电弧[3]等强迫点火器较为多见。② 燃料处理技术。燃料处理技术主要为燃料加热技术和燃料添加剂的应用。③ 燃料混合增强技术。燃料混合增强主要为应用混合性能较好的注油器，如气动斜坡和物理斜坡注油器、支杆注油器等以及液态燃料雾化技术[5-9]。④ 节流点火技术[10-11]。在流场相对下游位置进行一定程度的节流，能够在节流位置上游产生一定的激波串，减速和加热来流，对点火有利。节流装置一般有物理装置节流和气体节流（air

throttle）。Rasmussen 等人[1]在连续风洞设备上，在来流温度为 $640\,K$ 和 $560\,K$ 的条件下点燃了乙烯，测量了壁面压力曲线，研究了影响点火的因素；Mathur 等人[12]在总温低于 $1000\,K$ 条件下，利用雾化技术和点火器点燃了加热的 JP-7 燃料。国内也有低总温条件下的相关点火文献[13]。

国内外的低温点火文献中，均在连续式风洞上开展试验[14-15]，而本试验为脉冲风洞设备，相比之下，连续风洞的点火环境为热壁环境，相对有利于点火和燃烧，而脉冲风洞由于传热时间极短，保持冷壁状态，不利于点火和燃烧。本试验组合了点火器、引导氢气和节流技术 3 种辅助点火技术，通过调节注油位置和注油当量比，精确控制点火时序等，在总温为 $935\,K$ 的条件下成功点燃了气态乙烯，并维持了稳定燃烧。

1 试验设备简介

1.1 脉冲燃烧风洞设备

本试验在直连式脉冲燃烧风洞设备（见图 1）中进行，该设备能够模拟的飞行马赫数为 $4\sim7$，二元喷管出口马赫数为 $2\sim3$，总温为 $800\sim1900\,K$，总压为 $0.8\sim2.6\,MPa$，设备能够提供的稳定试验时间为 $200\sim270\,ms$。通过设备调试和流场校测，设备的设计参数和运行参数见表 1，试验时间约 $250\,ms$，满足脉冲试验的时间要求。

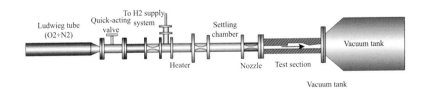

图 1 直连式脉冲燃烧风洞设备

Fig.1 Directly connected pulse combustion wind tunnel facility

表 1 风洞设备的试验参数

Table 1 Experimental parameters of wind tunnel facility

	飞行马赫数	二元喷管出口马赫数	总温/K	总压/MPa
设计参数	4	2	900	0.8
运行参数	4	2.1	935	0.8

1.2 火炬点火器

本次试验采用氢气-空气火炬点火器,见图2. 该点火器主要由3个子系统组成,氢气进气系统、

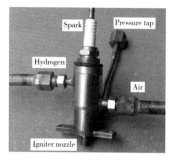

图2 火炬点火器

Fig. 2 Torch igniter

空气进气系统和点火花塞系统.单个点火器的输入功率调节范围为$0\sim70\,\mathrm{kW}$.试验模型上同时安装了两个独立控制的点火器,试验中点火器的功率调节范围为$0\sim140\,\mathrm{kW}$.

1.3 直连式超燃发动机燃烧室试验模型

发动机模型的入口截面为$50\,\mathrm{mm}\times100\,\mathrm{mm}$,燃烧室的具体尺寸见图3.凹槽深度为$18\,\mathrm{mm}$,长深比约为$10.8$,后缘角为$22.5^\circ$.压力测点主要布置在发动机的侧面中心位置,单侧测点数量为30个.试验模型中的凹槽、注油位置、点火器位置和节流位置见图4. A,B,C,D和E为注油位置,点火器F布置在凹槽底部,节流位置G布置在上壁面,离凹槽台阶为$430\,\mathrm{mm}$.节流喷气孔的当量直径约为$10.4\,\mathrm{mm}$,每排注油孔的当量直径为$3.46\,\mathrm{mm}$.

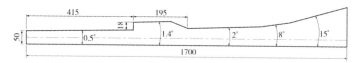

图3 超燃发动机燃烧室侧视图(单位:mm)

Fig. 3 Side-view of scramjet combustor (unit:mm)

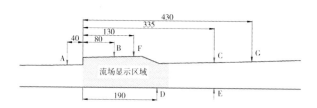

图4 凹槽、注油位置、点火器位置和节流位置(单位:mm)

Fig. 4 Positions of cavity,injectors,igniter and air throttle (unit:mm)

2 方法及理论

2.1 贫油极限和富油极限

在数值计算中,贫油点火极限可以根据生成的CO_2质量浓度精确判断.而实际试验中,只能通过冷流与热流试验的测量结果的比对进行判断,存在传感器测量范围、测量精度和误差等因素的影响,试验手段测量的贫油点火极限可能存在一定的误差.

受脉冲燃烧风洞试验时间的制约,不能比照连续风洞的研究方法,富油熄火极限研究困难.同

时,在试验中发现,在低马赫数条件下,发动机可能在还没有达到富油熄火极限时,燃烧室压力向上游传播至隔离段入口,达到超燃发动机的工作极限.在本试验中,富油极限主要为发动机富油工作极限.

2.2 稳定燃烧判据

脉冲燃烧风洞试验时间短,判断点火后是否维持稳定燃烧,主要判断依据为:①燃烧室测量点压力信号与冷流相比具有明显差别;②燃烧室测量的压力信号在试验时间内维持稳定.满足上述两个条件,判断为维持稳定燃烧.

3 试验内容及试验结果分析

在模拟飞行 $Ma=4$,总温为 $935\,K$ 的条件下,进行了气态乙烯燃料的点火试验.试验通过组合点火器、引导氢气和气体节流三种辅助点火方式成功点燃了乙烯燃料并使其维持了稳定燃烧.试验研究了引导氢气、点火器和气体节流辅助点火方式参数对点火的影响,通过试验确定了不同注油位置的贫油极限和富油极限.

3.1 前期的试验结果

前期开展了低温点火的部分内容研究,主要结果有:①仅靠点火器不能点燃乙烯燃料;②氢气从凹槽上游位置注入不能自点火燃烧;③利用点

火器加引导氢气的组合点火方式成功点燃了气态乙烯燃料.引导氢气辅助点火的最小当量比为 0.05,点火器的主要功能是点燃引导氢气,点火器功率在 $20\sim140\,kW$ 均能可靠地点燃引导氢气,进而点燃乙烯.

3.2 引导氢气自燃辅助点火方式

引导氢气自燃辅助点火试验的时序为:以流场稳定时刻为 $0\,ms$,$20\,ms$ 时注入氢气,$50\,ms$ 时注入乙烯,$80\,ms$ 时关闭氢气,$250\,ms$ 时试验结束.依据 $200\sim250\,ms$ 的压力测量数据与冷流压力数据比较判断点火是否成功.试验条件和试验参数见表 2(080804 试验数据测量时刻氢气正常注油).

表 2 引导氢气自燃辅助点火试验设置

Table 2 Ignition experiment conditions with self-combustion of pilot hydrogen

| No. | 氢气 | | | 乙烯 | 点火结果 |
	质量流量/(g/s)	当量比	注入位置	当量比	
080704	0.452	0.005	F	0.791	成功
080804	12.61	0.142	B	0	成功
081403	0.43	0.005	F	0.806	成功
081502	17.27	0.194	B	0.806	失败
081503	6.68	0.075	B	0.806	成功

从表 2 中的试验设置参数和试验结果可以总结出下列结果:①氢气以适当质量流量($0.43\sim6.68\,g/s$)从凹槽内两个位置注入,能够发生自点火燃烧并点燃乙烯.②氢气从凹槽注入的质量流量大于 $17.27\,g/s$ 时,不能发生自点火燃烧.这是因为氢气质量流量过大,以声速注入的氢气会迅速降低附近流场温度,氢气不能发生自点火燃烧.③利用适量氢气自燃的辅助点火方式,能够成功点燃乙烯燃料,有一定的工程应用价值.

图 5 为引导氢气自燃辅助试验压力曲线,从图 5 的压力曲线可以看出,针对相同的乙烯当量比,质量流量分别为 $0.452、0.43\,g/s$ 和 $6.68\,g/s$ 的氢气自燃点燃乙烯,燃烧室壁面压力与冷流压力差别显著,表明成功点火并维持稳定燃烧.

3.3 引导氢气加气体节流组合点火方式

引导氢气加气体节流点火试验的时序为:以流场稳定时刻为 $0\,ms$,$20\,ms$ 时注入氢气,$30\,ms$ 时开始节流,$50\,ms$ 时注入乙烯,$80\,ms$ 关闭节流

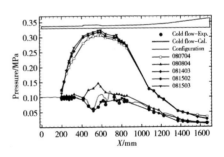

图 5 引导氢气自燃辅助试验压力曲线

Fig. 5 Pressure curves of ignition experiments with self-ignition hydrogen

和氢气,$250\,ms$ 时试验结束.依据 $200\sim250\,ms$ 的压力测量数据与冷流压力数据比较判断点火是否成功.试验条件和试验参数设置见表 3(080903 试验数据测量时刻为关闭节流后 $50\,ms$,氢气继续注油),表中以喷入的节流气体质量流率与隔离段入口来流质量流率之比来表征节流量.

表 3 气体节流点火试验设置

Table 3 Ignition experiment conditions with air throttle

No.	氢气			乙烯	气体节流		点火结果
	注入位置	注入压力/MPa	当量比	当量比	位置	$\dot{m}_{\text{throttle}}/\dot{m}_{\text{air}}$ /%	
080902	A	5.4	0.281	0.8	G	5	失败
080903	A	5.0	0.281		G	30	成功
081002	A	5.0	0.281	0.8	G	30	成功
081003	A	5.0	0.281	0.8	G	20	成功
081004	A	5.0	0.281	0.8	G	15	成功
081201	A	5.0	0.281	0.8	G	10	成功
081202	A	5.0	0.281	0.8	G	5	失败
081203	A	3.0	0.168	0.8	G	10	失败
081204	A	4.0	0.224	0.8	G	10	失败
081001	A	2.0	0.112	0.8	G	15	失败

从表 3 的试验参数和试验结果可以得出下列结论:①081002～081202 的试验结果表明,在引导氢气注油压力为 5.0 MPa 的条件下,能够成功点火的最小节流量为 10%.②081201～081204 和 081001 的试验结果表明,节流量和引导氢气的注油压力均是影响成功点火的重要因素,同在 10% 的节流量下,氢气的注油压力为 4 MPa 和 3 MPa 时,均不能成功点火.③适当提高氢气的注油压力至 5.4 MPa,5% 的节流量仍不能成功点火,该结果预示成功点火的极限节流量可能在 5%～10%.

图 6 的压力曲线均在引导氢气和节流完全关闭后 120 ms 的时刻测量得到,从压力曲线的分布可以看出,在乙烯当量比相同的条件下,节流量越大,燃烧越充分,压力越高,说明节流量对燃烧的

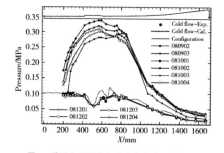

图 6 节流加引导氢气的点火试验压力曲线

Fig. 6 Pressure curves of ignition experiments with
air throttle and pilot hydrogen

性能有一定的影响.分析认为,节流量越大,通过激波串的作用,燃烧室内速度越低,温度相对较高,相同当量比的燃料在燃烧室内燃烧更充分,壁面压力更高,同时也会影响隔离段内的激波串位置向前移动.节流关闭后,激波串位置与燃烧情况保持一定的对应关系,继续影响燃烧室内的流场参数和燃烧情况.

3.4 凹槽上游注油和凹槽底部注油的贫油极限和富油极限

针对凹槽上游注油位置(A)和凹槽内注油位置(B),进行了贫油极限和富油极限的试验研究.脉冲风洞设备上贫油极限的试验是在辅助点火方式完全关闭后 120 ms 时刻,测量燃烧室壁面压力,将其与冷流压力进行比较,若测量压力与冷流压力曲线差别很小时,认为达到贫油熄火极限.对于富油极限,在试验过程中发现,随着燃料当量比的增加,还没有达到富油熄火极限之前,燃烧室压力已经扰出隔离段,超过了发动机的工作极限,因此脉冲风洞设备上将富油极限定义为富油工作极限.

3.4.1 凹槽上游注油的贫油极限和富油极限

乙烯从凹槽上游注入,利用点火器(功率为 40 kW)和引导氢气(当量比为 0.08)进行点火,改变乙烯的当量比,研究该位置注油的贫油极限和富油极限.乙烯当量比参数见表 4.

从图 7 中的曲线可以看出,当乙烯当量比较低时(如 073101 和 073102),壁面压力曲线除在凹槽段与冷流有差别之外,其他位置基本与冷流

表 4 凹槽上游注油贫油极限和富油极限的试验参数

Table 4 Experimental parameters of lean limit and rich
limit for ethylene injected from
upstream of cavity

No.	乙烯当量比
072801	0.184
072803	0.241
072804	0.327
073101	0.077
073102	0.076

一致.更低的当量比测量不出压力与冷流的差别,
认为已经达到了试验手段的贫油点火极限.随着
乙烯当量比的增加,壁面压力曲线升高.当乙烯当
量比达到 0.327 时,燃烧室内的压力已经向上游
传播至隔离段入口附近.更高的当量比会推出隔
离段,可能导致进气道不起动,超出了发动机的工
作范围,定义为发动机富油工作极限.

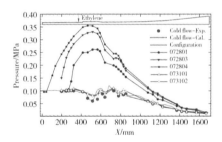

图 7 凹槽上游注油的贫油极限和富油极限

Fig.7 Lean limit and rich limit for ethylene injected
from upstream of cavity

3.4.2 凹槽底部注油的贫油极限和富油极限

乙烯从凹槽底部注入,利用点火器(功率为
40kW)和引导氢气(当量比为 0.08)进行点火,改
变乙烯的当量比,研究该位置注油的贫油极限和
富油极限.乙烯当量比参数见表 5.

从图 8 中的曲线可以看出,当乙烯当量比为
0.077 时(如 072401),壁面压力曲线与冷流压力
曲线差别不大,定义该状态下的贫油极限当量比
为 0.077;乙烯当量比为 0.535 时,燃烧室内压力
向上游传播,已经超出隔离段,超出了发动机的工
作范围;乙烯当量比为 0.471 时,燃烧室压力向上
传播至隔离段入口附近,接近发动机工作极限,定
义为该位置注油的发动机富油工作极限.

表 5 凹槽底部注油贫油极限和富油极限的试验参数

Table 5 Experimental parameters of lean limit and
rich limit for ethylene injected
from cavity floor

No.	乙烯当量比
072301	0.115
072302	0.143
072305	0.163
072401	0.077
072402	0.325
072403	0.535
072404	0.471

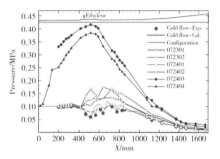

图 8 凹槽底部注油的贫油极限和富油极限

Fig.8 Lean limit and rich limit for ethylene injected
from cavity floor

对两个位置注油的贫油极限当量比进行比较可以
看出,除试验总压浮动的影响外(073101 总压偏
高),3 条曲线在凹槽区域的趋势基本一致.
073102 曲线在 700~1100mm 段比另外两条曲线
更低,更接近冷流压力曲线.从图 9 的比较可以看
出,0.076 的当量比接近于试验手段测量的贫油
点火极限.

比较图 7 和图 8 的曲线,能够得出这样的结
论:①在较高的燃料当量比时,凹槽上游注油比凹
槽内注油燃烧更加充分;②当燃料当量比降低到
一定程度时,两个位置注油的差异不大.通过计算
分析,当量比较高时,凹槽上游注油只有一部分燃
料能够进入凹槽,导致凹槽内的燃料当量比相对
有利于燃烧,而凹槽内注油仅有小部分燃料通过
扩散流出凹槽,大部分燃料富集在凹槽内,导致凹
槽内的富油程度稍高,不利于燃烧.而燃料当量比
很低时,喷注压力很低,凹槽上游注入的燃料基本
全部集中在剪切层内,大部分进入凹槽[1].此时的

注油效果与凹槽内注油基本一致，因此在相对较低的油气比时，两个位置的注油效果差别不大。

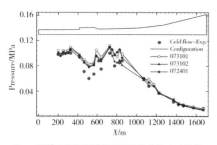

图 9　凹槽上游和凹槽底部注油的贫油点火极限

Fig. 9　Lean ignition limit for ethylene injected from upstream of cavity and cavity floor

4　结　论

在模拟马赫数为 4，总温为 935 K，隔离段入口马赫数为 2.1 的试验来流中，进行了气态乙烯燃料的点火试验，并测量了壁面压力数据。通过试验得到下列结论：

1) 利用点火器加引导氢气，引导氢气自燃辅助点火，节流加引导氢气三种组合点火方式在脉冲燃烧风洞设备上成功点燃了气态乙烯。关闭辅助点火机制后，乙烯维持了稳定燃烧。

2) 研究了引导氢气自燃辅助点火方式中氢气注油位置和氢气质量流量对点火的影响。氢气从凹槽内注入的位置对点火影响不大，氢气质量流量不能过大，17.27 g/s 的氢气不能发生自点火燃烧。

3) 研究了节流加引导氢气组合点火方式中节流量和氢气注油压力对点火的影响。在此试验中，能够成功点火的最小节流量为 10%，氢气的注油压力不能低于 5 MPa。

4) 通过试验确定了凹槽上游位置注油的贫油极限约为 0.076，富油工作极限当量比约为 0.327，凹槽内注油的贫油极限当量比约为 0.077，富油工作极限当量比约为 0.471。

5) 在较高燃料当量比时，凹槽上游注油比凹槽内注油燃烧效率更高，而燃料当量比很低时，注油位置对燃烧的影响不明显。这样的现象主要由不同注油方式对燃料的分布和混合的影响造成。

参考文献：

[1]　Rasmussen C C, Driscoll J F, Hsu K Y, et al. Stability limits of cavity-stabilized flames in supersonic flow[J]. Proceedings of the Combustion Institute, 2005, 30(2) : 2825-2833.

[2]　Watanabe J, Naoyuki A, Kenichi T. Effect of a rearward-facing step on plasma ignition in supersonic flow[J]. Journal of Spacecraft and Rockets, 2009, 46(3) : 561-567.

[3]　Aleksandrov A, Bychkov V, Chernikov V, et al. Arc discharge as a mean for ignition and combustion of propane-air mixture supersonic flow[R]. AIAA 2006-1462, 2006.

[4]　张弯洲，乐嘉陵，杨顺华，等. 超燃发动机混合效率评估方法探讨[J]. 航空动力学报，2012, 27(9) : 1958-1966.
ZHANG Wanzhou, LE Jialing, YANG Shunhua, et al. Evaluationof the fuel mixing efficiency for scramjet[J]. Journal of Aerospace Power, 2012, 27(9) : 1958-1966. (in Chinese)

[5]　Lin K C, Kirkendall K A, Keneddy P J, et al. Spray structures of aerated liquid fuel jets in supersonic cross-flows [R]. AIAA-99-2374, 1999.

[6]　Lin K C, Kenedy P J, Jackson T A. Spray penetration heights of angled-injected aerated-liquid jets in supersonic cross-flows[R]. AIAA 2000-0194, 2000.

[7]　Lin K C, Kenedy P J, Jackson T A. Penetration heights of liquid jets in high-speed cross-flows [R]. AIAA 2002-0873, 2002.

[8]　Segal C. The scramjet engine: processes and characteristics [M]. Cambridge, UK: Cambridge University Press, 2009.

[9]　Yu G, Li J G, Yue L J, et al. Characterization of kerosene combustion in supersonic flow using effervescent atomization[R]. AIAA 2002-5225, 2002.

[10]　Yang V, Li J, Choi J Y, et al. Ignition transient in an ethylene fueled scramjet engine with air throttling: Part I non-reacting flow development and mixing[R]. AIAA 2010-409, 2010.

[11]　Yang V, Li J, Choi J Y, et al. Ignition transient in an ethylene fueled scramjet engine with air throttling: Part II ignition and flame development[R]. AIAA 2010-410, 2010.

[12]　Mathur T, Lin K C, Kennedy P. Liquid JP-7 combustion in a scramjet combustor[R]. AIAA 2000-3581, 2000.

[13]　王靛，宋文艳，李建平，等. 超燃冲压发动机燃烧室碳氢燃料的点火和火焰稳定研究[J]. 固体火箭技术，2008, 31(5) : 449-452.
WANG Dian, SONG Wenyan, LI Jianping, et al. Investigation of hydrocarbon-fuel ignition and combustion in scramjet combustor[J]. Journal of Solid Rocket Technology, 2008, 31(5) : 449-452. (in Chinese)

[14]　Lin K C, Tam C J, Boxx I, et al. Flame characteristics and fuel entrainment inside a cavity flame holder in a scramjet combustor[R]. AIAA 2007-5381, 2007.

[15]　Mathur T, Streby G, Gruber M. Supersonic combustion experiments with a cavity-based fuel injector[R]. AIAA-99-2102, 1999.

Experimental investigation on heat transfer correlations of n-decane under supercritical pressure

L. Zhang [a,*], R.L. Zhang [b,1], S.D. Xiao [a,2], J. Jiang [b,1], J.L. Le [b,1]

[a] School of Mechanical Engineering, Southwest Jiaotong University, Chengdu, Sichuan 610031, China
[b] Airbreathing Hypersonic Technology Research Center of CARDC, Mianyang, Sichuan 621000, China

ARTICLE INFO

Article history:
Received 3 February 2013
Received in revised form 17 April 2013
Accepted 25 April 2013

Keywords:
Heat transfer correlation
Supercritical pressure
n-Decane
Laminar
Transition
Turbulent

ABSTRACT

A series of electrically heated tube experiments was conducted to investigate the heat transfer correlations of n-decane under conditions similar to a regeneratively cooled scramjet. The fuel was heated in a 1.5 mm inner diameter tube of 1Cr18Ni9Ti and the fuel pressure was varied from 4.0 to 4.3 MPa in the experiments. The Reynolds number ranged from about 800 to 70,000. The heat transfer correlations of n-decane in laminar, transition and turbulent flow regions were determined by using the method of least squares curve fitting. The calculated outer wall temperatures were compared with data in present and another electrically heated tube experiments, a fuel-cooled panel test, and a regeneratively cooled scramjet combustor test. The results of comparisons validated the applicability of the presented correlations.

© 2013 Elsevier Ltd. All rights reserved.

1. Introduction

Scramjet can achieve high specific impulse over a wide range of flight Mach numbers and altitudes [1]. It is well known that regenerative fuel cooling has been considered as one of the most effective and practical methods to solve the thermal protection of scramjets [2,3].

The typical working pressure of fuel in the cooling channel of scramjet is about 3–7 MPa and its critical pressure is around 2 MPa [2]. During the process of heating, the fuel temperature rises from ambient temperature to supercritical temperature. The heat transfer correlation is required for calculating the wall and fluid temperature [3], so it plays an important role in the design of cooling structure of a scramjet.

Isaev and Abdullaeva [4] presented a heat transfer correlation of n-heptane in laminar flow region under supercritical pressure. The Reynolds number changes within the limits 1087–2347 in the experiments and the fluid flow region in the channel was laminar. The laminar heat transfer correlation was used for describing the heat transfer intensity of n-heptane in small-size efficient heat-exchanging systems. Ghajar and Tam [5,6] developed an improved

correlation (in the form of laminar and turbulent flow Nusselt numbers) in transition flow region, and the data were correlated by using the traditional least squares method. These works investigated heat transfer correlation only on laminar or transition flow region.

Stiegemerier et al. [7] investigated the heat transfer of five common hydrocarbon fuels in transition and turbulent flow regions (Reynolds number larger than 5000). The tests were conducted at an electrically heated tube facility to simulate conditions encountered in regeneratively cooled rocket engines. The heat transfer correlation in transition and turbulent flow regions was determined using multiple linear regression method. This work developed a heat transfer correlation for transition and turbulent flow region. However, in a regeneratively cooled scramjet the Reynolds number may change from several hundreds to tens of thousands, and the flow region experiences firstly laminar, then transition, and finally turbulent flow. Furthermore, the transition behavior from laminar to turbulent should be also investigated.

The main motive of this paper is to develop experimental heat transfer correlations of n-decane under supercritical pressure valid for the three flow regions: laminar, transition and turbulent, and the experiment conditions were similar to those encountered in a regeneratively cooled scramjet.

2. Experiment facility and procedure

The experiments were conducted at an electrically heated tube facility. A simplified schematic of the facility is shown in Fig. 1. The

L. Zhang, R.L. Zhang, S.D. Xiao, J. Jiang, J.L. Le. Experimental investigation on heat transfer correlations of n-decane under supercritical pressure. International journal of heat and mass transfer, 2013, 64: 393-400.

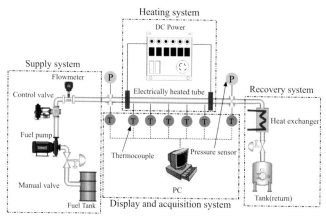

Fig. 1. Schematic diagram of an electrically heated tube facility.

tube was heated electrically by passing a current through it. The convective heat transfer between tube wall and fuel results in the fuel temperature rise along the tube. Steady temperature and flow parameters can be reached and maintained when the inner wall heat release (net heat flux) equals to the heat absorbed by fuel.

One direct current power was used for the heating power supply, and the uncertainty of measurements was less than ±0.01A. The fuel was stored in a supply tank rated for pressures up to 4.3 MPa. The fuel flowed from the supply tank through the test section and back through a heat exchanger to the return tank. Fuel pump was used for mass flux control, and the uncertainty of measurements was less than ±0.01 g/s. The fuel temperature and pressure were measured at the inlet and outlet of the test section by armoured thermocouples and pressure transducers. Seventeen type K thermocouples were spot welded directly to the outer surface of the tube to measure outer wall temperatures. The uncertainty of the wall and fuel temperatures measurements was less than ±3 K, and the uncertainty of the pressure measurements was less than ±0.2%F.S. The tube resistance was measured by a resistance bridge, and the uncertainty of measurements was less than ±0.001Ω (tube resistance is about 0.19 Ω).

The test section was composed of a 1.3 m long 1Cr18Ni9Ti stainless steel tube with an inner diameter of 1.5 mm and a wall thickness of 0.75 mm. The coolant fuel was n-decane (critical point: p_c = 2.1 MPa; T_c = 617.7 K). The fuel mass fluxes were 0.93 g/s, 1.24 g/s and 1.86 g/s.

The thermal loss from tube outer surface includes two parts, which were the natural convective heat transfer between tube wall and room temperature air, and the thermal radiation of tube wall. The calibrations of thermal loss rate q_{loss} at different temperatures were done adopting blank tube heating experiments, during which the tube had different currents and no fuel flowed inside. The thermal loss was equal to the heat generated by the tube electric resistant at steady state.

It was assumed the thermal loss rate q_{loss}(W/m) is a function of tube outer wall temperature and room temperature having the following form

$$q_{loss} = c_1(T_{wo} - T_{air}) + c_2\left(T_{wo}^4 - T_{air}^4\right) \tag{1}$$

where T_{wo} is outer wall temperature, T_{air} is room temperature, c_1 and c_2 are constants. The q_{loss} was different at different axial loca-

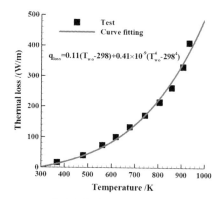

$$q_{loss} = 0.11(T_{wo}\text{-}298) + 0.41 \times 10^{-9}(T_{wo}^4\text{-}298^4)$$

Fig. 2. Profile of thermal loss at x = 0.9 m.

tion of the tube. The result for the blank tube test (T_{air} = 298 K) at the location of tenth thermocouple is shown in Fig. 2.

3. Analysis of heat transfer correlations

3.1. Correlations for laminar, transition and turbulent flow regions

It is generally stated that the flow mode of fluid in a circular tube can be determined by local Reynolds number (Re) [8,9]: $Re < 2300$, laminar flow; $2300 \leqslant Re \leqslant 10^4$, transition flow; $Re > 10^4$, turbulent flow.

It is well known that the Nusselt number (Nu) is a constant in a circular tube with constant properties, uniform surface heat flux and laminar, fully developed flow conditions [10]. The laminar Nusselt number (Nu_{l0}) can be expressed as

$$Nu_{l0} = 4.36 \tag{2}$$

When the property variations are large caused by differences between surface and mean fluid temperature, Nu calculated from the above equation contains errors. The viscosity variation can be particularly important for liquids, especially for oils. Kays et al. [8] suggest using the modified laminar Nusselt number (Nu_l) and it can expressed as

$$Nu_l = Nu_{l0}(\mu/\mu_w)^{0.14} \tag{3}$$

where μ and μ_w are the viscosities of the main fluid temperature and the surface temperature. This correction factor is valid both for fully developed and the entry length regions.

Two Gnielinski correlations are widely used for transition region [11,12]. For some liquid, the first Gnielinski correlation has been confirmed experimentally for the range of conditions: $0.5 \leqslant Pr \leqslant 2000$, $2300 \leqslant Re \leqslant 5 \times 10^6$, and can be expressed as

$$Nu_{tr1} = \frac{(f/8)(Re - 1000)Pr}{1 + 12.7(f/8)^{0.5}(Pr^{2/3} - 1)} \tag{4}$$

$$f = (1.82 \lg(Re) - 1.64)^{-2} \tag{5}$$

The second Gnielinski correlation can be used in the range of conditions: $1.5 \leqslant Pr \leqslant 500$, $3000 \leqslant Re \leqslant 10^6$, and can be expressed as

$$Nu_{tr2} = 0.012(Re^{0.87} - 280)Pr^{0.4} \tag{6}$$

Another correlation extensively used for transition region was Petukhov correlation [9,11], and it can be used in the same conditions as the above equation, and can be expressed as

$$Nu_{tr3} = \frac{(f/8)RePr}{C + 12.7(f/8)^{0.5}(Pr^{2/3} - 1)} \tag{7}$$

$$C = (1.82 \lg(Re) - 1.64)^{-2} \tag{8}$$

For turbulent flow in a circular tube Nu can be calculated by Dittus–Boelter (D–B) correlation [12,13], and its form is

$$Nu_{t1} = 0.023Re^{0.8}Pr^n \tag{9}$$

for heating, $n = 0.4$. This correlation can be used in the range of conditions: $0.7 \leqslant Pr \leqslant 160$, $Re \geqslant 10^5$.

Eq. (9) can be used only for small to moderate temperature differences between surface and main fluid. Sieder and Tate (S–T) correlation [13,14] is widely used for large property variations, and it can be expressed as

$$Nu_{t2} = 0.027Re^{0.8}Pr^{1/3}(\mu/\mu_w)^{0.14} \tag{10}$$

S–T correlation can be used in the range of conditions: $0.7 \leqslant Pr \leqslant 16,700$, $Re \geqslant 10^5$.

If the forms of correlations are determined for laminar, transition and turbulent flow regions, the experimental data can be correlated by using the least squares curve fitting method to develop new correlations.

3.2. Determination of heat transfer correlations

For a small cell of length Δl shown in Fig. 3, the flow and heating processes are equivalent to the forced convective heat transfer in a circular tube with heat generation. The heat generated by electric current and resistance minus q_{loss} equals to the heat absorbed by fuel at steady state.

The heat flux along the axial direction is so small compared with that in the radial direction, which can be ignored. Thus, the fuel flow in an electrically heated tube can be treated as one dimensional flow with friction and convective heat transfer.

The experimental Nusselt number (Nu_e) can be calculated by

$$Nu_e = hd_i/\lambda \tag{11}$$

where h is the convective heat transfer coefficient between fuel and inner wall, d_i is the inner diameter of the tube, and λ is thermal conductivity of the fuel. The definition of h is

$$h = q/(T_{wi} - T_l) \tag{12}$$

where q is the heat flux of the inner wall, T_{wi} is the inner wall temperature, and T_l is the mean fuel temperature. The heat flux q is related to current I, resistance R and q_{loss} of the tube wall per length, and can be expressed as

$$q = (I^2 R - q_{loss})/(\pi d_i) \tag{13}$$

For a circular tube which has heat generation inside tube wall, the relationship of T_{wi} and T_{wo} may be expressed as

$$T_{wi} = T_{wo} + \frac{I^2 R - q_{loss}}{4\pi \lambda_w} + \frac{1}{4\pi \lambda_w} \left[\frac{d_o^2 (I^2 R - q_{loss})}{d_o^2 - d_i^2} - q_{loss} \right] \ln \frac{d_i}{d_o} \tag{14}$$

where d_o is the outer diameter of the tube, and λ_w is the thermal conductivity of the tube wall.

The Re and Pr can be expressed as

$$Re = \rho u d_i/\mu \tag{15}$$

$$Pr = c_p \mu/\lambda \tag{16}$$

where c_p is fuel specific heat capacity at constant pressure, and ρ and u are fuel mean density and speed. The u can be determined from mass conservation law, and it can be expressed as

$$\dot{m} = \rho A u \tag{17}$$

where $\dot{m}$ is fuel mass flux, and A is the cross-section area of the tube.

The fuel density, speed, specific heat capacity, viscosity, and thermal conductivity can be calculated using a program for thermophysical properties of hydrocarbon mixtures according to the fuel temperature and pressure of the small cell. The fuel temperature rise and pressure loss should be determined in order to calculate the Nu_e, Re and Pr along the tube.

In the small cell above, fuel temperature rise ΔT_l can be calculated from energy conservation law, and it can be expressed as

$$\Delta T_l = \frac{\Delta h_T}{\dot{m} c_p} = \frac{I^2 R - q_{loss}}{\dot{m} c_p} \Delta l \tag{18}$$

where Δh_T is sensible enthalpy increase. It is assumed that no endothermic cracking happens in fuel, so that all absorbed heat is transferred to sensible enthalpy of fuel.

The pressure loss Δp of the small cell can be determined by

$$\Delta p = \frac{0.5 f_0 \rho u^2 \Delta l}{d_i} \tag{19}$$

where f_0 is friction factor, and can be defined as [8,10]

$$f_0 = \begin{cases} 0.316 Re^{-0.25} & \text{for } Re \leqslant 20,000 \\ 0.184 Re^{-0.20} & \text{for } Re > 20,000 \end{cases} \tag{20}$$

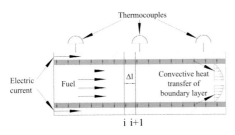

Thermocouples

Electric current

Fuel

Δl

Convective heat transfer of boundary layer

i i+1

Fig. 3. Schematic diagram of heat transfer in an electrically heated tube.

From above deduction, the heat transfer correlations can be determined through the measurements of the current I, fuel mass flux $\dot{m}$, resistance R, diameters and thickness of the tube, outer wall temperature T_{wo}, fuel temperatures and pressures at the inlet and outlet of the tube.

4. Results and discussion

4.1. Wall heat flux

The measured outer wall temperature at the outlet of the tube varied from 405 to 996 K. Fig. 4 shows the distribution of inner wall heat flux along tube in a run.

At the positions within 0–0.2 m and 1.2–1.3 m along tube, inner wall heat flux was smaller due to additional thermal loss by the joint metal blocks near the inlet and outlet. Especially within 1.2–1.3 m which was near the outlet, the tube temperatures were much higher than those within 0–0.2 m near the inlet, so the inner wall heat flux drops steeply down. In some following texts the data within 0.2–1.2 m along tube were chosen for study.

4.2. Reynolds number and Prandtl number

The Re of fuel ranged from about 800 to 70,000 in the experiments. Fig. 5 shows the Re and Pr along tube in a run. One can see that Re varied from about 1100 to 55000. The flow mode was firstly laminar near the inlet ($x = 0$–0.14 m), then changed into transition within $x = 0.14$–0.52 m, and lastly changed into turbulent at $x = 0.52$–1.3 m. Meanwhile, Pr changed from about 14 to 1.

The large range of Re is caused by μ, and the range of Pr is caused by three factors c_p, μ and λ. The distributions of c_p, μ and λ are shown in Fig. 6.

4.3. Calculation of heat transfer correlations

4.3.1. Laminar flow region
Fig. 7 shows the Nu_e and Re along tube in a run. At positions within 0.2–0.9 m, Re was less than 2300 and the flow was laminar.

It is found that Nu_l had a strong dependence on the inlet Reynolds number (Re_l) in laminar flow region in the experiment condition. Assuming Nu_l has a form of below

$$Nu_l = Nu_{l0}\left(\frac{Re_l}{300}\right)^a (\mu/\mu_w)^{0.14} \tag{21}$$

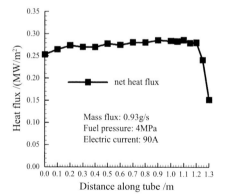

Fig. 4. Distribution of heat flux along tube.

where Re_i is the Reynolds number at the inlet. The heat transfer correlation in laminar flow region for n-decane can be determined using a least squares curve fitting program, and the result is

$$Nu_l = 5.13\left(\frac{Re_l}{300}\right)^{0.81}(\mu/\mu_w)^{0.14} \tag{22}$$

The above correlation is also shown in Fig. 7, about 20% of the experimental data were predicted within 10–20%, and about 5% with deviations greater than 20%. Eq. (22) and Nu_e in tests had values of 10–15 in laminar flow region. Several facts might be responsible for bigger Nu in tests compared with Eq. (2), which might include viscous dissipation of energy, surface roughness.

4.3.2. Transition flow region
As shown in Fig. 8, the Nu_{tr1} and Nu_{tr2} by Gnielinski, and Nu_{tr3} by Petukhov were compared with Nu_e in a run in which Re was between 2200 and 7500 at 0.2–0.9 m. It is clearly seen that Nu_{tr1} and Nu_{tr2} by Gnielinski are better than Nu_{tr3} in transition region.

In Fig. 8, the authors presented a correlation for n-decane in transition region, which is

$$Nu_{tr} = (1 - f(\theta))Nu_l + f(\theta)Nu_t$$
$$f(\theta) = (1 + sin(\theta))/2$$
$$\theta = 0.5\pi(Re - Re_m)/(Re_r - Re_m), Re_l < Re < Re_r \tag{23}$$
$$Re_m = (Re_l + Re_r)/2, \quad \begin{matrix} Re_l = 2100 \sim 2300 \\ Re_r = 6000 \sim 10,000 \end{matrix}$$

where Nu_l is the correlation used in laminar flow region (Eq. (22)), and Nu_t is the correlation used in turbulent flow region (see Eq. (24) in next part). This correlation was adopted because the result was better than the forms of Eqs. (4)–(7), and the curve of Nu is smooth from laminar to turbulent region, about 15% of the experimental data were predicted within 10–20%, and about 4% with deviations greater than 20%.

4.3.3. Turbulent flow region
As shown in Fig. 9, the Nu_{tr1} and Nu_{tr2} by Gnielinski, S–T and D–B correlations were compared with Nu_e in a run in which Re was between 10,000 and 35,000 at 0.62–1.2 m. It is clearly seen that S–T correlation is better than Nu_{tr1}, Nu_{tr2}, and D–B correlations in turbulent flow region.

According to the experimental data, the heat transfer correlation can be obtained in turbulent flow region for n-decane by using

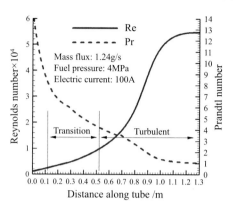

Fig. 5. Distributions of Reynolds and Prandtl numbers along tube.

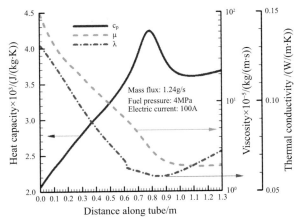

Fig. 6. Distributions of cp, μ and λ along tube.

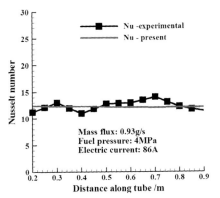

Fig. 7. Comparisons of Nusselt numbers in laminar flow region.

a least squares curve fitting program, which has a form of Eq. (10), and it can be expressed as

$$Nu_t = 0.020 Re^{0.82} Pr^{0.40} (\mu/\mu_w)^{0.16} \qquad (24)$$

The above correlation is also shown in Fig. 9, about 25% of the experimental data were predicted within 10–20%, and about 7% with deviations greater than 20%. One can see that Eq. (24) and S–T correlation gave close values in turbulent flow region, and both were better than other correlations.

It was found that Nu was very sensitive to outer wall temperature value, especially near the outlet where outer wall temperatures were high. In Figs. 8 and 9, it can be seen that deviations existed in high temperature places.

4.4. Comparisons of wall temperatures

4.4.1. Test cases of present electrically heated tube facility

The outer wall temperatures can be calculated by using the presented heat transfer correlations. Fig. 10 shows the comparison of calculated and experimental outer wall temperature in a different run of present electrically heated tube test.

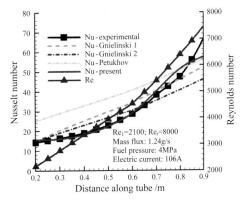

Fig. 8. Comparisons of Nusselt numbers in transition flow region.

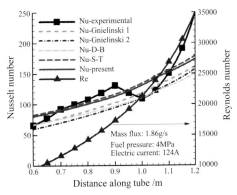

Fig. 9. Comparisons of Nusselt numbers in turbulent flow region.

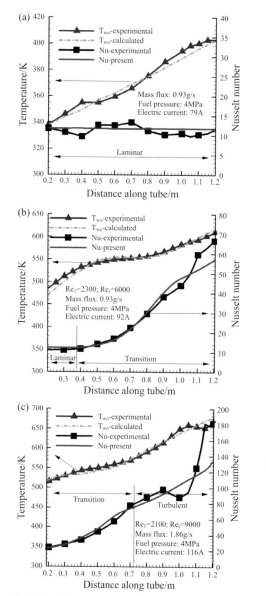

Fig. 10. Outer wall temperature comparisons of electrically heated tube test in present facility. (a) Laminar flow case. (b) Laminar and transition flow case. (c) Transition and turbulent flow case.

In Fig. 10, the outer wall temperature predictions adopting the presented heat transfer correlations were acceptable in all three flow regions.

Fig. 11. Electrically heated tube facility.

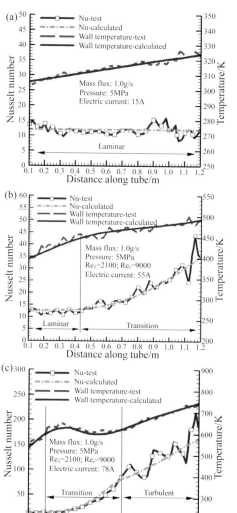

Fig. 12. Outer wall temperature comparisons of electrically heated tube test in another facility. (a) Laminar flow case. (b) Laminar and transition flow case. (c) Transition and turbulent flow case.

4.4.2. Test cases of another electrically heated tube facility

In order to verify the applicability of the presented correlations, similar experiment was conducted in another facility. Like part 4.4.1, the presented heat transfer correlations were adopted to calculate the outer wall temperature and compared with data of test cases of another electrically heated tube facility.

The test section was 1.3 meter long and mounted horizontally (Fig. 11). The tube was made of 0Cr18Ni9 stainless steel tube with an outer diameter of 3.0 mm and a wall thickness of 0.75 mm. The coolant fuel was also *n*-decane. The mass fluxes were 1.0 g/s and 1.5 g/s, and the outlet fuel pressure was about 5.0 MPa.

The calculated and experimental outer wall temperatures are shown in Fig. 12. It can be seen that the present correlations show a good agreement with the test data.

4.4.3. Fuel cooled panel test case

The presented heat transfer correlations were adopted to calculate the back wall temperatures in a fuel cooled panel test [15].

The fuel-cooled panel test was conducted in an arc-jet facility, as shown in Fig. 13. Inside the panel there was only one passage with an aspect ratio of close to 1, and the passage width was about 2 mm. The length of the channel was around 2 m. The high temperature gas flow side wall was made of GH3030, and the back side wall was made of stainless steel (1Cr18Ni9Ti). More details can be found in Ref. [15].

The calculated and experimental back wall temperatures are shown in Fig. 14. It can be seen that the present correlations in laminar and transition flow regions show better agreement with the measured data than S–T correlation.

4.4.4. Regeneratively cooled combustor test case

The presented heat transfer correlations were adopted to calculate the back wall temperatures in a regeneratively cooled combustor test. The combustor test was conducted in the same arc-jet facility (Fig. 15).

The calculated and experimental back wall temperatures in the center line are shown in Fig. 16. The calculation adopting the S–T correlation in laminar and transition flow regions underestimated back wall temperatures, which might be detrimental for cooling structure design. The presented *Nu* gave slightly larger predictions than test in laminar and transition flow regions, which might be caused by the errors of thermal environment and coolant flux between calculation and test.

From above stated wall comparisons, one can see that the presented heat transfer correlations are applicable for scramjet cooling structure design.

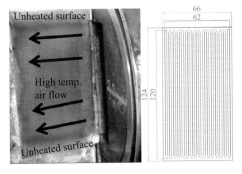

Fig. 13. Fuel cooled panel at nozzle exit and the cooling passage inside the panel.

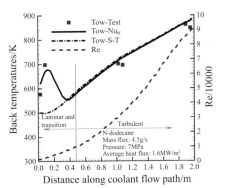

Fig. 14. Back temperature comparison of a fuel cooled panel test.

Fig. 15. The regeneratively cooled combustor in test.

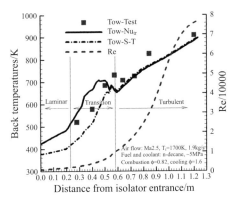

Fig. 16. Back temperature comparison of a combustor test.

5. Conclusion

The heat transfer behaviors of *n*-decane in laminar, transition, and turbulent flow regions under supercritical pressure were experimentally studied using an electrically heated tube facility and the method of least squares curve fitting.

The calculated outer wall temperatures were compared with experimental data of present and another electrically heated tube facilities, a fuel cooled panel test, and a regeneratively cooled scramjet combustor test. The results showed that the presented heat transfer correlations are applicable for scramjet cooling structure design.

References

[1] W.H. Heiser, D.T. Pratt, D.H. Daley, U.B. Mehta, Hypersonic Airbreathing Propulsion, AIAA Inc., Washington, 1994. pp. 1–15.

[2] F.Q. Zhong, X.J. Fan, G. Yu, J.G. Li, Heat transfer of aviation kerosene at supercritical conditions. J. Thermophys. Heat Transfer 23 (3) (2009) 543–550.

[3] J. Jiang, Investigation on the thermal-structural analysis and design of regeneratively cooled scramjet, Ph.D Thesis, Northwestern Polytechnical University, Xi'an, 2011.

[4] G.I. Isaev, G.K. Abdullaeva, Heat transfer to an ascending laminar flow of hydrocarbons under supercritical pressure conditions, J. Eng. Phys. Thermophys. 79 (4) (2006) 722–726.

[5] A.J. Ghajar, L.M. Tam, S.C. Tam, Improved heat transfer correlation in the transition region for a circular tube with three inlet configurations using artificial neural networks, Heat Transfer Eng. 25 (2) (2004) 3040.

[6] A.J. Ghajar, L.M. Tam, Heat transfer measurements and correlations in the transition region for a circular tube with three different inlet configurations, Exp. Therm. Fluid Sci. 8 (1994) 79–90.

[7] B. Stiegemeier, M.L. Meyer, R. Taghavi, A thermal stability and heat transfer investigation of five hydrocarbon fuels: JP-7, JP-8, JP-8+100, JP-10, and RP-1, in: 38th AIAA/ASME/SAE/ASEE Joint Propulsion Conference & Exhibit, Indianapolis, Indiana, July 2002.

[8] T.L. Bergman, A.S. Lavine, F.P. Incropera, D.P. Dewitt, Fundamentals of Heat and Mass Transfer, seventh ed., John Wiley & Sons Inc., New York, 2011. pp. 518–555.

[9] W.M. Rohsenow, J.P. Hartnett, Y.I. Cho, Handbook of Heat Transfer, third ed., McGraw-Hill Book Company, New York, 1998. pp. 5.18–5.36.

[10] S.M. Yang, W.Q. Tao, Heat Transfer, fourth ed., Higher Edition Press, Beijing, 2006. pp. 243–255.

[11] D. Huber, H. Walter, Forced convection heat transfer in the transition region between laminar and turbulent flow for a vertical circular tube, Latest Trends on Theoretical and Applied Mechanics, Fluid Mechanics and Heat & Mass Transfer, Greece, 2010.

[12] V. Gnielinski, New equations for heat and mass transfer in the turbulent flow in pipes and channels, Eng. Res. 41 (1) (1975) 7–16.

[13] D.L. Linne, M.L. Meyer, T. Edwards, D.A. Eitman, Evaluation of heat transfer and thermal stability of supercritical JP-7 Fuel, NASA TM-107485.

[14] T.A. Ward, Physical and chemical behavior of flowing endothermic jet fuels, Ph.D Thesis, University of Dayton, Ohio, 2003.

[15] J. Jiang, R.L. Zhang, J.L. Le, W.X. Liu, J. Wu, G.Z. Zhao, The investigation on heat transfer characteristic tests and thermal evaluation of fuel-cooled panels, J. Exp. Fluid Mech. 25 (1) (2011) 1–6.

超临界压力下正十烷流动传热的数值模拟[*]

赵国柱[1]，宋文艳[1]，张若凌[2]，乐嘉陵[2]

（1. 西北工业大学 动力与能源学院，陕西 西安 710072；

2. 中国空气动力研究与发展中心 高超声速冲压发动机技术重点实验室，四川 绵阳 621000）

摘　要：为了深入理解主动再生冷却过程中碳氢燃料的超临界传热特性，基于 SIMPLE 算法建立了数值模拟方法，考虑了碳氢燃料物性随温度的剧烈变化，并利用电加热管实验结果验证了计算方法。针对超临界压力下细管道内正十烷的流动传热现象进行了系统的数值计算研究，考察了计算网格无关性和超临界流动传热过程中的压力效应。结果表明：网格选择与正十烷的状态有关；在超临界压力下，较低的正十烷压力引起临界温度附近的努赛尔数减小，导致传热效率下降；目前常用的传热经验公式在正十烷临界区域附近与数值计算结果差别较大。

关键词：超临界压力；正十烷；再生冷却；数值模拟

中图分类号：V231.1　　　**文献标识码**：A　　　**文章编号**：1001-4055（2014）04-0537-07

Numerical Simulation on Flow and Heat Transfer of *n*-Decane Under Supercritical Pressure

ZHAO Guo-zhu[1], SONG Wen-yan[1], ZHANG Ruo-ling[2], LE Jia-ling[2]

（1. School of Power and Energy, Northwestern Polytechnical University, Xi'an 710072, China；

2. Science and Technology Laboratory on Scramjet, China Aerodynamics Research and Development Center, Mianyang 621000, China）

Abstract：In order to understand the heat transfer characteristic of hydrocarbon under supercritical pressure in regenerative cooling process, a numerical method was established based on SIMPLE algorithm. The method can model the drastic variations of hydrocarbon thermophysical properties with temperature. The results of an electrically heated tube test were adopted to verify the method. A systemic numerical investigation was then conducted to study the flow and heat transfer phenomena of *n*-decane under supercritical pressure in a tube. Grid independence and the pressure effects on supercritical flow and heat transfer were analyzed in detail. The results show that the state of *n*-decane has a great effect on grid selection. Nusselt number decreases with decreasing pressure in the vicinity of critical temperature under supercritical pressure. Thus the efficiency of heat transfer declines. There are obvious differences between the results of empirical expressions and that of numerical prediction due to the drastic variations of thermophysical properties near the critical region.

Key words：Supercritical pressure；*n*-decane；Regenerative cooling；Numerical simulation

赵国柱, 宋文艳, 张若凌, 乐嘉陵. 超临界压力下正十烷流动传热的数值模拟. 推进技术, 2014, 35(4): 537-543.

1 引　言

主动再生冷却技术是目前解决超燃冲压发动机热防护的重要研究方向[1,2]。再生冷却发动机工作时,碳氢燃料在喷入燃烧室前,首先经发动机壁面内的冷却通道吸热,冷却发动机壁面。为保证喷油效果,燃料系统工作压力通常超过燃料的临界压力,导致冷却通道内的流动传热在超临界压力下进行。在发动机壁面高热流密度条件下,随着燃料不断吸热,其温度逐渐升高至超过临界温度,经历了由高压液态到超临界态的转变过程,并可能出现裂解反应,这一过程中燃料的物理化学性质将发生剧烈变化。超燃冲压发动机再生冷却通道为矩形通道,单侧受热,且燃烧室热环境很不均匀,燃料流动传热呈现典型三维特征。因此,针对超临界压力下碳氢燃料的流动传热特性进行数值计算研究,对掌握主动再生冷却的流动传热机理具有重要意义。

国内外对超临界压力下的传热现象已有部分研究,早期主要针对二氧化碳和水两种介质。Duffey[3,4]系统总结了二氧化碳和水超临界传热的实验研究。Jiang et al[5]实验研究了流动雷诺数、热流密度、流动方向变化时对超临界压力下垂直管道内二氧化碳对流传热特性的影响,并通过数值计算分析了湍流模型的适用性。赵民富[6]通过数值计算与实验结果对比,研究了计算网格对模拟水的超临界传热现象时的影响。Eckart[7]对超临界压力下流体对流传热的数值研究进展进行了综述。基于航空航天应用的需求,近期关于碳氢燃料的超临界传热现象逐渐得到重视,国内部分研究机构和高校开展了相关研究。仲峰泉[8]在模拟飞行马赫数 6 的条件下对国内三号航空煤油 RP-3 的传热特性进行了实验研究和分析。张斌[9]基于 RP-3 的传热实验,提出了超临界压力下碳氢燃料的对流换热经验关系式。针对碳氢燃料超临界传热的数值计算研究也取得了一定进展[10-12]。但是,由于超临界压力下碳氢燃料的流动传热现象十分复杂,目前在高超声速技术领域内对其认识和应用仍处于探索阶段。

本文采用完整的 N-S(Navier-Stokes)方程与碳氢燃料物性计算程序耦合,建立了超临界压力下碳氢燃料的流动传热计算方法。以分子量及物性变化规律与超燃冲压发动机所用燃料接近的正十烷为工质进行了电加热管实验,通过实验验证了计算方法和网格无关性。采用验证后的方法详细研究了压力对正十烷流动传热的影响,并通过努赛尔数沿正十烷流向的变化规律,分析了正十烷在临界点附近流动传热的特殊性及目前常用传热经验公式的适用性。

2 数值计算方法

2.1 控制方程

连续方程

$$\frac{\partial}{\partial x_i}(\rho u_i) = 0 \tag{1}$$

动量方程

$$\frac{\partial}{\partial x_i}(\rho u_i u_j) = -\frac{\partial p}{\partial x_i} + \frac{\partial}{\partial x_i}\left(\mu_{\text{eff}}\frac{\partial u_j}{\partial x_i}\right) + \frac{\partial}{\partial x_i}\left(\mu_{\text{eff}}\left(\frac{\partial u_i}{\partial x_j} - \frac{2}{3}\delta_{ij}\frac{\partial u_k}{\partial x_k}\right)\right) \tag{2}$$

能量方程

$$\frac{\partial}{\partial x_i}(\rho u_i h) = \frac{\partial}{\partial x_i}\left(\alpha_{\text{eff}}\frac{\partial h}{\partial x_i}\right) + u_i\frac{\partial p}{\partial x_i} + \phi \tag{3}$$

式中 μ_{eff}、α_{eff} 分别为有效粘性系数和有效热扩散系数,Φ 为粘性耗散项。

由于再生冷却通道水力直径很小,通常小于 2mm,冷却燃料流速自入口处约 1~10m/s 逐渐增大,通道内格拉晓夫数和雷诺数平方的比值 Gr/Re^2 的量级约为 10^{-5}~10^{-4},表明重力对流动传热的影响很小,因此动量方程中忽略了重力项。

2.2 湍流模型

按照湍流边界层划分,近壁区域内可根据无量纲网格厚度 y^+ 分为粘性底层、过渡层和对数律层。紧邻壁面的粘性底层内湍流 Re 很低,与核心湍流区需采取不同的处理方式。y^+ 的表达式为

$$y^+ = \frac{y\sqrt{\tau_w\rho}}{\mu} \tag{4}$$

式中 τ_w 为壁面切应力,y 为附壁网格厚度,ρ 为密度,μ 动力粘度。

目前常用的双方程湍流模型有 $k\text{-}\varepsilon$ 模型、Wilcox[13] 的 $k\text{-}\omega$ 模型和 Menter[14] 的 $k\text{-}\omega$ SST 模型。标准 $k\text{-}\varepsilon$ 模型是高 Re 数模型,在计算有壁面约束的管道流动传热时需结合壁面函数方法应用;$k\text{-}\omega$ 模型在近壁处做了精确处理,是求解近壁区流动传热的较好选择;Menter 在有效结合以上两种模型,并对涡粘性进行修正后,提出 $k\text{-}\omega$ SST 模型,可应用于不同的湍流区域。综上,本文选择 $k\text{-}\omega$ SST 模型进行湍流传热模拟。

湍动能输运方程

$$\frac{\partial (\rho u_i k)}{\partial x_i} = P_k - D_k + \frac{\partial}{\partial x_i} \left((\mu + \alpha_k \mu_t) \frac{\partial k}{\partial x_i} \right) \quad (5)$$

比耗散率输运方程

$$\frac{\partial (\rho u_i \omega)}{\partial x_i} = \frac{\gamma \rho}{\mu_t} P_k + \frac{\partial}{\partial x_i} \left((\mu + \alpha_\omega \mu_t) \frac{\partial \omega}{\partial x_i} \right) -$$
$$\beta \rho \omega^2 + (1 - F_1) \frac{2 \rho \alpha_{\omega 2}}{\omega} \frac{\partial k}{\partial x_i} \frac{\partial \omega}{\partial x_i} \quad (6)$$

式中 P_k，D_k 分别为 k 的生成项和耗散项；F_1 为混合函数；α_k，α_ω 分别为 k 和 ω 的等效耗散系数。

SIMPLE(Semi-Implicit Method for Pressure Linked Equations) 算法自提出以来，经过不断发展完善，目前已比较成熟，能够较好满足求解低速对流传热问题的需求。本文采用 SIMPLE 方法作为压力-速度耦合方法，分离式迭代求解控制方程，直至满足收敛准则。

2.3 物性计算

超临界压力下碳氢燃料物性显著受到温度变化的影响，尤其在临界点附近，燃料物性变化十分剧烈。准确模拟燃料物性变化是求解碳氢燃料超临界传热的关键因素。

本文采用碳氢化合物的物性计算程序生成燃料的物性数据链表，在 CFD 程序中通过查表-插值的方法来更新每一步的物性值。相比直接调用物性计算程序，查表法在主程序循环外计算燃料物性，按照一定的压力、温度间隔生成数据链表，利于提高计算效率，尤其对多组分混合物提高更为明显。并且再生冷却通道内燃料的压降通常较小，温升低于 1000K，所需数据链表较为简洁，且具有良好的扩充性和重复性。

物性计算及与 CFD 程序的耦合见图 1。

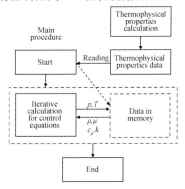

Fig. 1　Sketch of numerical calculation process

3　电加热管实验简述

电加热管实验具有热流密度高、燃料升温快和操作相对简单的特点，是研究燃料传热中广泛采用的方法[15,16]。

图 2 为实验示意图，实验中在电热管两端加直流电压，通过调节电压可控制壁面热流。电热管长 1.3m，内径 1.5mm，外径 3mm。管道外壁沿程焊有 17 个 K 型热电偶，用于测量外壁温度。管内燃料为正十烷（临界压力 p_c:2.11MPa，临界温度 T_c:617.7K），其物性与超燃冲压发动机所用燃料的物性随温度、压力变化趋势接近，且单一组分便于计算和分析。

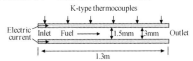

Fig. 2　Sketch of electrically heated tube

实验中正十烷流量 1.24g/s，入口温度 297K，压力约 4MPa。实验共进行五个热流恒定状态测量。基于能量守恒定律，对各实验状态进行传热分析[17]，可得到对应的管壁热流密度。图 3 给出了实验工况下热流恒定时的壁温测量结果，以及由传热分析得到的热流密度分布。

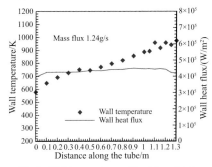

Fig. 3　Wall temperature and heat flux of test

4　网格无关性分析

传热计算时，无量纲附壁网格厚度 y^+ 对计算结果的影响较大，采用 k-ω SST 模型时应保证粘性底层内布有足够的网格点，需满足 $y^+ < 5$。y^+ 与网格质量相关，因此首先研究计算网格的影响。针对图 3 中所示的实验状态进行计算，采用二阶中心差分格式离散方程，实施 SIMPLE 算法时，压力松弛因子取 0.3，其

它变量统一取 0.7。计算时边界条件设置如下。

入口边界

$$u_{i,in} = \frac{\dot{m}}{\rho A} = u_{i,0}, T_{in} = T_0, \frac{\partial p}{\partial x} = 0 \qquad (7)$$

固壁边界

$$u_{i,w} = 0, q_w = q(x), \frac{\partial p}{\partial n} = 0 \qquad (8)$$

出口边界

$$\frac{\partial u_i}{\partial x} = 0, \frac{\partial T}{\partial x} = 0, p = p_0 \qquad (9)$$

式中 A 为管道截面积,$u_{i,0}$ 为入口速率,$q(x)$ 是图 3 中由传热分析得到的沿程热流密度。

正十烷分子量 142.287g/mol,图 4,图 5 分别给出了正十烷密度 ρ,定压比热 c_p,热导率 k 随温度和压力的变化,其他物性变化未一一列出。

生成计算网格时大致保证满足 $y^+ < 5$,最终轴对称面上的网格设置见表 1,其中 Number 1 ~ 4 用于验证 y^+ 的影响,Number 4 ~ 6 验证法向网格数的影响。

Table 1　Grids for calculation

Number	Grids	Size of nearest grid to wall/μm
1	1300 × 50	15
2	1300 × 50	10
3	1300 × 50	5
4	1300 × 50	1
5	1300 × 40	1
6	1300 × 30	1

计算结果见图 6、图 7。图 6 是壁面温度的实验和计算结果,以及正十烷流动 Re 的沿程分布。图 7 是不同网格的 y^+ 分布,以及正十烷与流动方向垂直截面上平均温度的沿程分布。正十烷截面平均温度 T_f 由下式计算[18]

$$T_f = \frac{\int_A c_p \rho u T dA}{\int_A c_p \rho u dA} \qquad (10)$$

按照文献[19]的划分,$Re < 2300$ 时为层流区;$2300 \leqslant Re \leqslant 10^4$ 时为过渡区;$Re > 10^4$ 时为旺盛湍流区。由图 6 可知,当电热管长度 $l > 0.44m$ 时,管内正十烷流动发展为充分湍流流动。在 $l \leqslant 0.44m$ 的层流和过渡区,传热效率低于湍流传热,因此这一区域实验测量的壁温较高。数值计算全部采用湍流模拟,本文主要对比分析湍流区的数据。计算时固壁边界条件为恒定热流,根据能量守恒定律,层流和过渡区正十烷温度与实际温度相符,在 $l \leqslant 0.44m$ 段流态的差别对燃料温升的影响较小,从而对后面湍流段的计算影响较小。

由图中可知,网格 Number 1 ~ 4 都满足 $y^+ < 5$,在 $l \geqslant 0.9m$ 的范围内计算结果差别较小。但在 $0.5m < l < 0.9m$ 的范围内,Number 1,Number 2 的壁温计算值与实验值相差约 40K,误差 $\Delta T/(T_w - T_0)$ 超过 8%。网格 Number 3 的 y^+ 不大于 1,此时壁温计算值与实验值的误差小于 1%,符合计算需求。继续减小 y^+(Number 4),计算结果基本保持不变。

y^+ 是近壁区流动规律的重要判据,y^+ 相同的位置流动规律相同。网格 Number 3 的计算结果最优,详细分析其 y^+ 分布,由图 7 知,y^+ 可明显分为 $l \geqslant 0.9m$,$0.7m \leqslant l < 0.9m$,$0.44m < l < 0.7m$ 三个区域。可以看出,$l \geqslant 0.9m$ 时正十烷具有相同的流动规律,此时 y^+ 近似为 1。$l = 0.7m$ 时,正十烷平均温度约 670K,对照图 5,该温度下正十烷的 c_p 处于峰值,而 k 取值基本为最小。$0.44m < l < 0.7m$ 的范围内,c_p,k 经历急剧的增大和减小过程,$0.7m \leqslant l < 0.9m$ 的范围内 c_p,k 的变化趋势相反。可见,正十烷的流动传热规律及对 y^+ 的要求与其所处的状态有关,生成网格时必须考虑这一因素的影响,以期获得最优计算结果。

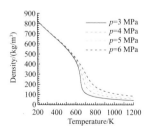

Fig. 4　Density of n-decane under various temperature and pressure

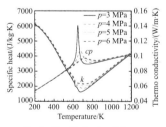

Fig. 5　Specific heat and thermo conductivity of n-decane under various temperature and pressure

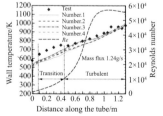

Fig. 6　Experimental and calculated wall temperature and Reynolds number along the tube

通过对网格 Number 4～6 进行对比计算,研究法向网格数对计算的影响。结果表明,当附壁网格尺寸保持不变时,三组网格的计算值之间误差小于 2%,法向网格数对计算结果的影响很小。

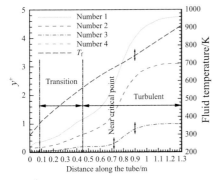

Fig.7　The y^+ variations of different grids and fluid temperature along the tube

5　压力对正十烷流动传热的影响

超燃冲压发动机工作时,常温碳氢燃料在超临界压力下注入冷却通道,经由燃烧室壁面加热,温度逐渐升高并超过临界温度至超临界态。超临界压力对流动传热的影响是主动冷却研究关心的问题之一。本节将通过数值计算,详细研究管道内正十烷处于不同压力时的流动传热特性。计算区域及设置与电加热管相同,但壁面热流取固定值 0.4MW/m²。按上节所述,生成计算网格时取 $y^+\sim1$,轴对称面网格数 1300×50。分别计算正十烷压力约为 3MPa,4MPa,5MPa 及 6MPa 的状态。

图 8 为不同压力下正十烷的 Re 数沿程分布,各 Re 数在 $l=0.5m$ 前都超过 10^4,流动发展为充分湍流流动。

图 9 为最终计算得到的壁温和正十烷平均温度。由图可知,在湍流区内,压力大于 4MPa 时,壁温基本一致;压力 3MPa 时,壁温在 0.45m≤l<0.8m 范围内突然升高;所有压力下正十烷平均温度基本不变。分析认为,压力 3MPa 已接近正十烷的临界压力,在壁温突然升高的范围内,正十烷平均温度大致处在 T_c ±70K 的区间。此区间对应的正十烷热导率最小,相比其它压力状态,固壁与流体间传热效率很低,流体带走的热量最少,因此出现壁温升高的现象。

正十烷物性变化对流动传热的影响可以从图 10

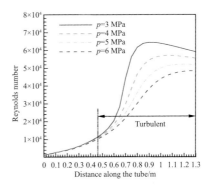

Fig.8　Variations of Reynolds number along the tube under different pressure

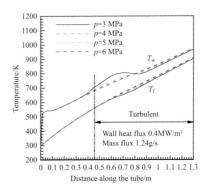

Fig.9　Wall temperature and fluid temperature under different pressure

中更为直观的加以说明。图 10(a)、(d)、(e) 分别为管道轴对称面上 T_f、c_p、k 的分布,c_p 在临界温度附近靠近壁面极窄的范围内突然升高,并逐渐延伸至主流区,随后迅速下降;而 k 在相应范围内先减少再增加,二者变化规律与上面的分析一致。(b) 中为正十烷密度分布,在临界温度附近,密度迅速减小,对照图 4 可知,此时密度随温度变化曲线的斜率极大。基于质量守恒定律,密度减少引起速度增加,图 11 给出了管道轴向不同位置截面上的速度分布,可明显看出($l=0.75m$)正十烷在临界点附近流动规律的特殊性。

进一步通过努赛尔数 Nu 分析正十烷在不同压力下的传热规律,并选取目前常用的对流传热经验公式 Gnielinski[11] 和 S-T(Sieder-Tate)[20] 公式对比。各 Nu 数计算公式如下。

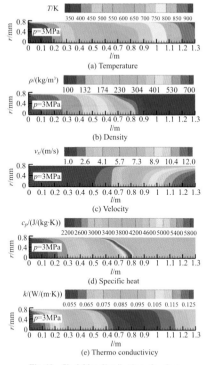

Fig. 10 Variables distribution of *n*-decane
on aix-symmetry plane

本文计算

$$Nu = \frac{hd}{k_f} = \frac{q}{(T_w - T_f)} \frac{d}{k_f} \qquad (11)$$

式中 k_f 根据平均温度 T_f 取值，h 为对流换热系数，d 为管道内径。

Gnielinski 公式

$$Nu = \frac{(f/8)(Re_f - 1000)Pr_f}{1 + 12.7(f/8)^{0.5}(Pr_f^{2/3} - 1)} \qquad (12)$$

$$f = (0.79 \ln Re_f - 1.64)^{-2}$$

适用范围为 $0.5 \leqslant Pr_f \leqslant 2000, 3000 \leqslant Re_f \leqslant 5 \times 10^6$。

S-T 公式

$$Nu = 0.027 Re_f^{0.8} Pr_f^{1/3} \left(\frac{\mu_f}{\mu_w}\right)^{0.14} \qquad (13)$$

努赛尔数计算结果见图 12。由图知，当压力为 3MPa 时，在 $0.5\mathrm{m} \leqslant l < 0.8\mathrm{m}$ 范围内，正十烷 Nu 数远低于压力 6MPa 时的值，表明此时管道内传热效率降低。根据计算结果，压力超过 4MPa 时，这一区域 Nu

数差别不大。由图中 Gnielinski 公式的计算结果可知，在远离临界温度的区域，其结果与本文计算值分布规律基本一致，而在临界温度附近，压力越接近临界压力，差别越大。S-T 公式的计算结果也符合这一规律，但差别更大。因此，对于超临界压力下正十烷的对流传热，Gnielinski 公式及 S-T 公式在临界点附近适用性可能较差。

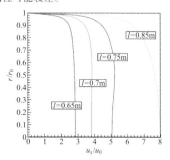

Fig. 11 Velocity distribution at different
position of the tube

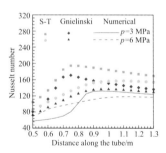

Fig. 12 Nusselt number in turbulent region
under different pressure

6 结 论

本文基于 SIMPLE 算法，通过数值计算模型与碳氢燃料物性程序的耦合，建立了模拟超临界压力下碳氢燃料流动传热的计算方法，并利用电加热管实验数据进行了验证。利用验证后的方法对超临界压力下管道内正十烷的流动传热进行了数值计算研究，主要结论如下：

（1）采用 $k\text{-}\omega$ SST 模型计算时，要求附壁网格在合理范围内取值。y^+ 的大小与正十烷所处的状态有关，不同位置的 y^+ 值不同。附壁网格法向尺寸不变时,法向网格数对计算结果的影响较小。

（2）在超临界压力下,较低的正十烷压力引起临界温度附近的 Nu 数减小,导致传热效率下降,壁温升高。

（3）临界点附近正十烷物性急剧变化,此时采用目前常用的 Gnielinski 和 S-T 对流传热经验公式计算得到的 Nu 数与数值计算结果差别较大。

参考文献:

[1] Jackson T A, Eklund D R, Fink A J. High Speed Prolusion Performance Advantage of Advanced Materials[J]. *Journal of Materials Science*, 2004,39:5905-5913.

[2] Verma A. Ameliorative Study of a Scramjet Engine by Regenerative Cooling Using Finite Element[J]. *International Journal of engineering and Technology*, 2010,2(6).

[3] Duffey R B, Pioro I L. Experimental Heat Transfer in Supercritical Carbon Dioxide Flowing Inside Channels[J]. *Nuclear Engineering and Design*, 2005, 235(8):913-924.

[4] Pioro I L, Duffey R B. Experimental Heat Transfer in Supercritical Water Flowing Inside Channels[J]. *Nuclear Engineering and Design*,2005,235(22):2407-2430.

[5] JIANG P X, Liu B, Zhao C R, et al. Convection Heat Transfer of Supercritical Pressure Carbon Dioxide in a Vertical Micro Tube From Transition to Turbulent Flow Regime[J]. *International Journal Heat and Mass Transfer*,2013,56:741-749.

[6] 赵民富,张国欣,陈玉宙. 竖直圆管内超临界水传热特性竖直模拟[J]. 原子能科学技术, 2011,45(2).

[7] Eckart L. Numerical Simulation of Flow and Heat Transfer of Fluids at Supercritical Pressure[EB/OL]. *University of Stuttgart*,IKE,URL:http://www.oecd-nea.org, 2008.

[8] Zhong F Q, Fan X J, Yu G, et al. Heat Transfer of Aviation Kerosene at Supercritical Conditions[R]. *AIAA* 2008-4615.

[9] 张 斌,张春本,邓宏武,等. 超临界压力下碳氢燃料在竖直圆管内换热特性[J]. 航空动力学报, 2012, 3:595-603.

[10] Qin J, Bao W, Zhou W X, et al. Flow and Heat Transfer Characteristics in Fuel Cooling Channels of a Recooling Cycle[J]. *International Journal of Hydrogen Energy*, 2010, 35:10589-10598.

[11] Hua Y X, Wang Y Z, Meng H. A Numerical Study of Supercritical Forced Convective Heat Transfer of n-Heptane Inside a Horizontal Miniature Tube[J]. *The Journal of Supercritical Fluids*,2010,52:36-46.

[12] 王亚洲,华益新,孟 华. 超临界压力下低温甲烷的湍流传热数值研究[J]. 推进技术,2010,31(4). (WANG Ya Zhou, HUA Yi Xin, MENG Hua. Numerical Investigation of Turbulent Heat Transfer of Cryogenic-Propellant Methane Under Supercritical Pressures[J]. *Journal of Propulsion Technology*,2010,31(4).)

[13] Wilcox D C. Reassessment of the Scale Determining Equation for Advanced Turbulence Models[J]. *AIAA Journal*, 1988,26(11):1299-1310.

[14] Menter F R. Two-Equation Eddy-Viscosity Turbulence Models for Engineering Applications[J]. *AIAA Journal*, 1994,32(8):1598-1605.

[15] Wishart D P, Fortin T, Guinan D, et al. Design Fabrication and Testing of an Actively Cooled Scramjet Propulsion System[R]. *AIAA* 2003-15.

[16] Chen A Y, Dang L. Characterization of Supercritical JP-7's Heat Transfer and Coking Properties[R]. *AIAA* 2002-0005.

[17] 张 磊,乐嘉陵,张若凌,等. 超临界压力下湍流区碳氢燃料传热研究[J]. 推进技术,2013,34(2). (ZHANG Lei, LE Jia Ling, ZHANG Ruo Ling, et al. Heat Transfer of Hydrocarbon Fuel in Turbulent Flow Region Under Supercritical Pressure[J]. *Journal of Propulsion Technology*, 2013,34(2).)

[18] 杨世铭,陶文铨. 传热学(第四版)[M]. 北京:高等教育出版社, 2006.

[19] 陶文铨. 数值传热学(第二版)[M]. 西安:西安交通大学出版社, 2001.

[20] Linne D L, Meyer M L, Edwards T, et al. Evaluation of Heat Transfer and Thermal Stability of Supercritical JP-7 Fuel[R]. *AIAA* 97-3041.

吸气式高超声速飞行器机体推进一体化技术研究进展

吴颖川*，贺元元，贺伟，乐嘉陵

中国空气动力研究与发展中心 超高速空气动力研究所 高超声速冲压发动机技术重点实验室，
绵阳 621000

摘　要：吸气式高超声速一体化飞行器最显著的特点是子系统之间的耦合较其他类型飞行器更加强烈，这使得其设计具有挑战性。所有的子系统之间部件相互干涉，包括气动、推进、控制、结构、装载和热防护等，特别是机体与超燃冲压发动机之间的耦合最为突出。飞行器的前体和后体下壁面既是主要的气动型面，又是超燃冲压发动机进气道外压缩型面和尾喷管的膨胀型面，在产生推力的同时也产生升力和俯仰力矩。机体与发动机的强耦合作用对飞行器的推力、升力、阻力、俯仰力矩、气动加热、机身冷却、稳定性和控制特性有直接的影响。本文介绍了国内外机体推进一体化技术的研究进展，重点介绍了中国空气动力研究与发展中心（CARDC）的相关研究工作，包括：密切曲锥曲面乘波进气道和基于双激波轴对称基准流场内转式进气道设计方法、独创的大尺度脉冲式燃烧加热风洞一体化飞行器带动力试验技术和高超声速内外流耦合数值模拟技术等。对高速飞行中激波边界层相互干扰、流动分离机理、可压缩湍流转捩及其控制、超燃冲压发动机燃烧流动机理等相关基础问题也进行了研究，强调了对高效高精度计算方法的迫切需求。

关键词：高超声速飞行器；超燃冲压发动机；机体推进一体化；乘波体；燃烧加热风洞；湍流燃烧；转捩

中图分类号：V475.2　　文献标识码：A　　文章编号：1000-6893(2015)01-0245-16

吸气式高超声速技术是研究飞行马赫数大于5，以吸气式发动机为动力，在大气层和跨大气层中实现高超声速远程飞行的飞行器技术。对其进行研究的目的是实现全球快速到达和低成本进入空间。

吸气式高超声速飞行器发展主要分为3个阶段：高超声速巡航飞行器——具有高速度、高能量、高生存和机动的特点；高超声速飞机——快速到达全球，用于远程侦察和轰炸；空天飞机——水平起降、快速进入空间、低成本、易于维护。

超燃冲压发动机（Scramjet）使来流经过斜激波压缩后仍然保持为超声速，燃料在燃烧室内进行超声速燃烧，这样可以有效地减小气流能量损失，降低对燃烧室热防护的要求，使飞行器在高马赫数飞行时能够获得较高的有效比冲。因此，超燃冲压发动机使吸气式高超声速飞行成为可能，其最适于作为高超声速飞行器的动力系统。

从20世纪80年代开始，美国开展了国家空天飞机（National Aero-Space Plane，NASP）计划[1]（见图1）、HyTech计划[2]（后来衍变为Hy-Set项目[2]）、HyFly项目[2]、X-43A项目[3-4]、X-51A项目[5]和FALCON（Force Application and Launch from CONUS）项目[6]等一系列直接或间接发展高超声速飞行器技术的计划或项目。这些

吴颖川，贺元元，贺伟，乐嘉陵. 吸气式高超声速飞行器机体推进一体化技术研究进展. 航空学报. 2015, 36(1): 245-260.

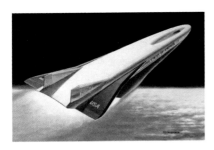

图 1　X-30 空天飞机示意图[1]

Fig 1　Schematic diagram of X-30 space plane[1]

计划或项目，有些虽然由于经费等原因被取消，但有些进行了整合并正在开展，具有很好的继承性和连续性。目前重点项目除 X-51A 外，还包括 FALCON 和 HyFly 等项目。

俄罗斯的高超声速计划[7]主要有冷计划、彩虹-D2 计划和鹰计划。法国[8]、澳大利亚[9]、德国[10]、印度[11]、日本[11]和韩国[11]等都开展了自己的高超声速研究计划。

NASP 计划由美国国家航空航天局（NASA）和国防部联合发展，计划用 11 年时间，投资 50 多亿美元，最后研制成两架 X-30 样机，这是单级入轨空天飞机的试验机。NASP 计划分 3 个阶段：第 1 阶段进行可行性研究，已于 1985 年完成，第

2 阶段（1986—1990 年）攻克关键技术，包括超燃冲压发动机和热防护材料等。第 3 阶段于 1990 年开始，拟研制两架 X-30 试验机进行试飞，并根据试飞情况作出研制实用空天飞机的决定。

NASP 计划虽然失败了，但奠定了美国高超声速技术的发展基础。计算流体力学（CFD）应用能力已扩展至能够处理三维几何，能提供复杂的流场细节，提高了代码效率，开发出一批新防热材料，建立了推进实验数据库，完善了地面试验能力（8 ft 高温风洞（HTT）等大型设备，1 ft=0.304 8 m）；发展了计算方法（三维全 Navier-Stokes 方程，有限速率化学反应和设计用先进工程代码等）。

NASA 的 X-43 计划的目的是演示、验证和发展高超声速飞行器机体推进一体化和发动机技术，包括试验技术、计算方法、设计工具和性能预测。X-43 的名义飞行弹道如图 2 所示。

X43-A 的飞行过程如下：一架经过改装的 B-52B 重型轰炸机，机翼下挂着一架 X-43A 飞机和一枚"飞马"助推火箭，从加州的爱德华兹空军基地起飞，很快 B-52B 上升至 12 km 高空，这时，和 X-43A 捆绑在一起的"飞马"火箭点火，它们脱离 B-52B 轰炸机，并由"飞马"火箭把 X-43A 推到大约 29 km 的高空；接下来，X-43A 脱离"飞马"火箭，自身发动机点火，开始以 10 000 km/h 的速度

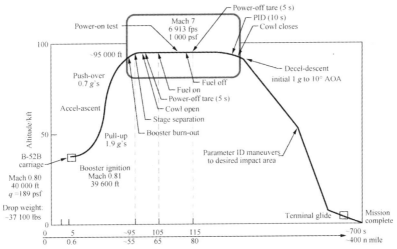

Notes: 1 psf = 47.848 6 Pa; 1 lbs = 0.453 6 kg; 1 n mile = 1 852 m; AOA—Angle of attack

图 2　X-43A 的名义飞行弹道[3]

Fig. 2　Nominal flight trajectory of X-43A[3]

独立飞行,约 $10s$ 后,燃料耗尽,飞机继续滑行了 $6 min$,经过 $1368 km$ 的距离坠入太平洋。

X-43A 飞行试验首次实现了以超燃冲压发动机为动力的升力体飞行器的高马赫数 $(Ma=7, 10)$ 自主飞行,具有里程碑意义。它验证了升力体分离、超燃发动机推进和一体化飞行器控制等关键技术的可行性。

美国 X-51A 项目的终极目标是要发展可以在 $1h$ 内进行远程飞行的飞行器,包括快速响应空间飞行器和高超声速巡航飞行器。X-51A(见图 3)采用碳氢燃料,设计 $12 min$ 内 Ma 从 4.5 增加到 6,其特点是固定几何进气道、乘波体前体外形和主动冷却发动机。X-51A 迄今共开展了 4 次飞行试验,目的是考验进气道起动、巡航加速、发动机模态转换、发动机点火/熄火、参数辨识以及机动性等能力,并在第 4 次取得成功。前 3 次飞行试验先后出现了尾喷管密封失效、进气道不起动和舵面失效等问题,第 4 次虽然取得了成功,但加速度远小于预期值。X51-A 的经验表明,吸气式高超声速飞行器技术难度大、周期长、投入大、风险高,必须循序渐进、坚持不懈地进行长期技术

图 3 X-51A 试飞器[5]

Fig 3 X-51A launch vehicle[5]

积累。

吸气式高超声速飞行器的主要关键技术有:发动机、结构、材料与热防护以及气动、推进、防热和控制的一体化。本文重点关注的是气动与推进的一体化,尤其关注推阻和升阻特性,也称为机体推进一体化。

1 吸气式高超声速技术面临的困难和挑战

吸气式高超声速飞行器最显著的特点是子系统之间的耦合较其他类型飞行器更加强烈,气动性能与发动机性能紧密耦合。机体推进一体化(见图 4)气动性能预测非常困难。

吸气式高超声速技术研究的三大手段分别是:CFD、风洞试验和飞行试验(见图 5)。首要任务是准确预测高超声速飞行器的机体推进一体化性能,在原理上验证高超声速技术的可实现性。

美国针对一体化飞行器推阻特性的预测是通过综合大量地面试验和分析研究结果获得的[12]。Navy 和 Air Force 负责整个研究计划(1976—1987 年),研究成果(见图 6)推动了 NASP 计划的建立。

有动力情况下,推阻和升阻特性仍然是当前最具挑战性和最紧迫的问题,亦是吸气式发动机性能预测中最困难的问题,主要原因在于地面试验设备尺寸太小,即使是类似 X-51A 的长约 $4.2 m$ 的飞行器,模拟仍然有很大困难。

解决的办法主要有 3 种:①大尺度飞行试验很困难、风险大;②建设比 APTU(Aerodynamic

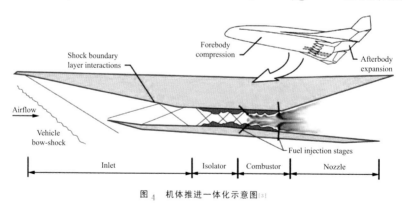

图 4 机体推进一体化示意图[3]

Fig 4 Schematic diagram of airframe-propulsion integration[3]

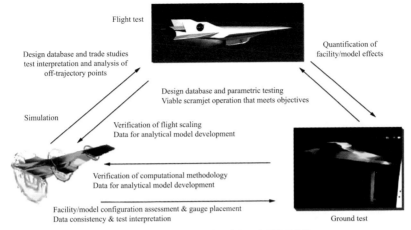

Flight test
Quantification of facility/model effects
Design database and trade studies test interpretation and analysis of off-trajectory points
Design database and parametric testing
Viable scramjet operation that meets objectives
Simulation
Verification of flight scaling
Data for analytical model development
Verification of computational methodology
Data for analytical model development
Facility/model configuration assessment & gauge placement
Data consistency & test interpretation
Ground test

图 5 吸气式高超声速技术研究的三大手段示意图

Fig. 5 Schematic diagram of three methods of air-breathing hypersonic technology research

Thrust
AIAA-1998-2506
Vehicle drag
Ma = 4
Design methods
Vehicle drag
Ma = 7
Ma = 4, experiment
Ma = 7
Fuel flow

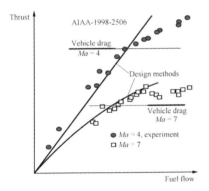

图 6 兰利中心超燃冲压发动机推阻性能试验结果(1976—1987)[12]

Fig. 6 Thrust drag performance summary of NASA Langley scramjet test results from 1976-1987[12]

and Propulsion Test Unit)大 10 倍的地面模拟设备——投入大、技术难度大;③在一些小型飞行试验的基础上,通过深入研究基本物理现象与尺度效应,综合试验与计算结果,进行一体化气动性能预测。

图 7 为对 X-43A 一体化气动性能预测[13]研究方案的总结,其中有基准进气道关闭的常规风洞试验、截断发动机试验和 8 ft 高温风洞的全流道发动机试验。

图 8 为 X-43A 进气道开/关、冷/热态的纵向力和力矩性能预测结果。由于小尺度飞行器模型不能进行进气道打开试验,以进气道关闭的常规高超风洞试验为基准,再结合 CFD 增量分析方法,预测飞行器进气道打开时带动力/不带动力的飞行器气动性能,包括马赫数、攻角和侧滑角的影响。并通过 8 ft 高温风洞全流道发动机试验(图 9)验证这一预测方法的准确性。

常规风洞缩比尺度很小,基准试验性能和 CFD 增量可能并不是简单的叠加,最终通过一体化综合分析得到全尺度飞行器的气动性能预测结果。

国内高超声速机体推进一体化技术研究起步较晚。主要是通过冷通气风洞试验进行一体化飞行器研究。

易军等[14]以美国 X-43A 和 X-51 两类高超声速飞行器为研究对象,对两类飞行器的气动性能进行了数值模拟,并以此为基础对比分析了两类高超声速飞行器的一体化气动特性。

张红英等[15]对一种类似于 X-43A 的吸气式高超声速一体化构形全流道开展了风洞试验和数值模拟研究,分析了不同来流 Ma、总压、飞行攻角下全流道的流场结构和气动力特性。

范晓樯等[16]以机体推进系统耦合、三维侧压式进气道为基本特征,设计了采用超燃冲压发动

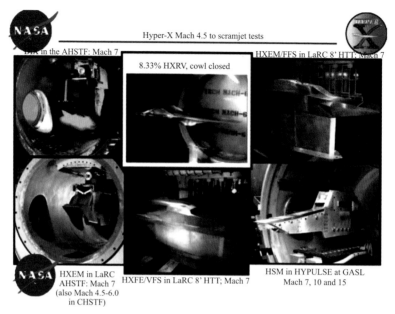

Notes: AHSTF is Langley arc-heated scramjet test facility; DFX—Dual-fuel experimental parametric engine; HXRV—Hyper-X research vehicle; HXEM—Hyper-X engine model; HXFE—Hyper-X flight engine; VFS—Vehicle flowpath simulator (used with HXFE); HSM—HYPULSE scramjet model; HYPULSE—NASA Langley hypersonic pulse facility at GASL., Inc., Ronkonkoma (NY); GASL—General Applied Science Laboratory

图 7　X-43A 风洞试验[13]

Fig 7　X-43A wind tunnel tests[13]

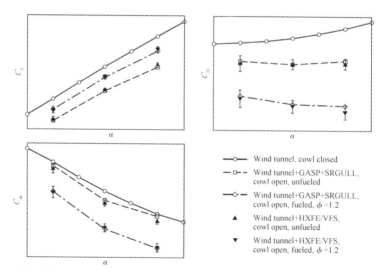

Notes: ϕ is fuel-air ratio; SRGULL is an advanced engineering model for the prediction of airframe scramjet engineering model for the predicton of airframe integrated scramjet cycle enformance; GASP—General Aviation Sythesis Program.

图 8　X-43A 气动性能增量法[13]

Fig 8　X-43A aerodynamic performance increment method[13]

(a) Ground demo engine 2

(b) X-51A flight clearance engine (FCE) SJX61-2

图 9   8 ft 高温风洞发动机试验

Fig 9   Eight-foot high temperature wind tunnel scramjet test

机为推进系统的高超声速一体化冷流通气实验模型，在高超声速炮风洞中完成了飞行器的整体气动测力试验。

金亮等[17]采用数值模拟和风洞试验方法，对高超声速一体化飞行器的缩比模型在发动机关闭以及发动机通流状态下的气动特性进行了研究。

无论是国外与国内，之前的机体推进一体化研究都是采用通气/不通气飞行器模型在低总温的常规风洞中获得飞行器冷态气动力数据，采用全流道发动机模型在高温推进风洞中获得发动机推力增量数据，通过数值计算修正和分析，综合出整个一体化飞行器的带动力性能。这样做的缺点是，常规风洞试验模型尺寸较小（目前国内最大的常规高超声速风洞口径是 1 m），不能模拟天上的总温条件，而内流道流态与尺度、温度等密切相关，通气模型不能简单地采用雷诺数相似模拟获得气动力数据；发动机在高温推进风洞中只能获得冷热态的推力增量数据，不能直接得到净推力，必须扣除冷态内阻，才能得到真正的推力性能，而目前冷态内阻的计算和试验测量都面临很大困难，有很大的不确定度，这给发动机真实性能的评估带来了不确定性。

综上所述，更好的办法是能够在短时大尺度高温风洞中直接开展飞行器的带动力一体化性能

试验，直接测量飞行器的推阻和升阻性能，然后通过风洞试验数据的相关性分析、尺度规律的影响研究，结合数值计算，得到全尺度飞行器的带动力一体化性能。

2   CARDC 机体推进一体化技术研究进展

美国机体推进一体化气动性能研究主要采用增量法，中国空气动力研究与发展中心（CARDC）在此基础上提出了结合脉冲燃烧风洞模型飞行器带动力一体化试验的综合分析方法。围绕预测飞行器一体化气动性能的目标，采用数值计算与风洞试验相结合的方法，直接预测飞行器气动性能，而不是采用增量法。

下面从气动布局设计、地面试验和数值模拟方法等 3 个方面介绍所取得的研究进展。

2.1   一体化气动布局设计技术

开发了乘波构型高超声速飞行器交互式参数化优化设计系统（Waverider derived Interactive Parametric Optimization and Design System of hypersonic vehicle，WIPODS），其具有参数化基线设置、快速的分析工具和图形化集成设计环境。

前体进气道压缩面采用曲面乘波压缩形式，基准流场由多段激波和曲面压缩轴对称流场组成，三维乘波面采用类密切锥（Osculating Cone Waverider，OCW）[18-19]方法由前缘线各点流线跟踪拟合构成流面。与国外密切锥方法采用单一直面或曲面激波构造基准流场不同，密切曲锥（Osculating Curved Cone Waverider，OCCW）[20-23]方法的基准流场由多个激波或等熵压缩流场组成，并且流线跟踪一直到进气道喉道内收缩段（见图 10）。其优点是可以灵活控制进气道的压缩量，并保持乘波体低阻力、高流量捕获和高总压恢复的特点。图 11 为密切锥方法与密切曲锥方法展向截面的对比。图 12 为采用密切曲锥方法生成的乘波进气道流场马赫数分布和壁面压力分布。

三维内转式进气道以其较高的压缩效率和较低的总压损失成为未来高超声速技术发展的一个重要方向，而流线追踪技术的引入拓宽了三维内转式进气道的设计方法，更在一定程度上克服了纯内压缩进气道的起动问题。

利用特征线理论提出了一种基于设计状态消

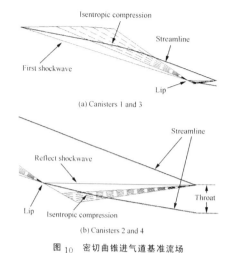

(a) Canisters 1 and 3

(b) Canisters 2 and 4

图 10　密切曲锥进气道基准流场

Fig. 10　Basic flow field of OCCW inlet

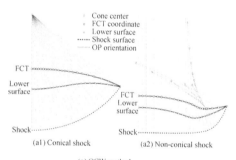

(a1) Conical shock　　　(a2) Non-conical shock

(a) OCW method

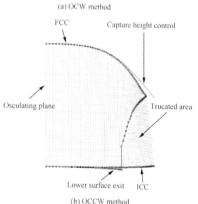

(b) OCCW method

Notes: OP—Osculating plane; FCT—Flow
capturetube; ICC—Inlte capture curve.

图 11　密切锥与密切曲锥方法对比

Fig. 11　Comparison of OCW and OCCW methods

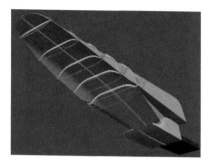

图 12　密切曲锥乘波进气道流场 YZ 截面马赫数分布和
壁面压力分布

Fig. 12　YZ cross section Mach number and wall pressure
distributions of OCCW inlet

波的双激波轴对称基准流场的设计方法．流场只
包含入射激波和反射激波．入射激波终止于唇口
前缘．反射激波入射至肩点并实现消波．不仅可实
现压缩面上流动参数的优化．还在最大程度上实
现了设计状态下流场结构的优化．基于这种基准
流场的流线追踪进气道设计状态下只存在两个激
波面(入射激波面和反射激波面，隔离段内完全消
除了激波反射)，如图 13 和图 14 所示．图 13 中：

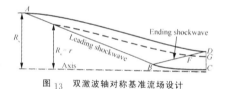

图 13　双激波轴对称基准流场设计

Fig. 13　Basic flow field design of dual shockwaves

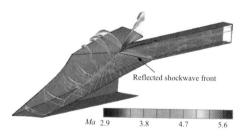

图 14　基于消波流场的流线追踪设计的内转式进气道
CFD 数值验证

Fig. 14　CFD results of inward turning inlet designed by
streamtracing method based on eliminating theory
of shockwave

R_s 为轴对称基准流场的最大半径，r 为流线追踪起始点的半径。图 15 为按照上述方法设计的内转式进气道试验模型和不同堵塞比（0%～80%）情况下进气道起动性能的考核结果。试验表明，进气道在设计状态下能够起动。

(a) Test model

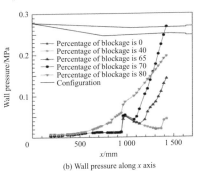

(b) Wall pressure along x axis

图 15　内转式进气道试验模型及起动性能

Fig. 15　Test model and start performance of inward turning inlet

2.2　一体化带动力试验技术

除了 ⌀1 m 等常规气动力高超声速风洞之外，CARDC 成功研制和改造了脉冲式和连续式两种类型的燃烧加热设备。在国际上首次独立提出了挤压式脉冲燃烧高超声速推进风洞原理和总体技术方案，建成了最大口径为 2.4 m，试验时间为 300～600 ms 的脉冲燃烧风洞[24]（见图16）。

⌀2.4 m 脉冲风洞能够直接获得 4～5 m 量级飞行器的带动力一体化气动性能，结合连续风洞和 CFD 结果，分析预测飞行器整机的推阻与力矩特性。

图 16　脉冲燃烧风洞照片

Fig. 16　Picture of pulsed combustion heated wind tunnel

1.5 m 飞行器[25]是开展机体推进一体化技术研究的一个基准模型，最早采用氢燃料，后逐步过渡到煤油燃料。基于 1.5 m 飞行器发展的一体化飞行器带动力试验技术为未来有可能在 ⌀2.4 m 脉冲燃烧风洞开展大尺度模型飞行器的一体化试验进行技术准备。

图 17 为 1.5 m 飞行器的试验模型，图 18 为发动机工作状态风洞试验典型测力时序曲线。发动机点火后，天平信号出现了反号，表明飞行器获得了正推力。表 1 为试验测力与 CFD 计算结果的比较，冷态结果基本一致，热态试验推力收益比计算预测约大 100 N。

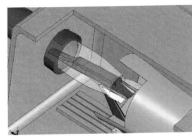

(a) Model installation diagram

(b) Model photo

图 17　1.5 m 飞行器机体推进一体化风洞试验模型

Fig. 17　Airframe-propulsion integration wind tunnel test model of 1.5 m vehicle

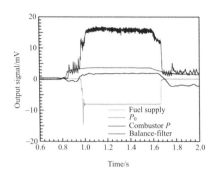

图 18　1. 5 m 飞行器机体推进一体化试验中典型试验参数随时间变化曲线

Fig. 18　Curves of typical test parameters vs time of 1. 5 m vehicle airframe-propulsion integration test

2. 3　高超声速内外流耦合数值模拟技术

自主开发了面向应用、功能完善的高超声速内外流耦合数值模拟软件系统——AHL3D[23]，主要用于一体化飞行器和发动机的设计和性能评估。

基于 AHL3D 建立并发展了一体化气动性能 3 级数值预测体系（如图 19 所示）：第 1 级快速预测，采用工程算法及发动机性能一维计算[26]（如图 19(a) 所示），快速评估一体化飞行器的气动性能，主要用于设计选型阶段；第 2 级高效预测，采用三维外流、二维内流（如图 19(b) 所示）相结合的方法，高效评估飞行器的一体化气动性能，主要用于详细设计阶段；第 3 级精细预测，采用三维内外流耦合数值模拟的方法（如图 19(c) 所示），精细评估飞行器的性能及流场结构，主要用于飞行器的性能分析阶段。图 19 中，P_t 为总压；T_t 为总温；P 为静压；T 为静温；γ 为比热比。

另外，在风洞试验数据修正中，CFD 也发挥了重大作用。例如，图 20 中针对有背部支撑的飞行器试验模型，通过计算得到支架干扰的影响量，从而对风洞试验数据进行支架干扰修正；图 21 中通过飞行器在真实风洞试验流场中的数值模拟对试验数据进行流场非均匀性修正等。

表 1　地面试验与计算结果对比

Table 1　Comparison of numerical and test results

Method	Cold flow (no fuel)		Hot flow (with fuel)	
	F_x/N	F_y/N	F_x/N	F_y/N
CFD	−596. 01	955. 25	−3. 13	1 056. 97
Experiment	−593. 07	1 102. 56	−106. 82	1 103. 87

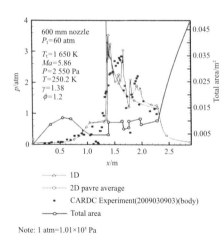

600 mm nozzle
P_t=60 atm
T_t=1 650 K
Ma=5.86
P=2 550 Pa
T=250.2 K
γ=1.38
ϕ=1.2

- 1D
- 2D pavre average
- CARDC Experiment(2009030903)(body)
- Total area

Note: 1 atm=1.01×10⁵ Pa

(a) One-dimensional engine performance calculations

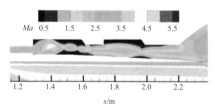

(b) Two-dimensional propulsive path calculation with combustion

(c) Three-dimensional large-scale parallel computational results

图 19　机体推进一体化性能三级预测体系

Fig. 19　Three stages of numerical predictions of airframe-propulsion integraion

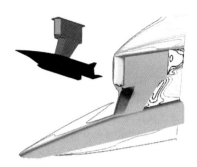

图 20 支架干扰修正

Fig 20 Support interference correction

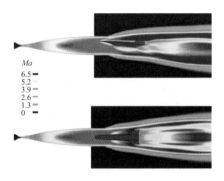

图 21 风洞非均匀流场修正

Fig 21 Non-uniform flow field correction of wind tunnel

3 机体推进一体化相关基础与机理研究

高超声速技术目标高、难度大，多数技术没有可借鉴的基础，需要结合应用进行大量的基础研究，澄清与高超声速飞行相关的物理、化学、流动力和热等各方面的机理性问题，因而基础研究应该贯穿研究的全过程。

目前最需要关注的是以下 3 个方面的问题：高速飞行中激波边界层相互干扰、流动分离机理，可压缩湍流、转捩及其控制，超燃冲压发动机点火与燃烧流动机理。对于数值模拟来说，发展更加先进的高效高精度计算方法及实验验证是当前最为迫切的需求。

3.1 高速飞行中激波边界层相互干扰、流动分离的机理

激波边界层干扰对高超声速飞行器及其推进系统的性能有至关重要的影响。激波引起的强逆压梯度影响边界层流动结构，甚至产生流动分离，对飞行器的热流和阻力产生很大影响。特别是对于高超声速进气道，激波/边界层干扰几乎贯穿于所有现象之中，如边界层分离、总压损失、不起动/重起动、溢流、抗反压、边界层转捩、内通道激波串和激波震荡等，直接影响进气道和发动机能否正常起动和稳定工作。开展相关研究有助于更加深入理解进气道中各种复杂流动的流动机理，为进气道的设计、性能评估和控制奠定更加坚实的技术基础。

激波振荡是当进气道没有起动时有可能发生的一种周期性的激波和分离区吞进/吐出现象。进气道发生激波振荡时，进气道内的气流发生振荡，速度和压强会发生强烈的脉动，引起进气道性能严重下降，导致发动机推力损失，严重时会导致燃烧室熄火。图 22 为典型高超声速进气道激波振荡现象研究的试验模型，采用 LES-DES(Large Eddy Simulation-Detached Eddy Simulation)方法对该进气道的激波振荡进行了数值模拟。LES-DES 方法作为一类典型的 RANS (Reynold-Averaged Navier Stokes equations)/LES 混合方法，适用于包含大分离流动的问题。计算所得振荡周期为 6.5 ms，试验所得振荡周期为 7.5 ms，计算结果和实验结果基本吻合。

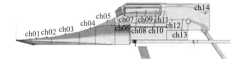

图 22 进气道激波振荡特性研究试验模型

Fig 22 Study on shockwave oscillation characteristics at inlet unstart

图 23 为测压点 ch10 处计算结果和试验所得的无量纲静压变化比较，二者的压力峰值和变化规律基本一致。

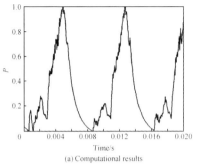

(a) Computational results

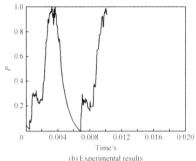

(b) Experimental results

图 23　隔离段上壁面后端测点 ch10 处的压强变化

Fig 23　Pressure change at upper wall test point ch10 in isolator

3.2　高速飞行中可压缩湍流、转捩及其控制

近年来虽然在湍流结构、层次结构模型、湍流直接模拟和大涡模拟等方面取得了长足进展，但是，对于可压缩湍流，特别是高超声速湍流，还缺乏研究。由于缺乏描述大、小旋涡相互作用的定量关系，大涡模拟也遇到困难，仍需要对湍流机理作深入了解。研究湍流的多尺度特征及其尺度间的相互作用，从中建立相对普遍适用的、反映多尺度特征的湍流模式理论。转捩和湍流紧密相关，飞行器推迟转捩减阻、进气道强制转捩增强自起动性能，都需要搞清转捩机理，提出控制转捩的途径，研究可压缩层流到湍流转捩的新特征，揭示影响转捩的主要因素，提出控制转捩的理论和方法。

转捩控制的研究主要分为两个方面：① 为了利用层流摩阻比湍流低的特性，通过修改气动构型，或者采用边界层抽吸的方法，尽可能推迟转捩，保持层流构型；② 为了利用湍流抗反压能力比层流强的特性，通过在进气道上添加强制转捩装置，促进从层流到湍流的转捩，减少激波与边界层相互作用的分离区，提高进气道的起动能力。

2004 年，美国 X-43A 的 2 次飞行试验表明没有转捩装置的进气道保持层流流动状态[27]。当进气道为层流状态时，风洞试验表明[28]，在进气道的拐角处产生了比较大的分离区，严重时将导致进气道的不起动，甚至可能导致飞行试验失败。为此，X-43A 进气道在前体上加装了斜坡型强制转捩装置，确保在隔离段入口的流动为湍流，便于进气道的起动。X-43A 的 2 次飞行试验[27] 表明强制转捩装置是可行的，可以在弹道的关键区域实现强制转捩，确保隔离段入口是湍流状态。X-43A 之后，美国的 X-51A 和 Hyfly(见图 24) 等都在进气道上安装了强制转捩装置。

(a) X-51A inlet

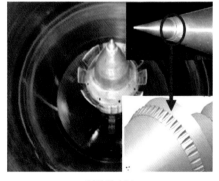

(b) HyFly inlet

图 24　X-51A 和 HyFly 进气道强制转捩装置的构型

Fig 24　Forced-transition trip of X-51A and HyFly inlets

对强制转捩装置的设计一般有 2 个要求：① 在进气道上实现强制转捩；② 满足沿弹道的热结构要求。

强制转捩装置的设计主要包括以下 4 个方面：① 强制转捩装置的转捩机理；② 强制转捩装置的安装位置；③ 强制转捩装置的选型和几何参数的优选；④ 强制转捩装置的天地相关性研究。其中第④项天地相关性是最困难的。

关于强制转捩装置的转捩机理，目前还没有完全搞清楚。Schneider[29]认为：没有一个通用的机理来说明粗糙带在什么条件下可以引起转捩。目前至少有 3 种解释：粗糙带后尾迹中的流向涡不稳定性的增加导致转捩；横流和 Görtler 不稳定性的失稳导致转捩；边界层外扰动和粗糙带相互作用导致转捩。Choudhari 等[30]对 X-43A 进气道涡流发生器的转捩机理研究表明：涡流发生器诱导的转捩一般是几种不稳定模式共同作用下的结果，绝不是一种模式的结果。

关于强制转捩装置的安装位置，X-43A 进气道强制转捩装置[27]的位置在第 1 个压缩面的中点，这里的层流边界层外缘的 $Ma<4$，根据流动稳定性理论，此时流动第 1 模式在转捩过程中起主导作用，转捩装置在这里可以激发流动第 1 模式的不稳定性，促进转捩。但是这个位置是否最优的，安装在 $Ma>4$ 的位置，促进第 2 模式的不稳定性是否可行，尚不得而知。对于 HyFly 进气道构型，除了第 1，2 模式不稳定性外，还存在横流不稳定性，此时转捩装置的安装位置在哪里比较合适，公开的文献中没有说明。

关于转捩装置的选型，虽然在地面风洞试验中，强制转捩装置有很多种构型，但是在飞行试验中，主要采用钻石型和斜坡型（见图 25）。这些构型又称为涡流发生器构型。Berry 等[28]指出根据以前的工程经验，这类构型的尾流中存在反向旋转的涡流，可以有效地促进转捩。关于转捩装置几何参数的优选，目前认为转捩装置的高度 k 是最主要的几何参数。美国 X-43A 项目组和气动中心都针对高度进行了详细的风洞试验研究[28,31]，结论都是随着转捩带高度 k 的增加，转捩区域逐步前移（见图 25，$Ma=6$，CARDC 实验）。清华大学的 Xiao 等[32]通过转捩预测方法对斜坡型转捩带的 4 个几何参数对转捩区域的影

(a) $k=0$ mm

(b) $k=0.5$ mm

(c) $k=1.0$ mm

图 25　涡流发生器高度对转捩位置的影响

Fig 25　Influence of vortex generator height on transition position

响规律开展了研究，他认为，4 个参数对转捩区域的影响顺序从高到底依次为高度、间距、底边长度和角度。目前很少有报告研究过转捩装置几何参数对进气道性能的影响。

强制转捩装置的天地相关性研究包括以下内容：① 地面试验和飞行试验的差别在哪里，这些差别对转捩区域会带来多大的影响；② 如何根据风洞试验验证成功的转捩装置设计飞行试验强制转捩装置的安装位置和几何参数。由于缺乏足够的飞行试验数据，天地相关性是转捩装置设计最困难的一项内容。一般认为地面试验和飞行试验的差别主要在于来流扰动（包括湍流度和噪声）、壁温/总温比和模型尺度。Berry 等[28]认为 X-43A 进气道转捩装置的天地相关性主要根据经验判断，认为飞行试验的 k/δ（δ 为当地层流边界层厚度）大于风洞试验的有效 k/δ 时，可以确保在强制转捩装置后实现转捩。但是这样简单关系的理论依据及其对其他进气道构型是否都适用尚待探索。

3.3 Scramjet 燃烧流动机理

超燃冲压发动机工作过程中，由于燃料的加入，在流道内形成了湍流、激波与燃烧相互作用的高速燃烧流场。在超声速流动的条件下，流体的可压缩性、激波和详细化学反应机理等的影响都很显著，流场中的波系、湍流对燃料的雾化和混合、燃烧室的点火、火焰的结构及其演变、火焰的传播特性等起着举足轻重的作用。

对于类似 X-51A 的发动机，起动过程中液体燃料的雾化、蒸发及点火延滞的评估与分析是一项关键技术。CARDC 研究组通过典型的小尺度发动机自由射流试验，于 2002 年第一次获得了液体煤油在 $Ma=5$、总温为 1 500 K 的条件下，燃料引射后约 5 ms 时间内的自点火性能[33]。

由于这一问题涉及到多相流、湍流与化学反应的相互作用，目前仍然是燃烧室和引射系统设计中的挑战性问题，需要通过物理建模和数值计算研究其复杂过程。物理建模包括采用粒子脉冲激光全息技术，对超声速气流中射流的雾化过程进行测量分析[34]。实验观测到射流表面的不稳定波结构，揭示了射流柱表面不稳定波的增长是导致超声速气流中射流破碎的主要原因（见图 26 和图 27）。通过实验分析建立的基于 Rayleigh-Taylor 和 Kelvin-Helmhotz 相结合、适用于超声速横流中的液滴破碎模型与计算结果，与 $Ma=2$ 来流、液气动量比为 15～30 的实验基本一致。关于液粒蒸发和化学反应延滞的整个点火延滞特性分析，采用 CARDC 的 AHL3D 软件并行计算中两相流非定常双时间方法[31-35]。计算结果表明（见图 28），在实验条件下液滴破碎时间为 0.2 ms、蒸发时间为 0.8 ms、化学反应延滞时间为 4 ms，十分接近实验获得的约为 5 ms 的液体燃料从注入到稳定的点火时间。

除了点火特性外，从工程应用上，两类问题需特别引起关注，一类是来流 $Ma=4～5$ 情况下亚燃强燃烧流场研究，即既要获得强燃烧又要使燃烧产生的激波串不致造成进气道的不起动，在这种工况下，提出在飞行轨道、姿态角、油气比条件下激波串不能超越 50% 隔离段长度的建议，这亦是 X-51A 下一步要研究的重点；另一类是 $Ma>7$ 的湍流混合与燃烧效率问题，X-43A 在 $Ma=$

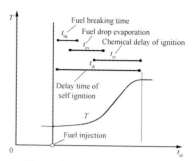

图 26　液体碳氢燃料延滞点火特性

Fig 26　Ignition time delay characteristic of liquid hydro-carbon fuel

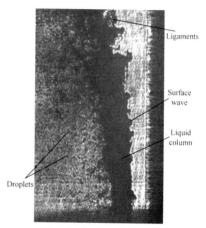

图 27　射流表面不稳定波结构

Fig 27　Jet surface instability wave structure

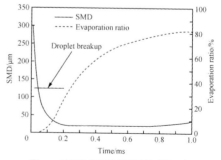

图 28　液滴粒度及其蒸发比例时间历程

Fig 28　Time history of sauter mean diameter (SMD) and evaporation ratio

10 条件下没有获得正推力,表明这一问题的重要性。高超声速冲压发动机中的超声速燃烧现象十分复杂,总体上它涉及宽范围复杂反应动力学、受限空间内复杂湍流和燃烧相互作用及高超声速条件下燃烧及其稳定性等问题,亦涉及到高分辨率多场多组分燃烧流场测量及超级计算机的科学计算,是物理、化学和数学力学等多科的交叉,亦是国际上正在突破的前沿[36]。

4 结束语

中国空气动力研究与发展中心主要突破了以下 3 项机体推进一体化关键技术:

1) 在一体化飞行器气动布局研究中,自主创新发展了密切曲锥乘波体设计方法,基准流场由多个激波或等熵压缩流场组成,流线跟踪一直到进气道喉道内收缩段,优点是可以灵活控制进气道的压缩量,并保持乘波体低阻力、高流量捕获和高总压恢复的特点。

2) 自主开发了高超声速内外流耦合数值模拟软件 AHL3D,并以此为基准,建立了一体化气动性能三级数值预测体系。

3) 突破了脉冲燃烧风洞带动力一体化试验技术,建立了计算与试验、脉冲与连续式燃烧风洞相结合的机体推进一体化气动性能预测体系。

高超声速机体推进一体化技术涉及到激波边界层干扰、分离流动、湍流转捩和超声速燃烧等大量流动机理问题,必须从物理机理层次搞清上述问题,才能进一步深化机体推进一体化技术研究,从而设计出工程上实用的吸气式高超声速飞行器。

致 谢

感谢唐志共研究员的关心与支持。感谢课题组的刘伟雄、倪鸿礼、郑忠华、杨顺华、李向东、余安远、贺旭照以及赵慧勇等同事的支持和贡献。

参 考 文 献

[1] Tank M H. National Aero-Space Plane (NASP) Program, N1991-28214[R]. Washington, D. C.: NASA Space Transportation Propulsion Technology Symposium, 1991, 2: 383-407.

[2] Foelsche R O, Leylegian J C, Betti A A. Progress on the development of a free flight atmospheric scramjet test technique, AIAA-2005-3297[R]. Reston: AIAA, 2005.

[3] Peebles C. Road to Mach 10: lessons learned from the X-43A flight research program[M]. Reston: Library of Flight Series, AIAA, 2008: 36-78.

[4] Shelly F, Charles M, Kenneth R, et al. Hyper-X Mach 7 scramjet design, ground test and flight results, AIAA-2005-3322[R]. Reston: AIAA, 2005.

[5] Hank J M, Murphy J S, Mutzman R C. The X-51A scramjet engine flight demonstration program, AIAA-2008-2540[R]. Reston: AIAA, 2008.

[6] Walker S H, Sherk J, Shell D. The DARPA/AF Falcon Program: the hypersonic technology vehicle #2 (HTV-2) flight demonstration phase, AIAA-2008-2539[R]. Reston: AIAA, 2008.

[7] Liu T L. Hypersonic technology flight test program in Russia (I)[J]. Aerodynamic Missile Journal, 2000(4): 23-30 (in Chinese).
刘桐林. 俄罗斯高超声速技术飞行试验计划(一)[J]. 飞航导弹, 2000(4): 23-30.

[8] Duveau P, Hallard R, Novelli P, et al. Aerodynamic performance analysis of the hypersonic airbreathing vehicle Japhar[C]//ISABE 1999. Florence: ISABE Congress, 1999.

[9] Neuenhahn T, Olivier H. Development of the HyShot stability demonstrator, AIAA-2006-2960[R]. Reston: AIAA, 2006.

[10] Steelant J. Sustained hypersonic flight in Europe: technology drivers for LAPCAT II, AIAA-2009-7240[R]. Reston: AIAA, 2009.

[11] Chen Y Y, Ye L, Su X X. Current situation of air-breathing hypersonic vehicle abroad[J]. Aerodynamic Missile Journal, 2008(12): 25-32 (in Chinese).
陈英硕, 叶蕾, 苏鑫鑫. 国外吸气式高超声速飞行器发展现状[J]. 飞航导弹, 2008(12): 25-32.

[12] Rogers R C, Capriotti D P, Guy R W. Experimental supersonic combustion research at NASA langley, AIAA-1998-2506[R]. Reston: AIAA, 1998.

[13] Engelund W C, Holland S D, Cockrell C E. Jr. Aerodynamic database development for the hyper-X airframe-integrated scramjet propulsion experiments[J]. Journal of Spacecraft and Rocket, 2001, 38(6): 803-810.

[14] Yi J, Xiao H, Shang X S. Aerodynamic performance research of two integrated hypersonic configurations[J]. Advances in Aeronautical Science and Engineering, 2011, 2(3): 305-311 (in Chinese).
易军, 肖洪, 商旭升. 两种高超声速一体化构型的气动性能对比分析[J]. 航空工程进展, 2011, 2(3): 305-311.

[15] Zhang H Y, Cheng K M, Wu Y Z. A study on the flowpath and the aerodynamic characteristic of a hypersonic vehicle[J]. Acta Aerodynamica Sinica, 2009, 27(1): 119-123 (in Chinese).

张红英，程克明，伍贻兆．某高超飞行器流道冷流特征及气动力特性研究[J]．空气动力学学报，2009，27（1）：119-123．

[16] Fan X Q, Li H, Yi S H, et al. Experiment of aerodynamic performance for hypersonicvehicle integrated with sidewall compression inlet[J]. Journal of Propulsion Technology, 2004, 25(6): 499-502 (in Chinese).
范晓檐，李桦，易仕和，等．侧压式进气道与飞行器机体气动一体化设计及实验[J]．推进技术，2004，25（6）：499-502．

[17] Jin L, Liu J, Luo S B, et al. Aerodynamic characterization of an integrated hypersonic vehicle[J]. Journal of Experiments in Fluid Mechanics, 2010, 24(1): 42-45 (in Chinese).
金亮，柳军，罗世彬，等．高超声速一体化飞行器冷流状态气动特性研究[J]．实验流体力学，2010，24（1）：42-45．

[18] Jones K D, Sobieczky H, Seebass A R, et al. Waverider design for generalized shock geometries[J]. Spacecraft and Rockets, 1995, 32(6): 957-963.

[19] Sobieczky H, Zores B, Wang Z, et al. High speed flow design using osculating axisymmetric flows[C] // PICAST' 3. Beijing: Aviation Industry Press, 1997: 1-5.

[20] He X Z, Le J L, Wu Y C. Design of a curved cone derived waverider forebody. AIAA-2009-7423[R]. Reston: AIAA, 2009.

[21] He X Z, Ni H L. Osculating curved cone (OCC) waverider: design methods and performance analysis[J]. Chinese Journal of Theoretical and Applied Mechanics, 2011, 43(6): 1077-1082 (in Chinese).
贺旭照，倪鸿礼．密切曲面锥乘波体——设计方法和性能分析[J]．力学学报，2011，43（6）：1077-1082．

[22] Wu Y C, He Y Y, Yu A Y, et al. Aerodynamic layout of spanwise truncated curved waverider compression inlet [J]. Journal of Aerospace Power, 2013, 28(7): 1570-1575 (in Chinese).
吴颖川，贺元元，余安远，等．展向截断曲面乘波压缩前体进气道气动布局研究[J]．航空动力学报，2013，28（7）：1570-1575．

[23] Wu Y C, He Y Y, He W, et al. The design of osculating curved cone waverider based hypersonic vehicle[J]. Acta Aerodynamica Sinica, 2014, 32(1): 8-13 (in Chinese).
吴颖川，贺元元，贺伟，等．基于密切曲锥的乘波构型一体化飞行器设计方法研究[J]．空气动力学学报，2014，32（1）：8-13．

[24] Mao X B. ∅2.4 m impulse combustion wind tunnel[EB/OL]. (2012-7-12)[2014-6-15]. http: // www. cardc. cn/ html/Facility /cgs/ Cumbustion/44. html.
毛雄兵．∅2.4 m 脉冲燃烧风洞[EB/OL]．（2012-7-12）[2014-6-15]．http: // www. cardc. cn/ html/ Facility /cgs/

[25] He Y Y, Le J L, Ni H L. Numerical and experimental study of airbreathing hypersonic airframe/propulsion integrative vehicle[J]. Journal of Experiments in Fluid Mechanics, 2007, 21(2): 29-34 (in Chinese).
贺元元，乐嘉陵，倪鸿礼．吸气式高超声速机体/推进一体化飞行器数值和试验研究[J]．实验流体力学，2007，21（2）：29-34．

[26] Wang L, Xing J W, Zheng Z H, et al. One-dimensional evaluation of the scramjet flowpath performance[J]. Journal of Propulsion Technology, 2008, 29(6): 641-646 (in Chinese).
王兰，邢建文，郑忠华，等．超燃发动机内流性能的一维评估[J]．推进技术，2008，29（6）：641-646．

[27] Berry S, Daryabeigi K, Wurster K, et al. Boundary layer transition on X-43A, AIAA-2008-3736[R]. Reston: AIAA, 2008.

[28] Berry S, Auslender A H, Dilley A D, et al. Hypersonic boundary-layer trip development for hyper-X[J]. Journal of Spacecraft and Rockets, 2001, 38(6): 853-864.

[29] Schneider S P. Effects of roughness on hypersonic boundary-layer transition, AIAA-2007-0305[R]. Reston: AIAA, 2007.

[30] Choudhari M, Li F, Edwards J. Stability analysis of roughness array wake in a high-speed boundary layer. AIAA-2009-0170[R]. Reston: AIAA, 2009.

[31] Zhao H Y, Zhou Y, Ni H L, et al. Test of forced boundary-layer transition on hypersonic inlet[J]. Journal of Experiments in Fluid Mechanics, 2012, 26(1): 1-6 (in Chinese).
赵慧勇，周瑜，倪鸿礼，等．高超声速进气道边界层强制转捩试验[J]．实验流体力学，2012，26（1）：1-6．

[32] Xiao Z X, Zhang M H, Xiao L H, et al. Studies of roughness-induced transition using three-equation $k-\omega-\gamma$ transition/turbulence model. AIAA-2013-3111[R]. Reston: AIAA, 2013.

[33] Le J L, Liu W X, He W, et al. Impulse combustion wind tunnel and its application in rocket and scramjet research [J]. Journal of Experiments in Fluid Mechanics, 2005, 19(1): 1-10 (in Chinese).
乐嘉陵，刘伟雄，贺伟，等．脉冲燃烧风洞及其在火箭和超燃发动机研究中的应用[J]．实验流体力学，2005，19（1）：1-10．

[34] Yang S H, Le J L. Numerical simulation of liquid fuel atomization in supersonic cross flow[J]. Journal of Propulsion Technology, 2008, 29(5): 519-522 (in Chinese).
杨顺华，乐嘉陵．超声速气流中液态燃料雾化数值模拟[J]．推进技术，2008，29（5）：519-522．

[35] Le J L, He W, Yang S H, et al. Investigation of ignition characteristics for kerosene fueled scramjet, ISABE-2009-

1322[R]. Montreal: The International Society Engines of Airbreathing, 2009.

[36] The National Natural Science Fund Committee Office. The annual directory of the major research plan 2014 of "Basic Research of Turbulent Combustion for the Engine" [R]. Beijing: The National Natural Science Fund Committee Office, 2014 (in Chinese).
国家自然科学基金委员会办公室. "面向发动机的湍流燃烧基础研究"重大研究计划 2014 年度项目指南[R]. 北京: 国家自然科学基金委员会办公室, 2014.

Progress in airframe-propulsion integration technology of air-breathing hypersonic vehicle

WU Yingchuan*, HE Yuanyuan, HE Wei, LE Jialing

Science and Technology on Scramjet Laboratory, Hypervelocity Aerodynamics Institute, China Aerodynamics Research and Development Center, Mianyang 621000, China

Abstract: Air-breathing hypersonic vehicle is highly integrated making its design challenging. All vehicle parts and functions interact including aerodynamics, propulsion, control, structure, tank and thermal protection, especially for airframe and scramjet engine coupling. The lower wall of the aircraft forebody and afterbody is either compression part of the engine inlet or expansion part of the engine nozzle and it produces lift and pitching moment as well as thrust. The strong coupling of the airframe and engine has direct influence to the thrust, lift, drag, pitching moment, aerodynamic heating, airframe cooling, stability and control characteristics of the vehicle. The research developments of airframe-propulsion integration technology are introduced and the related works of China Aerodynamics Research & Development Center (CARDC) are emphasized. These works included osculating curved cone waverider inlet design, double shockwave axissymetric flow field-based inward turning inlet design, airframe-propulsion integrated vehicle tests in pulsed combustion heated hypersonic high-temperature wind tunnels and hypersonic large-scale parallel numerical simulation platform (AHL3D). The related fundamental researches of hypersonic shock-boundary layer interaction, compressible turbulent transition of flow separation mechanism and its control, scramjet combustion study on flow mechanism and other related basic issues are introduced. The urgent need of efficient high-precision calculation method is emphasized.

Key words: hypersonic vehicle; scramjet; airframe-propulsion integration; waverider; combustion heated wind tunnel; turbulence combustion; transition

Investigation of combustion and flame stabilization modes in a hydrogen fueled scramjet combustor

Ye Tian [a,*], Shunhua Yang [a], Jialing Le [a], Tie Su [b], Maoxiong Yue [b], Fuyu Zhong [a], Xiaoqiang Tian [a]

[a] Science and Technology on Scramjet Laboratory of Hypervelocity Aerodynamics Institute, CARDC, Mianyang, China
[b] Facility Design and Instrumentation Institute of CARDC, Mianyang, China

ARTICLE INFO

Article history:
Received 24 May 2016
Received in revised form
21 July 2016
Accepted 23 July 2016
Available online 11 August 2016

Keywords:
Combustion mode
Flame stabilization
Hydrogen
Scramjet mode
Ramjet mode

ABSTRACT

The combustion and flame stabilization modes in a hydrogen fueled scramjet combustor were investigated by experiments and numerical simulations in the present paper. The results were obtained with the inflow conditions of Mach number of 2.0, static temperature of 656.5 K and static pressure of 0.125 MPa, respectively. The equivalence ratio (ER) of hydrogen which was used as fuel changed from 0.04 to 0.30. Wall pressure measurement, schlieren, differential interferometry, high-speed framing of flame luminosity and OH-PLIF (planar laser-induced fluorescence) were introduced to characterize the combustion flow. Two typical combustion modes (scramjet mode and ramjet mode) could be attained by changing the ER of hydrogen. The combustion mode was scramjet mode when the ER of hydrogen was less than 0.23, while it was ramjet mode when the ER of hydrogen was greater than 0.23. When the combustion mode was scramjet mode, the flame stabilization mode was cavity shear layer stabilized combustion, and when the combustion mode was ramjet mode, the flame stabilization mode was combined cavity shear-layer/recirculation stabilized combustion. The flame was unstable and oscillated with the flow during the ramjet-mode operation.

© 2016 Hydrogen Energy Publications LLC. Published by Elsevier Ltd. All rights reserved.

Introduction

Scramjet engines have been expected to be one of the propulsion systems for air-breathing systems in hypersonic flight and receive a considerable amount of interest from many researchers. Hydrogen is widely used as fuel due to its high energy release, high reactivity, wide flammability limits and high diffusivity [1–12]. The residence time in a scramjet combustor is very short (of the order of milliseconds), so achieving flame stabilization is a different problem. This has aroused the attention of many researchers. Micka and Driscoll

[13] used CH-PLIF to investigate the reaction zone in a dual-mode scramjet combustor and demonstrated two different reaction zone structures which corresponded to cavity-stabilized and fuel-jet-wake stabilized modes. Rasmussen et al. [14,15] investigated stability limits and flameholding mechanism in cavity-stabilized mode. When the fuel was injected from the aft wall, primary combustion occurred under the shear layer and in the aft region of the cavity. On the contrary, when fuel was injected from the down wall, a jet-driven recirculation zone of hot products near the upstream wall of the cavity served as a flameholder, and the reaction

Nomenclatures

Ma   Mach number
P   wall pressure of combustor
t   time after facility starts
ER   equivalence ratio
PLIF   planar laser-induced fluorescence
ICCD   intensified charge-coupled device
x   distance from combustor entrance

then occurred on the underside of the shear layer. Performance and operating limits of an ethylene-fueled recessed cavity flameholder with various cavity lengths were investigated both experimentally and numerically by Lin et al. [16]. They found that the L/D = 6 cavity exhibited the poorest performance among the three cavities in terms of the lean blowout limit at cavity-only operation. With back pressurization, both the lean ignition limit and the lean blowout limit increased as compared to the case without back pressurization. Flame characteristics and flame stabilization mechanism in a supersonic combustor with hydrogen injection upstream of cavity flame holders were studied by Sun et al. [8]. They found that hydrogen fuels were transported into the cavity shear layer and an approximately steady partially-premixed flame front exists in the cavity shear layer. The physical process of the flame stabilization demonstrated the similarity with triple flame characteristics, which indicted that triple flame theory might be the basic flame stabilization mechanism of the cavity flame holders. The characteristics of cavity assisted hydrogen jet combustion in a supersonic flow were investigated by Wang et al. [6]. Three combustion modes were observed for the cavity assisted hydrogen jet combustion in their results: cavity assisted jet-wake stabilized combustion, cavity shear-layer stabilized combustion, and combined cavity shear-layer/recirculation stabilized combustion. They also found stable combustion could not be obtained without a cavity. Combustion mode transition in a hydrogen fueled scramjet was investigated by Chang et al. [17−19]. A thermodynamic cycle analysis was made to identify the differences between Scramjet-mode and Ramjet-mode cycles, and three typical combustion modes (scramjet-mode, transitional mode, and ramjet-mode) could be obtained by changing the total amount of fuel added or adjusting the fuel distribution between two injectors.

In the present paper, the combustion and flame stabilization modes of hydrogen injection downstream of a cavity-based flame holder in a scramjet combustor were investigated by using schlieren, differential interferometry, high-speed framing of flame luminosity and OH-PLIF system in the experiment, and numerical simulations.

Experimental and numerical simulation methods

Facility and scramjet configuration

The experiments were carried out at China Aerodynamics Research and Development Center (CARDC)'s direct-connect supersonic combustion facility (Fig. 1). A H_2/O_2 vitiated air heater was used to generate high enthalpy airflow supplied into the combustor to simulate Mach 4 flight conditions. The vitiated air flow entering the test article had a molar composition of 21% O_2, 12% H_2O, and 67% N_2 and a mass flow rate of 2.68 kg/s. Stagnation conditions were 950 K and 0.82 MPa and the Mach number at the isolator entrance was 2.0. The test sequence of the facility is shown in Fig. 2, when hydrogen entered into the facility heater at t = 0.08 s, the cold flow then generated. The running time of the facility was about 440 m (0.26s−0.70s), when the supply pressure of hydrogen was kept as constant. The hydrogen which was used as fuel was injected into the combustor from t = 0.27 s to t = 0.55 s, and the ER of hydrogen of the studying cases is shown in Table 1. The spark in the combustor cavity was working from t = 0.0 s to t = 0.40 s.

A schematic illustration of the combustor model is shown in Fig. 3, the model consisted of an isolator and a combustor. The cross-sectional area was 30 × 150 mm² at the isolator entrance. The isolator length was 0.43 m, and it contained two parts: a 0.35 m length straight section and a 0.08 m length expansion section with the top wall divergent angle being 1.4°. The combustor length changed from 0.43 m to 1.07 m, which included a cavity (Depth: 0.011 m, Length: 0.121 m) and a four-part expansion section (Range: 0.551 m−1.07 m). The first part was from 0.551 m to 0.656 m, the wall divergent angle was 1.4°; the second part was from 0.656 m to 0.801 m, the wall divergent angle was 2.0°; the third part was from 0.801 m to 0.943 m, the wall divergent angle was 8.0°; and the last part was from 0.943 m to 1.07 m, the wall divergent angle was 15.0°. The hydrogen was directly injected into the airflow from the injector which was located 25 mm downstream of the cavity step, and there were 10 fuel injection holes (1.0 mm in diameter, 10 mm interval in span wise direction) in the injector. In tests, room temperature (295 K) hydrogen was injected at sonic speed at an angle of 90° to the airflow.

Instrumentation

In the test, wall pressure measurements were made along stream wise direction by pressure transducers (range: 0−700 kPa) at a sampling frequency of 1 kHz. Also schlieren,

Fig. 1 − Photo of the CARDC's supersonic combustion facility.

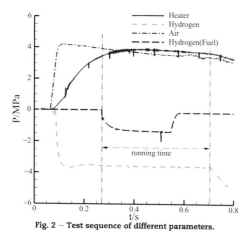

Fig. 2 — Test sequence of different parameters.

Table 1 — The studying cases for this study.

Case	ER
1	0.0
2	0.10
3	0.30
4	0.04
5	0.17
6	0.23

differential interferometry, high-speed framing of flame luminosity and OH-PLIF were introduced to characterize the combustion flow. Three quartz glass windows (two windows were in sidewalls and one window was in bottom wall) as shown in Fig. 3 were mounted on the combustor to allow optical access. The exposure time and frame rate of flame luminosity and differential interferometry were 0.5 m and 2000 frame/s, respectively, while Schlieren images were collected at 0.2 m exposure and 5000 frame/s.

The PLIF system, which is shown in Fig. 4, consisted of an Nd:YAG laser, a dye laser, a frequency doubler, and an intensified charge-coupled device (ICCD) system. The output beam from the Nd:YAG laser was at 532 nm, which pumped the dye laser to generate the laser beam at 567.106 nm. Then the output from the dye laser was transferred into the doubler, which transformed it into a UV laser beam at 283.03 nm. This wavelength was located at the transition Q1(8) of the OH

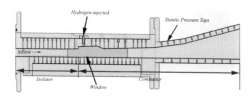

Fig. 3 — Schematic illustration of a scramjet combustor.

radical, which was not sensitive to the flow temperature (1400 K–2500 K). The laser sheet was positioned at the spanwise centerline in the combustor, and the thickness and width of the plane beam were about 1.0 mm and 130 mm, respectively. The PLIF image was photographed by an ICCD camera. Two sets of filters which were made up of UG11 and FF02-320/40-25 were mounted in front of the camera, and the isochronous controller was used to control the laser and camera. Since the frequency of laser was 10 Hz, and the running time of the facility was 440 m, so only three or four valid pictures could be obtained in a single test.

Numerical methods

In this study, the in-house CFD code AHL3D was used for computation. The physical and chemical models of the code had already been validated in Refs. [20,21].

For our present simulation [22], a fully coupled form of species conservation equations and Reynolds averaged Navier–Stokes equations were used as a governing equation set for a chemically reacting supersonic viscous flow. Cell-averaged finite volume techniques were used to discrete the governing equations. LU-SGS method was used for solving linearized equations. Third order MUSCL interpolation method and AUSMPW + scheme were used for inviscid fluxes construction, central difference method was used for viscous fluxes. Kok's modified k–ωTNT two-equation turbulence mode [23] was used for turbulence simulations. Although three-dimensional effect could not be ignored for several reasons [22,24], such as the corner flow effect, the boundary layer of the side wall and fuel injected methods, considering the calculating costs, two-dimensional numerical simulation was used in this paper.

Results and discussion

Discussion of combustion modes

The experimental wall pressure results of all studying cases are shown in Fig. 5. The plot also contains a representation of the flowpath for spatial reference. We found the wall

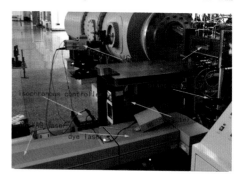

Fig. 4 — The PLIF system.

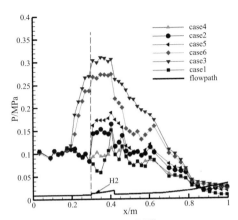

Fig. 5 — Wall pressures of different cases.

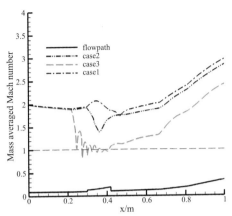

Fig. 7 — Mass averaged Mach number of two different combustion mode cases and non-reacting flow.

pressures increased with increasing ER of hydrogen because more heat released at a higher ER of hydrogen. The combustion-induced back pressure of case3 (ER = 0.30) and case6 (ER = 0.23) had disturbed the flow in the isolator, and the disturbance distance of case3 and case6 was both 0.1 m from the isolator exit. The significant pressure rise points of case3 and case6 were upstream of the hydrogen injection site, but that of the other studying cases were not, this was because the ER of hydrogen of these cases was lower. As discussed by Cabell et al. [25], the relative position of significant pressure rise point and the primary injector can be used as a combustion mode identification method. Scramjet mode operation was characterized by supersonic combustion flow with no large-scale flow separation and minimal combustion-induced pressure rise upstream of the primary fuel injection site [25].

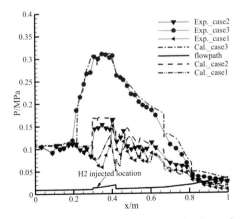

Fig. 6 — Wall pressures of two different combustion mode cases and non-reacting flow.

To the opposite, ramjet mode operation was characterized by subsonic combustion flow with large-scale flow separation and maximal combustion-induced pressure rise upstream of the primary fuel injection site. So we considered the combustion modes of case3 and case6 to be ramjet mode, and that of the other reacting cases to be scramjet mode. In order to provide more in-depth analysis of the combustion mode, the numerical results of two different combustion modes (case2 and case3) will be discussed in the following. The wall pressures of case2 and case3 are shown in Fig. 6, and also that of the non-reacting flow was added in the Figure which was used for comparing with the two cases. Comparing the numerical results with the experimental results of the same ER of hydrogen in the Figure, we found the numerical results well matched the experimental results, so we considered the numerical methods were applicable and the computational results were credible. The mass averaged Mach number distributions of the three cases are shown in Fig. 7. The mass averaged Mach number of case2 (ER-0.10) was a little less than that of case1 (Non-reacting flow) downstream of the isolator exit, but it was much higher than the sonic speed (Mach = 1.0). The mass averaged Mach number of case3 was much lower than that of the other two cases, which was even lower than

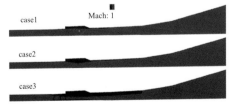

Fig. 8 — Mach number contours of two different combustion mode cases and non-reacting flow.

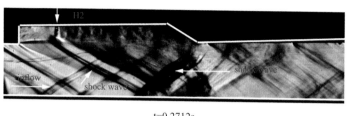

t=0.2712s

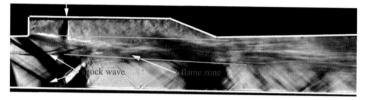

t=0.2740s

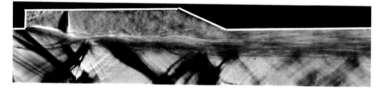

t=0.2810s

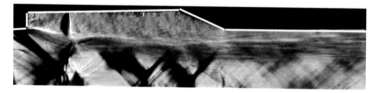

t=0.2916s

t=0.3352s

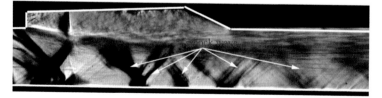

t=0.4678s

Fig. 9 – The high-speed schlieren images of case2 at different times.

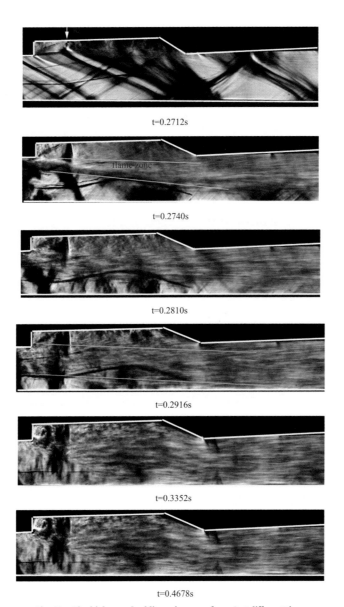

t=0.2712s

t=0.2740s

t=0.2810s

t=0.2916s

t=0.3352s

t=0.4678s

Fig. 10 – The high-speed schlieren images of case3 at different times.

the sonic speed from $x = 0.27$ m to $x = 0.41$ m. As discussed by Heiser and Pratt [26], they considered the velocity on the cross section of the isolator exit as a mode identification method. When the velocity was subsonic, the combustion mode was ramjet mode; when the velocity was supersonic, the combustion mode was scramjet mode [27]. So based on this identification method, the same conclusion of combustion mode of case2 and case3 could be summarized as described

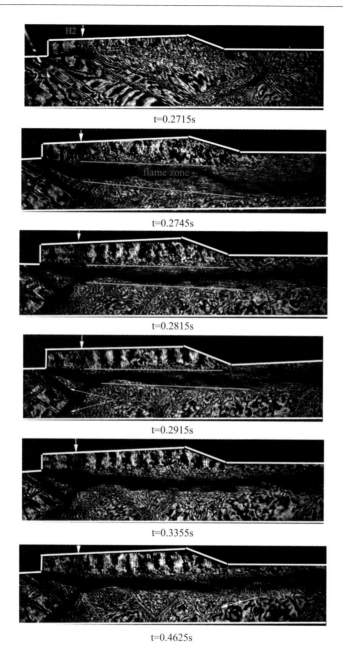

t=0.2715s

t=0.2745s

t=0.2815s

t=0.2915s

t=0.3355s

t=0.4625s

Fig. 11 – The differential interferometry pictures of case2 at different times.

before: the combustion mode of case2 was scramjet mode, and that of case3 was ramjet mode. The Mach number contours of the three cases are shown in Fig. 8. The Mach number of the main flow of scramjet mode was higher than 1.0, only which of the cavity region was less than 1.0. But the velocity of the core flow of ramjet mode was much less than the sonic

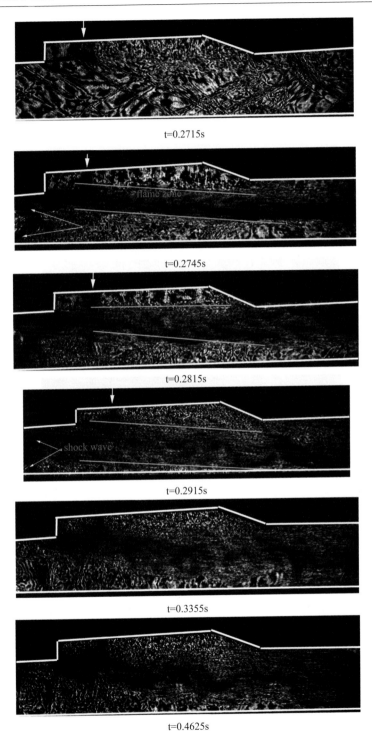

Fig. 12 − The differential interferometry pictures of case3 at different times.

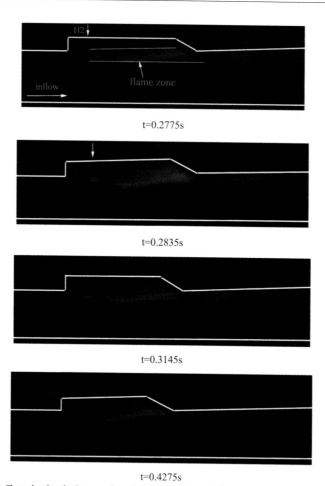

t=0.2775s

t=0.2835s

t=0.3145s

t=0.4275s

Fig. 13 — Flame luminosity images of case2 at different times (Taken through the side wall window).

speed and we could also see a large separation in the ramjet mode combustor.

Discussion of flame stabilization modes

As we know, the combustion heat release would raise the static temperature of the flow field, along with changing the density of the flow. So we could use Schlieren and differential interferometry pictures of different times to show the flame development process of the combustion flow field. First, the schlieren pictures of different times are shown in Fig. 9 (case2) and Fig. 10 (case3). The flow direction is from the left to the right. The pictures show the hydrogen being injected into the flow field from the cavity wall at the time of 0.2712s, a shock wave generated by the hydrogen injection in the supersonic

flow and a reflected wave off of the bottom wall. The hydrogen was burning at t = 0.2740 s in the shear layer between the cavity flow and the core flow, and the shock wave moved into the isolator due to the increased combustion back pressure. The flame and flow structure were stable in the shear layer during the whole test, and a shock train was generated below the shear layer, which could be seen at t = 0.4678 s. Like the discussion in Fig. 9, the hydrogen had been injected into the combustor at t = 0.2712 s in Fig. 10. Also, the flame was in the shear layer, and a shock wave was generated by the high combustion-induced pressure at t = 0.2740 s. But a different phenomenon came into being after t = 0.2810 s in case3 compared with that of case2. The shock wave moved forward into the isolator, and finally disappeared after t = 0.3352 s. This was because the higher back pressure was generated by

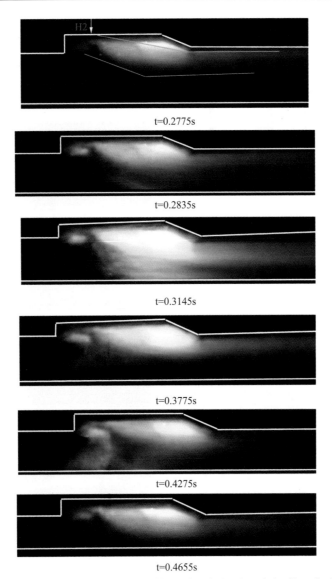

t=0.2775s

t=0.2835s

t=0.3145s

t=0.3775s

t=0.4275s

t=0.4655s

Fig. 14 − Flame luminosity images of case3 at different times (Taken through the side wall window).

the larger ER of hydrogen and pushed the shock train into the isolator, which also could be explained by the wall pressure in Fig. 6.

The differential interferometry pictures at different times from the two cases are shown in Figs. 11 and 12. The same

flow structures are observed in Figs. 11 and 12 when the hydrogen was injected into the combustor at t = 0.2715 s including the shock wave near the injected location. The same flame development process revealed by the Schlieren images is also observed using interferometry. However, the

Fig. 15 – Flame luminosity image taken through the bottom wall window of case2 and case3 (Left: case2; Right: case3).

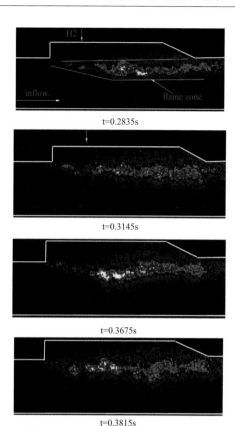

t=0.2835s

t=0.3145s

t=0.3675s

t=0.3815s

t=0.4355s

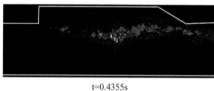

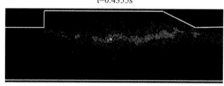

t=0.4675s

Fig. 16 – OH-PLIF images of side view in case2 at different times.

interferometry revealed new information concerning flow stability in these cases. The interference fringe had been found in the cavity region after t = 0.2815 s in Fig. 11, but that could not be found in Fig. 12. As we know, the interference fringe responded to the density of the flow field, so we considered the flow structure of case2 (ER = 0.10) was stable after t = 0.2815 s, but that of case3 was unstable, which would also be discussed in the following paragraph.

The flame luminosity images of case2 and case3 are shown in Figs. 13 and 14, respectively. The flame of case2 was stable in the shear layer, as concluded in the previous paragraphs. However, the flame luminosity images of case3 in Fig. 14 reveal this flame zone to be unstable with the hydrogen burning violently and the flame spread across the combustor to the bottom wall. The flame propagated upstream along the bottom wall from t = 0.3145 s to t = 0.4275 s, and then moved downstream again at t = 0.4655 s. The phenomenon of unstable flow structure of case3 had also been summarized by the differential interferometry method. The hydrogen was burning more violently in case3 than that of case2, which could also be seen through the bottom window in Fig. 15.

In order to clearly show the location of the flame, the OH-PLIF images of side view of case2 and case3 are shown in Figs. 16 and 17, respectively. As the OH free radical is an intermediate product during the reaction of hydrogen and air, the OH-PLIF images well describe the location of the flame. The OH-PLIF signals of case2 mainly distribute in the cavity shear layer, which means that the majority of fuel entered into the cavity shear layer, mixed with air, and burned in the shear layer (consistent with the previous observations). The OH-PLIF signals of case3 distributed not only in the shear layer but also in the main flow, which means the hydrogen was burning both in the shear layer and in the main flow. This was because the higher combustion-induced back pressure had been generated by a larger ER, and the shock waves moved into the isolator, then the interactions of the shock waves and boundary layer would cause the wall boundary layers to separate, which would enhance the fuel/air mixing, so the flame could propagate into the main flow. Also the shape of OH-PLIF signals changed largely in case3, which indicates the

flame structure is unstable (also consistent with the previous observations).

As discussed by Micka [28] and Sun et al. [8], the flame of case2 was stable and located in the shear layer, which meant

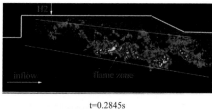

t=0.2845s

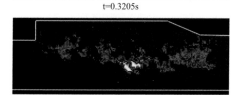

t=0.3205s

t=0.3660s

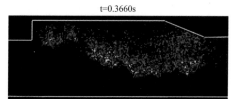

t=0.3800s

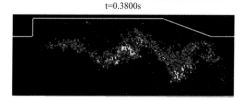

t=0.4340s

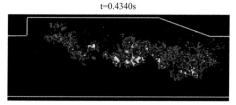

t=0.4715s

Fig. 17 – OH-PLIF images of side view in case3 at different times.

the flame stabilization mode was cavity shear layer stabilized combustion. The flame of case3 was not only in the cavity shear layer but also in the main flow, which meant the flame stabilization mode was combined cavity shear layer/recirculation stabilized combustion. As discussed earlier, the reacting cases (case2~case6) in this paper could be divided into two typical combustion modes according to the wall pressure distribution: scramjet mode (case2, case4 and case5) and ramjet mode (case3 and case6). So the flame stabilization mode of scramjet mode in this paper was cavity shear layer stabilized combustion, and that of ramjet mode was combined cavity shear layer/recirculation stabilized combustion.

Conclusions

The combustion and flame stabilization modes in a hydrogen fueled scramjet combustor were investigated by experiments and numerical simulations. The combustion mode of the reacting cases could be divided into two typical modes, the combustion mode was scramjet mode when the ER of hydrogen was less than 0.23, and that was ramjet mode when the ER of hydrogen was not less than 0.23. The flame stabilization mode of scramjet mode was cavity shear layer stabilized combustion, and that of ramjet mode was combined cavity shear layer/recirculation stabilized combustion. The flame was unstable and oscillated with the flow, when the flame stabilization mode was combined cavity shear layer/recirculation stabilized combustion mode.

Acknowledgments

The project supported by FengLei Youth Innovation Fond of CARDC and National Natural Science Foundation of China (No. 51376194, No. 51376193, No. 51406222, No. 91441204).

REFERENCES

[1] Cecere D, Ingenito A, Giacomazzi E, Romagnos L, Bruno C. Hydrogen/air supersonic combustion for future hypersonic vehicles. Int J Hydrogen Energy 2011;36:11969e84.
[2] Tsujikawa Y, Northam GB. Effects of hydrogen active cooling on scramjet engine performance. Int J Hydrogen Energy 1996;21(4):299e304.
[3] Choudhuri AR, Gollahalli SR. Combustion characteristics of hydrogen-hydrocarbon hybrid fuels. Int J Hydrogen Energy 2000;25:451e62.
[4] Huang W. Design exploration of three-dimensional transverse jet in a supersonic crossflow based on data mining and multi-objective design optimization approaches. Int J Hydrogen Energy 2014;39:3914e25.
[5] Lu S, Fan J, Luo K. High-fidelity resolution of the characteristic structures of a supersonic hydrogen jet flame with heated co-flow air. Int J Hydrogen Energy 2012;7:3528–39.
[6] Wang HB, Wang ZG, Sun MB, Wu HY. Combustion modes of hydrogen jet combustion in a cavity-based supersonic combustor. Int J Hydrogen Energy 2013;38:12078–89.

[7] Wang HB, Wang ZG, Sun MB, Qin N. Combustion characteristics in a supersonic combustor with hydrogen injection upstream of cavity flameholder. Proc Combust Inst 2013;34:2073–82.

[8] Sun MB, Wu HY, Fan ZQ, Wang HB, Bai XS, Wang ZG, et al. Flame stabilization in a supersonic combustor with hydrogen injection upstream of cavity flame holders: experiments and simulations. Proc. IMechE Part G J Aerospace Eng, Vol. 225 1351–1365.

[9] Qin J, Bao W, Zhou WX, Yu DR. Flow and heat transfer characteristics in fuel cooling channels of a recooling cycle. Int J Hydrogen Energy 2010;35:10589–98.

[10] Qin J, Zhou WX, Bao W, Yu DR. Thermodynamic analysis and parametric study of a closed Brayton cycle thermal management system for scramjet. Int J Hydrogen Energy 2010;35:356–64.

[11] Bao W, Qin J, Zhou WX, Yu DR. Parametric performance analysis of multiple re-cooled cycle for hydrogen fueled scramjet. Int J Hydrogen Energy 2009;34:7334–41.

[12] Bao W, Qin J, Zhou WX, Yu DR. Effect of cooling channel geometry on re-cooled cycle performance for hydrogen fueled scramjet. Int J Hydrogen Energy 2010;35:7002–11.

[13] Micka DJ, Driscoll JF. Reaction zone imaging in a dual-mode Scramjet combustor using CH-PLIF. AIAA; 2008. 2008-5071.

[14] Rasmussen CC, Driscoll JF, Hsub KY. Stability limits of cavity-stabilized flames in supersonic flow. Proc Combust Inst 2005;30:2825–34.

[15] Rasmussen CC, Dhanuka SK, Driscoll JF. Visualization of flameholding mechanisms in a supersonic combustor using PLIF. Proc Combust Inst 2007;31:2505–12.

[16] Lin KC, Tam CJ, Jackson K. Study on the operability of cavity flameholders inside a scramjet combustor. AIAA 2009-5028.

[17] Yang QC, Chang JT, Bao W, Deng J. A mechanism of combustion mode transition for hydrogen fueled scramjet. Int J Hydrogen Energy 2014;39:9791–7.

[18] Cao RF, Chang JT, Bao W, Guo ML, Qin J, Yu DR, et al. Analysis of combustion mode and operating route for hydrogen fueled scramjet engine. Int J Hydrogen Energy 2013;38:5928–35.

[19] Cao RF, Chang JT, Tang JF, Wang ZQ, Yu DR. Study on combustion mode transition of hydrogen fueled dual-mode scramjet engine based on thermodynamic cycle analysis. Int J Hydrogen Energy 2014;39(36):21251–8. http://dx.doi.org/10.1016/j.ijhydene.2014.10.082.

[20] LE JL, Yang SH, Liu WX, Xing JW, Massively parallel simulations of Kerosene-fueled scramjet. AIAA 2005-3318.

[21] Zheng ZH, LE JL. Massively parallel computation of three-dimensional scramjet combustor. Proceedings of the 24th international symposium on shock waves, Beijing, China, July 11–16, 2004, 2005, pp 897–902.

[22] Noh J, Choi JY, Byun R, Lim JS, Yang V. Numerical simulation of auto-ignition of ethylene in a scramjet combustor with air throttling, AIAA 2010-7036.

[23] Kok JC, Resolving the dependence on free-stream values for the K-omega turbulence model, NLR-TP-99295.

[24] Tian Y, Yang SH, Le JL. Numerical study on effect of air throttling on combustion mode formation and transition in a dual-mode scramjet combustor52; 2016. p. 173–80.

[25] Hass N, Cabell K, Storch A, Gruber M. HiFiRE direct-connect rig (HDCR) phase I, Scramjet test results from the NASA langley arc-heated scramjet test facility, AIAA-2011-2248.

[26] Heiser WH, Pratt DT. Hypersonic airbreathing propulsion, AIAA education series. USA: AIAA Pub; 1994.

[27] Tian Y, Xiao BG, Zhang SP, Xing JW. Experimental and computational study on combustion performance of a kerosene fueled dual-mode scramjet combustor46; 2015. p. 451–8.

[28] Micka DJ. Combustion stabilization, structure, and spreading in a laboratory dual-mode scramjet combustor. University of Michigan; 2010.

强制转捩对高超声速进气道性能影响

易淼荣，赵慧勇，乐嘉陵

（中国空气动力研究与发展中心 超高速空气动力学研究所

高超声速冲压发动机技术重点实验室，四川 绵阳 621000）

摘　　要：针对一个高超声速进气道，设计了不同高度的钻石形和斜坡形转捩带，在来流马赫数为 6，攻角为 1°工况下，通过风洞试验与三维数值模拟相结合的方法，研究了转捩带对进气道壁面热流密度，压强，隔离段入口处总压和马赫数分布的影响，数值计算与风洞试验在壁面压强，进气道总压和马赫数上吻合较好，在壁面热流密度上相差约 35%。研究结果表明，钻石形和斜坡形转捩带都可以有效地实现强制转捩，此外，随着转捩带高度的增加，转捩区域逐渐前移，直至转捩带后缘。对于已经起动的进气道，转捩带对进气道下壁面静压几乎不产生影响，但会使得隔离段下壁面附近的总压和马赫数有所下降，从而导致从隔离段入口直至整个隔离段的质量加权总压和流量均下降 3.5% 左右，而质量加权马赫数的下降量则并不明显，在 1% 左右。

关　键　词：高超声速；进气道；强制转捩；钻石形；斜坡形

中图分类号：V211.3　　　　文献标志码：A

Effect of forced-transition on performance of hypersonic inlet

YI Miao-rong，ZHAO Hui-yong，LE Jia-ling

（Science and Technology on Scramjet Laboratory，

Hypersonic Aerodynamics Research Institute，

China Aerodynamics Research and Development Center，Mianyang Sichuan 621000，China）

Abstract：Diamond and ramp forced-transition trips with different heights were designed for a hypersonic inlet. On the condition of freestream Mach number 6 and the angle of attack 1 degree，the effects of forced-transition on wall heat flux distribution，pressure，total pressure and Mach number of isolator inlet were investigated by both wind-tunnel experiment and three-dimensional numerical simulation. The computational data agreed well with wind-tunnel experiment results in wall pressure，inlet total pressure and Mach number. And there was at most 35% difference in wall heat flux between simulation and experiment results. The results show that both the diamond and ramp forced-transition trips can promote forced-transition effectively. The transition location moves forward with the increasing height of the trips. For the started inlet，the forced-transition trips have little effect on the static pressure of the inlet bottom surface，but the total pressure and Mach number near the bottom surface of the isolator will decline；in such case，the mass-averaged total pressure and the mass flow rate in the isolator decline about 3.5%. The decline of the mass-averaged Mach number is only about 1%.

Key words：hypersonic；inlet；forced-transition；diamond；ramp

在吸气式一体化高超声速飞行器的设计中，若进气道入口流动为层流，在前体压缩面拐角和隔离段入口，流动容易产生层流分离，导致横向溢流，从而使发动机进口捕获的流量降低，影响发动

易淼荣，赵慧勇，乐嘉陵. 强制转捩对高超声速进气道性能影响. 航空动力学报，2016，31(8)：1380-1387.

机的性能 严重时甚至可能导致进气道不起动 而若进气道入口流动为湍流 则可有效抑制进气道内的流动分离 提高进气道流量 也有助于发动机内燃料喷射时的混合 增加燃烧效率 但是在高空中 来流湍流度和噪声都很低[1] 飞行器在飞行试验时进气道入口前的边界层很难发生自然转捩 如 X-43A 飞行器的飞行试验表明[2] 没有安装转捩带的进气道背面保持层流状态 因此美国从 X-43A 飞行器到 X-51A 飞行器 均在进气道入口之前的压缩面上加装了强制转捩装置 确保进入进气道的空气为湍流状态 X-43A 飞行器的两次飞行试验也均证明了其强制转捩装置的有效性

强制转捩装置又称为涡流发生器或转捩带 一般为沿展向分布的一排或多排离散型的粗糙颗粒 常见的粗糙颗粒有圆柱形 三棱柱 钻石形 后掠斜坡形(简称斜坡形)等[3] 关于转捩带的转捩机理 目前有多种解释 Schneider[4] 在其综述中提出 高超声速情况下 目前研究得较多的机制是涡流发生器产生的尾流中有流向涡和不稳定的剪切层 尾流不稳定性的增加导致转捩 清华大学的 DUAN Zhiwei 等[5] 用直接数值模拟对来流马赫数为 6 下单个圆柱形粗糙颗粒诱导平板转捩的模拟表明 粗糙单元上游形成不稳定分离泡 随主流向下游发展形成复杂尾迹涡及马蹄涡 在尾迹区形成较强的剪切层 它们在尾迹区相互缠绕 发展 失稳并造成了粗糙单元尾迹区的转捩 Choudhari 等[6] 通过对 X-43A 飞行器前体/进气道转捩机理的研究 则认为转捩带诱导的转捩一般是多种不稳定模式共同作用下的结果 其中包括转捩带之前的 Rayleigh 和 Gortler 不稳定模式和转捩带之后的条带(streak)不稳定模式 在转捩带的几何参数中 一般认为粗糙颗粒的高度是影响转捩的最主要因素[7] 随着粗糙颗粒高度(即转捩带高度)的增加 达到起始高度后 强制转捩位置从自然转捩位置开始前移 当移到靠近粗糙颗粒的位置时 达到有效高度 之后转捩位置不再随着转捩带高度的增加而前移 也不再随着雷诺数的增加而前移 此外 粗糙颗粒的展向宽度和间距也影响转捩效果

迄今为止 对边界层控制装置进行了全面系统研究的高超声速飞行器是美国的 X-43A 为了开展 X-43A 进气道转捩带的选型 美国历经 3 年时间 在 3 座高超风洞内 开展了 366 次不同涡流发生器外形的风洞试验[8] 涡流发生器构型包括钻石形 圆柱形 三角形和斜坡形等 其风洞试验

结果表明 钻石形和斜坡形是比较好的两种构型 虽然钻石形的转捩效率稍优于斜坡形 但由于斜坡形在热防护上的优势 X-43A 的飞行试验最终采用了斜坡形转捩带 美国海军和 DARPA(Defense Advanced Research Projects Agency)联合研究的 HyFly 计划在 LENS(Large Energy National Shock)风洞中对全尺寸进气道模型开展了转捩带的设计和筛选试验[9] 研究了不同钝度 攻角和雷诺数对强转捩的影响 最终选取了斜坡形的涡流发生器构型 涡流发生器的高度接近当地层流边界层厚度 此外 X-51A 也分别在 LENS 激波风洞[10]和 Purdue 大学的马赫数为 6 的静音风洞[11]中研究了斜坡形和钻石形转捩带的转捩效果 值得一提的是其在静音风洞中的试验表明 风洞的来流噪声降低了转捩雷诺数 X-51A 最终也选取了斜坡形的转捩带

目前针对转捩带的研究主要集中在转捩位置和转捩效果上 如 Berry 等[12] 采用磷光热图技术得到了 X-43A 下壁面的壁面热流密度 从而判断出其具体转捩位置 Borg[13] 采用温敏漆技术测量壁面温度从而获得转捩位置 周玲等[13] 采用符松 王亮提出的三方程转捩模型对进气道的下壁面摩擦阻力进行了三维数值模拟 得到了转捩位置 针对加装转捩带之后对进气道性能的影响的研究却很少 美国的 Cockrell 等[14] 比较过转捩带对隔离段入口总压的影响 但是超声速来流状态下 测量的总压是皮托管前面正激波后的总压 与激波前的总压并不能等同 国内的赵慧勇等[15]则针对钻石形转捩带 通过风洞试验测量了进气道压缩面的壁面热流密度 并通过二维计算研究了进气道内的流量捕获系数和总压恢复系数 蔡巧言等[16]通过二维转捩模式计算指出 小的粗糙体对进气道的性能几乎没有影响 大的粗糙体会导致进气道的总压恢复系数下降 22% 赵慧勇和蔡巧言等针对强制转捩对进气道性能影响的研究均只采用了二维计算 而且计算结果并没有与试验数据进行比较 因此其结论是否能够推广到三维情况还有待进一步研究

本文针对一种升力体高超声速进气道 设计了不同高度的钻石形和斜坡形转捩带 通过风洞试验与三维数值计算相结合的方法 研究了在来流马赫数为 6 攻角为 1° 工况下转捩带对进气道性能的影响 比较了安装钻石形和斜坡形转捩带的进气道在转捩区域 隔离段入口总压和马赫数分布的差异 风洞试验与数值模拟在定量数据上

吻合良好.证明本文提出的方法可以应用于转捩带对进气道影响及转捩带筛选研究.研究结果还表明.本文研究的转捩带有效实现了强制转捩.对于已经起动的进气道.转捩带对进气道下壁面静压几乎不产生影响.但会使得隔离段下壁面附近的总压和马赫数有所下降.从而导致从隔离段入口直至整个隔离段的质量加权总压和流量均下降 3.5% 左右.而质量加权的马赫数的下降量则并不明显.在 1% 左右.

1 进气道模型及试验测量方法

进气道采用三维鸭嘴形状（图 1）.多波系平面顶压.工作范围.来流马赫数为 $5\sim7$.

转捩带分为钻石形和斜坡形两种.分别是一排钻石形和斜坡形的粗糙颗粒.转捩带高度 H 分为 $0.5\,mm$ 和 $1\,mm$ 两种.如图 2、图 3 所示.安装在进气道第 1 道压缩面上.

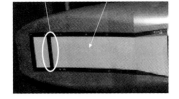

图 1 进气道模型示意图

Fig. 1 Sketch of inlet model

图 2 钻石形转捩带

Fig. 2 Diamond forced-transition trip

图 3 斜坡形转捩带

Fig. 3 Ramp forced-transition trip

试验在中国空气动力研究与发展中心的直径为 $0.5\,m$ 的高超声速风洞中进行.来流马赫数为 6.攻角为 $1°$.总温为 $473.2\,K$.总压为 $2\,MPa$.试验分为测热试验和测压试验.测热试验采用红外热

技术.详细的测量方法可以参见文献[13].根据以往的经验.热流密度测量精度大约为 $20\%\sim25\%$.由于红外热测量要求模型表面采用非金属材料.这里在进气道压缩面上采用聚四氟乙烯材料（如图 1）.但是由于转捩带采用聚四氟乙烯容易受热变形.所以采用不锈钢材料.这会导致转捩带处的红外热测量结果并不准确.因此本文将不对转捩带上的热流密度进行讨论.测压试验采用压力传感器测量进气道壁面中心线的壁面压强.采用总压耙测量隔离段入口的总压分布.总压耙安装位置如图 4 所示.

图 4 总压耙安装位置

Fig. 4 Installation position of the total pressure rakes

2 数值计算方法

计算模型与试验模型完全一致.由于偏航角为 $0°$.为了减少计算量.所以计算采用一半的进气道模型.转捩带处的网格进行了局部加密.网格细节如图 $5\sim$ 图 6 所示.壁面 y^+ 控制在 1 以内.

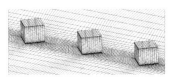

图 5 钻石形转捩带网格细节

Fig. 5 Detail grid of the diamond forced-transition trips

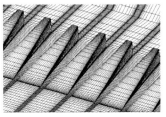

图 6 斜坡形转捩带网格细节

Fig. 6 Detail grid of the ramp forced-transition trips

为了消除网格不同引起的误差,无转捩带时和安装钻石形转捩带时网格仅在转捩带位置处有差别,其余部分完全一致,网格量约为3 000万,安装斜坡形转捩带后,由于斜坡形颗粒数为钻石形颗粒数的2倍,网格量约为6 000万.

数值计算采用中国空气动力研究与发展中心研制的AHL3D软件[17],该软件经过十多年在超燃冲压发动机和一体化飞行器计算中的应用,具有较高的可靠性,关于该计算软件的验证可参考文献[17-18].本文通过有限体积法求解黏性NS(Navier-Stokes)方程,时间推进采用LU-SGS(lower-upper symmetric Gauss-Seidel)方法,无黏通量采用Steger-Warming格式,黏性项采用Gauss定理计算,湍流模型采用SST(shear stress transport)双方程湍流模型,壁面采用无滑移等温壁条件,壁面温度根据试验结果来给定,转捩位置根据试验热流密度数据给定,在转捩位置之前采用全层流计算,转捩位置之后采用全湍流计算.

3　结果讨论

3.1　进气道下壁面热流密度

将红外热测得的进气道压缩面热流密度分布与计算所得结果进行对比如图7~图9所示.从图中可以看出进气道下壁面热流密度在4道压缩面上呈现阶梯形增长分布,这是由于每一道压缩面之前都会有一道斜激波,斜激波的压缩作用使空气的压强增加,压强与热流密度成正比关系,从而导致了壁面热流密度的增加.壁面热流密度在展向分布比较均匀.

为了更好地定量比较壁面热流密度值,按照图10方法分别取出靠近对称面的S_1和S_2切线的热流密度计算值与试验值比较(由于无转捩带

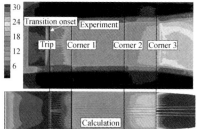

图8　钻石形转捩带进气道壁面热流密度
($H=1$ mm)

Fig. 8　Heat flux of inlet wall with diamond
forced-transition trip ($H=1$ mm)

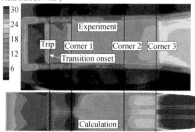

图9　斜坡形转捩带壁面热流密度($H=1$ mm)

Fig. 9　Heat flux of inlet wall with ramp
forced-transition trip ($H=1$ mm)

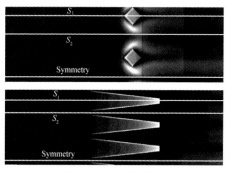

图10　S_1,S_2切线示意图

Fig. 10　Sketch of tangentlines of S_1,S_2

时试验测量值在展向上分布很均匀,因此只取了一条切线)如图11~图15所示,图中L表示进气道长度.

结合图7和图11可以得到,对于无转捩带的进气道,在第2道压缩拐角附近出现了一个热流

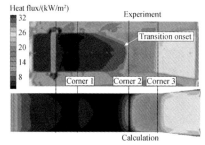

图7　无转捩带进气道壁面热流密度

Fig. 7　Heat flux of inlet wall without
forced-transition trip

密度迅速增加的区域,这一区域正是由于空气由层流转捩为湍流造成的,因此无转捩带进气道的转捩区域起始位置即为第 2 道压缩拐角附近的位置。

对于安装了 $H=1\,\mathrm{mm}$ 的斜坡形和钻石形转捩带的进气道,热流密度值均在转捩带之后就马上明显增加,结合图 8~图 9 和图 12~图 13 可以看出,转捩带之后气流马上由层流转捩为湍流,因此 $H=1\,\mathrm{mm}$ 的斜坡形和钻石形转捩带均有效促进了进气道的强制转捩,使得转捩位置提前到转捩带后缘。热流密度值在转捩带之后到第 1 道压缩拐角之前出现条带形分布,这是由于转捩带拖出的条带形尾涡导致的,但是这种条带形尾涡在被第 1 个压缩拐角的斜激波压缩后逐渐消失,之

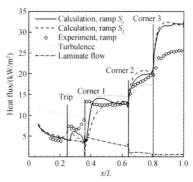

图 13 安装斜坡形转捩带时进气道壁面热流密度曲线
($H=1\,\mathrm{mm}$)

Fig. 13 Heat flux curves of inlet wall with ramp forced-transition trip ($H=1\,\mathrm{mm}$)

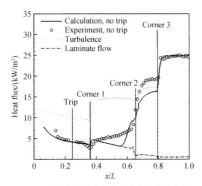

图 11 无转捩带时进气道壁面热流密度曲线

Fig. 11 Heat flux curves of inlet wall without forced-transition trip

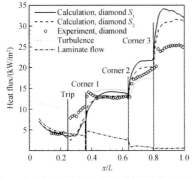

图 12 安装钻石形转捩带时进气道壁面热流密度曲线
($H=1\,\mathrm{mm}$)

Fig. 12 Heat flux curves of inlet wall with diamond forced-transition trip ($H=1\,\mathrm{mm}$)

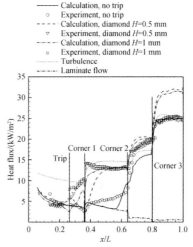

图 14 安装不同高度钻石形转捩带时进气道壁面热流密度曲线

Fig. 14 Heat flux curves of inlet wall with diamond forced-transition trip in different heights

后热流密度在展向上的分布都比较均匀。从 S_1 和 S_2 切线的热流密度对比也可以看出,两条切线的差别更多的体现在第 1 道压缩拐角之前,而对于之后的热流密度值,差别都比较小。

图 14~图 15 则表示了钻石形和斜坡形转捩带高度对进气道热流密度的影响规律(由于 S_1 和 S_2 切线的热流密度差别较小,因此在这里只显示了其中一条切线的热流密度值)。从图中可

以看到,安装了 $H=1\,\mathrm{mm}$ 转捩带进气道的转捩位置要稍微早于安装了 $H=0.5\,\mathrm{mm}$ 转捩带的进气道,再与无转捩带的情况进行对比,可以得出转捩位置随 H 的增加而前移的结论.在本文的模型中,$H=1\,\mathrm{mm}$ 的转捩带已经达到了有效高度.

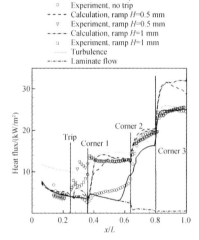

图 15　安装不同高度斜坡形转捩带时进气道
壁面热流密度曲线

Fig. 15　Heat flux curves of inlet wall with
ramp forced-transition trip in
different heights

定量对比试验和计算值,可得前 3 道压缩面上热流密度值均符合良好,在第 4 道压缩面上,计算值普遍比试验值高出 35% 左右,计算偏高的原因还不太清楚,需要进一步的研究.

3.2　进气道下壁面压强

从图 16 中可以看出中心线上壁面静压沿流向分布的计算与试验值吻合良好.静压在进气道下壁面出现台阶形分布,与热流密度结果相吻合.在唇口附近和隔离段中,由于激波与激波,激波与边界层的相互干扰,导致静压的分布沿流向变化剧烈.加装转捩带对进气道壁面静压并没有多大影响,仅在转捩带处出现一个小的静压突起.

从对称面静压分布云图(图 17)可以看出安装转捩带的进气道在转捩带处形成一道斜激波,但由于转捩带高度不是很高,斜激波在远离壁面后迅速变弱,并在与第 1 道压缩波相交后消失.同时可以从图 17 中看处,后 3 道压缩波均相交于唇口处,证明进气道均处于起动状态.

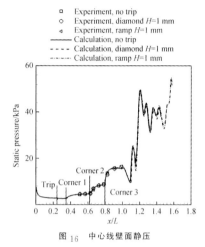

图 16　中心线壁面静压

Fig. 16　Wall static pressure on the centerline

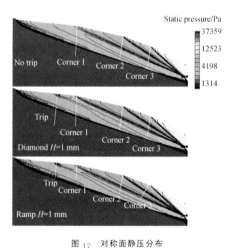

图 17　对称面静压分布

Fig. 17　Distribution of static pressure on
symmetry plane

3.3　隔离段入口的总压和马赫数

隔离段入口的总压和马赫数的试验和计算结果如图 18~图 20 所示,其中 y 表示总压管与进气道下壁面距离,h 表示唇口高度.计算与试验结果在 $y/h=0.44,0.61$ 处吻合较好,误差在 5% 以内,但在 $y/h=0.27$ 处试验值则普遍比计算值小,但对于总压和马赫数随转捩带高度变化的规律,试验与计算结果仍然是一致的,结合图 18~图

20可以得到,安装转捩带后隔离段入口靠近下壁面处的总压有所下降,钻石形的总压比斜坡形略小.总压下降量并不随转捩带高度的增加而线性

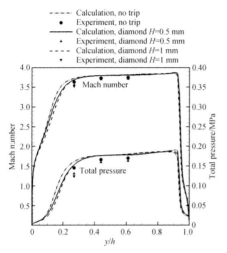

图 18 安装不同高度钻石形转捩带后隔离段入口处的总压和马赫数分布

Fig.18 Distribution of total pressure and Mach number at the entrance of isolator with diamond forced-transition trip of different heights

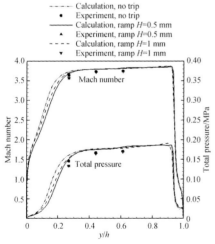

图 19 安装不同高度斜坡形转捩带后隔离段入口处的总压和马赫数分布

Fig.19 Distribution of total pressure and Mach number at the entrance of isolator with ramp forced-transition trip of different hights

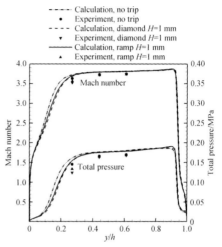

图 20 安装不同转捩带后隔离段入口处的总压和马赫数分布

Fig.20 Distribution of total pressure and Mach number at the entrance of isolator with different forced-transition trips

增加.无论是对于斜坡形还是钻石形转捩带,$H=0.5\ mm$ 时的总压与 $H=1\ mm$ 时的总压都比较接近.马赫数分布与总压遵循同样的规律.

隔离段内流量(mass flow rate)、质量加权总压($p_{t,m}$)和质量加权马赫数(Ma_m)如图 21 所示,可以看出,由于安装转捩带后进气道入口下壁面附近的总压和马赫数的下降,造成整个隔离段内的质量加权总压和质量加权马赫数也有所下降,其中质量加权总压和流量在隔离段出口处下降约 3.5%,质量加权马赫数的下降量则在 1% 左右.

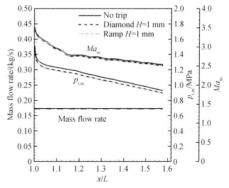

图 21 隔离段内进气道性能参数

Fig.21 Performance parameters in the inlet of isolator

4 结 论

针对一种升力体高超声速进气道模型,设计了不同高度的斜坡形和钻石形转捩带,利用风洞试验和三维数值模拟相结合的方法,研究了进气道壁面热流密度、中心线壁面压强和隔离段入口处的总压和马赫数分布,结果表明:

1) 数值计算与风洞试验在壁面压强、隔离段入口处总压和马赫数上吻合较好,在壁面热流密度上相差 35% 左右,因此本文的计算方法可用于对高超声速进气道转捩带的进一步研究、筛选和设计。

2) 所设计的钻石形和斜坡形转捩带都有效促发了强制转捩,转捩位置随着转捩带高度的增加向前移动。针对本进气道模型,$H=1$ mm 的斜坡形和钻石形转捩带均达到了有效高度。

3) 对于已经起动的进气道,转捩带使得隔离段下壁面处的总压有所下降,造成隔离段出口处质量加权总压也随之下降了 3.5% 左右,马赫数遵循跟总压相同的规律,但下降量在 1% 左右,流量的下降量也在 3.5% 左右。

4) 相同高度的斜坡形和钻石形转捩带的转捩区域基本相同,钻石形的隔离段入口处的总压损失略高于斜坡形,因此推荐采用斜坡形转捩带。

参考文献:

[1] Schneider S P. Effects of high-speed tunnel noise on laminar-turbulent transition [J]. Journal of Spacecraft and Rockets,2001,38(3):323-333.

[2] Scott B,Kamran D,Kathryn W. Boundary layer transition on X-43A[R]. AIAA-2008-3736,2008.

[3] 赵慧勇,易淼荣. 高超声速进气道强制转捩装置设计综述[J]. 空气动力学学报,2014,32(5):623-627.
ZHAO Huiyong, YI Miaorong. Review of design for forced-transition trip of hypersonic inlet[J]. Acta Aerodynamica Sinica,2014,32(5):623-627. (in Chinese)

[4] Schneider S P. Effects of roughness on hypersonic boundary-layer transition[J]. Journal of Spacecraft and Rockets,2008,45(2):193-209.

[5] DUAN Zhiwei,XIAO Zhixiang,FU Song. Direct numerical simulation of hypersonic transition induced by ramp roughness elements[R]. AIAA-2014-0237,2014.

[6] Choudhari M, LI Fei, Edwards J. Stability analysis of roughness array wake in a high-speed boundary layer[R]. AIAA-2009-0170,2009.

[7] Vermeersch O,Arnal D. May transient growth theory explain isolated roughness induced transition? [R]. Lisbon, Portugal:European Conference on Computational Fluid Dynamics,2010.

[8] Berry S A,Auslender A H,Dilley A D,et al. Hypersonic boundary-layer trip development for hyper-X[J]. Journal of Spacecraft and Rockets,2001,38(6):853-864.

[9] Holden S M,Wadhams T P,Maclean M. Experimental studies in the LENS supersonic and hypersonic tunnels for hypervelocity vehicle performance and code validation[R]. AIAA-2008-2505,2008.

[10] Wadhams T P,Maclean M G,Holden M S. A review of transition studies on full-scale flight vehicles at duplicated flight conditions in the LENS tunnels and comparisons with prediction methods and flight measurement [R]. AIAA-2010-1246,2010.

[11] Borg M P. Entitled laminar instability and transition on the X-51A[D]. West Lafayette,US:Purdue University,2009.

[12] Berry S A,Nowak R J,Horvath T J. Boundary layer control for hypersonic airbreathing vehicles[R]. AIAA-2004-2246,2004.

[13] 周玲,阎超,孔维萱. 高超声速飞行器前体边界层数值模拟[J]. 航空学报,2014,35(6):1487-1495.
ZHOU Ling,YAN Chao,KONG Weixuan. Numerical simulation of forced boundary layer transition on hypersonic vehicle forebody[J]. Acta Aeronautica et Astronautica Sinica,2014,35(6):1487-1495. (in Chinese)

[14] Cockrell C E,Jr,Auslender A H,White J A,et al. Aeroheating predictions for the X-43 cowl-closed configuration at Mach 7 and 10[R]. AIAA-2002-0218,2002.

[15] 赵慧勇,周瑜,倪鸿礼,等. 高超声速进气道边界层转捩试验[J]. 实验流体力学,2012,26(1):1-6.
ZHAO Huiyong,ZHOU Yu,NI Hongli,et al. Test of forced boundary-layer transition on hypersonic inlet[J]. Journal of Experiments in Fluid Mechanics,2012,26(1):1-6. (in Chinese)

[16] 蔡巧言,谭慧俊. 前体边界层状态对高超声速进气道流动结构及性能的影响[J]. 航空动力学报,2008,23(4):699-705.
CAI Qiaoyan,TAN Huijun. Effects of the forebody boundary layer state on flow structure and performance of hypersonic inlets[J]. Journal of Aerospace Power,2008,23(4):699-705. (in Chinese)

[17] 赵慧勇. 超燃冲压发动机并行数值研究[D]. 四川 绵阳:中国空气动力研究与发展中心,2005.
ZHAO Huiyong. Parallel numerical study of whole scramjet engine [D]. Mianyang,Sichuan:China Aerodynamics Research and Development Center,2005. (in Chinese)

[18] YANG Shunhua,LE Jialing. Computational analysis of a kerosene-fueled scramjet[R]. ISABE-2005-1195,2005.

亚声速横向气流中液体射流破碎过程的直接模拟[*]

刘日超[1,2]，乐嘉陵[2]，杨顺华[2]，郑忠华[2]，宋文艳[1]，黄　渊[2]

（1. 西北工业大学 动力与能源学院，陕西 西安 710072；

2. 中国空气动力研究与发展中心 超高速空气动力研究所 高超声速冲压发动机技术重点实验室，四川 绵阳 621000）

摘　要：为研究亚声速横向气流中液体射流柱的变形、弯曲以及其破碎过程，利用LES结合VOF的方法，对射流破碎过程进行了直接模拟。通过计算观察得到射流柱进入到横流气体中后由于Rayleigh-Taylor（RT）不稳定性和Kelvin-Helmholtz（K-H）不稳定性的共同作用迅速发生变形，并形成表面波，同时伴随着细小液滴的脱落，气流在射流柱后方形成涡状结构。在射流柱的迎风面上同时存在两种表面波，即RT表面波和KH表面波。模拟所得液柱纵向初始表面波波长为0.22mm，与K-H表面波波长理论公式计算所得相接近。通过对比射流柱的背风面和迎风面两侧，发现在横向来流中背风面上的大量细小液滴主要是由横向来流对液柱的不稳定性扰动造成迎风面上的液体向背风面方向挤压并剪切，最终脱落造成。

关键词：RT不稳定性；射流柱；首次破碎；雾化

中图分类号：V231.2　　　**文献标识码**：A　　　**文章编号**：1001-4055（2016）11-2135-07

DOI：10.13675/j.cnki.tjjs.2016.11.018

Direct Numerical Simulations of Atomization Processes of Liquid Jet in Subsonic Cross-Flow

LIU Ri-chao[1,2]，LE Jia-ling[2]，YANG Shun-hua[2]，ZHENG Zhong-hua[2]，SONG Wen-yan[1]，HUANG Yuan[2]

（1. School of Power and Energy，Northwestern Polytecnical University，Xi'an 710072，China；

2. Science and Technology on Scramjet Laboratory，Hypervelocity Aerodynamics Institute，

CARDC，Mianyang 621000，China）

Abstract：In order to deeply research the mechanisms of column distortion，bend and breaking up when a liquid jet was injected into subsonic cross-flow，the method of VOF（volume of fluid）and LES（large eddy simulation）was used to direct simulation of the atomization processes. The spray column begin distortion after spray was injected into cross-flow and yield surface wave due to Rayleigh-Taylor（RT）instability and Kelvin-Helmholtz（K-H）instability respectively. At the same time，Vortex was formed behind the spray column accompany with a lot of tiny droplets. There are two types of surface wave at the windward of liquid column：K-H wave and R-T wave. The initial wavelength of column wave is 0.22mm，and it is equal to expressions of K-H surface wave approximately. Droplets after leeward of spray were mostly because liquid of windward was pushed to leeward due to RT instability by cross flow by compare of windward and leeward of spray column.

Key words：Rayleigh-Taylor（RT）instability；Spray column；Primary breakup；Atomization

刘日超，乐嘉陵，杨顺华，郑忠华，宋文艳，黄渊. 亚声速横向气流中液体射流破碎过程的直接模拟. 推进技术，2016，37(11): 2135-2141.

1　引　言

横向气流中的射流喷射应用广泛,如超燃冲压发动机、航空发动机燃烧室、液体火箭发动机、农业喷洒等。特别是在环境问题日益受到重视后,对商用飞机的污染物排放要求也越来越严格。国际民用航空组织(ICAO)对飞机提出了更高的排放标准,如在报告[1]中明确规定了减少50%的二氧化碳和80%的NO_x排放目标。因此要达到排放标准,必须采用更先进的燃烧室技术,当前主要的方法有富燃-淬熄-贫燃(Rich burn-Quence-Lean burn,RQL)、贫油预混预蒸发(Lean Premixed Prevaporized LPP)、贫油直接喷射燃烧(Lean Direct Injection LDI)。其中LPP技术就是将燃油喷射进预混与蒸发设备,形成油气混合。因此射流在横向气流作用下的雾化质量直接影响到最终燃烧室性能、污染物的排放等。

当前国内外对横向气流下的射流做了广泛的研究。试验方面,国外主要将重点集中在横向射流的穿透深度、射流轨迹、混合程度等特性上。如Pei-Kuan Wu[2]对横向射流的过程进行了试验研究,并得出了不同射流速度和穿透深度之间的关系,认为气液动量比是影响射流轨迹的重要参数。Prakash R Surya[3]对横向气流作用下压力旋流喷雾进行了研究,分别对流速度不变,依次增加来流速度的雾化和气流速度不变,增加射流速度两种情况进行了研究,并将后一种情况下的雾化过程分为了五个阶段:崩溃剪切破碎、双袋状破碎、单袋状破碎、袋状/剪切复合破碎、剪切支配破碎。Mazallon J[4]对不同液体进行了横向射流的试验研究,认为射流柱在横向来流作用下发生剪切破碎的过程类似于液滴的二次破碎,并将射流在不同Weber数下的雾化状态进行了分类。国内朱英、黄勇等[5]采用高速摄像仪对横向气流中的液体圆柱射流破碎进行了研究,并给出了射流破碎位置坐标与液气两相动量通量比之间的关系式以及射流液柱在破碎点之前类似抛物线的轨迹曲线公式。张海滨、孙慧娟[6~8]等对压力旋流喷嘴在横向来流下的掺混情况进行研究,并得到了不同射流角度,不同液气动量比情况下的流场结构。此外,刘静、林宇震、郭婷婷[9~11]等分别对横向射流的研究现状进行了总结。尽管对横向射流问题展开了深入的研究,但对横向射流破碎过程的机理认识却远远不够,这是因为射流破碎过程中在射流的近场区域液相分布浓密,射流柱在气动力、表面张力、粘性力等的相互作用下发生变形、弯曲并破碎产生大量液滴,

试验采用的光学设备在近场区域因光线无法穿透液雾分布极密的区域,很难对其进行测量。因此人们想到了用数值计算的方法对横向射流进行研究。Alireza Mashayek等[12·13]对横向射流的模拟大多采用欧拉-拉格朗日方法,即将液相视为离散颗粒,对颗粒在流场中的破碎、混合过程采用经验和半经验的模型,如Reitz等提出来的波致破碎模型[14]。这种方法的缺陷在于无法直接模拟射流柱的变形、弯曲、脱落等物理过程,采用的物理模型的经验常数往往取决于试验条件或者喷嘴类型。近来发展起来的捕捉自由界面方法恰好可以弥补离散相模型的这一缺陷,可以直接对射流破碎过程进行直接模拟。当前比较常用的是VOF方法和Level set方法。如Sina Ghods[15]采用Level set方法对不同气液密度比的横向射流的雾化过程进行了模拟,并将计算结果与实验结果进行对比,通过线性稳定性理论和非线性稳定性理论对雾化过程进行了分析。但Level set方法则是把界面看作零等值面,通过定义流场中各点的符号距离函数得到一个连续的标量场,零等值面的变化过程就是界面的变化过程,但是在函数输运以及初始化的过程中,方程无法保证质量守恒,因此这是Level set方法存在最大的问题。VOF方法在每个时间步内都通过界面形态构造体积分数的输运量,因此这种VOF方法具有很好的守恒性。VOF的界面重构[16~18]当前主要有施主-受主格式(Donor-Acceptor Scheme)、几何重构格式(Geo-Reconstruction Scheme)和可压缩界面捕捉格式(Compressive Interface Capturing Scheme for Arbitrary Meshes,CICSAM)等。其中CICSAM根据可压缩性构造思想提出,该方法采用半隐格式处理,通过C-N格式离散,并对边界上的流通量修正,适合于非结构网格。之后研究人员对CICSAM进行了改进,如采用了显示的输运方程以及多维连续流格式,适用于四边形、三角形网格。Chee[19]采用VOF方法对横向射流的柱状破碎、袋状破碎还有剪切破碎状态进行了模拟,模拟得到了各状态下射流柱的变形过程。但其模拟结果并没有观察到很细致的表面波结构。研究人员[20~23]采用这两种方法对射流雾化过程中射流柱脱落、液滴破碎,以及表面波的形成与发展进行了深入的研究,并取得了一定的成果。但是在国内采用界面捕捉方法对横向射流进行模拟却鲜见报道。

本文通过大涡模拟和VOF相结合的方法,对横向气流下液体圆柱射流的破碎过程进行了数值模拟,其中液相介质为煤油。分析射流柱初次雾化的

细部结构、射流柱的演变过程以及射流柱表面脱落过程,为进一步深入研究射流柱的雾化机理奠定基础。

2 数学模型

2.1 大涡模拟理论

大涡模拟认为湍流流动由不同尺度的漩涡组成,其中大尺度涡旋对流动影响较大,而小尺度涡旋则主要起耗散作用。基于此,对大涡进行直接模拟,而小涡的影响则通过近似模型处理,即亚网格尺度模型。其中大尺度的可解分量通过控制方程直接求解,而亚网格尺度量则通过模型化求解

$$\bar{\phi}(x) = \int_D \phi(x') G(x, x') dx' \tag{1}$$

式中 D 为流动区域,$G(x, x')$ 是滤波函数,x' 为实际流动区域内的空间坐标,而 x 就是经过滤波后的大尺度空间上的空间坐标。

亚网格尺度应力 τ_{ij} 体现了小尺度涡对运动方程的影响

$$\tau_{ij} = \rho \overline{u_i u_j} - \rho \overline{u_i} \overline{u_j} \tag{2}$$

亚格子尺度模型

$$\tau_{ij} - \frac{1}{3} \delta_{ij} \tau_{kk} = -2\mu_t \overline{S_{ij}} \tag{3}$$

式中 τ_{kk} 是亚格子尺度应力的各项同性部分,μ_t 为亚网格尺度湍流粘性系数,$\overline{S_{ij}}$ 是分辨尺度的应变率张量。对于 μ_t 的计算,本文采用 Smagorinsky-Lilly 模型。

2.2 VOF方法

VOF方法通过求解单独的动量方程并计算区域每种流体的体积分数来模拟不同流体。如 γ 为网格单元内流体与网格体积的比值。当 $\gamma = 0$,表示该网格单元内全部为气体,当 $\gamma = 1$,则网格单元内全为液相,当 $0 < \gamma < 1$ 时,则认为网格单元内气液两相同时存在,为气液混合物。体积分数的输运方程为

$$\frac{\partial}{\partial t}(\tilde{\gamma}\bar{\rho}) + \nabla(\tilde{\gamma}\bar{\rho}\tilde{u}) = 0 \tag{4}$$

式中 $\tilde{\gamma}$ 表示网格单元内液相的体积分数,$\bar{\rho}$ 为密度。在气液混合区内,流体密度和粘性可以表示为

$$\bar{\rho} = (1 - \tilde{\gamma})\rho_g + \tilde{\gamma}\rho_l \tag{5}$$

$$\bar{\mu} = (1 - \tilde{\gamma})\mu_g + \tilde{\gamma}\mu_l \tag{6}$$

式中 ρ_g 和 ρ_l 分别表示气相密度和液相密度。

刘儒勋[24]通过对比上述三种界面捕捉方法,认为CICSAM方法精度较高。CICSAM方法控制体内流体体积为连续流构造的流通量和可压缩性思想构造

的流通量的平均值,利用精细网格内界面的迁移方法得到界面在网格内的运动位置和法向。本文采用的CICSAM方法。该方法在文献[25]中对直流喷射首次破碎的形变过程进行了研究,模拟结果与实验结果能够很好地吻合。

3 数值计算

3.1 计算域与参数设置

计算区域如图1所示,喷嘴直径为0.3mm,垂直喷入一个长、宽、高分别为3.9mm,3.9mm,12.5mm的方腔内。算例采用六面体结构网格,网格总数为2400万,在轴向方向上均匀分布,径向则由中心向四周越来越稀疏,在射流柱区域附近网格单元间距小于1μm,能够捕捉到射流柱的变化过程。方腔底部一个直径为0.3mm的小孔为液相速度入口,左侧为空气速度入口,右侧为压力出口条件,其余均为无滑移壁面。在VOF模型中令气相为第一相,液相为第二相。表1为各种参数的设置。

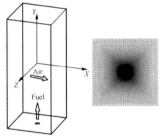

Fig. 1 Calculate domain

Table 1 Liquid properties and test conditions

Nozzle diameter/mm	0.3
Injector velocity/(m/s)	22
Air velocity/(m/s)	21
Liquid density/(kg/m³)	780
Air density/(kg/m³)	1.225
Liquid viscosity/(m²/s)	2.4×10⁻³
Air viscosity/(m²/s)	1.78×10⁻⁵

3.2 计算结果

3.2.1 射流柱整体变形过程

图2显示了横向射流情况下射流柱周围的三维流线图。图中显示了当气体遇到横向射流的射流柱的阻挡,在液柱后方形成了两个反向的尾涡。

图3(a)显示出了射流柱在横向气流作用下不断变形的过程,并伴随着在射流柱的背风面产生大量

的细小液滴。图3(b),(c)分别为为射流柱的 Z 截面和 X 截面的液相体积分数分布图,本文后面出现的云图无特殊说明均表示液相体积分数。图中显示了射流柱在气流作用下不断变扁平,表现为从 Z 截面看射流柱沿射流方向越来越窄,而从 X 截面看射流柱沿射流方向变宽。随着射流柱的迎风面不断变大,射流柱本身受到来流气体的冲击也越大,因此射流柱会向来流气体的下游方向弯曲,直至破碎。图中虚线表示射流柱离开喷嘴后在距离喷嘴3.2mm处发生断裂,产生液带和较大的液块。

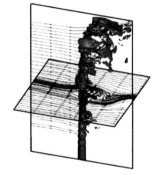

Fig. 2   Stream traces of air around liquid column

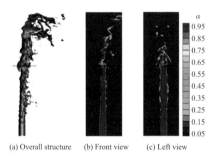

(a) Overall structure   (b) Front view   (c) Left view

Fig. 3   Spray overall structure

3.2.2   射流柱表面变形过程

图4显示了射流柱在气流作用下与气流相互作用产生表面波现象。图中1处为射流柱沿 Y 方向运动与周围气流存在 Y 方向上的速度差而相互剪切产生的KH表面波,射流柱在表面波的作用下向四周突起并脱落。箭头2处指示的为射流柱变形后的突起部分与横向气流在 X 方向上存在的速度差导致了相互作用产生的KH表面波。箭头3指出了射流柱在横向气流的挤压下变扁平的过程中与来流相互作用并在其液膜表面上形成表面波。表面波的形成原因将

在后面的小节进行阐述。图4(b)显示了射流柱的背风面液相的脱落过程。

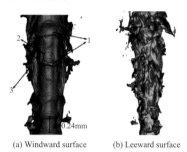

(a) Windward surface   (b) Leeward surface

Fig. 4   Spray column

根据Reitz将色散方程的解拟合得到的表面波波长公式

$$\frac{\Lambda_{KH}}{a} = 9.02\frac{(1+0.45Oh^{0.5})(1+0.4T_p^{0.7})}{(1+0.87We_g^{1.67})^{0.6}} \tag{7}$$

式中 Λ_{KH} 为对应波长, a 表示液滴半径, We_g 表示气体Weber数,表征惯性力和表面张力的比值, Oh 数物理含义为粘性力与表面张力的比值, $T_p = Oh \times We_g^{0.5}$。本算例取气液相对速度为22m/s, a 值取射流柱出口半径,为0.3mm。根据计算公式求得波长 $\Lambda_{KH} = 0.24mm$。与模拟结果初始表面波波长的0.22mm基本一致。进一步验证了射流柱的纵向表面波是由K-H不稳定性引起的。

图5显示了射流柱表面由于射流速度在射流方向上的速度差产生的表面波波长的变化,图中横坐标表示波数,如第1个波为射流柱离开喷嘴后射流柱表面较为明显的第一个波,以此类推。从图中看出表面产生的表面波波长从0.22mm随着射流柱的喷射增长到0.74mm。并在波长为0.74mm的波长处发生断裂。

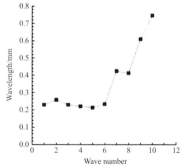

Fig. 5   Liquid column surface wave evolvement

3.2.3 射流柱射流方向截面变形过程

图6为射流柱的Y截面液相体积分数分布随着射流方向的演变过程，每两张图在Y方向上相距0.2mm。图7为图6(d)的放大图。从图中可以看出射流柱在横流中的雾化过程类似于液滴在气流中的二次破碎过程。射流柱在受到来流的挤压下发生变形，射流柱的正迎风面与来流方向相互垂直，气液密度差导致产生RT不稳定性的作用，RT波将射流柱前端液体向两侧推移，射流柱表面的波状结构也越来越明显。当液膜向两侧弯曲到一定程度后，气流方向与射流柱表面不再垂直，RT不稳定性减弱，气液间速度差的相互作用开始成为支配地位，也就是说射流柱经历了由RT不稳定波主导变形到KH不稳定波主导变形的过渡过程。同时在图中可以看到在射流柱的后方形成两个明显的反向漩涡对，漩涡的存在促进了射流柱两侧因气液剪切作用而脱落下的液块进一步破碎成较小的液块。

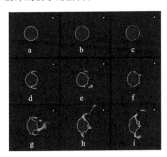

Fig. 6　Spray column evolvement along Y-direction

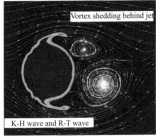

Fig. 7　Air flow streamline

图8显示了射流柱截面的上下游流场的变化，其中白色实线为液柱的轮廓线。射流柱上游气体压力沿液柱侧向方向有不断减小的趋势。液柱的迎风面上形成一个滞留区，在滞留区内流速为零，同时在液柱的背风面产生了一个低压区。前后压差的存在以及液体表面张力的作用使液柱截面从圆形不断变形为椭圆。随着液柱变形，压力增大，加速液柱的变形。在液柱两侧的气体存在高速区，气液间剪切力增大，从而引发了K-H表面不稳定性，导致液柱表面液体的突起并脱落。

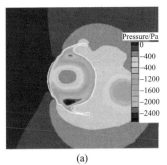

(a)

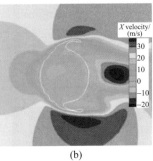

(b)

Fig. 8　Pressure and velocity contour around of spray column

图9对比了射流柱两侧X方向和Y方向的速度变化。图中白色实线为射流柱$Z=0$截面的轮廓线。从图中射流柱两侧的速度云图显示无论是X方向还是Y方向，其射流柱左侧的液相速度与周围气体的速度差都要明显大于右侧，因此再一次证明了在射流柱后方观察到的小液滴更多的是由于射流柱的迎风面遭遇横流气体后液体向两侧挤压并向后方脱落而形成。而在射流柱的背风面上，由于气液相对速度较小，因此射流的不稳定性较弱，所以脱落现象并不明显。

3.2.4 不同横流速度下的射流柱变形对比

图10，图11对比了相同横流速度下，不同空气来流速度下射流的变形过程，从图中可以看出，横流速度增大，液柱变形过程也加快，并且加速了射流柱的弯曲。但是如图所示，$\lambda_1=0.22mm$，$\lambda_2=0.21mm$射流柱表面的纵向表面波波长均在0.2mm左右，表明射流

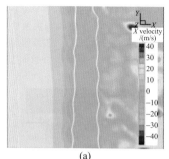

(a)

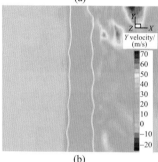

(b)

Fig. 9　Velocity contour beside of spray

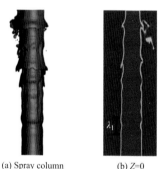

(a) Spray column　　　(b) Z=0

Fig. 10　Spray column ($V_1 = 22$m/s, $V_a = 21$m/s)

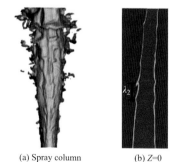

(a) Spray column　　　(b) Z=0

Fig. 11　Spray column ($V_1 = 22$m/s, $V_a = 60$m/s)

柱表面纵向扰动主要取决于射流柱的射流速度,而非横向来流的速度。图11中射流柱表面横向扰动明显强于图10中射流柱的横向扰动,这是由于横向气体流速增大,导致射流柱表面由气液密度比引起的R-T不稳定性加强。

4　结　论

通过采用大涡模拟和VOF相结合的方法对横向气流中液体射流的破碎过程进行研究,得出以下结论:

(1)大涡模拟和VOF相结合的方法能够有效地捕捉到横向气流中射流柱的变形、弯曲、断裂等物理过程。

(2)射流柱在气流作用下发生变形,在射流柱表面形成RT不稳定性并使射流柱表面具有波状结构,其横截面向扁平化过渡,并在射流柱后方形成气涡。整个过程类似于液滴的二次破碎过程。射流柱表面波在射流柱迎风面到两侧的过程中经历了由R-T表面波主导到K-H表面波主导的过渡过程。

(3)射流柱整体也在横向气流作用下向下游发生弯曲,并带有明显的波状,表面波振幅不断变大,波长由0.22mm增大到0.74mm,并在达到0.74mm处射流柱最终发生断裂,形成较大的液带,液带在气动力作用下进一步发生破碎形成小液滴。

(4)射流柱在相同射流速度,不同来流速度的条件下液柱表面纵向表面波波长没有很大的变化,表明液柱表面纵向表面波的决定性因素是射流速度,而非横流速度。

此次模拟因重点放在射流柱柱体上,因此在流场的远场区域网格较为稀疏,所以不能很好地反映远场液滴颗粒的二次破碎过程。从计算结果的液相等势图还可以看出,因网格量的原因,仍不能非常细致地体现射流柱表面的细部结构,只能做定性分析。因此对射流柱细部结构做定量分析可以作为下一步的工作。

参考文献:

[1]　Busquin P, Arguelles P. European Aeronautics: A Vision for 2020-A Synopsis[R]. *Air & Space European*, 2001, 3(s 3-4):16-18.

[2]　Pei-Kuan Wu, Kevin A. Breakup Processes of Liquid Jets in Subsonic Crossflows[R]. *AIAA 96-3024.*

[3]　Prakash R Surya, Hrishikesh Gadgil. Breakup Processes of Pressure Swirl Spray in Gaseous Cross-Flow[J]. *International Journal of Multiphase Flow*, 2014, 66(7):

79-91.

[4] Mazallon J, Dai Z. Aerodynamic Primary Breakup at the Surface of Nonturbulent Round Liquid Jets in Cross-flow[R]. *AIAA* 1998-716.

[5] 朱 英, 黄 勇. 横向气流中的液体圆形射流破碎实验[J]. 航空动力学报, 2010, 25(10): 2216-2266.

[6] 孙慧娟, 刘 利. 横流中喷雾掺混流场结构研究 [J]. 推进技术, 2012, 33(2): 221-226. (SUN Hui-juan, LIU Li. Investigation of Mixing Flow Field of Spray Droplets in Crossflow[J]. *Journal of Propulsion Technology*, 2012, 33(2): 221-226.)

[7] Haibin Zhang, Bofeng Bai. Experimental Study of the Mixing of Two Impinging Pressure-Swirl Sprays in Cross-flow [J]. *Experimental Thermal and Fluid Science*, 2013, 49(9): 67-74.

[8] Haibin Zhang, Bofeng Bai. Droplet Dispersion Characteristics of the Hollow Cone Sprays in Crossflow[J]. *Experimental Thermal and Fluid Science*, 2013, 45 (2): 25-33.

[9] 刘 静, 徐 旭. 高速气流中横向液体射流雾化研究进展[J]. 力学进展, 2009, 39(3): 273-283.

[10] 林宇震, 李 林. 液体射流喷入横向气流混合特性研究进展[J]. 航空学报, 2014, 35(1): 46-57.

[11] 郭婷婷, 李少华. 横向紊动射流的数值与实验研究进展[J]. 力学进展, 2005, 35(2): 211-220.

[12] Alireza Mashayek. Experimental and Numerical Study of Liquid Jets in Crossflows[D]. *Toronto: University of Toronto*, 2006.

[13] Liu F Smallwood. Numerical Study of Breakup Processes of Water Jet Injected into a Cross Air Flow[C]. *USA: ICLASS* 2000 *Proceedings of the Eighth International Conference on Liquid Atomization and Spray Systems*: 67-74.

[14] Kuo K K. Recent Advances in Spray Combustion: Spray Atomization and Drop Burning Phenomena[R]. *AIAA* 96-866418.

[15] Sina Ghods. Detailed Numerical Simulation of Liquid Jet in Crossflow Atomization with High Density Ratios [D]. *USA: Arizona State University*, 2013.

[16] Hirt C W, Nichols B D. Volume of Fluid (VOF) Method for the Dynamics of Free Boundaries[J]. *Journal of Computational Physics*, 1981, 39(81): 201-225.

[17] Youngs D L. Time-Dependent Multi-Material Flow with Large Fluid Distortion [M]. *Aldermaston: Atomic Weapons Research Establishment*, 1982.

[18] Ubbink O. Numerical Prediction of Two Fluid Systems with Sharp Interfaces[D]. *London: Imperial College of Science, Technology and Medicine*, 1997.

[19] Chee-Loon NG. Deformation, Wave Phenomena, and Breakup Outcomes of Round Nonturbulent Liquid Jets in Uniform Gaseous Crossflow[D]. *USA: Oklahoma State University*, 2006.

[20] Martinez J M, Chensneau X, Zeghmati B. A New Curvature Technique Calculation for Surface Tension Contribution in PLIC-VOF Method[J]. *Computational Mechanics*, 2006, 37(2): 182-193.

[21] Dexjardins O, Moureau V, Pitsch H. An Accurate Conservative Level Set/Ghost Fluid Method for Simulating Turbulent Atomization[J]. *Journal of Computational Physics*, 2008, 227(18): 8395-8416.

[22] Shinjo J, Umemura A. Simulation of Liquid Jet Primary Breakup: Dynamics of Ligament and Droplet Formation [J]. *International Journal of Multiphase Flow*, 2010, 3 (8): 513-532.

[23] 刘 娟, 李清廉, 刘卫东. 离心式喷嘴液膜破碎过程实验[J]. 推进技术, 2011, 32(4): 539-543. (LIU Juan, LI Qing-lian, LIU Wei-dong. Experiment on Liquid Sheet Breakup Process of Pressure Swirl Injector [J]. *Journal of Propulsion Technology*, 2011, 32(4): 539-543.

[24] 刘儒勋, 刘晓平, 张 磊. 运动界面的追踪和重构方法[J]. 应用数学和力学, 2004, 25(3): 279-290.

[25] 刘日超, 乐嘉陵. 直流喷射首次破碎的形变过程研究 [J]. 推进技术, 2016, 37(7). (LIU Ri-chao, LE Jia-ling. Investigation of Deformation of Primary Breakup in Direct Inject[J]. *Journal of Propulsion Technology*, 2016, 37(7).)

电加热圆管内流动的自然转捩过程研究

张若凌*，乐嘉陵

（中国空气动力研究与发展中心超高速空气动力研究所 高超声速冲压发动机技术重点实验室，四川 绵阳 621000）

摘要：分析了在一个电加热圆管内的自然转捩流动和对流传热。对于圆管内的流动，提出在径向脉动速度不随流动模式变化的假设下，自然转捩流动是充分发展的层流与湍流流动按照比例的合成。采用合成比例来描述该合成流动，合成比例在转捩区间会发生振荡。根据最小熵产生准则得到自然转捩发展演化的方程，其中转捩发展演化的控制因素，是合成比例的振荡。给出了一个与测量结果一致的合成比例的振荡函数，包括圆管内转捩过程的传热实验测量和速度及其脉动统计特性的实验测量。指出圆管内层流向湍流的转捩过程，可以与热力学平衡系统的连续相变过程进行比较，并且在电加热圆管内的流体，其速度和温度可以有相似和独立的转捩演化过程。

关键词：转捩流动；振荡；流动合成；最小熵产生

中图分类号：V211.1　**文献标识码**：A

Natural laminar-to-turbulent transition inside an electrically heated circular tube

Zhang Ruoling*, Le Jialing

（Science and Technology on Scramjet Laboratory，Hypervelocity Aerodynamics Institute of China Aerodynamics Research and Development Center，Mianyang Sichuan 621000，China）

Abstract：The natural laminar-to-turbulent transitional flow and convective heat transfer inside an electrically heated circular tube are analyzed. It is proposed that the transitional flow can be decomposed into the fully developed laminar flow and the turbulent flow，under the assumption that the fluctuating velocity in the radial direction does not change with varying flow modes. The composite ratios are adopted to define the composite flow，and they fluctuate during the flow transition. The minimum entropy production criterion is used to derive an equation which can describe the evolution of the transitional flow. It is deduced that the transitional behavior is governed by the fluctuations of the composite ratios. One fluctuation function is given to attain agreements with measurements including those obtained in heat transfer and flow experiments. It is pointed out that the process of the laminar-to-turbulent transition inside the tube can be compared with continuous phase transitions in a thermodynamic equilibrium system，and similar and separate processes for the transitions of the velocity and temperature inside the tube can be allowed.

Keywords：transitional flow；fluctuations；composition of flows；minimum entropy production

0 引 言

不可压缩和可压缩流体的层流向湍流的转捩过程，对许多应用来说十分重要。对于圆管内的低速轴对称流动，详细的测量研究被不断重复了100多年，文献[1]对其中一些进行了综述。

为进行超燃冲压发动机的冷却设计，多年来碳氢化合物在电加热圆管内的流动和传热引起了一些兴趣[2-3]。如图1所示，当电流在管壁内流过时，电流和管子电阻相互作用会产生热量。处于超临界压力下的一定流量的正十烷（n-Decane）在管内流动时，会与内壁面发生近似恒定热流的对流传热，能够模拟碳氢燃料在一根燃烧室冷却通道内的流动和吸热。正十烷吸收热量后温度升高，导致密度ρ和粘性系数μ

降低。在合适的流量、管径 d_i 和电流条件时,处于超临界压力下的正十烷在电加热圆管中建立定常流动,并且由图 2 看出,流动的雷诺数 Re(即 $\rho U d_i/\mu$,此处 ρ、U 和 μ 为截面平均值)可以自入口的 886 增加至出口的 15 000。

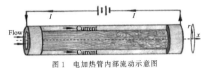

图 1　电加热管内部流动示意图
Fig. 1　Flow inside an electrically heated tube

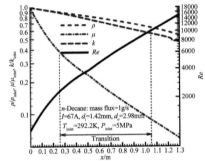

图 2　管内密度、粘性系数、热传导系数和 Re 的变化
Fig. 2　Density, viscosity, thermal conductivity and Re inside the tube

通过调节管子的电流来控制管壁的热流,可以使流体的温度自室温升高至 800K 左右的高温。流体的一些热物理性质,例如密度 ρ、粘性系数 μ 和热传导系数 k,在出口处其数值较入口处会下降较多。在图 2 中,不同温度和压强下正十烷的 ρ、μ 和 k 采用一个高温碳氢化合物的物性程序计算得到。

众所周知,圆管内流体的流动在雷诺数 Re 约为 2300 时会发生转捩[4],因此在管内可以发生完整的自然转捩过程。管内的层流是线性稳定的,即使 Re 很高时也需要有限振幅的扰动才能引发转捩。转捩的起始 Re 和结束 Re 与实际的流动条件相关,取决于管内流动的扰动,并且强迫转捩的起始 Re 和结束 Re 与自然转捩大不相同。

1883 年雷诺在转捩流动实验中发现湍流的闪斑(Flash)。随后在圆管内的流动转捩研究中,长期以来人们观测到 2 种扰动流动(Disordered flow),分别被称为 Puff 和 Slug[1,5]。在这 2 种扰动流动内部,既不是层流也不是完全湍流,用间歇性或者间歇因子来描述转捩流动是不够的。在很宽的雷诺数范围内,这些特殊状态的扰动流动和典型湍流之间的关系,至今不明。自 1883 年雷诺初始的实验至今没有确立公认的理论来解释层流向湍流的转捩过程[1,5],圆管内流动转捩的本质仍然是流体力学中的未解之谜。对于转捩过程中的对流传热特性也没有理论解释。

层流向湍流的转捩,很早就被看作是非平衡热力学系统的一种相变[6]。序参数(Order parameter)常用来描述相变(如正常导体变为超导体),接近相变点时序参数发生振荡。层流向湍流的转捩过程中发现了很大的振荡(或脉动、涨落,Fluctuations)[5,7],但是采用非平衡相变观点来进行的讨论不多。

在圆管内的转捩流动中,速度的随机振荡具有一定的统计特性[5]。这种振荡的统计特性随着 Re 数增加的演变过程十分重要,即使已知转捩流动的起始 Re 和结束 Re。本文尝试解释在电加热圆管内流动转捩的起始和终止之间振荡统计特性的发展演化,并解释这种振荡统计特性对流动和对流传热的影响。采用的方法包括 3 个步骤。在第一步中,先给出求解管内层流和湍流的方程,然后假定在转捩流动中沿半径方向脉动速度的数值与湍流流动一样,把转捩流动分解成层流和湍流成分,即转捩流动是相同雷诺数下成层流和湍流 2 种流动的合成,合成比例用来定义这种合成流动。在第二步中,对于转捩流动引入合成比例的振荡,采用最小熵产生准则给出一个方程,来描述合成比例振荡的统计特性和转捩的发展演化。在最后一步中,在给出一些层流向湍流的转捩和相变的相似性后,给出一个合成比例的振荡函数,并与传热和流动测量实验结果进行对比。

1　求解速度和温度的定常层流方程

从图 2 可以看出,在出口处密度 ρ 和热传导系数 k 下降至大约为入口数值的 50%。在出口处粘性系数 μ 下降至大约为入口数值的 6%。因为质量流量和管子内径 d_i 不变,造成在出口 Re 增加至约 17 倍的入口数值。

在如图 2 所示的条件下,Re 自入口的 886 连续增加至出口的 15 000。自然转捩起始于位置 $x \approx 0.26$m,约等于 180 倍管子内径 d_i(1.42mm),在该处 $Re \approx 2300$。对于对流传热(温度)来说,自然转捩终止于位置 $x \approx 1.05$m,在该处 $Re \approx 10000$[8-9],转捩长度约等于 $740d_i$。文献[5]的测量结果说明,流动(速度)的转捩区间(终止雷诺数与起始雷诺数之差)要小得多。对于本文感兴趣的情形,流动的转捩长度大约 $100d_i$。在管子入口 $x=0$ 之前,有比 $500d_i$ 还长的一段管子,与图 1 所示的加热部分属于同一根管子。在下文中,从转捩开始直至出口的流动,均认为在流动和传热上得到了充分发展。

对于该圆管内充分发展的轴对称流动,每个截面上的参数可以由平均速度、平均温度和边界条件(如管壁处的热流和零速度)来确定。本文研究的电加热圆管长度约为 $1000d_i$,且管壁热流数值不是很高。当研究一个截面上的流向速度 u 和流体温度 T 分布时,可以不考虑一些物理量沿着轴向坐标 x 方向的变化。理由是:与 u 和 T 沿着半径 r 方向的变化相比,这些物理量沿着 x 方向的变化很小。下文中,在每一个截面位置进行求解时,u 和流体物性如 μ 和 k 等均假定不随 x 方向变化,仅考虑 T 和压强 p 随 x 的变化。在每一截面,u 和 ρ 是沿半径方向位置 r 的函数。整个管内的流动通过求解每一个截面位置的流动参数而确定。

实际上,在管子的入口和出口流体物性 μ 和 k 的差别很大。尽管如此,μ 和 k 沿着 x 方向的变化对于 u 和 T 的解影响不大,因此在下文中没有用到 μ 和 k 对 x 的导数。这与管子的长径比很大并且管壁热流不很高是一致的。在求解每一个截面位置的流动参数时,采用的是与当地 T 和 p 对应的当地 μ 和 k。

对于管内的轴对称层流,采用图 1 所示轴向坐标为 x 和径向坐标为 r 的坐标系,速度 u 满足[4,10]:

$$\frac{\mathrm{d}p}{\mathrm{d}x} = \frac{1}{r}\frac{\mathrm{d}}{\mathrm{d}r}\left(\mu r \frac{\mathrm{d}u}{\mathrm{d}r}\right) \tag{1}$$

有传热时的热充分发展的管内定常层流流动,传热引起的温升远大于粘性摩擦引起的温升。按照文献[4]的做法,忽略粘性摩擦和轴向传热引起的温度变化,有:

$$\alpha_p u \frac{\partial T}{\partial x} = \frac{1}{r}\frac{\partial}{\partial r}\left(kr \frac{\partial T}{\partial r}\right) \tag{2}$$

对于恒定壁温的情形,可以使用变量分离法把温度表示成两部分的乘积,一部分随 x 变化,一部分随 r 变化。对于近似适用于本加热管道的恒定热流的情形,可以令:

$$T = X(x) + \Theta(r) \tag{3}$$

随 x 变化的部分代表管壁加热引起的沿 x 方向的平均温升,随 r 变化的部分代表管壁加热引起的径向温度分布(剖面)。(2)式变为:

$$\frac{\mathrm{d}X}{\mathrm{d}x} = \frac{1}{\alpha_p u}\frac{1}{r}\frac{\mathrm{d}}{\mathrm{d}r}\left(kr \frac{\mathrm{d}\Theta}{\mathrm{d}r}\right) \tag{4}$$

这里忽略了径向速度,根据质量平衡方程,乘积 ρu 不随 x 变化,因此(4)式可以分为 2 个可以积分求解的方程。在壁面有传热发生时,传热会引起 μ、k 和 c_p 变化,因此求解速度和温度分布的方程(1)和(2)是耦合的。这种耦合是通过温度和压力影响物性(μ、k

和 c_p)而发生的,属于较弱的耦合。

流体由大量分子组成,分子间不停发生碰撞。这些碰撞不会使动量或者能量产生或者消失,因此采用守恒变量时动量和能量平衡方程没有源项。但是在处于非平衡态的流体内部,与分子频繁碰撞相伴而生的是耗散过程,流体通过耗散过程来改变非平衡分布从而向平衡态靠近。因此对于处于非平衡的流体流动系统,熵的平衡方程有源项,表征着耗散过程中熵的产生。由于温度和速度梯度引起的熵产生为[6,10,11]:

$$\sigma_{\mathrm{Lam}} = \frac{k}{T^2}(\nabla T)\cdot(\nabla T) + \frac{2\mu}{T}(\nabla v)^s:(\nabla v)^s \tag{5}$$

式中:σ 是熵产生,下标 Lam 表示层流,v 是速度矢量,$(\nabla v)^s$ 是 ∇v 的对称零阵迹部分(对角线各项之和为 0),: 表示二次缩并。在柱坐标系中 $(\nabla v)^s$ 为:

$$(\nabla v)^s = \frac{1}{2}[\nabla v + (\nabla v)^T] - \frac{1}{3}\mathrm{Tr}(\nabla v)\overline{\overline{U}}$$

$$= \begin{pmatrix} \frac{2}{3}\frac{\partial v_r}{\partial r} - \frac{1}{3}\frac{v_r}{r} & 0 & \frac{1}{2}\frac{\partial v_x}{\partial r} \\ 0 & -\frac{1}{3}\frac{\partial v_r}{\partial r} + \frac{2}{3}\frac{v_r}{r} & 0 \\ \frac{1}{2}\frac{\partial v_x}{\partial r} & 0 & -\frac{1}{3}\frac{\partial v_r}{\partial r} - \frac{1}{3}\frac{v_r}{r} \end{pmatrix}$$

$$\tag{6}$$

式中:$\overline{\overline{U}}$ 是二阶单位张量,上标 T 表示转置,Tr 表示张量的迹(对角线各项之和),v_r 是沿半径方向的速度,v_x 是沿 x 方向的速度。对于所研究的情形,有 $v_x = u$ 和 $v_r = 0$。

实际上,(2)式忽略了粘性摩擦和轴向传热引起的温度变化,因此(5)式中第二项可以忽略,第一项中可以忽略沿轴向的微分,以此来与(2)式一致。尽管如此,下面在讨论熵产生时仍然保留这些项。将会看到这种保留不影响结论。

(5)式第一项在柱坐标系中的表达式为:

$$\sigma_{\mathrm{Lam},T} = \frac{k}{T^2}(\nabla T)\cdot(\nabla T)$$
$$= \sigma_{\mathrm{Lam},X} + \sigma_{\mathrm{Lam},\Theta} = \frac{k}{T^2}\left(\frac{\mathrm{d}X}{\mathrm{d}x}\right)^2 + \frac{k}{T^2}\left(\frac{\mathrm{d}\Theta}{\mathrm{d}r}\right)^2 \tag{7}$$

(5)式第二项的表达式为:

$$\sigma_{\mathrm{Lam},V} = \frac{2\mu}{T}(\nabla v)^s:(\nabla v)^s = \frac{\mu}{T}\left(\frac{\mathrm{d}u}{\mathrm{d}r}\right)^2 \tag{8}$$

在(7)和(8)式中,下标 T、X、Θ 和 V 分别表示由对应物理量的梯度引起的熵产生,V 表示速度。可以看出,熵产生包含温度和速度梯度的平方项。

流体的流动方程也称为 NS 方程。在所考虑的方程中,广义流(如应力张量)与广义力(如张量 ∇v)

的关系是线性的[6]。层流中的流体微元处于非平衡状态，其中的耗散输运过程引起熵增。对于非平衡系统，在广义流与广义力关系是线性的区域内，Prigogine 最早指出最小熵产生对应的状态是一个定态[6]。最小熵产生准则与适用于平衡系统的最小自由能准则相对应。最小自由能准则常用于处理平衡系统的相变特性，著名的 Ginzburg-Landau 相变理论就是利用这个准则来处理连续相变的一个平均场理论，认为自由能是序参数的解析函数[6,12-13]。

2 求解速度和温度的定常湍流方程

对于充分发展的定常轴对称湍流，同样可以忽略速度和密度沿 x 的变化，每个截面上 ρ 和 u 是 r 的函数。在每一截面上，x 方向动量方程为[1]：

$$\frac{\mathrm{d}\bar{p}}{\mathrm{d}x} = \frac{1}{r}\frac{\mathrm{d}}{\mathrm{d}r}\left(\mu r \frac{\mathrm{d}u}{\mathrm{d}r}\right) - \frac{1}{r}\frac{\mathrm{d}}{\mathrm{d}r}\left(\rho r \overline{\tilde{u}\tilde{v}}\right) \quad (9)$$

式中：符号上方有"$-$"表示物理量的时间平均值，符号上方"$\sim$"表示物理量的瞬时脉动值（瞬时值减去平均值）。由于径向平均速度为 0，$\tilde{v}$ 是 r 方向的瞬时速度分量。

对于热充分发展的管内定常湍流流动，有：

$$\rho c_p u \frac{\partial T}{\partial x} = \frac{1}{r}\frac{\partial}{\partial r}\left(kr \frac{\partial T}{\partial r}\right) - \frac{1}{r}\frac{\partial}{\partial r}\left(\rho c_p r \overline{\tilde{T}\tilde{v}}\right) \quad (10)$$

(9)和(10)式等号右边第一项与层流流动的耗散项一致。对于充分发展湍流，求解速度和温度的方程可以改写为：

$$\frac{\mathrm{d}\bar{p}}{\mathrm{d}x} = \frac{1}{r}\frac{\mathrm{d}}{\mathrm{d}r}\left(\mu r \frac{\mathrm{d}\bar{u}}{\mathrm{d}r} - \rho r \overline{\tilde{u}\tilde{v}}\right) \quad (11)$$

$$\bar{\rho} c_p \bar{u} \frac{\partial \bar{T}}{\partial x} = \frac{1}{r}\frac{\partial}{\partial r}\left(kr \frac{\partial \bar{T}}{\partial r} - \rho c_p r \overline{\tilde{T}\tilde{v}}\right) \quad (12)$$

由于乘积 ρu 是不随 x 变化的，对于管内定常充分发展的湍流，求解速度和温度分布的方程之间具有由温度和压力影响物性带来的弱耦合关系。

由于 u 和 ρ 均为 r 的函数，在恒定热流情形下对(12)式可以采取类似(3)式的方式处理，只有随 r 变化的部分具有湍流脉动量，即：

$$T_{\text{Turb}} = X(x) + \Theta_{\text{Turb}}(r) = X(x) + [\bar{\Theta}(r) + \tilde{\Theta}(r)] \quad (13)$$

$$\tilde{T}_{\text{Turb}} = \tilde{\Theta}(r) \quad (14)$$

下标 Turb 表示湍流。这里的处理方法与(3)和(4)式类同，对于层流和湍流情形，如果流量和热流相同则 X 和 $\mathrm{d}X/\mathrm{d}x$ 相同。如果流量和热流反向变为相反值，对于层流情形根据(1)、(2)和(4)式，u、X 和 Θ 的梯度就改变为相反值。对于湍流应该有相同结论。

这要求流量和热流的反向不影响 $\tilde{v}$ 的数值。因此，对于湍流情形根据(11)、(12)和(13)式，流量和热流的反向导致 u、X 和 Θ 的梯度变为相反值。根据(9)、(10)和(14)式，$\tilde{u}$ 和 $\tilde{\Theta}$ 也变成相反值。

由于层流和湍流均遵从 NS 方程，因此对于充分发展湍流，分别以瞬时温度和速度代替(5)式中的层流温度和速度，可以得到由于温度和速度梯度引起的熵产生的平均值为：

$$\overline{\sigma_{\text{Turb}}} = \overline{\sigma_{\text{Turb},T}} + \overline{\sigma_{\text{Turb},V}}$$
$$= \overline{\frac{k}{T^2}(\nabla T) \cdot (\nabla T)} + \overline{\frac{2\mu}{T}(\nabla \boldsymbol{v})^s : (\nabla \boldsymbol{v})^s} \quad (15)$$

式中：下标 Turb 表示湍流。下标 T 和 V 的意义与(7)和(8)式中的相同。忽略分子和分母统计相关性的影响时，(15)式第一项在柱坐标系中的表达式为：

$$\overline{\sigma_{\text{Turb},T}} = \overline{\frac{k}{T^2}(\nabla T) \cdot (\nabla T)} = \overline{\sigma_{\text{Turb},X}} + \overline{\sigma_{\text{Turb},\Theta}} \quad (16)$$

其中：

$$\overline{\sigma_{\text{Turb},X}} = \frac{k}{T^2}\left(\frac{\mathrm{d}X}{\mathrm{d}x}\right)^2 \quad (17)$$

$$\overline{\sigma_{\text{Turb},\Theta}} = \frac{k}{T^2}\left(\frac{\mathrm{d}\bar{\Theta}}{\mathrm{d}r}\right)^2 + \frac{k}{T^2}\overline{\left(\frac{\mathrm{d}\tilde{\Theta}}{\mathrm{d}r}\right)^2} \quad (18)$$

(15)式第二项在柱坐标系中表达式为：

$$\overline{\sigma_{\text{Turb},V}} = \overline{\frac{2\mu}{T}(\nabla \boldsymbol{v})^s : (\nabla \boldsymbol{v})^s} = \overline{\sigma_{\text{Turb},u}} + \overline{\sigma_{\text{Turb},v}} \quad (19)$$

其中：

$$\overline{\sigma_{\text{Turb},u}} = \frac{\mu}{T}\left[\left(\frac{\mathrm{d}\bar{u}}{\mathrm{d}r}\right)^2 + \overline{\left(\frac{\mathrm{d}\tilde{u}}{\mathrm{d}r}\right)^2}\right] \quad (20)$$

$$\overline{\sigma_{\text{Turb},v}} = \frac{4}{3}\frac{\mu}{T}\left[\left(\frac{\mathrm{d}\bar{v}}{\mathrm{d}r}\right)^2 + \overline{\left(\frac{\tilde{v}}{r}\right)^2} - \overline{\frac{\mathrm{d}\bar{v}}{\mathrm{d}r}\frac{\bar{v}}{r}}\right] \quad (21)$$

在(16)~(18)式中，下标 X 和 Θ 的意义与(7)和(8)式中的相同。在(19)~(21)式中，下标 u 和 v 表示对应速度分量引起的熵产生。

3 在定常转捩区间求解速度和温度的方程

比较层流和湍流的求解速度和温度的方程(1)、(2)、(9)和(10)式，如果认为对于层流流动，流向速度和温度的脉动量为 0，则 2 种流动遵从形式上完全相同的方程。

在定常自然转捩区间的速度和温度方程，形式上与(9)和(10)式相同。如果在转捩区间径向脉动速度 $\tilde{v}$ 的数值与完全湍流的一样，则每一点的自然转捩流动可以认为是充分发展的层流和湍流 2 种流动模式的合成。此时在每一点 u 和 Θ 的变化，一部分由层流贡献，一部分由湍流贡献，贡献比例分别为 $1-\eta$ 和 η，η 是流体微元的运动属于湍流模式的比例。对于 u 的变化，有：

$$(\mathrm{d}u)_{\mathrm{Tran}} = (1-\eta)\,(\mathrm{d}u)_{\mathrm{Lam}} + \eta\,(\mathrm{d}u)_{\mathrm{Turb}}$$
$$= (1-\eta)\,\mathrm{d}u + \eta\,\mathrm{d}(u+\tilde{u}) \tag{22}$$

其中，下标 Tran 表示转捩。对于 x 方向的动量有：

$$(1-\eta)\frac{\mathrm{d}\bar{p}}{\mathrm{d}x} + \eta\frac{\mathrm{d}\bar{p}}{\mathrm{d}x} = (1-\eta)\frac{1}{r}\frac{\mathrm{d}}{\mathrm{d}r}\left(\mu r\frac{\mathrm{d}u}{\mathrm{d}r}\right)$$
$$+ \eta\frac{1}{r}\frac{\mathrm{d}}{\mathrm{d}r}\left(\mu r\frac{\mathrm{d}u}{\mathrm{d}r} - \rho r\overline{\tilde{u}\tilde{v}}\right) \tag{23a}$$

$$\frac{\mathrm{d}\bar{p}}{\mathrm{d}x} - \frac{1}{r}\frac{\mathrm{d}}{\mathrm{d}r}\left(\mu r\frac{\mathrm{d}u}{\mathrm{d}r}\right) = \frac{\mathrm{d}\bar{p}}{\mathrm{d}x} - \frac{1}{r}\frac{\mathrm{d}}{\mathrm{d}r}\left(\mu r\frac{\mathrm{d}u}{\mathrm{d}r} - \rho r\overline{\tilde{u}\tilde{v}}\right) = 0 \tag{23b}$$

（23b）式是层流与湍流的动量方程的结果。（23）式中，变量上方无"—"的表示层流量，变量上方有"—"的表示湍流量。

在一个截面位置且在一个时刻仅允许一个压力梯度，因此在一个截面位置 η 数值相同，u 和 Θ 以及它们对 r 的导数包含层流和湍流 2 种成分。湍流具有 r 方向的脉动速度 $\tilde{v}$。（23）式的成立，要求在转捩区间 $\tilde{v}$ 具有和全湍流一样的数值。因此在转捩区间速度矢量增量是 $\mathrm{d}v = (\mathrm{d}v_r, \mathrm{d}v_\theta, \mathrm{d}v_x) = (\mathrm{d}\tilde{v}, 0, (\mathrm{d}u)_{\mathrm{Tran}})$。把 $\tilde{v}$ 和该速度矢量增量代入完整的动量方程（9）式可以导出（23a）式（将利用（26）式进行进一步说明）。对于传热（温度）来说转捩长度约为几百倍的 d_i，对于流动（速度）来说转捩长度大约为一百倍的 d_i。在（23）式中，在每一截面位置求解时忽略 η 沿着 x 方向的变化。

对于在转捩区间流体的温度，可以写出类似（22）和（23）式的方程。与纯粹的层流或湍流情形一样，由于乘积 ρu 和 $\rho\bar{u}$ 不随 x 变化，在转捩区间温度和速度方程之间也具有由温度和压力影响物性带来的弱耦合关系。层流、转捩或湍流分别具有确定的 u 和 Θ 或者 $\bar{u}$ 和 Θ 以及 $\tilde{u}$ 和 $\tilde{\Theta}$ 等变量的剖面类型，它们影响 μ、k 和 c_p 等物性的分布剖面。由于是雷诺数 Re 影响着变量的剖面类型，物性参数本身不改变变量的剖面类型或者流动/传热模式。温度和速度的弱耦合方程具有这种性质，即速度（和其脉动）的剖面类型不影响温度（和其脉动）的剖面类型，反过来也一样。由于温度和速度方程之间的这种弱耦合关系，并且根据（5）和（15）式，在熵产生表达式中温度梯度项和速度

梯度项是独立的，因此对于温度可以采取另外一个不同的 η。为了简明起见，下面对于温度仍然采用相同的 η。Θ 的变化为：

$$(\mathrm{d}\Theta)_{\mathrm{Tran}} = (1-\eta)\,(\mathrm{d}\Theta)_{\mathrm{Lam}} + \eta\,(\mathrm{d}\Theta)_{\mathrm{Turb}}$$
$$= (1-\eta)\,\mathrm{d}\Theta + \eta\mathrm{d}(\Theta+\tilde{\Theta}) \tag{24}$$

恒定热流情形下的温度方程为：

$$\frac{\mathrm{d}X}{\mathrm{d}x} = (1-\eta)\frac{1}{\rho c_p u}\frac{1}{r}\frac{\mathrm{d}}{\mathrm{d}r}\left(kr\frac{\mathrm{d}\Theta}{\mathrm{d}r}\right)$$
$$+ \eta\frac{1}{\rho c_p u}\frac{1}{r}\frac{\mathrm{d}}{\mathrm{d}r}\left(kr\frac{\mathrm{d}\Theta}{\mathrm{d}r} - \rho c_p r\overline{\tilde{\Theta}\tilde{v}}\right) \tag{25a}$$

$$\frac{\mathrm{d}\Theta}{\mathrm{d}r} = \frac{\mathrm{d}\Theta}{\mathrm{d}r} - \frac{\rho c_p}{k}\overline{\tilde{\Theta}\tilde{v}} \tag{25b}$$

（25b）式是层流与湍流的壁面热流相同的结果。与（23a）式一样，在（25a）式中对每一截面位置处求解时忽略 η 沿着 x 方向的变化。

在转捩区间动量方程形式上与（9）式相同，由于在一个截面位置 η 数值相同，通过（22）式沿着 r 方向积分，可以得到流向合成速度 u_{Tran} 和 $\tilde{u}_{\mathrm{Tran}}$。其中，流向的合成平均速度 $u_{\mathrm{Tran}} = (1-\eta)u + \eta\bar{u}$，流向的合成脉动速度 $\tilde{u}_{\mathrm{Tran}} = (1-\eta)0 + \eta\tilde{u} = \eta\tilde{u}$。把它们代入（9）式，在径向脉动速度 $\tilde{v}$ 具有和全湍流一样数值的假设下，得到：

$$(1-\eta)\frac{\mathrm{d}\bar{p}}{\mathrm{d}x} + \eta\frac{\mathrm{d}\bar{p}}{\mathrm{d}x} = \frac{1}{r}\frac{\mathrm{d}}{\mathrm{d}r}\left\{\mu r\left[(1-\eta)\frac{\mathrm{d}u}{\mathrm{d}r} + \eta\frac{\mathrm{d}u}{\mathrm{d}r}\right]\right\}$$
$$- \frac{1}{r}\frac{\mathrm{d}}{\mathrm{d}r}\left\{\rho r\,\overline{[(1-\eta)0+\eta\tilde{u}]\,[(1-\eta)\tilde{v}+\eta\tilde{v}]}\right\} \tag{26a}$$

$$(1-\eta)\frac{\mathrm{d}\bar{p}}{\mathrm{d}x} + \eta\frac{\mathrm{d}\bar{p}}{\mathrm{d}x} = (1-\eta)\frac{1}{r}\frac{\mathrm{d}}{\mathrm{d}r}\left(\mu r\frac{\mathrm{d}u}{\mathrm{d}r}\right)$$
$$+ \eta\frac{1}{r}\left[\frac{\mathrm{d}}{\mathrm{d}r}\left(\mu r\frac{\mathrm{d}u}{\mathrm{d}r}\right) - \frac{\mathrm{d}}{\mathrm{d}r}(\rho r\overline{\tilde{u}\tilde{v}})\right] \tag{26b}$$

其中，层流的流向脉动速度为 0，这里对 η 不考虑振荡（下节将考虑 η 的振荡）。由于在一个截面位置 η 数值相同且无振荡，很容易得出（26）和（23a）式是等价的。由（26）式可以看出，在转捩区间流动是层流和湍流的合成运动，因为 $(1-\eta)u + \eta\bar{u}$ 是流向的合成平均速度，$(1-\eta)0 + \eta\tilde{u} = \eta\tilde{u}$ 是流向的合成脉动速度，$(1-\eta)\tilde{v} + \eta\tilde{v} = \tilde{v}$ 是径向的合成脉动速度。用 Θ 替代 u，$\tilde{\Theta}$ 替代 $\tilde{u}$，Θ 替代 $\bar{u}$，温度方程（25a）式也可以改写成类似（26）式的形式。这种合成运动对于恒定壁温的情形也成立。

在转捩区间 $\tilde{v}$ 具有和全湍流一样数值的假设，可以进行进一步解释。众所周知，即使 Re 很高时也需

要有限振幅的扰动才能引发转捩,因为圆管内的层流是线性稳定的。在所关心的圆管内能够发生完整的转捩过程,其原因是 Re 的增加和管内有限振幅的扰动。对于管内确定的流动,这个假设相当于说,无论流动状态是层流、转捩还是湍流,$\bar{v}$ 的数值均相同,而且这个 $\bar{v}$ 的数值是由存在于层流之中的扰动带来的。在管内转捩过程中 $\bar{v}$ 的数值相同不违反质量平衡方程,因为对应于一个有限振幅的扰动,在另外 2 个方向还有 2 个脉动速度分量。有限振幅的扰动决定层流区内三维脉动速度的数值,脉动速度 $\bar{v}$ 的大小在流动模式的转捩过程中保持不变。在上述推导中忽略了这个三维脉动速度对层流流动的影响,但是转捩是否起始和结束取决于该扰动。在强迫转捩中往往利用不同振幅的扰动,造成很不相同的转捩的起始和终止 Re 数值。

尽管物性参数本身不改变变量的剖面类型或者流动/传热模式,对于管内有加热的流动,在一个站位上不同的流动模式有不同的温度分布剖面,导致 c_p、μ 和 k 等物性参数的分布剖面不同。不难理解,不同的 c_p、μ 和 k 分布剖面也会导致不同的速度和温度分布剖面。因此,在转捩区间的流动是层流和湍流 2 种流动模式的合成,层流部分 u 和 Θ 的数值与纯粹层流对应的数值不同。如果忽略这种由不同流动模式引起的物性参数差异(即不同流动模式下物性参数的分布剖面相同),那么在转捩区间,层流部分 u 和 Θ 的数值与纯粹层流对应的数值相同。这种分析对于转捩区间湍流部分 $\bar{u}$ 和 $\bar{\Theta}$ 的数值也一样成立。

显然,(23)、(25)和(26)式对于任意的 η 均成立,因此在转捩区间 $\bar{v}$ 具有和全湍流一样数值的假设,使得这种流动合成在力学上是可能的。由(26)式,利用某种假设可以推出,即使在层流区存在非 0 的流向脉动速度,这种流动合成也是可以成立的。这里讨论的是流体 2 种运动的合成,即使对 $\eta<0$ 或 $\eta>1$,这种合成也是允许的。η 或者 $1-\eta$ 是负的意味着反方向的流量、压力梯度和热流的贡献,对应着相反数值的 x 方向速度梯度和温度(X 和 Θ)梯度以及相反数值的 $\bar{u}$ 和 $\bar{\Theta}$。

在本节里,假定在转捩流动中径向脉动速度 $\bar{v}$ 与湍流流动一样,将转捩流动分解成层流和湍流成分。$\eta=0$ 时流动是全层流的,$\eta=1$ 时流动是全湍流的。这里多了一个变量,通过(23)和(25)式无法直接确定自然转捩过程。

4 圆管内自然转捩流动中的振荡

要确定自然转捩如何发展演化,必须给出 η 的动力学方程。记转捩起始雷诺数为 Re_L,记转捩终止雷诺数为 Re_R。对流传热实验发现,对于充分发展的圆管流动,转捩在 $Re_L \approx 2300$ 时开始,在 $Re_R \approx 10\,000$ 时结束[8-9]。关于流速振荡的测量结果说明,流动的转捩区要小得多[5]。转捩开始和终止的精确位置,与实际的流动条件相关。这里仅对圆管内的转捩流动在起始和终止雷诺数之间的发展演化以及这种发展演化对流动和对流传热特性的影响感兴趣。

在写出转捩区的熵产生表达式前,引入 η 的振荡 $\tilde{\eta}$。认为在自然转捩区间 η 由 2 部分组成:统计平均值 $\bar{\eta}$ 和振荡值 $\tilde{\eta}$,即:

$$\eta = \bar{\eta} + \tilde{\eta} \tag{27}$$

代入(22)和(24)式得到:

$$(du)_{Tran} = (1-\eta)\,du + \eta d\bar{u} \\ + \tilde{\eta} d\tilde{u} + \bar{\eta} d(\bar{u} - u + \tilde{u}) \tag{28}$$

$$(d\Theta)_{Tran} = (1-\eta)\,d\Theta + \eta d\bar{\Theta} \\ + \tilde{\eta} d\tilde{\Theta} + \bar{\eta} d(\bar{\Theta} - \Theta + \tilde{\Theta}) \tag{29}$$

式中:$\eta = \bar{\eta}(Re)$,是取值 0 和 1 之间的单调递增函数。引入 $\bar{\eta}$ 和 $\tilde{\eta}$ 可以描述转捩过程中某一时刻流动瞬时状态对应的合成比例。文献[1]和[5]报道了在转捩过程中,瞬时流态包含典型的与常规湍流不同的扰动运动(Disordered motions),其他文献亦有报道。常常采用的间歇因子 γ 不能给出某一时刻流动的瞬时状态,也不能区分这种扰动运动接近湍流的程度。因此引入 $\bar{\eta}$ 和 $\tilde{\eta}$ 可以更好地描述转捩流动。(28)和(29)式说明,在转捩区间 u 和 Θ 变化的脉动量由两部分组成:湍流脉动量和与 η 的振荡对应的量。$\tilde{u}$、$\tilde{v}$ 和 $\tilde{\Theta}$ 之间具有统计相关性。这里认为湍流脉动量和 η 的振荡之间是统计无关的,即存在如下关系:

$$\overline{(\tilde{\eta})^n (\tilde{\phi})^m} = \overline{(\tilde{\eta})^n}\,\overline{(\tilde{\phi})^m}, \\ \overline{(\tilde{\eta})^n \left(\frac{d\tilde{\phi}}{dr}\right)^m} = \overline{(\tilde{\eta})^n}\,\overline{\left(\frac{d\tilde{\phi}}{dr}\right)^m} \tag{30}$$

式中:n 和 m 均为正整数,$\tilde{\phi}$ 表示 $\tilde{u}$、$\tilde{v}$ 或 $\tilde{\Theta}$。(30)式的成立,其原因在于脉动或者振荡的不同产生机制。湍流脉动量统计相关性的机制在于湍流这种流动模式,而 $\tilde{\eta}$ 的统计特性源于转捩区间内流动模式的振荡机制。

在一个截面位置且在一个时刻仅允许一个压力梯度,因此在一个截面位置和一个时刻 η 和 $\bar{\eta}$ 各有相同的数值。可以证明 η 和 $\bar{\eta}$ 的引入不影响(26b)式的成立。通过积分(28)式可以得到在径向任意一点处的流向合成速度 u_{Tran} 和 $\bar{u}_{\text{Tran}}$，其中流向的合成平均速度 $u_{\text{Tran}} = (1-\eta)u + \eta u$，流向的合成脉动速度 $\bar{u}_{\text{Tran}} = \eta\bar{u} + \bar{\eta}(u - \bar{u} + \bar{u})$。径向的合成脉动速度永远是 $\bar{v}$。把它们和 η 与 $\bar{\eta}$ 代入(9)式，并利用(30)式，可以得到与(26b)相同的方程，只是 $\bar{\eta}$ 代替了 η。因此，η 和 $\bar{\eta}$ 的引入不影响流动合成的成立。

在转捩区间流动遵从全 NS 方程，因此其熵产生的方程与(15)式相同。(28)和(29)式代入(15)式，应用(30)式整理后可得：

$$\overline{\sigma_{\text{Tran}}} = \overline{\sigma_{\text{Turb}.X}} + \overline{\sigma_{\text{Turb}.v}}$$
$$+ (1-\eta)(\sigma_{\text{Lam}.\Theta} + \sigma_{\text{Lam}.V}) + \eta(\overline{\sigma_{\text{Turb}.\Theta}} + \overline{\sigma_{\text{Turb}.u}})$$
$$+ [\bar{\bar{\eta}} - (1-\eta)\eta]\left\{\frac{k}{T_{\text{Tran}}^2}\left[\left(\frac{d\Theta}{dr} - \frac{d\Theta}{dr}\right)^2 + \overline{\left(\frac{d\Theta}{dr}\right)^2}\right]\right.$$
$$\left. + \frac{\mu}{T_{\text{Tran}}}\left[\left(\frac{du}{dr} - \frac{du}{dr}\right)^2 + \overline{\left(\frac{d\bar{u}}{dr}\right)^2}\right]\right\}$$

$$(31)$$

式中：$\sigma_{\text{Lam}.\Theta}$ 与(7)式中相同，$\sigma_{\text{Lam}.V}$ 为(8)式，$\overline{\sigma_{\text{Turb}.X}}$ 为(17)式，$\overline{\sigma_{\text{Turb}.\Theta}}$ 为(18)式，$\overline{\sigma_{\text{Turb}.u}}$ 为(20)式，$\overline{\sigma_{\text{Turb}.v}}$ 为(21)式，只是分母中的 T 和 T^2 要用 $\overline{T_{\text{Tran}}}$ 和 $\overline{T_{\text{Tran}}^2}$ 代替。在分析熵产生时，忽略了分子和分母统计相关性的影响。η 和 $\bar{\bar{\eta}}$ 仅是 Re 的函数，与 r 无关。

(31)式右边除 $\overline{\sigma_{\text{Turb}.X}}$ 和 $\overline{\sigma_{\text{Turb}.v}}$ 外的其他项，令人联想到两元混合物的自由能公式。对于两元混合物，在利用最小自由能准则分析相变特性时的处理方法，在文献[6]和[12]中有详细的论述。仿照 Brian Cowan 在文献[12]中对处于平衡态的二元混合物的混合自由能定义，去掉(31)式中前四项，对于管内非平衡转捩流动系统(一个横截面上的流体整体)定义流体微元的合成熵产生为

$$\overline{\sigma_{\text{Tran}.c}} = [\bar{\bar{\eta}} - (1-\eta)\eta]$$
$$\left\{\frac{k}{T_{\text{Tran}}^2}\left[\left(\frac{d\Theta}{dr} - \frac{d\Theta}{dr}\right)^2 + \overline{\left(\frac{d\Theta}{dr}\right)^2}\right]\right. \qquad (32)$$
$$\left. + \frac{\mu}{T_{\text{Tran}}}\left[\left(\frac{du}{dr} - \frac{du}{dr}\right)^2 + \overline{\left(\frac{d\bar{u}}{dr}\right)^2}\right]\right\}$$

式中：下标 c 表示合成。根据 Prigogine 最早提出的最小熵产生准则，η 的取值应使 $\overline{\sigma_{\text{Tran}.c}}$ 取极小值。这要求(32)式对 η 求导为 0。当 $Re = Re_L$ 时，$\eta = 0$ 和 $\bar{\eta} = 0$；当 $Re = Re_R$ 时，$\eta = 1$ 和 $\bar{\eta} = 0$。因此，(32)式右边的 η 项，即右边第一个方括号内几项之和，在 $Re = Re_L$ 和 $Re = Re_R$ 时为 0。(32)式右边其它项，即大括号内各项之和随 r 变化而变化，但是对于每一截

面位置 η 和 $\bar{\bar{\eta}}$ 均相同，η 的取值与 r 无关。令(32)式右边的 η 项对 η 的导数为 0，即

$$\frac{d(\bar{\eta}^2)}{d\eta} - 1 + 2\bar{\eta} = 0 \qquad (33)$$

该式的成立使得(32)式右边 η 项在 $Re_L \leqslant Re \leqslant Re_R$ 时恒为 0，此时 $\overline{\sigma_{\text{Tran}.c}}$ 对 η 的导数为 0。

这种 η 的取值，对应于在转捩区间维持流体微元运动需要的最小熵产生。η 与 r 无关，仅是 Re 的函数，因此(33)式决定了圆管内层流向湍流自然转捩的发展演化。

对于恒定壁温的情形，也可以推导出(33)式。在温度方程中忽略的摩擦项和轴向传热项，无论在熵产生方程中是否保留，均不影响(33)式的成立。

5　一个振荡函数和与实验现象的对比

层流向湍流的过渡和平衡系统的连续相变(Landau 所谓的第二类相变[13])，二者宏观特性非常相似。它们均具有一个振荡剧烈的临界区间。在此区间前后，系统的行为显著不同。在此区间内系统的微观特性对其行为的影响不大，不同系统的表现大致相似。在层流向湍流的过渡这个过程中，系统的行为特性发生了显著转变，正如文献[6]中所述，这种流动转捩属于一种非平衡系统的相变。相变可以采用序参数来描述。通常认为，很多关于平衡相变的认识可以推广到非平衡情形[14]。如果用序参数来描述圆管内层流向湍流转捩这一非平衡流动系统的相变，则这里自然转捩过程的序参数应当是 η 或其线性函数。

对于热力学平衡系统中发生的连续相变过程，文献[13]讨论了序参数的振荡。靠近相变点时存在一个很窄的温度区间，其间热力学函数的物理本质主要来源于序参数振荡的反常增加。这个区间被称为振荡区间，序参数的振荡占据统治地位。

由于不考虑序参数的振荡，Ginzburg-Landau 相变理论不适用于振荡区间[12,13]。在此区间，热力学势不能仅仅展开成序参数(和其空间导数)其他热力学变量的函数。在(31)式中引入了 η 的振荡。推导(33)式的过程，在对待序参数的处理方式上与 Ginzburg-Landau 理论的精神一致，也就是说，在通过对热力学势求极值来确定序参数时把它看成是与热力学变量地位相同的独立变量[13]。

(33)式是采用最小熵产生准则得到的结果，其中不包含任何细致的层流或时间平均湍流的剖面信息。这与相变的临界现象一致。靠近临界点时，系统的微观性质不影响系统的热力学行为，完全不同的系统可

以有许多相似的性质[12]。(33)式表明,在层流向湍流的转捩过程中,转捩的发展演化,与层流剖面和时间平均的湍流剖面均无直接关系。

对于所考察的管内流动,η 仅随雷诺数变化。$\overline{\tilde{\eta}^2}$ 在 $\bar{\eta}=0$ 或 $\bar{\eta}=1$ 时均为 0。这里假定(33)式中的 $\overline{\tilde{\eta}^2}$ 能够展开成 $\bar{\eta}$ 随雷诺数导数的级数形式。如果认为 η 的振荡函数为:

$$\tilde{\eta}=C\frac{\mathrm{d}\bar{\eta}}{\mathrm{d}Re}\tilde{g}(t),\ \overline{[\tilde{g}(t)]^2}=1 \tag{34}$$

此时,$\overline{\tilde{\eta}^2}$ 中仅包含最低阶导数的平方项,且系数为正。(34)式代入(33)式得到:

$$2C^2\frac{\mathrm{d}^2\bar{\eta}}{\mathrm{d}Re^2}-1+2\bar{\eta}=0 \tag{35}$$

转捩开始于 Re_L 并且结束于 Re_R,因此对 $\bar{\eta}$ 的约束条件为:

$$\bar{\eta}(Re_L)=0,\bar{\eta}(Re_R)=1 \tag{36}$$

由于 $\bar{\eta}$ 是 Re 的单调递增函数,在完全层流区和完全湍流区 η 的振荡为 0,因此对 $\bar{\eta}$ 导数的约束为:

$$\frac{\mathrm{d}\bar{\eta}}{\mathrm{d}Re}\geq 0,\frac{\mathrm{d}\bar{\eta}}{\mathrm{d}Re}\Big|_{Re_L}=\frac{\mathrm{d}\bar{\eta}}{\mathrm{d}Re}\Big|_{Re_R}=0 \tag{37}$$

满足这些条件的解为:

$$\bar{\eta}=\frac{1+\sin\theta}{2},\theta=\frac{2Re-Re_R-Re_L}{Re_R-Re_L}\frac{\pi}{2} \tag{38}$$

并且,$C=(Re_R-Re_L)/\pi$。(38)式代入(34)式得到 $\tilde{\eta}=\frac{\cos\theta}{2}\tilde{g}(t)$。

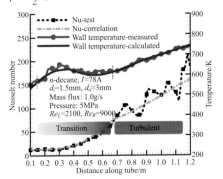

图 3 电加热管实验中的外壁温和 Nu 数对比
Fig. 3 Outer wall temperature and *Nu* comparisons in an e-lectrically heated tube test

这与文献[15]由传热实验得到的结果一致。文献[15]指出,圆管内定常自然转捩区间的 Nu 数,形式上为相同雷诺数下层流传热 Nu 数和湍流传热 Nu 数的权重叠加,权重因子与(38)式一样。图 3 取自文献[15]中图 12(c)的数据,其中转捩区间的 Nu 关系式采用了这种权重叠加规律($Nu_{\mathrm{Tran}}=(1-\bar{\eta})Nu_{\mathrm{Lam}}$

$+\bar{\eta}Nu_{\mathrm{Turb}}$),并由此计算了圆管外壁温度分布。显然,相同的权重叠加规律适用于转捩区间的摩擦因子。

$\bar{\eta}$、$1-\bar{\eta}$ 和 $\sqrt{\overline{\tilde{\eta}^2}}$ 作为 θ 的函数在图 4 中显示。可以看出,在转捩区间 η 的振荡是非常大的。$\tilde{\eta}$ 的方均根值在 $\theta<0$ 侧超过了 $\bar{\eta}$,在 $\theta>0$ 侧超过了 $1-\bar{\eta}$,造成在某些时刻的流体微元,$\bar{\eta}$ 至少在 $\theta<0$ 一侧、$1-\bar{\eta}$ 至少在 $\theta>0$ 一侧为负值。由于讨论的是流体流动的合成,不是 2 种物质的混合,因此负值是允许的。

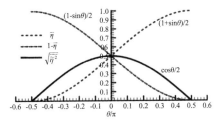

图 4 $\bar{\eta}$、$1-\bar{\eta}$ 和 $\sqrt{\overline{\tilde{\eta}^2}}$ 作为 θ 的函数
Fig. 4 $\bar{\eta}$、$1-\bar{\eta}$ and $\sqrt{\overline{\tilde{\eta}^2}}$ as functions of θ

这里给出的转捩区间内大的振荡,能够解释文献[7]中提到的"奇怪效应"(Strange effect)。在管道流动实验中,流速达到某个数值时,雷诺数也达到一个临界数值,压力计的读数开始剧烈震荡。这种现象在一个速度范围内一直持续,直到流速达到某个数值压力计的读数重新回到平稳状态,随后一直保持平稳。

由于在一个截面上不同点的 $\bar{\eta}$ 相同,在中心线上的轴向速度为同一 Re 下层流和湍流对应值的权重叠加。可以通过沿一条半径积分(28)式来得到 $\bar{u}_{\mathrm{Tran}}$ 和 $\tilde{u}_{\mathrm{Tran}}$,即:

$$\bar{u}_{\mathrm{Tran}}=(1-\bar{\eta})u+\bar{\eta}\tilde{u} \tag{39a}$$

$$\tilde{u}_{\mathrm{Tran}}=\bar{\eta}\tilde{u}+\tilde{\eta}(\tilde{u}-u+\tilde{u}) \tag{39b}$$

在流动实验中可以沿着中心线对它们进行测量,并与上式的预测值进行对比。

文献[5]、[16]和[17]均利用常温水为介质,在圆管中进行了大量强迫转捩和少量自然转捩实验。根据 $Re=\rho Ud_i/\mu$,改变水的流量可以进行不同雷诺数的实验,从而测量在转捩起始和终止 Re 之间轴向速度及其脉动量的统计特性的变化。不难理解,可以用本文的推导来分析这种无传热发生时的强迫或自然转捩过程。

在同一 Re 下,与湍流脉动值 $\tilde{u}$ 相比,层流和湍流平均的速度对应值之差 $u-\bar{u}$ 在数值上是很大的。由(39b)式可以看出,在层流向湍流转捩的过程中,在中心线上轴向速度脉动的方均根数值应该有过冲

(Overshoot)现象。Durst 和 Ünsal 清楚地测量了这种现象[5]。为了与测量值进行对比,可以计算出中心线上轴向速度的脉动强度($I = u'/\bar{u}_{Tran}$)。不考虑完全湍流时 $u'_{r=0}/u_{r=0}$ 随 Re 的变化,采用文献[5]图 3 中的数据,即对于湍流 $\bar{u}_{r=0} = 1.4U_{mean}$ 和 $u'_{r=0} = 0.035u_{r=0}$,对于层流 $u_{r=0} = 2.0U_{mean}$(在层流区轴线速度脉动强度约为 0.002)。如图 5 所示,(39)式给出在 $\theta/\pi = 0.07$ 时中心线上的轴向速度脉动强度达到峰值 20.7%,而在 Re_L 大于 4000 时文献[5]的测量值约为 20%。根据(39b)式,在图 5 中 u' 的数值由下式计算得到:

$$u' = \sqrt{\overline{u_{Tran}^2}} = \sqrt{\bar{\eta}^2 \, \overline{u^2} + \overline{\bar{\eta}^2 [(u-u)^2 + \overline{u^2}]} + \overline{u_{Lam}^2}}$$

(40)

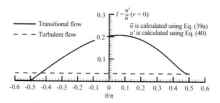

图 5 中心线上轴向速度脉动强度的大的过冲
Fig. 5 The large overshoot of I at the center line in transition region

图 6 是图 5 的不同形式。在图 6 中,y 轴变为对数坐标且 x 轴进行了平移和收缩处理,使得容易与文献[5]的测量结果进行对比。图 6 中曲线的形状和数值,均与文献[5]中图 3、5 和 10 在 Re_L 大于 4000 时的测量结果相接近(实验数据难以取出与计算值进行比较)。轴向速度脉动过冲的测量,在文献[16]和[17]中也有详细报道。文献[5]还报道,这种轴向速度脉动的过冲现象,没有导致摩阻系数(应是时间平均值)的过冲。本文推导也能够解释这种现象。

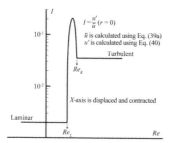

图 6 中心线上轴向速度脉动强度的大的过冲,可以与文献[5]中的测量结果进行对比
Fig. 6 The large overshoot of I at the center line in transition region(both the shape and values can be compared with the measurements in reference[5])

文献[17]在第 50 页给出了在强迫转捩过程中圆管中心线上轴向速度及其脉动量统计平均的测量值,图 7 和 8 是本文预测值与它们的比较。计算中未考虑完全湍流时 $u'_{r=0}/u_{r=0}$ 随 Re 的变化。从图 7 可以看出,对于在转捩过程中圆管中心线上轴向速度的统计平均值,(39a)式能够得到与实验测量一致的预测结果。从图 8 可以看出,对于在转捩过程中圆管中心线上轴向脉动速度的统计平均值,(40)式在 $Re_L \approx 2120$ 时仅能够给出与实验测量的趋势大致一致的预测结果。本文预测的脉动速度统计平均值的峰值较高,并且在转捩快要开始前或刚刚结束后的位置,本文的计算与测量的数值甚至趋势不同。从文献[5]的图 10 可以看出,中心线上轴向速度脉动强度的统计平均值(见图 6)在 Re_L 小于 3000 时,其过冲幅度的测量值小于在 Re_L 大于 4000 时的数值,其原因有待于进一步研究。在 Re_L 大于 4000 时,文献[5]、[16]和[17]给出的所有测量结果(包括中心线上轴向速度及其脉动统计平均值),均不适于取出数据与计算进行比较。在转捩开始前或结束后的较远位置,图 8 中的计算与文献[5]、[16]和[17]测量的数值和趋势一致。

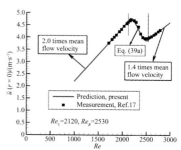

图 7 中心线上轴向速度统计平均值的计算与测量对比
Fig. 7 The calculated and measured mean values of longitudinal velocity at the center line

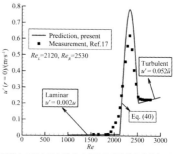

图 8 圆管中心线上轴向脉动速度统计平均值的计算与测量对比
Fig. 8 The calculated and measured mean values of the fluctuations of longitudinal velocity at the center line

(33)式的成立使得(32)式为 0,此时(31)式为:

$$\overline{\sigma_{Tran}} = \overline{\sigma_{Turb,X}} + \overline{\sigma_{Turb,v}}$$
$$+ (1 - \overline{\eta})(\sigma_{Lam,\Theta} + \sigma_{Lam,V}) + \overline{\eta}(\overline{\sigma_{Turb,\Theta}} + \overline{\sigma_{Turb,u}})$$

(41)

对比(31)与(41)式(二者的差别是(32)式)可以看出,自然转捩的演化以这种方式进行,单纯合成引起的负的熵产生与 η 的振荡引起的正的熵产生,二者相互抵消。这使得在自然转捩过程中 η 能够连续变化,并且熵产生在形式上为相同雷诺数下层流和湍流熵产生的权重叠加。

在转捩区间合成流动由 η 描述。(34)式确定的是在转捩区间 η 的振荡,由(34)式得到的(38)式确定的是具体的转捩行为。不同的(34)式可以得到不同的(38)式。从上述与实验现象的对比可以看出,转捩过程中流体的振荡统计特性和流动及传热行为,可以用运动合成、合成比例的振荡和最小熵产生准则进行解释,并且 η 的振荡控制着层流向湍流过渡的自然转捩行为。

最后,如果对于温度采用不同的 η,可以得到相应的熵产生方程,其形式与(31)式相似。由于温度和速度方程之间仅仅具有弱耦合关系,并且在熵产生表达式中温度梯度的平方项和速度梯度的平方项是独立的,因此这种处理对上述讨论不会带来大的改变。对于速度和温度来说,不同的 η 允许相似和独立的转捩过程。

6 结　论

流体在电加热圆管内自然转捩过程中的振荡统计特性和流动及传热行为,可以用层流和湍流 2 种流动模式的运动合成、合成比例的振荡和最小熵产生准则进行解释。

(1) 在自然转捩区间流动是充分发展的层流与湍流 2 种模式的流动按比例的合成。合成流动由 2 种流动模式的合成比例描述。

(2) 在层流向湍流过渡的转捩过程中,合成比例会发生振荡,并控制层流向湍流过渡的自然转捩行为。

(3) 在电加热圆管内的流体,其速度和温度可以有相似和独立的转捩过程。

(4) 圆管内层流向湍流的转捩过程,可以与热力学平衡系统中发生的连续相变过程进行比较。描述后者的序参数和振荡区间等概念,可以在流动转捩研究中采用。

参考文献:

[1] Mullin T. Experimental studies of transition to turbulence in a pipe[J]. Annu Rev Fluid Mech, 2011, 43: 1-24.

[2] Linne D L, Meyer M L, Edwards T, et al. Evaluation of heat transfer and thermal stability of supercritical JP-7 fuel[R]. AIAA-97-3041, 1997.

[3] Huang H, Sobel D R, Spadaccini L J. Endothermic heat-sink of hydrocarbon fuels for scramjet cooling[R]. AIAA-2002-3871, 2002.

[4] Eckert E R G, Drake R M JR. Analysis of heat and mass transfer[M]. McGraw-Hill Kogakusha, Ltd, 1972.

[5] Durst F, Ünsal B. Forced laminar-to-turbulent transition of pipe flows[J]. J Fluid Mech, 2006, 560: 449-464.

[6] Reichl L E. A modern course in statistical physics[M]. A Wiley-Interscience Publication, John Wiley & Sons Inc, 1998.

[7] McComb W D. The physics of fluid turbulence[M]. Claredon Press, 1992.

[8] Bergman T L, Lavine A S, Incropera F P, et al. Fundamentals of heat and mass transfer[M]. 7th ed. John Wiley & Sons Inc, 2011.

[9] Rohsenow W M, Hartnett J P, Cho Y I. Handbook of heat transfer[M]. 3rd ed. McGraw-Hill Book Company, 1998.

[10] Landau L D, Lifshitz E M. Fluid mechanics[M]. 2nd ed. Pergamon Press, 1987.

[11] Lifshitz E M, Pitaevskii L P. Statistical physics[M]. Part 2, 2nd ed. Pergamon Press, 1980.

[12] Cowan B. Topics in statistical mechanics[M]. Imperial College Press, 2005.

[13] Landau L D, Lifshitz E M. Statistical physics[M]. Part 1, 3rd ed. Pergamon Press, 1980.

[14] Henkel M, Hinrichsen H, Lübeck S. Non-equilibrium phase transitions, Volume I: Absorbing phase transitions[M]. Canopus Academic Publishing Limited, 2008.

[15] Zhang L, Zhang R L, Xiao S D, et al. Experimental investigation on heat transfer correlations of n-decane under supercritical pressure[J]. International Journal of Heat and Mass Transfer, 2013, 64: 393-400.

[16] Nishi M, Ünsal B, Durst F, et al. Laminar-to-turbulent transition of pipe flows through puffs and slugs[J]. J Fluid Mech, 2008, 614: 425-446.

[17] Nishi M. Laminar to turbulent transition in pipe flow through puffs and slugs[D]. Der Technischen Fakultät der Friedrich-Alexander-Universität Erlangen-Nürnberg, 2009.

Experimental and simulation study of aeroengine combustor based on CARS technology and UFPV approach

Xiong Moyou[1,*]，Le Jialing[2]，Huang Yuan[2]，Song Wenyan[1]，
Yang Shunhua[2]，Zheng Zhonghua[2]

(1. School of Power and Energy，Northwestern Polytechnical University，Xi'an　710072，China；2. Science and Technology on Scramjet Laboratory of Hypervelocity Aerodynamics Institute，China Aerodynamics Research and Development Center，Mianyang Sichuan　621000，China)

Abstract：Based on the Unsteady Reynolds Averaged Navier Stokes(URANS) method，a three-dimensional two-phase turbulent combustion numerical software for aeroengine combustor has been developed. The physical and chemical processes taking place in the liquid fuel are simulated completely，including liquid film formation，breakup，evaporation and combustion. LISA and KH-RT are used as the primary and second atomization model respectively，and also the standard evaporation model is used to simulate the evaporation process. Besides，detailed chemical mechanism of kerosene is used for reaction kinetics，and the Unsteady Flamelet/Progress Variable (UFPV) approach in which the unstable combustion characteristics of the flame could be simulated is used as the combustion model. The temperature and species of the flow field and the diameter of fuel droplets in the aeroengine combustor are obtained. At the same time，the Coherent Anti-stokes Raman Scattering (CARS) technology is used to measure the temperature in the primary zone of the aeroengine combustor. Then the temperature of the simulation is compared with that measured by CARS technology，and the calculation error of numerical results is less than 7.3%. The studies have shown that the numerical method in this paper and UFPV approach can simulate the two-phase turbulent combustion process appropriately in the aeroengine combustor.

Keywords：aeroengine combustor；two-phase combustion；UFPV approach；CARS technology

采用 CARS 试验技术与 UFPV 数值方法研究航空发动机燃烧室

熊模友[1,*]，乐嘉陵[2]，黄　渊[2]，宋文艳[1]，杨顺华[2]，郑忠华[2]

(1. 西北工业大学 动力与能源学院，西安　710072；2. 中国空气动力研究与发展中心超高速空气动力研究所 高超声速冲压发动机技术重点实验室　四川 绵阳　621000)

摘要：在自主开发的软件平台上，采用基于 URANS 的方法计算航空发动机燃烧室的三维两相燃烧流动，考虑了液态燃油从液膜、液滴、燃气、燃烧的完整物理化学过程。其中，颗粒相采用 LISA 一次破碎模型、KH-RT 二次破碎模型和标准的蒸发模型，湍流燃烧模型采用可以考虑非稳态燃烧特性的非稳态火焰面/反应进度变量方法，得到了航空发动机燃烧室中温度、组分浓度和燃油液滴的颗粒直径分布规律。同时，采用 CARS 光学手段测量燃烧室主燃区的温度分布，并将数值计算结果与光学试验测量值进行比较，数值计算结果和试验值吻合较好，数值计算误差小于7.3%。说明了本文的数值计算方法和 UFPV 方法在计算航空发动机燃烧室的两相燃烧流动时具有较高的精度。

关键词：航空发动机燃烧室；两相燃烧；UFPV 方法；CARS 技术

中图分类号：V231.2　　　　文献标识码：A

熊模友，乐嘉陵，黄渊，宋文艳，杨顺华，郑忠华. 采用 CARS 试验技术与 UFPV 数值方法研究航空发动机燃烧室. 实验流体力学，2017，31(5): 15-23.

0 Introduction

As the demands for the new generation aero-engine increase, higher requirements are put for the combustor, including combustion efficiency, outlet temperature distribution, combustion stability, emissions, ignition/blowout limits, and so on[1]. The studies of the new combustion organization, chemical reaction, and the interaction between turbulence and combustion are carried out. The aviation kerosene is a macromolecular hydrocarbon fuel. During its combustion process, thousands of elementary reactions take place in the combustor. The combustion mode is a coexistence of the non-premixed and premixed combustion[2-5], and thus the combustion processes are quite complicated. Numerical simulation is one of the important methods in the study of combustion mechanism and aeroengine combustor design, and the combustion model is a key technical problem in the numerical simulation.

At present, the majority studies of the numerical simulation about turbulent combustion in aeroengine combustor adopt simplified combustion models. Some disadvantages of these models are as follows: there are too many assumptions, and more assumptions lead to worse accuracy. These combustion models ignore the interaction between the turbulence and chemical reactions, and detailed chemical reaction kinetics can not be simulated. Additionally, the calculation efficiency is low, the processes of calculation are complex, intensive and expensive. As a result, they are not suitable for practical engineering applications. It is especially important to develop a combustion model that is applicable to the numerical simulation of aeroengine combustor at present and in the future.

Peters[6-7] has first proposed a steady laminar flamelet model, referred to SLFM, in which mixture fraction is used as an independent variable to solve the flamelet equation, and the mixture fraction is used to describe the mixing process of the fuel and the oxidant in combustion. The model has been used widely lately. Owing to the unstable characteristics of local extinguishing and reigniting of turbulent combustion, Pierce and Moin[8-9] proposed a steady flamelet/progress variable approach on the basis of SLFM, referred to as SFPV, which replaces the scalar dissipation rate in the SLFM with the progress variable, and it could represent the intermediate process of the local extinguishing, thus some unstable properties of the turbulent combustion, such as local extinguishing and reigniting phenomenon can be captured correctly. According to the results of DNS early, Pitsch and Ihme[10] first found that the steady flamelet library could not capture the unsteady structure of the flame, because the transient process of the local extinguishing and reigniting of the actual turbulent combustion is far from the steady combustion state, so unsteady flamelet/progress variable approach, referred to as UFPV was proposed based on SFPV by Pierce, and then it was studied and verified systematically in doctoral thesis by Sadasivuni[11], and the lift-off height of partially premixed flame was calculated accurately with the UFPV approach and LES. Auto-ignition process of methane/air lifted flame was studied by Ihme and See[12], who found that the traditional combustion model based on the steady combustion can not simulate the variation of the thermodynamic scalar parameter in space-time during the transient ignition process. Auto-ignition process of the lifted flame in the diesel engine was simulated using the UFPV approach based on RANS by Bajaj[13], and the combustion model was proved more efficient than the finite rate model which solves the multi-step chemical reaction directly in the flow field[14-15]. Based on RANS, the UFPV approach was used to simulate the lifted flame by Naud and Novella[16], and the numerical simulation results are in good agreement with the experimental data.

The temperature and pressure are high, and meanwhile the turbulence is strong in the aeroengine combustor. The true combustion process is to be understood urgently. As a result, the temperature is to be measured correctly with the optical measurement technology, and CARS is an advanced measurement, and the measurement of the temperature in the primary zone with CARS technology is necessary.

Based on the above statement, the UFPV approach has not been used to simulate the aeroengine combustor currently. For the reason, the UFPV approach combined with liquid atomization evaporation models is used to simulate the true turbulent combustion of the liquid kerosene in the aeroengine combustor, and the numerical results are compared with the measured temperature by CARS technology so as to provide the basis for the design and evaluation of aeroengine combustor.

1 Physical and mathematical models

1.1 Transport equations of mixture fraction and mixture fraction variance

The transport equation of the mean mixture fraction $\tilde{f}$ in UFPV approach is as follows:

$$\frac{\partial(\rho\tilde{f})}{\partial t} + \frac{\partial}{\partial x_j}(\rho\tilde{f}u_j) = \frac{\partial}{\partial x_j}\left(\left(\rho D + \frac{\mu_t}{\sigma_f}\right)\frac{\partial\tilde{f}}{\partial x_j}\right) \quad (1)$$

The transport equation of the variance of the mixture fraction $\tilde{f}^{\prime\prime 2}$ in UFPV approach is as follows:

$$\frac{\partial(\rho\tilde{f}^{\prime\prime 2})}{\partial t} + \frac{\partial}{\partial x_j}(\rho\tilde{f}^{\prime\prime 2}u_j) = \frac{\partial}{\partial x_j}\left(\left(\rho D + \frac{\mu_t}{\sigma_f}\right)\frac{\partial\tilde{f}^{\prime\prime 2}}{\partial x_j}\right)$$
$$+ 2\frac{\mu_t}{\sigma_f}\left(\frac{\partial\tilde{f}}{\partial x_j}\right)^2 - \rho\tilde{\chi} \quad (2)$$

where the coefficients[17] are $\sigma_f = 0.7$, $\tilde{\chi} = C_\chi \frac{\varepsilon}{k}\tilde{f}^{\prime\prime 2}$, $C_\chi = 2.0$。

1.2 Transport equation of progress variable

In the UFPV approach, the progress variable represents the degree of reaction, and there is a one-to-one correspondence between the degree of the reaction and the progress variable, so it must be a non-conserved quantity when defining the progress variable. Linear combination of the mass fraction of the species is adopted usually. The sum of the mass fraction of carbon dioxide, carbon monoxide, water and hydrogen is used to define the progress variable in this paper.

$$C = Y_{H_2O} + Y_{CO_2} + Y_{CO} + Y_{H_2} \quad (3)$$

The corresponding source term for the equation of progress variable is defined as:

$$\dot{\omega}_C = \dot{\omega}_{H_2O} + \dot{\omega}_{CO_2} + \dot{\omega}_{CO} + \dot{\omega}_{H_2} \quad (4)$$

The transport equation of the progress variable is as follows:

$$\frac{\partial(\rho\tilde{C})}{\partial t} + \frac{\partial}{\partial x_j}(\rho\tilde{C}u_j) = \frac{\partial}{\partial x_j}\left(\left(\rho D + \frac{\mu_t}{\sigma_C}\right)\frac{\partial\tilde{C}}{\partial x_j}\right) + \rho\tilde{\dot{\omega}}_C \quad (5)$$

where σ_C represents schmidt number of the progress variable, and $\sigma_C = 0.7$ here; $\tilde{\dot{\omega}}_C$ is obtained by looking up flamelet library with mixture fraction, progress variable and stoichiometric scalar dissipation rate.

1.3 Flamelet equations

By the coordinate transformation, the space coordinates of the species equations and the energy equation are replaced by the mixed fraction f, which can be obtained by the unsteady flamelet equations (6) and (7)[17], as follows:

$$\rho\frac{\partial Y_i}{\partial \tau} = \frac{1}{Le_i}\rho\frac{\chi}{2}\frac{\partial^2 Y_i}{\partial f^2} + \rho\dot{\omega}_i \quad (6)$$

$$\rho c_p\frac{\partial T}{\partial \tau} = \frac{1}{2}\rho c_p\chi\frac{\partial^2 T}{\partial f^2} + \sum_{i=1}^{m}h_i\rho\dot{\omega}_i + q_R + \frac{\partial p}{\partial t} \quad (7)$$

where ρ, τ, T, p and c_p refer to the density, time, temperature, pressure and specific heat capacity respectively; Y_i, $\dot{\omega}_i$, h_i and Le_i refer to the corresponding species mass fraction, chemical reaction rate, enthalpy and Lewis number of species i; q_R refers to heat loss due to thermal radiation; and χ is scalar dissipation rate.

The unsteady laminar flamelet library $\phi(f, C, \chi_{st})$ is obtained by means of the chemical reaction kinetics and unsteady flamelet equation(6) and (7) and detailed chemical kinetics of kerosene in this paper.

Turbulent flamelet library is obtained by the way of integrating the probability density function with the laminar flamelet library:

$$\tilde{\phi} = \iiint\phi(f, C, \chi_{st})\tilde{P}(f, C, \chi_{st})\mathrm{d}f\mathrm{d}C\mathrm{d}\chi_{st} \quad (8)$$

The size of turbulent flamelet library about UFPV approach is $100 \times 50 \times 12 \times 28$ here.

1.4 Atomization model

The whole breakup and evaporation process of fuel particles is considered. LISA model[18-19] is used to simulate the primary atomization of liquid fuel, KH-RT model[20] is used for secondary breakup, and standard evaporation model is used for simulating the evaporation of fuel particles. Detailed description about atomization model is given in Ref.[18-20].

2 Experimental and numerical settings

2.1 Study objects and flow conditions

The study object is a model combustor, as shown in Fig.1. The air flows in the X direction, and the vertical direction of the primary air jet holes are in the Y direction, and the lateral flow is in the Z direction. The liquid kerosene is injected from the fuel inlet through the swirl atomizer nozzle into the combustor, and turbulent flame is formed when the fuel steam and the air flow are mixed after atomization and vaporization. The burnt mixing gas interacts with the fresh air from primary air jet holes, diluted air jet holes and air film holes, and then flows out from the combustor exit.

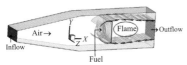

Fig.1 Schematic diagram for aeroengine combustor
图1 航空发动机燃烧室构型

Due to experimental limitations, the experimental condition is a reduced pressure model state. As shown in Fig.1, the air flow rate from "Inflow" to the combustor is 0.44kg/s, and the temperature and the pressure are 861K and 0.55MPa respectively. The flow rate of the liquid kerosene from "Fuel" into the combustor is 0.0115kg/s, and the temperature is 300K. The calculated side wall boundary condition is also wall, which is the same with the experiment. Moreover, k-ε two-equation model is used as the turbulent model in simulation, and detailed chemical reaction mechanism of the kerosene is used for kinetics. Then 203 species are included in the chemical reaction mechanism. In this paper, the boundary conditions of experimental measurements and numerical simulation are shown in Table 1.

Table 1 Boundary inlet condition of combustor
表1 燃烧室边界条件

Flow rate of fuel /(kg·s⁻¹)	Flow rate of air /(kg·s⁻¹)	Excess air coefficient	Inlet total temperature /K	Inlet total pressure /MPa
0.0115	0.440	2.6	861	0.55

2.2 Experimental platform and measurement method

Fig.2 is a photograph of the aeroengine combustor measuring platform. The experimental system mainly includes resistance heater, high pressure air source, fuel supply system, test control, data acquisition system and aeroengine combustor. Using resistance heater to heat the air is the main feature of the measuring platform. Compared to the pollution air by combustion, it provides clean and dry air for the entrance of combustor.

Fig.2 Photograph of the aeroengine combustor measuring platform
图2 航空发动机燃烧室测量平台照片

Fig.3 shows a photograph of model combustor. To measure the temperature in the primary zone of the combustor, rectangular windows of 0.1m length and 0.08m width are placed on each side in the primary zone, and quartz glass and sealed material are used for filling and sealing.

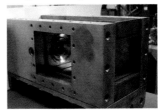

Fig.3 Photograph of model combustor
图3 模型燃烧室实物照片

CARS is used to measure the temperature of the primary zone in the combustor. A pump and a broadband Stokes beam are selected on the basis of measuring molecular Raman displacement, and the non-linear effect resonance CARS signal is formed by the way of phase matching approach focuses on the combustion flame zone and measured molecules. The linear profile of the CARS signal spectrum is related to the temperature, species concentration and pressure of the measured medium. When the concentration of the detected species is greater than

30% and the pressure is constant, the line profile of the CARS spectrum is determined by temperature. As the hydrocarbon fuel is usually used in the aeroengine combustor, the nitrogen concentration in the combustion field is generally higher than 30%, and nitrogen is chosen as the detected species for temperature measurement. The temperature measurement range of CARS technology is $200 \sim 3000K$. Fig.4 shows a picture of the CARS test integration system for aeroengine combustor, the system mainly includes laser module, signal receiver module, spectral receiver module, computer controller module and optical fiber.

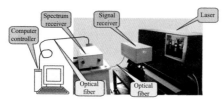

Fig.4 Picture of CARS test integration system for aeroengine combustor
图 4 CARS 测量集成系统

Fig.5 shows the CARS measurement location of temperature in the zone near the primary air jet holes in the combustor. Ten scattered points are measured, and all of these points are in the section of the primary air jet hole in the combustor.

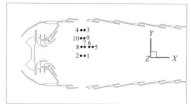

Fig.5 Schematic diagram for CARS measurement location of temperature in the zone near the primary air jet holes
图 5 航空发动机主燃孔附近 CARS 测温位置示意图

2.3 Numerical setting

2.3.1 Grid of combustor

Considering the extremely complex configuration of the combustor, the hybrid grid is selected. For example, hexahedral grid is used for the diffuser, and tetrahedral grid is used for inside of swirler. The computational grid completely simulates the true geometry of combustor, and in particular, small structures such as air film holes are not sim-

plified and authenticity of numerical simulation is improved. The grid number of combustor is about 6. 21 million, while the local mesh of the wall and primary recirculation zone is locally encrypted, reaching 10^{-2} millimeter magnitude. The grid of combustor is shown in Fig.6.

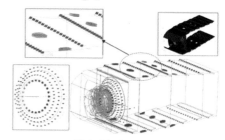

Fig.6 Grid of combustor
图 6 燃烧室网格划分

2.3.2 Generation of flamelet library

Fig.7 shows the S curve of the maximum temperature with the variation of the stoichiometric scalar dissipation rate when the flamelet library is generated under the boundary conditions given in Table 1 above. In Fig.7, stable burning branch is the red segment, partial extinguished unsteady burning branch is the blue segment and pure mixing limit branch is brown segment. Each green point corresponds to an unsteady flamelet. All unsteady flamelets are used to generate flamelet library of UFPV approach.

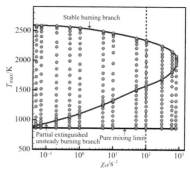

Fig.7 S curve of the maximum temperature with the variation of the stoichiometric scalar dissipation rate
图 7 最大温度随当量标量耗散率变化的 S 型曲线

Fig.8 is the temperature T distribution in the mixture fraction space when the stoichiometric scalar dissipation rate is $100s^{-1}$. Each line corresponds to a transient laminar flamelet. The black line at bottom is referred to "T-extinguish", which are extinguished flamelets. The red upper line is referred to "T-stable", which are stable burning flamelets. The middle blue line is referred to "T-unstable", which are partial extinguished unsteady burning flamelets. And all the flamelets between "T-extinguish" and "T-stable" are the laminar unsteady flamelets in transient state.

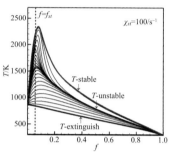

Fig.8 Distribution of temperature T in the mixture fraction space when the stoichiometric scalar dissipation rate is 100s^{-1}
图 8 当量标量耗散率为 100s^{-1} 时，温度在混合分数空间的分布

3 Experimental and numerical results

3.1 Comparison study of experiment and numerical simulation

Table 2 shows the temperature of the CARS technology measurements and numerical simulation at the locations in Fig.5. As seen from Table 2, the second column temperature is CARS measured value. Because the points 1, 2, 7 and 8 are inside the primary high temperature zone, and the chemical reaction is intense and the temperature is high here. The temperature at point 5 and 6 is relatively low due to the low temperature fresh air from primary air jet holes diluting the burnt gas. Overall, the CARS measurement values are consistent with the true combustion characteristics of the primary zone, and it has been proved that the uncertainty of the CARS measured value is less than 4%.

The temperature of the UFPV approach based on the atomization model and measurement temperature of the CARS is in good agreement, and the error is usually within 5%. There is only one point "6", and the maximum error is up to 7.3%. There-

fore it shows the high precision of the UFPV approach combing atomization models to simulate the two-phase turbulent combustion in the combustor, because the combustion model contains a large number of unsteady flamelets, and thus the unsteady effects can be simulated accurately in the zone close to the nozzle. Due to the fact that the measurement uncertainty of the CARS is about 4 percent, overall the simulation values of temperature in this paper are consistent with the CARS measurement results.

Table 2 Temperature of the CARS technology measurements and numerical simulation
表 2 数值模拟与 CARS 测量温度分布结果

Position	CARS measured value/K	Simulation value/K	Error /%
1	2350	2263	-3.70
2	2300	2281	-0.08
3	1915	1951	1.88
4	1896	1936	2.11
5	1870	1772	-5.24
6	1907	1768	-7.29
7	2235	2258	1.03
8	2188	2255	3.00
9	2135	2168	1.54
10	2100	2165	3.10

3.2 Numerical results and analysis

Fig.9 shows that the iso-surface of liquid particle axial velocity is 5m/s (red surface) and -1m/s (blue surface) respectively. Because the distribution of liquid particles is not continuous, the contour distribution of liquid particles velocity is also discontinuous. As a result of the action of the swirler, the liquid phase particles are attached to the combustor wall of the primary zone of the combustor (the red iso-surface in Fig.9 indicates that the axial velocity is positive). Under the action of air in recirculation zone, some of the liquid particles flow back (the blue iso-surface in Fig.9 represents that the axial velocity is negative).

Fig.10 shows that the iso-surface distribution of fuel mass fraction is 0.015. The liquid fuel is vaporized and the gas fuel is gradually formed in around the wall surface of venture. Under the action of swirler, the gas fuel has a spiral motion, then turbulent flame is formed when mixing with air. From the head of the combustor, the fuel flows into the combustor counter clockwise.

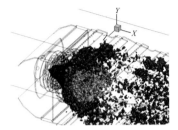

Fig.9 Iso-surface of liquid particle axis velocity is 5m/s
(red surface) and -1m/s (blue surface) respectively
图 9　燃油颗粒速度为 5m/s(红色表面)和-1m/s(蓝色表面)的等值面图

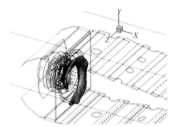

Fig.10 Iso-surface distribution of fuel mass fraction of 0.015
图 10　燃料质量分数为 0.015 的等值面图

Two legends are given in Fig.11 ~ 12 of the flow field parameters contour distribution. One is the liquid particle diameter, and the other is gas phase flow field parameters, including velocity and mixture fraction.

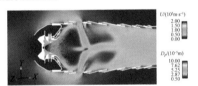

Fig.11 Velocity and diameter of fuel droplets in combustor
图 11　燃烧室速度及燃油液滴直径分布

Fig.11 shows the velocity of the combustion flow field and the diameter distribution of fuel particles in the liquid phase. The velocity is the sum of the three directions. The velocities at the swirler exit and the primary air jet holes are high, because a large amount of air enters the combustor from the swirler and the primary air jet holes. But the velocity is low in the primary recirculation zone.

Fig.12 shows the mixture fraction of combustion flow field and the diameter distribution of fuel particles in the liquid phase. The value of mixture fraction in primary zone from fuel nozzle to primary

air jet holes is high, and it is higher in the rotational shear layer, because there is the evaporation zone of fuel particles.

Fig.12 Mixture fraction and diameter of fuel droplets in combustor
图 12　燃烧室混合分数及燃油液滴直径分布

Fig.13 to Fig.15 show the temperature, progress variable and carbon monoxide mass fraction contour of the two-phase combustion.

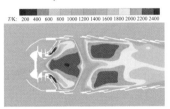

Fig.13 Contour of temperature in combustor
图 13　燃烧室温度云图

Fig.13 shows the temperature of the two-phase combustion flow field. The high temperature zone is formed in the primary zone where the gas fuel is obtained after breakup and evaporation of liquid fuel. Then turbulent combustion flame is generated when gas fuel and fresh air are mixed. A large area of recirculation is formed in the primary zone, and stable flames are generated by the combustion of fuel in the shear layer near the boundary of recirculation. The burnt gas flows back to the upstream of the recirculation zone, which is a ignition source for the fresh combustible mixture. At the same time, intermediate species, such as carbon monoxide, don't burn completely in the primary zone, and continue to flow downstream and mix with the fresh air which is from primary air jet holes. As a result, two high temperature zones are formed in the up and down side of combustor axis behind the primary air jet holes.

Fig.14 shows the progress variable of the two-phase combustion flow field. Progress variable is defined with the sum of mass fraction of carbon monoxide, carbon dioxide, water and hydrogen in this paper. According to the definition of the progress variable, it is characterized with the degree of the reaction, so it is noted that from the progress variable in the flow field, it is high in primary zone, and so the degree of reaction in the zone is high and the combustion is intense. In the zone after the primary air jet holes, because of the injection of fresh air from primary air jet holes, intermediate products of incomplete combustion, such as carbon monoxide, continue to burn with fresh air to produce a large amount of final combustion products, carbon dioxide and water. As a result, the value of progress variable is up to 0.33 in this zone.

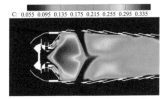

Fig.14　Contour of progress variable in combustor
图 14　燃烧室反应进度变量云图(后附彩图)

Fig.15 shows the mass fraction of carbon monoxide of the two-phase combustion flow field. As seen from the above analysis, combustion process occurs mainly in the primary zone of combustor, a large amount of fuel vapor is formed in this zone due to the evaporation of fuel particles. Because of a small amount of air from the head of the combustor, rich flame is formed in this zone, and a lot of carbon monoxide is obtained because of combustion is not complete. In Fig. 15, carbon monoxide is found in large quantities in the zone. And the rest carbon monoxide flows to the downstream where a plenty of fresh air coming from the primary air jet holes mixes and burns with it. So lean flame is already formed in the secondary zone, as carbon

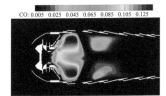

Fig.15　Contour of mass fraction of CO in combustor
图 15　燃烧室一氧化碳云图(后附彩图)

monoxide is depleted further. Therefore there is almost no carbon monoxide at the exit of combustor.

4　Conclusions

Detailed chemical reaction kinetics of the kerosene, the atomization model, the evaporation model and the UFPV approach based on URANS are adopted in this paper, and the three-dimensional two-phase turbulent combustion in the combustor is simulated. At the same time, the temperature of the primary zone in the combustor is measured using the CARS optical technology, and numerical results are compared with the experimental results. The results are as follows:

(1) Scattered temperature of the primary zone in the combustor is measured with the CARS technology, and then comparative study of numerical results and experimental results is conducted. The maximum error between the calculated results and the experimental values is less than 7.3 percent, and the UFPV approach and numerical method adopted in this paper are verified. They are suitable for simulating the turbulent combustion process in combustor.

(2) The combustor is simulated with the UFPV approach and the atomization model. The distribution features of fuel droplet diameter, mass fraction of fuel, velocity, mixture fraction, temperature, progress variables and mass fraction of carbon monoxide are obtained, and the atomization, evaporation and combustion processes of liquid fuel are analyzed. The UFPV approach and numerical calculation method adopted in this paper can provide reference for the design and optimization of the combustor.

References:

[1] Jin J, Liu D H. Recent advances in turbulent two-phase combustion models[J]. Journal of Nanjing University of Aeronautics and Astronautics, 2016, 48(3): 304-309.

[2] Legier J P, Poinsot T, Veynante D. Dynamically thickened flame LES model for premixed and non-premixed turbulent combustion[C]//Center for Turbulence Research Proceedings of the Summer Program, 2000.

[3] Selle L, Lartigue G, Poinsot T, et al. Large-eddy simulation of turbulent combustion for gas turbines with reduced chemistry [C]//Center for Turbulence Research Proceedings of the Summer Program, 2002.

[4] Yang J H, Liu F Q, Mao Y H, et al. A partially premixed combustion model and its validation with the turbulent bunsen flame calulation[J]. Journal of Engineering Thermophysics, 2012, 33(10): 1793-1797.

[5] Xiao H H, Shen X B, Sun J H. Experimental study and three-dimensional simulation of premixed hydrogen/air flame propagation in a closed duct[J]. International Journal of Hydrogen Energy, 2012, 37(15): 11466-11473.

[6] Peters N. Laminar diffusion flamelet models in non-premixed turbulent combustion[J]. Progress in Energy and Combustion Science, 1984, 10(3): 319-339.

[7] Peters N. An asymptotic analysis of nitric oxide formation in turbulent diffusion flames[J]. Combust Sci and Tech, 2007, 19 (1-2): 39-49.

[8] Pierce C D. Progress-variable approach for large-eddy simulation of turbulent combustion[D]. Stanford: Stanford University, 2001.

[9] Pierce C D, Moin P. Progress-variable approach for large-eddy simulation of non-premixed turbulent combustion[J]. Journal of Fluid Mechanics, 2004, (504): 73-97.

[10] Pitsch H, Ihme M, Nevada R. An unsteady/flamelet progress variable method for LES of nonpremixed turbulent combustion [R]. AIAA-2005-557, 2005.

[11] Sadasivuni S K. LES modelling of non-premixed and partially premixed turbulent flames[D]. Loughborough: Loughborough University, 2009.

[12] Ihme M, See Y C. Prediction of autoi̇̇̇̇̇̇̇̇̇̇ted methane/air flame using an UFPV model[J]. Combustion and Flame, 2010, 157(10): 1850-1862.

[13] Bajaj C, Ameen M, Abraham J. Evaluation of an unsteady flamelet progress variable model for autoignition and flame lift-off in diesel jets[J]. Combustion Science and Technology, 2013, 185(3): 454-472.

[14] Van Oijen J, De Goey L. Modelling of premixed counterflow flames using the flamelet-generated manifold method[J]. Combustion Theory and Modelling, 2002, 6(3): 463-478.

[15] Ribert G, Gicquel O, Darabiha N, et al. Tabulation of complex chemistry based on self-similar behavior of laminar premixed flames[J]. Combustion and Flame, 2006, 146(4): 649-654.

[16] Naud B, Novella R, Pastor J M, et al. RANS modelling of a lifted H_2/N_2 flame using an unsteady flamelet progress variable apporach with presumed PDF[J]. Combustion and Flame, 2014, 162(4): 893-906.

[17] Xing J W. Applications of chemical equilibrium and flamelet model for the numerical simulation of scramjet[D]. Mianyang: China Aerodynamics Research and Development Center, 2007.

[18] Fung M C, Inthanvong K, Yang W, et al. Experimental and numerical modelling of nasal spray atomisation[C]. Ninth International Conference on CFD in the Minerals and Process Industries CSIRO, Melbourne, 2012.

[19] Senecal P K, Schmidt D P. Modeling high-speed viscous liquid sheet atomization [J]. International Journal of Multiphase Flow, 1999, 25(6-7): 1073-1097.

[20] Reitz R D. Modeling atomization processes in high pressure vaporizing sprays[J]. Atomization Spray Technoology, 1987, 3 (309): 309-337.

Investigation of combustion process of a kerosene fueled combustor with air throttling

Ye Tian*, Shunhua Yang, Jialing Le, Fuyu Zhong, Xiaoqiang Tian

Science and Technology on Scramjet Laboratory of Hypervelocity Aerodynamics Institute, CARDC, Mianyang 621000, China

ARTICLE INFO	ABSTRACT
Article history: Received 19 April 2016 Revised 14 October 2016 Accepted 24 January 2017 *Keywords:* Combustion process Flame stabilization Air throttling Kerosene Ignition Blown off	An experimental and numerical study was carried out to investigate the combustion process of a kerosene fueled combustor with air throttling. The results were obtained with the inflow conditions of Mach number of 2.0, total temperature of 953 K and total pressure of 0.82 MPa, respectively. The air throttling was located 0.575 m downstream the combustor entrance, and the mass flux of air throttling was 27.2% inflow mass flux. The pilot flame was blown off by the room temperature kerosene when the kerosene supply pressure was 0.25 MPa, but the kerosene was ignited successfully when the throttling air was injected into the combustor, and the flame stabilization was achieved even when the pilot hydrogen was removed. The combustion process could be divided into four parts based on changes in the pressure monitored near the cavity: kerosene was ignited successfully by the pilot flame, and the mixture flame was stable during part-a. As the kerosene supply pressure was increasing, the flame was blown off by the room temperature kerosene in part-b. Successful ignition and flame stabilization had been achieved with the aid of air throttling in part-c, and the combustion mode was subsonic combustion. The flame was stable even after the pilot hydrogen was removed in part-d, but the combustion mode was supersonic combustion. © 2017 The Combustion Institute. Published by Elsevier Inc. All rights reserved.

1. Introduction

Achieving flame stabilization is a difficult problem in scramjet combustor, because in supersonic combustor, the time available for fuel injected, vaporized, mixed with air, and combustion is very short, of the order of milliseconds [1]. This problem applies especially to hydrocarbon fuels such as kerosene that are often used in the scramjet, Which consists of long chains of hydrogen and carbon molecules with longer reaction times than smaller molecules (such as hydrogen and ethylene) and thus has long ignition delay times, often exceeding a millisecond [2]. So the flameholders should be used in the scramjet combustor in order to achieve flame stabilization. Some different kinds of flame holders, such as cavity [3–5], strut [6–9], step [10] and air throttling [11–14], have been investigated by many researchers.

Ignition transients in a scramjet engine with air throttling were investigated by Li and coworkers [13,14]. In their paper, a pre-combustion shock train was generated in the isolator due to the increased back pressure by the throttling air. The resultant increase in the temperature and pressure of the airstream in the combustor, along with the decrease in the flow velocity, lead to smooth and reliable ignition. The incidentally formed separated flows adjacent to the combustor sidewall improved fuel/air mixing as a result of enhanced flow distortion and increased residence time. Successful ignition could only be achieved with the aid of air throttling under the present flow conditions. Chemical reactions were intensified and produced sufficient heat release to maintain a flow environment conducive to flame stabilization. A self-sustaining mechanism was thus established between the flow and flame development. Stable flames were achieved even after the deactivation of air throttling. Our previous work studied the effect of air throttling on flow structure, fuel/air mixing, ignition transients and flame stabilization in the scramjet combustor [15–19]. Mathur et al. [20] conducted an experiment using air throttling to initiate combustion in a scramjet combustor. Their results showed that once the air throttling was removed after the flame establishment, the shock train was retained leading to sustained combustion if heat release was sufficient. Conversely, insufficient heat release might result in an unstable shock train and caused flame blowout. Donbar et al. [21] tested the operation sequence of ignition in an ethylene fueled scramjet combustor. Air throttling was used after a stable fuel condition was reached. Once ignition occurred by activating spark igniters, the air throttling was removed, after which sustained combustion proceeded.

From the above discussion, we found the effect of air throttling on flame stabilization had been investigated by several researchers, but most of them focused on gas (ethylene or hydrogen) fueled

Tian Ye, Yang Shunhua, Le Jialing, Zhong Fuyu, Tian Xiaoqiang. Investigation of combustion process of a kerosene fueled combustor with air throttling. Combustion and Flame. 2017, 179: 74-85.

Fig. 1. Photo of the supersonic combustion facility in CARDC.

Nomenclatures

Ma	Mach number
ER	equivalence ratio
D	depth of cavity
L	length of cavity
P	wall pressure of combustor
t	time
x	distance from combustor entrance
T	mass averaged temperature

combustor. Few published papers investigated the effect of air throttling on room temperature liquid (kerosene) fueled combustor, which was our main purpose of the present paper. We used a small scaled cavity as a flameholder in the combustor in order to reduce the cavity drag, also the pilot hydrogen was used to ignite the room temperature kerosene, and air throttling was used to achieve flame stabilization after pilot hydrogen was removed.

2. Experimental and numerical simulation methods

2.1. Facility and combustor configuration

Experimental investigations were conducted on a direct-connected supersonic combustion facility (Fig. 1) in China Aerodynamics Research and Development Centre (CARDC). Hydrogen fueled heater was used to heat the air up to 1000 K and additional oxygen was added to maintain a 21% O_2 mole fraction in the

vitiated air, and the mole fraction of H_2O and N_2 were 12%, and 67%, respectively. A pulse Mach 2.0 airflow was supplied via a two-dimensional nozzle which was connected to the upstream of the combustor. The total temperature and total pressure of the inflow were 953 K and 0.82 MPa, respectively.

The lab-scale combustor [19] was made of stainless steel and divided into two sections, which was shown in Fig. 2, the first section was a rectangular isolator with the length of 430 mm (350 mm straight section and 80 mm expansion section with upwall divergent angle being 1.4°) and the cross section area was 30 × 150mm². The second section was the combustor which included a cavity (D: 11 mm, L/D: 11) and a four-part expansion section (range: 551mm-1070 mm). There were two fuel injected positions shown in Fig. 2, the first injector was designed for injecting room temperature kerosene and the second injector was designed for introducing pilot hydrogen, and the locations of the two injectors were 410 mm and 440 mm from the isolator entrance, respectively. The kerosene was injected at sonic speed at an angle of 90° to the airflow by fifteen 0.3 mm in diameter fuel injection holes, and the hydrogen was injected by ten 1.0 mm in diameter fuel injection holes, that of the throttling air were twenty 3 mm in diameter injection holes. The sampling frequency of pressure transducer was 1 kHz, which was used for measuring the wall pressure. Schlieren images were captured by a CCD camera, and the exposure time was 1μs and the frame rate was 10,000 fps. The chemiluminescence of CH* was used to mark the flame zones in the combustor. The luminosity from CH* was imaged by a CCD camera with ± 5 nm bandwidth interference filters centered at 430 nm and the exposure time was 1/2000 s.

The test sequence of the studying case was shown in Fig. 3 and Table 1, when hydrogen entered into the facility heater at $t = 1.80$ s, the cold flow then generated. The running time of the facility was about 600 ms (1.95s-2.57 s), when the supply pressure of hydrogen was kept as constant. The kerosene was injected into the combustor from $t = 1.95$ s to $t = 2.57$ s, and the supply pressure of kerosene was increased to 2.0 MPa (equivalence ratio: 0.3) from $t = 1.95$ s to 2.30 s. The spark in the combustor cavity was working all the test time, so the pilot hydrogen was ignited at once when it was injected into the combustor. The air throttling was started to be injected into the combustor at $t = 2.35$ s, just 0.04 s before pilot hydrogen was off. The location of air throttling was 0.575 m from the combustor entrance, and the mass flux of air throttling was about 27.2% inflow mass flux. The equivalence ratio (ER) of kerosene was 0.3, and the ER of pilot hydrogen was 0.08.

2.2. Numerical methods

In this study, the inhouse CFD code AHL3D [19,23] software which had been introduced in reference [18,19,23] was used for computation. A fully coupled form of species conservation

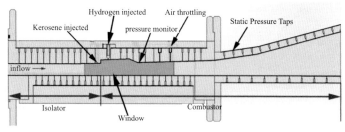

Fig. 2. Schematic illustration of the combustor.

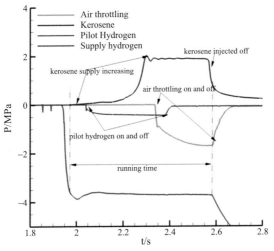

Fig.3 Test sequence of different parameters

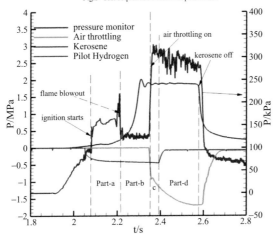

Fig. 4. Wall pressures of pressure monitor, kerosene supply and pilot hydrogen.

Table 1
Operation sequence of the test.

Operation sequence	Cold flow starts	Kerosene injected on	Hydrogen injected on	Air throttling on	Hydrogen injected off	Kerosene and Air throttling off
Time/s	1.80	1.95	2.05	2.35	2.39	2.57

equations and Reynolds averaged Navier–Stokes equations were used as a governing equation set for a chemically reacting supersonic viscous flow. The governing equations expressed in conservative vector form using the Cartesian coordinate system are:

$$\frac{\partial Q}{\partial t}+\frac{\partial F}{\partial x}+\frac{\partial G}{\partial y}+\frac{\partial E}{\partial z}=\frac{\partial F_v}{\partial x}+\frac{\partial G_v}{\partial y}+\frac{\partial E_v}{\partial z}+S$$

where, $Q = (\rho,\rho u,\rho v,\rho w,\rho E_t,\rho C_i)^{\mathrm{T}}$, C_i was the mass concentration for species i. E_t was the total energy, including kinetic energy and internal energy. E,F,G were the inviscid fluxes, F_v,G_v,E_v were the viscous fluxes. S was the source term, u,v,w were the velocity components in Cartesian coordinate(x,y,z). ρ was the density.

Cell-averaged finite volume techniques were used to solve the conservative form governing equations. *LU-SGS* method was used

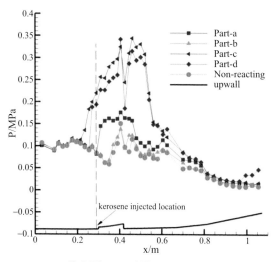

Fig. 5. Wall pressures of different combustion processes.

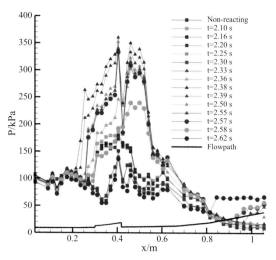

Fig. 6. Wall pressures at different times.

in time-marching. In space terms difference, third order *MUSCL* interpolation method and *AUSMPW+* scheme were used in inviscid fluxes construction, central difference method was used in viscous fluxes. Kok's modified k-ω TNT two-equation turbulence mode [22] was used in turbulence simulations. The kerosene reaction mechanism modified version of CARDC's chemistry mechanism [8], involving 12 elementary reaction steps and 10 reaction species was used in this simulation. A 2D structured grid with a size of 100,000 grid points was used in this simulation. Considering the calculating costs, the

kerosene atomization process [19,23] would not be considered and two-dimensional numerical simulation was used in the present calculation.

3. Results and discussion

In this section, the combustion process of the combustor with air throttling is going to be presented. Time evolution of the pressure at a monitoring station, spatial distribution along the

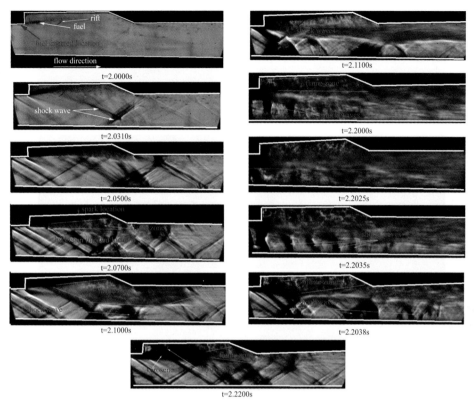

Fig. 7. The high-speed schlieren images of part-a and part-b combustion process.

streamwise coordinate of the wall-mounted average pressure over the different stages of the combustion process, schlieren and CH* luminosity images of experiments, also non-reacting and reacting simulation data are discussed. Non-reacting and reacting cases are tested and will be compared to assess the relative pressure drop in the combustor as the flame is found to nearly extinguish before air throttle was enabled for stabilization.

3.1. The whole combustion process

The monitor pressure with the pressure of kerosene and pilot hydrogen is shown in Fig. 4. During the whole test time (1.95s–2.60 s), the whole combustion process could be divided into four parts according to the monitor pressure changing. The pressure inside the combustor was monitored at a single location immediately downstream of the cavity flameholder. part-a indicated that the ignition started at $t = 2.07$ s, 0.02 s after the pilot hydrogen was injected into the combustor. And the flame was nearly blown off at $t = 2.22$ s, when the kerosene supply pressure was about 0.25 MPa. The kerosene flame was blown off and the pilot flame only existed in the cavity ramp from $t = 2.22$ s to $t = 2.35$ s. But the kerosene burnt intensively again when the throttling air was injected into the combustor at $t = 2.35$ s at the beginning

of part-c, the kerosene kept burning intensively during the whole part-c. The monitor pressure was around 0.28 MPa from $t = 2.35$ s to $t = 2.39$ s when the pilot hydrogen was removed (beginning of part-d and ending of part-c). But the pressure went down slightly from $t = 2.39$ s to $t = 2.45$ s and then kept steady around 0.25 MPa until $t = 2.57$ s when the throttling air was removed. The wall pressures of the four parts and non-reacting flow are shown in Fig. 5. The wall pressure of part-a was higher than that of non-reacting flow, especially in the cavity part, which meant the combustion mainly occurred in the cavity. The wall pressure of part-b was almost equal to the non-reacting flow except for the pressure near the cavity ramp, which meant the flame was nearly blow out and only existed near the cavity ramp where the pressure and temperature was higher and the velocity was lower. The wall pressures of part-c and part-d were higher obviously than that of the non-reacting flow and combustion-generated high backpressure had spread into the isolator, but the isolator entrance had not been disturbed. The disturbing distance of part-c was 0.10 m (0.20 m from the isolator entrance) and that was 0.05 m (0.25 m from the isolator entrance) of part-d. The precombustion shock train of part-c extended farther upstream in the isolator than that of part-d, this was because the pilot hydrogen had not been removed, and more heat released in part-c.

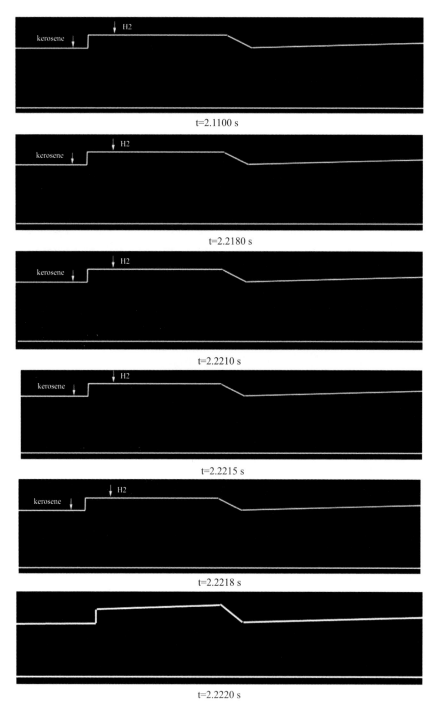

Fig. 8. The CH⁺ luminosity images of part-a and part-b combustion process.

Wall pressures at different time in the whole combustion process are shown in Fig. 6, Wall pressures in part-a ($t = 2.10$ s, $t = 2.16$ s, $t = 2.20$ s) were higher than that of the non-reacting case, which meant combustion existed in the combustor. But wall pressures in part-b ($t = 2.25$ s, $t = 2.30$ s, $t = 2.33$ s) were almost equal to that of non-reacting case, which meant the flame was nearly blown off in the combustor. Wall pressures in part-c ($t = 2.38$ s and $t = 2.39$ s) were higher obviously than that of the other parts. Wall pressures in part-d ($t = 2.50$s– $t = 2.58$ s) were almost the same, which meant the combustion was stable in the combustor with air throttling.

3.2. The combustion process of part-a and part-b

The high-speed schlieren images of part-a and part-b combustion process are shown in Fig. 7. The kerosene located upstream the cavity was gradually injected into the combustor from $t = 2.0000$ s, and the shock waves generated by the cavity ramp and the flow separation zone could be firstly seen at $t = 2.0310$ s. The kerosene had been filled in the cavity at $t = 2.0500$ s, and the pilot hydrogen had been injected into the cavity at $t = 2.0700$ s, then the ignition occurred and the flame was in the cavity near the cavity ramp. The flame propagated into the core flow rapidly at $t = 2.1000$ s, this was because the kerosene had already mixed with the air well before the pilot hydrogen was injected into the combustor. The supersonic core flow went through the narrow zone caused by the combustion heat release, then the shock waves generated. The shock waves moved upstream rapidly from $t = 2.1000$ s to $t = 2.2025$ s, this was because more heat had been released as the mass flux of kerosene increased. And almost no shock waves could be seen in the image at $t = 2.2025$ s, the shock waves had moved upstream from the cavity region into the isolator, because of the higher combustion backpressure caused by the increasing kerosene. But the shock waves could be seen again when $t = 2.2035$ s, and the shock waves moved downstream from the isolator into the cavity region, also the flame zone was reduced from the whole cavity to a small area near the cavity ramp at $t = 2.2200$ s. This was because more room temperature kerosene had been injected into the combustor, and then the flame was nearly blown off, only a small portion of the flame was anchored in the rearward of the cavity. So we could see the wall pressure of part-b was almost equal to the non-reacting flow except for the cavity ramp position.

CH* luminosity images are shown in Fig. 8, which was used to better analysis the flame distribution and to know if the flame was nearly blown off. The CH* signal was only present in the figure from $t = 2.110$ s to $t = 2.220$ s during part-a and part-b, and the CH* signal distribution area was rather small. When the kerosene supply pressure was increased to 0.25 MPa at $t = 2.2220$ s, no CH* signal could be seen in the figure. Which meant the kerosene flame was blown off at this time.

3.3. The combustion process of part-c and part-d

The flame was nearly blown off by the room temperature kerosene after $t = 2.2200$ s, which can also be seen at $t = 2.3500$ s in Fig. 9. But the shock wave had been generated by the throttling air at $t = 2.3500$ s. And the shock waves kept moving upstream from the cavity region into the isolator, and then the flame near the cavity ramp propagated upstream along the cavity wall. The flame was filled in the cavity at $t = 2.3600$ s, and almost all the shock waves had moved into the isolator at $t = 2.4000$ s. But the shock waves moved downstream from the isolator at $t = 2.4500$ s, this was because the combustion heat release was less than before after the pilot hydrogen was removed. Finally the shock waves were stable under the cavity after $t = 2.4500$ s, which

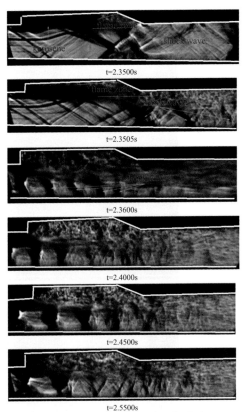

t=2.3500s

t=2.3505s

t=2.3600s

t=2.4000s

t=2.4500s

t=2.5500s

Fig. 9. The high-speed schlieren images of part-c and part-d combustion process.

meant the flame stabilization was achieved by the throttling air after the pilot hydrogen was removed.

CH* luminosity images of part-c and part-d are shown in Fig. 10. The CH* signal was present in the figure from $t = 2.360$ s to $t = 2.550$ s during the whole part-c and part-d, which distributed in the cavity and downstream the cavity near the top wall. From the CH* signal configuration, the mixture flame was stable and mainly distributed in the cavity shear layer.

3.4. The effect of air throttling

Based on the above discussion, we found air throttling was an effective method to achieve flame stabilization after pilot hydrogen was removed. The wall pressures of the non-reacting flow with and without air throttling are shown in Fig. 11, and the mass averaged temperature and Mach number of the non-reacting flow are shown in Fig. 12. From the figures, the effect of air throttling increased in the temperature and pressure of the flow in the combustor, along with the decrease in the flow velocity, which would lead to smooth and reliable ignition. In the shock train region, the mass averaged Mach number changed from 1.8 to 0.9, the average static temperature changed from 600 K

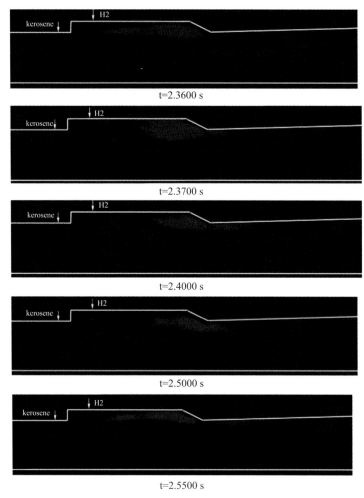

t=2.3600 s

t=2.3700 s

t=2.4000 s

t=2.5000 s

t=2.5500 s

Fig. 10. The CH⁺ luminosity images of part-c and part-d combustion process.

to 770 K, the average static pressure changed from 103 kPa to 285 kPa. The shock waves interacted with boundary layer, which would cause the wall boundary layers to separate, so the vorticity could spread into the core flow (seen in Fig. 13), which would improve the fuel/air mixing in the core flow (seen in Fig. 14) as a result of enhanced flow distortion and increased residence time.

The reacting simulations of the combustion field in the combustor with air throttling (part-c and part-d) are shown in Figs. 15–19, which were calculated by solving steady Navier–Stokes equations. This was because the combustion flow in part-c and part-d was stable, which could also be explained by the monitor pressure in Fig. 4. The numerical results of wall pressure was used

to compare with the experimental results to show the validity of the simulations in Fig. 15. From the figure, we found the simulated results better matched the experimental results, especially on the disturbing distance of combustion induced backpressure in the isolator.

The disturbing distances of part-c and part-d are 0.1 m ($x = 0.2$ m) and 0.05 m ($x = 0.25$ m) in Fig. 16, which was in accordance with the information in Fig. 5. The separation zone has been propagated into the isolator from the cavity, but it did not cause the inlet unstart. The core flow of part-d was supersonic, but that of part-c was subsonic. This was because the pilot hydrogen was still injected into the combustor, and more heat released from the mixture. The mass averaged Mach numbers of part-c and part-d were

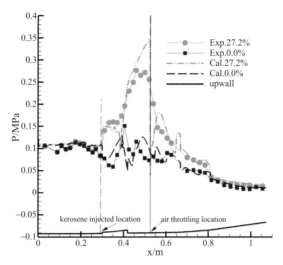

Fig. 11. Wall pressures of the non-reacting flow with and without air throttling (Exp.: experimental results, Cal.: calculative results, 27.2%: results with air throttling, 0.0%: results without air throttling).

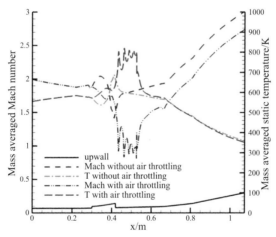

Fig. 12. Mass averaged Mach number and temperature of the non-reacting flow with and without air throttling.

shown in Fig. 17. The mass averaged Mach number was lower than 1.0 from $x = 0.25$ m to $x = 0.58$ m, the combustion mode of pat-c was subsonic combustion, and that of part-d was supersonic combustion. The static pressures of part-c and part-d in the cavity region were shown in Fig. 18, and shock waves generated due to the combustion back pressure. Mass fraction of carbon dioxide of part-c and part-d was shown in Fig. 19, the carbon dioxide of part-c had propagated into the isolator along the separation zone near the top wall.

4. Conclusions

The combustion process of a kerosene fueled combustor with pilot hydrogen and air throttling was investigated by experiments and numerical simulations. Time evolution of the pressure at a monitoring station, spatial distribution along the streamwise coordinate of the wall-mounted average pressure over the different stages of the combustion process, schlieren and CH* luminosity

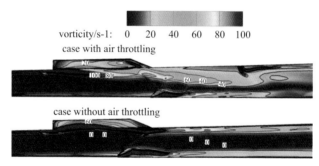

Fig. 13. Vorticity contours in the non-reacting flow with and without air throttling.

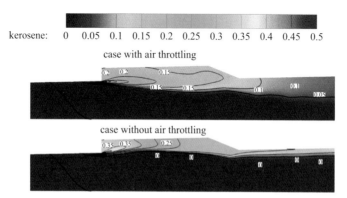

Fig. 14. The mass fraction of kerosene in thenon-reacting flow with and without air throttling.

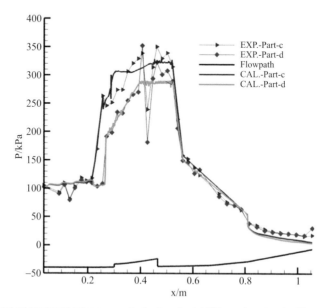

Fig. 15. Numerical and experimental wall pressure results of part-c and part-d. (EXP.: experimental results; CAL.: numerical results).

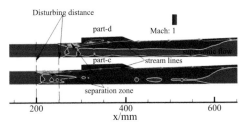

Fig. 16. Mach number contours in the reacting flow of part-c and part-d.

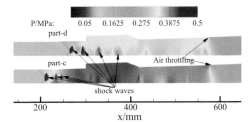

Fig. 18. Static pressure contours in the reacting flow of part-c and part-d.

images of experiments, also non-reacting and reacting simulation data were discussed.

We found the combustion process could be divided into four parts according to the monitor pressure changing. The kerosene was ignited by the pilot hydrogen successfully, and the flame was stable during part-a. As the kerosene supply pressure was increasing, the flame was blown off by the room temperature kerosene in part-b. Successful ignition and flame stabilization had been achieved with the aid of air throttling in part-c, and the flame was stable even after the pilot hydrogen was removed in part-d. Compared with previous works, the flame was either stable or blown off in their results, the interesting phenomenon of the flame was from stable and blown off to stable again was found out in this paper, which strengthened the effect of air throttling on flame stabilization.The wall pressure was often used in the previous works to show the flame stable or blown off, but we used pressure monitor, schlieren and CH* luminosity images to show the flame development and flow structure, which was easier to understand the combustion process. Finally we discussed the different combustion modes in the combustion process of part-c and part-d, the combustion mode was subsonic combustion when the pilot hydrogen existed (part-c), but it was supersonic combustion when the pilot hydrogen was removed (part-d).

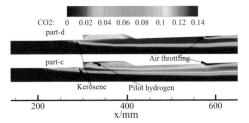

Fig. 19. Mass fraction of carbon dioxide contours in the reacting flow of part-c and part-d.

Conflicts of interest

The authors declare there are no conflicts of interest regarding the publication of this paper.

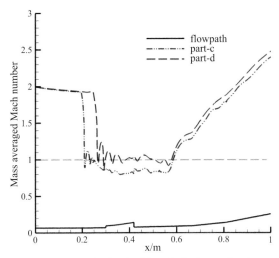

Fig. 17. Mass averaged Mach number of the reacting flow of part-c and part-d.

Acknowledgments

The project supported by Feng Lei Youthx Innovation Fond of CARDC 20150013 and National Natural Science Foundation of China (Nos. 51376194, 51376193, 51406222 and 91441204).

References

[1] A. Ben-Yakar, R.K. Hanson, Cavity flame-holders for ignition and flame stabilization in scramjets: an overview, J. Propuls. Power 17 (4) (2001) 869–877.

[2] B.J. Tatman, R.D. Rockwell, C.P. Goyne, et al., Experimental study of vitiation effects on flameholding in a cavity flameholder, J. Propuls. Power 29 (2) (2013) 417–423.

[3] M.R. Gruber, J.M. Donbar, C.D. Carter, K.Y. Hsu, Mixing and combustion studies using cavity-based flameholders in a supersonic flow, J. Propuls. Power 20 (No.5) (2004) 769–778.

[4] A. Ben-Yakar, R.K. Hanson, Cavity flame-holders for ignition and flame stabilization in scramjets-Review and experimental study, AIAA/asme/sae/asee Joint Propulsion Conference & Exhibit 68 (3) (2013) 210–222.

[5] T. Mathur, M. Gruber, K. Jackson, J. Donbar, W. Donaldson, T. Jackson, F. Billig, Supersonic combustion experiments with a cavity-based fuel injector, J. Propuls. Power 17 (6) (2001) 1305–1312.

[6] T. Mitani, N. Chinzei, T. Kanda, Reaction and mixing controlled combustion in scramjet Engines, J. Propuls. Power 17 (No.2) (2001) 308–314.

[7] Wen Bao, Jichao Hu, Youhai Zong, Qingchun Yang, Meng Wu, Juntao Chang and Daren Yu. "Combustion characteristic using O2-pilot strut in a liquid-kerosene-fueled strut-based dual-mode scramjet". Proc. IMechE PartG: J Aerosp. Eng. 227(12) 1870–1880.

[8] Jichao Hu, Jiang Qin, Juntao Chang, Wen Bao, Youhai Zong, Qingchun Yang, Combustion stabilization based on a center flame strut in a liquid kerosene fueled supersonic combustor, J. Therm. Sci. 22 (5) (2013) 497–504.

[9] Jichao Hu, Juntao Chang, Wen Bao, Qingchun Yang, John Wen. "Experimental study of a flush wall scramjet combustor equipped with strut/wall fuel injection". Acta Astronaut., 104(2014)84–90.

[10] Mohieldin, T.O., Tiwari, S.N., Olynciw M.J., "Asymmetric flow-structures in dual mode scramjet combustor with significant upstream interaction", AIAA Paper, 2001-3296.

[11] Bao, W., Hu, J., Zong, Y., et al "Ignition characteristic of a liquid kerosene fueled scramjet by air throttling combined with a gas generator" J. Aerosp. Eng. doi: 10.1061/(ASCE)AS.1943-5525.0000329 **34**. No. 12.

[12] Y. Tian, S.H. Yang, J.L. Le, Study on the effect of air throttling on flame stabilization of an ethylene fueled scramjet combustor, Int. J. Aerosp. Eng. 2015 (2015) 10 Article ID 504684, doi:10.1155/2015/504684.

[13] Y Vigor, J. Li, Y.C. Jeong, et al. "Ignition transient in an ethylene fueled scramjet engine with air throttling. Part 1: non-reacting flow development and mixing". AIAA 2010-409

[14] Y. Vigor, J. Li, Y.C. Jeong, et al. "Ignition transient in an ethylene fueled scramjet engine with air throttling. Part 2: ignition and flame development". AIAA 2010-410

[15] Ye Tian, Jialing Le, Shunhua Yang, et al., Numerical study on air throttling influence of ignition transient in the scramjet combustor, J. Aerosp. Power 28 (7) (2013) 1495–1502 (in Chinese).

[16] Ye Tian, Jialing Le, Shunhua Yang, et al., Study on the flow structure of an ethylene-fueled scramjet combustor with air throttling, J. Propuls. Technol. 34 (6) (2013) 795–801 (in Chinese).

[17] Ye Tian, Jialing Le, Shunhua Yang, et al., Study on the effects of air throttling on combustion performance of a kerosene-fueled scramjet combustor, J. Astronaut. 36 (12) (2015) 1421–1427 (in Chinese).

[18] Ye Tian, Jialing Le, Shunhua Yang, Study on the effect of air throttling on flame stabilization of an ethylene fueled scramjet combustor, Int. J. Aerosp. Eng., 2015 (2015) 10 Article ID 504684, doi:10.1155/2015/504684.

[19] Ye Tian, Jialing Le, Shunhua Yang, Numerical study on effect of air throttling on combustion mode formation and transition in a dual-mode scramjet combustor, Aerosp. Sci. Technol. 52 (5) (2016) 173–180.

[20] Mathur, T., Lin, K.C., Kennedy, P., Gruber, M., Donbar, J., Jackson, T., and Billig, F., "Liquid JP-7 combustion in a scramjet combustor," AIAA Paper 2000-3581, July 2000.

[21] Donbar, J., Powell, O., Gruber, M., Jackson, T., Eklund, D., and Mathur, T., "Post-test analysis of flush-wall fuel injection experiments in a scramjet combustor," AIAA Paper 2001-3197, July 2001.

[22] J.C. Kok. "Resolving the dependence on free-stream values for the K-omega turbulence model", NLR-TP-99295.

[23] Y. Tian, B.G. Xiao, S.P. Zhang, et al., Experimental and computational study on combustion performance of a kerosene fueled dual-mode scramjet engine, Aerosp. Sci. Technol. 46 (2015) 451–458. http://dx.doi.org/10.1016/j.ast.2015.09.002.

Experimental research of air-throttling ignition for a scramjet at *Ma* 6.5

Weixin DENG *, **Jialing LE**, **Shunhua YANG**, **Wanzhou ZHANG**, **Ye TIAN**

Science and Technology on Scramjet Lab, Hypersonics Aerodynamics Institute, China Aerodynamics Research and Development Center, Mianyang 621000, China

Received 24 October 2015; revised 2 December 2016; accepted 30 December 2016
Available online 8 May 2017

KEYWORDS

Air-throttling;
Combustion;
Flame stabilization;
Ignition;
Scramjet;
Shock

Abstract An experimental investigation on ignition characteristics with air-throttling in an ethylene-fueled scramjet under flight *Ma* 6.5 conditions was conducted. The dynamic process of air-throttling ignition was explored systematically. The influences of throttling parameters, i.e., throttling mass rate and duration, were investigated. When the throttling mass rate was 45% of the inflow mass rate, ambient ethylene could be ignited reliably. The delay time from ignition to throttling was about 45–55 ms. There was a threshold of throttling duration under a certain throttling mass rate. It was shorter than 100 ms when the throttling mass rate was 45%. While a 45% throttling mass rate would make the shock train propagate upstream to the isolator entry in about 10–15 ms, four lower throttling mass rates were tested, including 30%, 25%, 20%, and 10%. All of these throttling mass rates could ignite ethylene. However, combustion performances varied with them. A higher throttling mass rate made more ethylene combust and produced higher wall pressure. Through these experiments, some aspects of the relationships between ignition, flame stabilization, combustion efficiency, and air-throttling parameters were brought to light. These results could also be a benchmark for CFD validation.

© 2017 Chinese Society of Aeronautics and Astronautics. Production and hosting by Elsevier Ltd. This is an open access article under the CC BY-NC-ND license (http://creativecommons.org/licenses/by-nc-nd/4.0/).

1. Introduction

The resident time of fuel in a scramjet combustor decreases remarkably under a flight *Ma* 6.5 condition than that of a *Ma* 4.5 condition.[1] Although hydrocarbon fuels have greater fuel densities and endothermic cooling capabilities than hydrogen,[2] it is more difficult for hydrocarbon fuels to mix sufficiently and combust with inflow air because of their long ignition delay time.[3] Therefore, some aided ignition techniques must be employed to increase resident time and facilitate local mixing.[4,5] Air-throttling is a widely used ignition method, which injects high-pressure air into a scramjet combustor to produce a throttling effect.[6] Throttling can slow down the flow speed of inflow air and increase the mixing time for the fuel. A pre-combustion shock train can be formed and a local environment with a high temperature and a low flow speed can be

Weixin Deng, Jialing Le, Shunhua Yang, Wanzhou Zhang, Ye Tian. Experimental research of air-throttling ignition for scramjet at *Ma* 6.5. Chinese Journal of Aeronautics, 2017, 30(3): 932-938.

established. Furthermore, shock/boundary interaction could produce flow separation near the wall and distortion of the main flow, which would ease the difficulty of ignition.[7, 10]

Previous investigations on ignition and combustion in scramjets with air-throttling are briefly summarized in Table 1. T_0 is total temperature of inflow, p_0 is total pressure of inflow, Ma_i is Mach number at isolator entry. Mathur et al. conducted an experiment to research air-throttling aided combustion.[11] It was shown that if the heat release from combustion was strong enough to maintain a pre-combustion shock train, combustion could be stabilized; otherwise, combustion couldn't persist. Donbar et al. studied the influence of throttling timing.[12] Viacheslav et al. ignited preheated methane with air-throttling and a methane/air torch. Flame-out happened after throttling off if the heat release was not enough to stabilize combustion.[13] Li et al. investigated the transient dynamics of air-throttling ignition. The dynamic optimization of air-throttling was achieved by implementing a genetic algorithm in the quasi-one dimensional code.[14, 15] Noh et al. studied the ignition process of ethylene with air-throttling numerically under a Ma 3 condition.[16] Li et al. used the computational fluid dynamics (CFD) method to study the cold and reacting flow field in a combustor with/without air-throttling.[17, 18] Bao et al. ignited liquid kerosene with air-throttling and a kerosene/oxygen torch.[19] Tian et al. studied the flow-field of a scramjet and flame stabilization with air-throttling using the CFD method.[20]

Although a series of excellent results has been gotten, most of them were accomplished by CFD. The other experimental results were mainly conducted under flight Ma 5 or less conditions. So far there is almost no experimental investigation on ignition characteristics under flight Ma 6.5 conditions, when air-throttling is the unique aided ignition technique. However, the re-ignition ability offered by air-throttling is important for a scramjet cruising at a high Mach number. A higher flow speed increases the difficult of ignition and makes the ignition process different from that under a lower speed. Meanwhile, the throttling mass rate and duration must be precisely controlled and optimized. An excess throttling mass rate or a too long throttling duration would result in the shock train propagating upstream to the isolator entry and make the inlet un-start. However, insufficient throttling could not produce a

low-speed region with a high local pressure and temperature to facilitate ignition.

In this study, the dynamic process of air-throttling ignition in an ethylene fueled scramjet is explored systematically. The influences of throttling parameters, i.e., throttling mass rate and duration, are investigated. The inflow conditions simulate those of flight Ma 6.5. The development of flame with air-throttling is exposed by wall pressure measurements and high-speed photographs. Through these experiments, some aspects of the relationships between ignition, flame stabilization, combustion efficiency, and air-throttling parameters are brought to light.

2. Experimental setup

2.1. Pulse combustion wind tunnel

The pulse combustion wind tunnel is a directly-connected experiment facility. It consists of several subsystems, including an oxygen-rich-air supply system, a hydrogen supply system, a combustion heater, a facility nozzle, and a vacuum tank. The wind tunnel uses hydrogen as fuel combusting with oxygen-rich-air to produce experiment inflow. Meanwhile, the mole fraction of oxygen after combustion remains 21% just as the same as that in atmosphere.[21, 23]

Once stable combustion is established in the combustion heater, the high-temperature gas would be accelerated by the two-dimensional facility nozzle to achieve a needed Mach number. Combined with the currently available Ma 2, 2.6, and 3 facility nozzles, the combustion heater is fine-tuned to simulate flight conditions from Ma 4 to 6.5. For the presented study, the experiment inflow conditions simulate those of flight Ma 6.5. The gas parameters at the scramjet isolator entry are listed in Table 2. The total temperature is measured by a platinum-rhodium Type-B thermocouple. The total pressure is calculated by the wall static pressure and the Mach number with the Rayleigh Pitot tube formula. The ratios of specific heat and combustion efficiency of the heater are supposed to be 1.36 and 0.85, respectively. The mass flows of hydrogen and oxygen-rich-air are computed by their supply pressures and the area of the sonic throat.

Table 1 Survey of investigations of ignition and combustion in scramjets using air-throttling.

Reference	Fuel	Condition	Injector	Flame holder	Igniter	Research method	Year
Mathur et al.[11]	Ethylene	$Ma_\infty = 4$–5	Flush-wall orifice	Cavity	Spark plugs/plasma torch	Experiment	2000
Donbar et al.[12]	Ethylene	$Ma_\infty = 4$–5	Flush-wall orifice	Cavity	Spark plugs	Experiment	2001
Viacheslav et al.[13]	Methane	$Ma_i = 2.0$ $p_0 = 0.7$ MPa $T_0 = 910$ K	Transverse orifice	Pylon	CH_4/air torch	Experiment	2003
Li et al.[14,15]	Ethylene	$Ma_\infty = 3.5$–5	Transverse orifice	Cavity		Experiment and CFD	2006
Noh et al.[16]	Ethylene	$Ma_\infty = 5$	Transverse orifice	Cavity		CFD	2010
Li et al.[17,18]	Ethylene	$Ma_\infty = 5$	Transverse orifice	Cavity	Spark plugs	CFD	2014
Bao et al.[19]	Kerosene	$Ma_i = 3$	Struts		Kerosene/oxygen torch	Experiment	2012
Tian et al.[20]	Ethylene	$Ma_\infty = 2$	Flush-wall orifice	Cavity	Pilot hydrogen	CFD	2015

Table 2 Gas parameters at isolator entry.

Parameter	Value
Total temperature T_0 (K)	1610
Total pressure p_0 (MPa)	2.77
Mach at isolator entry Ma_i	3
Mole fraction of H_2O (%)	30
Mole fraction of O_2 (%)	21
Mole fraction of N_2 (%)	49
Mass flow rate of inflow, $\dot{m}_{air}$ (kg/s)	2.43

2.2. Scramjet model

The scramjet model consists of a heat-sink constant area isolator, a rectangular combustor featuring a recessed cavity flame holder and several flush wall perpendicular injectors, and a diverging nozzle. A 2D schematic of this scramjet model is shown in Fig. 1. The height and width of the isolator entry are 50 mm and 100 mm. The length of the isolator is 415 mm. The total length of the scramjet model is 1700 mm. The cavity flame holder is located at the divergent upper wall. The cavity geometry is fixed with a depth of 30 mm and a length of 300 mm, which is measured from the step to the end of the aft ramp, so the ratio of length to depth equals to 10.

Ambient ethylene is used as fuel. Based on the CFD results, four banks of injectors are designed to provide various fueling options. There are twelve orifices at each injector bank, which

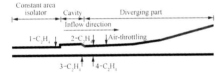

Fig. 1 2D schematic of scramjet model.

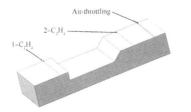

Fig. 2 3D schematic of scramjet model.

are distributed symmetrically span-wise. The diameters of the sonic fuel injector orifices are $\varnothing 1.0$ mm. The location of air-throttling in the scramjet combustor is 870 mm from the isolator entrance. The location is chosen according to the CFD results. Throttling at an upstream location makes it easy for the inlet to un-start. However, throttling at a downstream location could not produce a strong enough shock train. Throttling air could be injected into the combustor from the twelve sonic orifices at the upper wall. The diameters of the air-throttling orifices are $\varnothing 3.0$ mm. The first fuel injector bank is located upstream 10 mm to the cavity leading edge. The third fuel injector bank is located at the lower wall just opposite to the aft end of the cavity. The second and fourth injector banks have the same flow-wise coordinate, and their distances to the air-throttling point are 110 mm (shown in Fig. 2). The mass flow rates of ethylene and throttling air are computed by their pressures, densities, and the area of orifices.

The research conclusions of post-air-throttling combustion performance are related to the initial air-throttling mass. In order to shorten the opening or closing time of the valve, an electromagnetic valve assembled with a pneumatic executor is employed to inject throttling air. This assembly could open or close in 20 ms. It is located just near the throttling orifices, so the residual mass in the air-throttling line could be reduced as little as possible.

2.3. Measurement

The dynamic process of ignition and combustion with air-throttling is obtained by wall pressure measurements. The measurement system consists of an amplifier, a recorder, a controlling computer, and a 128-channel electronic pressure scanning system. Pressure taps are strategically positioned throughout the entire rig for instrumentation and health monitoring. The locations of the pressure taps are shown in Fig. 3, including five important pressure taps. Based on the value of pressure, three transducer scales are used at the same time, including 0.3 MPa, 0.7 MPa, and 15 MPa. The measurement errors are 0.2% of their scales. The sampling frequency of pressure measurements is 10 kHz.

An optical high-speed camera is placed perpendicular to the flow-field. This camera records light emitted within the visible spectrum at a capture rate of 5000 frames per second. The intensity of each pixel is the product of line integration across the span of the cavity. The luminosity of an image is assumed to correlate to the combustion efficiency.

2.4. Test procedure

Different ignition methods are tested to make clear the ignition ability of this scramjet model and prove the necessity of air-

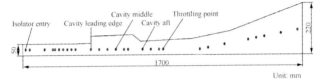

Fig. 3 Locations of pressure taps.

Table 3 Parameters of air-throttling.

Test No.	Fuel	Throttling duration (ms)	Throttling mass rate (%)
1	C₂H₄	160	45
2	C₂H₄	100	45
3	C₂H₄	200	30
4	C₂H₄	200	25
5	C₂H₄	250	20
6	C₂H₄	360	10

throttling. Both pilot hydrogen and a torch are used to ignite ethylene. However it couldn't be ignited reliably. These results are not listed here because of the paper length. Therefore, air-throttling is introduced as the unique aided technique to solve the ignition problem in the next section.

The starting time of the measurement system is set as the zero point of the test time. The measurement system worked about 1 s earlier than the wind tunnel to ensure the integrality of recording. After the test inflow is steady at $t = 1118$ ms, ethylene and throttling air are injected into the combustor at the same time of $t = 1240$ ms. Ethylene is supplied until the end of experiment at $t = 1500$ ms. The mass rates of ethylene at the four injectors are controlled by their pressures independently. In the presented paper, the fuel equivalence ratios at the four injectors are constant, which are $\Phi_1 = 0.3$, $\Phi_2 = 0.3$, $\Phi_3 = 0.2$, and $\Phi_4 = 0.2$, respectively. According to the volume of the air-throttling system, five levels of air-throttling mass flow rate are tested as shown in Table 3.

3. Results and discussion

3.1. Dynamic process of ignition

The wall pressure profiles and high-speed photographs are used to analyze the dynamic process of ignition with air-throttling. Results of Test 1 are shown in Figs. 4 and 5. The frequency of the photos is 5000 Hz. The throttling mass rate is 45% and the duration is 160 ms. Because the research focuses on the combustor wall pressure development, only one ethylene supply pressure is drawn in Fig. 4. The throttling mass rate is 45% and the duration time is 160 ms.

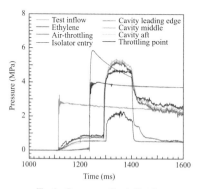

Fig. 4 Pressure profiles in Test 1.

(a) Combustion taken place in cavity, t=1295 ms

(a) Ethylene combusted strongly aided by air-throttling, t =1320 ms

(c) Ethylene combusted solely while throttling was taken off, t=1440 ms

Fig. 5 High-speed photographs of dynamic process of combustion in scramjet model, Test 1.

All pressures at the five locations in the combustor are magnified ten times. According to their importance, attention is paid to the pressures at five special locations in the scramjet model, named as isolator entry, cavity leading edge, cavity middle, cavity aft, and throttling point (see Fig. 3). The pressures at the five locations are magnified ten times in Fig. 4 for better exhibition.

Previous experiments showed that ethylene couldn't be auto-ignited in the present condition without any aided ignition technique, because a high flow speed shortens the fuel resident time to achieve efficient mixing. Under the throttling condition shown in Fig. 4, ethylene and throttling air were injected into the combustor at $t = 1240$ ms simultaneously, and it delayed about 55 ms before wall pressures increased in the combustor (see Fig. 4). Then ethylene was ignited by throttling in the cavity (see Fig. 5(a)). Pressures at the cavity downstream locations rose almost at the same time and achieved their peaks in about 20 ms. High-speed photos showed throttling making ethylene combust strongly (see Fig. 5(b)). The shock train produced by air-throttling and combustion propagated upstream to the isolator entry in about 10 ms. The pressure at the isolator entry formed a step while throttling on, so the 45% throttling mass rate was too high to keep the inlet from un-starting. After the throttling air was closed at $t = 1400$ ms, pressures at all the five locations dropped to the normal level of combustion. The effect of the isolator resumed. It indicated that ethylene was combusting steadily with the test inflow (see Fig. 5(c)). Although the location of the shock train moved downstream, it was maintained by the heat release from combustion.

Note that pressures at the five locations attenuated somehow. Reason for this is the decrease of the wind tunnel test inflow. The oxygen-rich air of this wind tunnel is supplied by a tube, the volume of which is fixed and finite, so the pressure of the test inflow decreases about 10% during the experiment.

This research showed that air throttling could generate a pre-combustion shock train in the combustor. The pre-combustion shock train could decrease flow velocity and increase temperature and pressure in the combustor section, which would improve the ignition characteristics and strengthen the flame stabilization process.

3.2. Different throttling durations

Results of Test 2 are drawn in Fig. 6. The throttling mass rate is 45% and the duration time is 100 ms. All pressures at the five locations in the combustor are magnified ten times.

Pressures at the five locations in the combustor are also magnified ten times for the same reason as in Fig. 4. The throttling mass rate is also 45% and the duration time is shortened to 100 ms. Ethylene and throttling air were injected into the combustor at $t = 1240$ ms simultaneously, and the delay from ignition to throttling was about 45 ms. The difference between the delay times in Tests 1 and 2 was only 10 ms. This difference didn't change the physical nature and could be thought to be equal, considering the error in such a complicated experimental system. This phenomenon also validated the reliability of air-throttling ignition. Pressures at the five locations rose and achieved their peaks in the same way. Pressure disturbing propagated upstream to the isolator entry in about 15 ms. They dropped to the normal level of combustion when air-throttling was closed, and the steady combustion was maintained to the end of the experiment.

Based on the results mentioned above, the duration time must be longer than the ignition delay time. There must be enough time for the shock train produced by throttling to stabilize and promote fuel mixing. Therefore, a deduction can be made that the throttling duration has a threshold under a given throttling mass rate. If the throttling duration is longer than the threshold, fuel could be ignited by air-throttling. In the present experiment condition, the threshold is shorter than 100 ms and longer than 55 ms when the throttling mass rate is 45%.

However, an accurate value of threshold for each throttling mass rate is difficult to get without a great deal of experiments.

It is more practical to analyze qualitatively from the trend. A reasonable estimation can be made that the threshold of duration is longer with a lower throttling mass rate.

3.3. Different throttling mass rates

A 45% throttling mass rate makes the shock train propagate too upstream, which produces disturbance at the isolator entry, so a lower throttling mass rate is needed to keep the shock train at the proper location and produce an enough low-speed and high-temperature region to ease ignition. Four different throttling mass rates are tested in this section to study its influence on ignition.

The influence embodies in two aspects. Firstly, a lower throttling mass rate needs a longer delay time to ignite ethylene. This can be explained by the fact that a lower throttling mass rate produces a weaker shock train. The main flow doesn't slow down so sharply. It is comparatively difficult for ethylene mixing. More time is needed to achieve the lean ignition limit. The delay time for each throttling mass rate is obtained from the measurement system and drawn in Fig. 7. It is up to 350 ms when the throttling mass rate is as low as 10%.

Secondly, different throttling mass rates would offer different ignition energy, which would change combustion performance. Wall pressures in the scramjet with different throttling mass rates are drawn in Fig. 8. While a 45% throttling mass rate would produce pressure disturbance, a 30% or lower throttling mass rate could keep the isolator in effect. Although all the five throttling mass rates from 45% to 10% could ignite ethylene reliably, the wall pressures vary with them. A higher throttling mass rate would make more ethylene combust, which means a higher combustion efficiency and wall pressure. Notice that the wall pressures in Fig. 8(a) are extracted at $t = 1320$ ms, so they are produced by both combustion and throttling. Not all the ethylene could combust in the scramjet model after throttling was taken off, which is the reason of the wall pressure variation in Fig. 8(b). Obviously, there is an optimum throttling mass rate, which could maximize the combustion performance and prevent the inlet from un-starting. Under the presented experimental conditions, a throttling mass rate between 20%-30% may be the rational choice.

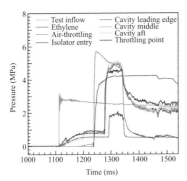

Fig. 6   Pressure profiles in Test 2.

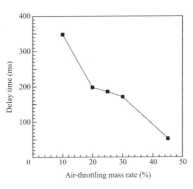

Fig. 7   Delay time vs air-throttling mass rate.

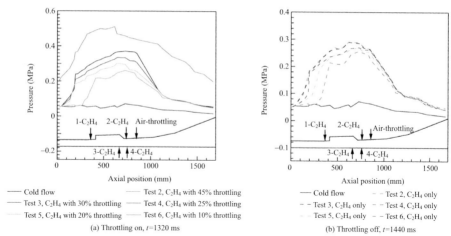

(a) Throttling on, t=1320 ms　　　　　　　　　　(b) Throttling off, t=1440 ms

Fig. 8　Wall pressures with different throttling mass rates.

4. Conclusions

This study presents experimental research on air-throttling ignition in a scramjet at *Ma* 6.5. The dynamic process of ignition is investigated by wall pressure measurements and high-speed photographs. Main conclusions include:

(1) Under present experimental conditions, when the throttling mass rate is 45% of the inflow mass rate, ethylene could be ignited reliably.

(2) The delay time from ignition to throttling changes with the throttling mass rate. There is a threshold of throttling duration under a certain throttling mass rate.

(3) While a 45% throttling mass rate would make the shock train propagate upstream to the isolator entry, four other throttling mass rates are tested, i.e., 30%, 25%, 20%, and 10%. All of these throttling mass rates can ignite ethylene. However, combustion performances vary with them. A higher throttling mass rate makes more ethylene combust and produces higher wall pressure. A lower throttling mass rate needs a longer delay time to ignite ethylene.

(4) There are optimum throttling mass rate and duration, which could maximize the combustion performance and prevent the inlet from un-starting. These values intensively depend on the inflow conditions, scramjet geometry, and fueling scheme. However, the design principles for throttling discovered in this paper could be a general reference. These results could also be a benchmark for CFD validation.

Acknowledgements

This work was supported by the National Natural Science Foundation of China (No. 51406222 and No. 51376194). The authors would like to thank Mr. Lairong ZENG and Mr. Xiaoqian DU for their supports in experiments.

References

1. Curran ET. Scramjet engines: the first forty years. *J Propul Power* 2001;**17**(6):1138–48.
2. Bao W, Yang QC, Chang JT. Dynamic characteristics of combustion mode transitions in a strut-based scramjet combustor model. *J Propul Power* 2013;**29**(5):1244–8.
3. Corin S. *The scramjet engine: processes and characteristics*. Cambridge: Cambridge University Press; 2009.
4. Gruber MR, Donbar J, Jackson K. Newly developed direct-connect high-enthalpy supersonic combustion research facility. *J Propul Power* 2001;**17**(6):1296–304.
5. Guo P, Chen Z. Ignition enhancement of ethylene/air by NO$_x$ addition. *Chin J Aeronaut* 2013;**26**(4):876–83.
6. Tian Y, Le JL, Yang SH. Numerical study on air throttling influence of flame stabilization in the scramjet combustor. *J Propul Tech* 2013;**34**(6):54–61 [Chinese].
7. Lin KC, Tam CJ, Jackson KR. Characterization of shock train structures inside constant-area isolators of model scramjet combustors. Reston: AIAA; 2006. Report No.:AIAA-2006-0816.
8. Deng WX, Le JL, Wang XY. Air-throttling effect of scramjet combustion characteristics. *J Aero Power* 2013;**28**(2):316–23 [Chinese].
9. Deng WX, Le JL, Yang SH. Air-throttling ignition experiments of ethylene fueled scramjet at *Ma* 3. *J Propul Tech* 2013;**34**(9):1240–7 [Chinese].
10. Wu ZN, Bai CY, Li J. Analysis of flow characteristics of hyper flow. *Acta Aeronaut Astronaut Sin* 2015;**36**(1):58–85 [Chinese].
11. Mathur T, Lin KC, Kennedy P. Liquid JP-7 combustion in a scramjet combustor. Reston: AIAA; 2000. Report No.:AIAA- 383 2000-3581.
12. Donbar J, Powell O, Gruber M. Post-test analysis of flush-wall fuel injection experiments in a scramjet combustor. Reston: AIAA; 2001. Report No.:AIAA-2001-3197.
13. Viacheslav AV, Yuri MS, Ruslan VA. Experimental research of pre-injected methane combustion in high speed subsonic airflow. Reston: AIAA; 2003. Report No.:AIAA-2003-6940.
14. Li J. Ignition transient in an ethylene fueled scramjet engine with air throttling [dissertation]. University Park (PA): Pennsylvania State University; 2009.

15. Li J, Ma FH, Yang V. Control and optimization of ignition transient in scramjet engine using air throttling. Reston: AIAA; 2006. Report No.:AIAA-2006-1028.

16. Noh JY, Choi JY, Byun JR. Numerical simulation of auto-ignition of ethylene in a scramjet combustor with air throttling. Reston: AIAA; 2010. Report No.:AIAA-2010-7036.

17. Li J, Zhang L, Choi JY, Yang V, Lin KC. Ignition transients in a scramjet engine with air throttling part 1: nonreacting flow. *J Propul Power* 2014;**30**(2):438–48.

18. Li J, Zhang L, Choi JY, Yang V, Lin KC. Ignition transients in a scramjet engine with air throttling part II: reacting flow. *J Propul Power* 2015;**31**(1):79–88.

19. Bao W, Hu JC, Zong YH. Ignition characteristic of a liquid kerosene fueled scramjet by air throttling combined with a gas generator. *J Aerosp Eng* 2014;**27**(5):1–27.

20. Tian Y, Yang SH, Le JL. Study on the effect of air throttling on flame stabilization of an ethylene fueled scramjet combustor. *Int J Aerosp Eng* 2015;**2015**:1–10.

21. Le JL, Liu WX, He W. Preliminary study of integrated aero-propulsive performance of vehicle with CFD and experiments. Reston: AIAA; 2006. Report No.:AIAA-2006-7982.

22. Le JL, Liu WX, He W. Preliminary aero-propulsive performance study for vehicle in short duration facility. *Proceeding of international conference on methods of aerophysical research*; 2008 June 9–13; Glasgow, Scotland. ICMAR; 2008.

23. Le JL, Yang SH, Deng WX. Ignition and combustion characteristics of an ethylene fueled scramjet combustor. *Proceeding of international symposium on air breathing engines*; 2013 Sep 9–13; Busan, Korea. Berlin: ISABE; 2013.

径向双旋流燃烧室流场结构大涡模拟研究*

周　瑜[1]，乐嘉陵[1,2]，陈柳君[1]，黄　渊[2]

（1. 西北工业大学 动力与能源学院，陕西 西安 710072；
2. 中国空气动力研究与发展中心 超高速空气动力研究所 高超声速冲压发动机技术重点实验室，四川 绵阳 621000）

摘　要：为深入了解真实航空发动机燃烧室内流场结构，在自有CFD平台上采用动态亚网格湍流模型对一种径向双旋流环形燃烧室的单个头部构型冷态流场进行了大涡模拟。为保证模拟精度，没有对模型进行常规简化处理，对包括全部气膜冷却小孔在内的所有精细结构均进行了完全仿真。计算验证了程序对高度复杂流场的模拟能力，结果表明，大涡模拟能较为全面地反映燃烧室内复杂流场从静止启动到统计定常的非定常发展过程，并成功捕捉到流场中心回流区等各种大尺度结构及涡旋破碎泡等旋流特征；大涡模拟所获得的时间平均流场结构与已有PIV试验结果定性一致，与RANS计算相比更接近试验测量值。

关键词：航空发动机；燃烧室；径向双旋流；大涡模拟；动态亚网格模型

中图分类号：V231.3　　**文献标识码**：A　　**文章编号**：1001-4055（2017）04-0909-09
DOI：10.13675/j.cnki.tjjs.2017.04.023

Large Eddy Simulation of Aeroengine Combustor with Counter-Rotating Swirler

ZHOU Yu[1], LE Jia-ling[1,2], CHEN Liu-jun[1], HUANG Yuan[2]

（1. School of Power and Energy, Northwestern Polytechnical University, Xi'an 710072, China;
2. Science and Technology on Scramjet Laboratory, Hypervelocity Aerodynamics Institute of CARDC, Mianyang 621000, China）

Abstract：To gain an improved understanding of the flow and turbulence structures in practical aeroengine combustor, large eddy simulation with dynamic Smagorinsky sub-grid model was used to explore the complex flow field in a single sector of a typical aeroengine combustor with counter-rotating swirler. The complex geometric configuration including all film cooling holes was fully simulated without any conventional simplification in order to reduce the modeling errors. The capability of our CFD code to simulate very complicated flow was validated by the calculation. The unsteady process that turbulent swirling flow developing from static to statistically stationary status was totally reproduced. Large-scale coherent structures like central recirculation zone （CRZ） and swirling flow characteristics like vortex breakdown were well captured. Time-averaged flow field predicted by LES shows qualitative agreement with PIV measurement and performed better than RANS.

Key words：Aeroengine; Combustor; Counter-swirler; Large eddy simulation; Dynamic smagorinsky model

1 引　言

燃烧室是航空发动机的核心部件之一，其内部流场结构、油气掺混情况直接影响燃烧室点火、火焰稳定、燃烧效率、出口温度分布均匀性等性能参数。由于发动机试验开销大，周期长，且难以获得内部流场重要结构和关键信息，无法形成对发动机内流动机理和规律的深入理解认知，因此CFD已逐渐发展成为燃烧室流场研究、指导设计的有力工具。尤其是近年来随着计算机性能的大幅进步,大涡模拟(Large Eddy Simulation, LES)以其能分辨流场细观结构、可接受的计算量,越来越多应用于复杂流场的研究。

在工业实践中通常会对燃烧室进行冷吹风试验以在燃烧室不加温条件下评估燃烧室的气流结构、流道设计、流量分配和压力损失等是否满足设计要求,因此冷态流场计算是燃烧室数值研究中的一个重要环节。目前国外已在这方面开展了大量相关工作。然而,真实的航空发动机燃烧室结构和流场的复杂性给计算带来了巨大困难,现有研究工作大多针对简单构型开展机理性研究,或对几何结构进行了大幅简化。如文献[1~3]对简单小型单级旋流工业燃气涡轮燃烧室进行了大涡模拟。对于更复杂的两级旋流燃烧室,Widenhorn A 等[4]采用 CFX 软件,应用 URANS 和 RANS/LES 混合模型对一种燃气涡轮双旋流模型燃烧室无反应冷态流动进行了数值研究并与激光多普勒测速(LDA)结果作了对比分析,验证了位于内剪切层的频率为 1514Hz 的进动涡核(PVC)的存在。Yee Chee See 和 Matthias Ihme[5]重点考察了网格疏密、网格划分类型和流量分配对两级旋流燃烧室流场大涡模拟获得的速度分布的影响,发现该算例存在较明显的网格敏感性,但具体影响因素还有待进一步研究。Jong-Chan Kim 等[6]对 GEAE 的 LM6000 型贫油预混低污染轴向双旋流燃烧器进行了大涡模拟,清楚地再现了主流场剪切层强烈的涡旋破碎以及中心和角回流区。然而上述研究针对的模型都是较为简单的研究型旋流燃烧器,特别是没有横向掺混射流结构,不能反映真实燃烧室内流动状态。

国内燃烧室数值模拟起步较晚,尤其是对复杂工业燃烧室的大涡模拟研究还很少,以采用 Fluent 等商业软件和 RANS 方法研究燃烧室稳态流动为主[7~11],或是采用大涡模拟方法研究简单射流燃烧器,徐宝鹏等[12]基于 Openfoam 对同轴模型燃烧室内燃料喷射、蒸发和混合过程进行了大涡模拟,认为可将其应用到真实的燃气涡轮燃烧室中。张济民[13]利用大涡模拟方法研究了一种中心钝体分层旋流燃烧器的冷态流场,获得了旋流强度对回流区、涡旋破碎和进动特征的影响规律。对于真实燃烧室的大涡模拟,颜应文[14]采用两种亚网格模型计算了一个单头部模型燃烧室冷态湍流流场,与 PIV 测量数据比较表明计算较好地模拟了旋流器后面回流区的瞬态变化,然

而该模型网格仅有 134 万,计算规模太小,流场结果比较粗糙。

总体而言,由于真实燃烧室结构十分复杂,对其进行准确建模、网格划分和数值求解难度较大,而 LES 方法计算量又相对较高,因此现有研究工作大多针对较为简单的构型如单级旋流燃烧室,或者实验室研究用简化燃烧器,数值模拟也较多地采用商业软件,并行计算规模不大。而对真实工业燃烧室的常规数值研究通常都进行了一定的简化处理,例如去除次要结构,按等面积原则将气膜冷却孔简化为单条槽道或直接采用多孔边界近似处理等。这种简化在降低计算难度的同时,带来的模拟误差难以量化评估,一定程度上影响了数值模拟的准确可靠性。文献[15]在燃气涡轮模型燃烧室研究中指出,真实燃烧室复杂构型和工况的 LES 计算迄今还未被充分广泛应用的一大难点就在于其几何复杂度和高质量贴体网格生成,及其相应带来的计算复杂度和对程序计算能力的较高要求。本文针对一种真实的径向双旋流环形燃烧室的单头部模型,采用大涡模拟方法对其冷态流场进行了数值研究。模型几何与文献[14]相似,为尽量减少模拟误差,对模型包括所有气膜孔在内的精细结构均作了完全仿真,在深入了解获得真实燃烧室多旋流湍流流场结构的同时,也进一步验证在自有 CFD 平台 AHL-3D 上开发的基于非结构有限体积离散的不可压流动计算程序面对高度复杂流场的模拟能力。

2 计算方法和模型

对 N-S 方程进行过滤得到不可压缩冷态湍流场的 LES 控制方程

$$\frac{\partial \widetilde{u_i}}{\partial x_i} = 0 \tag{1}$$

$$\frac{\partial \widetilde{u_i}}{\partial t} + \frac{\partial (\widetilde{u_i u_j})}{\partial x_j} = \frac{1}{\rho}\left(\mu \frac{\partial^2 \widetilde{u_i}}{\partial x_j \partial x_j} - \frac{\partial \widetilde{P}}{\partial x_i} - \frac{\partial \tau_{ij}}{\partial x_j}\right) \tag{2}$$

其中亚网格应力 $\tau_{ij} = \widetilde{u_i u_j} - \widetilde{u_i}\widetilde{u_j}$,采用涡粘模型封闭有

$$\tau_{ij} = -2\mu_t \widetilde{S_{ij}} + \frac{1}{3}\tau_{ij}\delta_{ij} \tag{3}$$

其中 $\widetilde{S_{ij}} = \frac{1}{2}\left(\frac{\partial \widetilde{u_i}}{\partial x_j} + \frac{\partial \widetilde{u_j}}{\partial x_i}\right)$ 为可解尺度的变形率张量。粘性系数 μ_t 需要采用亚网格模型进行模化,标准的 Smagorinsky 模型为

$$\mu_t = \rho\left(C_s \widetilde{\Delta}\right)^2 \sqrt{2\widetilde{S_{ij}}\widetilde{S_{ij}}} \tag{4}$$

式中 C_s 为 Smagorinsky 常数。$\widetilde{\Delta}$ 为过滤尺度,

$\widetilde{\Delta} = 2|\Omega|^{1/3}$，$|\Omega|$ 为网格单元体积。本文采用的动态模型（Dynamic Smagorinsky Model，DSM）则是应用两种尺度对方程进行两次过滤而动态地确定系数 C_s。假设两次过滤尺度均在惯性子区范围内，两次过滤的亚格子应力系数相等，则应用 Germano 等式

$$L_{ij} = (u_i)_{1,2}(u_j)_{1,2} - \left[(u_i)_1(u_j)_1\right]_2 = (\tau_{ij})_1 - \left[(\tau_{ij})\right]_2 \tag{5}$$

得到

$$L_{ij} = -\frac{1}{3}L_{kk}\delta_{ij} = C_D M_{ij}$$
$$M_{ij} = 2\left[\Delta_2^2\widetilde{|\widetilde{S}|}\widetilde{S}_{ij} - \Delta_1^2\overline{|\overline{S}|\overline{S}_{ij}}\right] \tag{6}$$

式中上标"—"为尺度 Δ_1 的过滤，上标"～"为尺度 Δ_2 的过滤。最终得到动态确定的系数 C_D

$$C_D = \frac{L_{ij}M_{ij}}{\langle M_{ij}M_{ij}\rangle} \tag{7}$$

3　燃烧室流场大涡模拟

本文研究对象为航空发动机环形燃烧室的单头部矩形试验模型，带有径向两级旋流器以及火焰筒壁上下各 6 道开缝式气膜冷却孔阵列和数个主燃孔/掺混孔，试验模型前方布置有长度约 250mm 矩形扩压段。为提高网格质量，在生成壁面网格之后，采用法向外推方式建立多层三棱柱/六面体结构网格，以更好地处理近壁面流动。对各种大小孔隙生成六面体纯结构网格，空间则采用各向同性正四面体非结构网格进行划分。在开展正式模拟前，通过多次小规模试验对网格进行了反复迭代调整，保证小孔及壁面首层网格位于程序计算判断的边界层粘性子层以内，且多处监控点流动变量的变化趋于网格无关。最终建立的计算网格中每个气膜冷却孔包含约 500 个网格单元，火焰筒内部空间约为 600 万网格单元，网格总量为 1700 万如图 1 所示。计算采用基于非结构有限体积离散的 SIMPLE 算法求解，时间二阶隐式，空间二阶迎风格式。

基于上述计算设置对该燃烧室模型流场进行了大涡模拟。来流条件如表 1 所示。由于 LES 计算中各种进出口条件给法都是近似的，判断哪种提法更快更准确，所需额外的数值实验对本文研究算例规模而言，计算开销难以承受，因此在出口按照常规方式直接给定变量的梯度为零。对于入口，由于试验中来流的湍流度未知，无法确定在入口如何叠加人工脉动/人工涡更为合适。考虑到计算对象在旋流器头部之前已经存在一个较长的扩压段，因此选择根据试验流量直接给定扩压段入口的平均法向速

值，这样对于旋流器而言等同于一个经过了自然转捩发展的湍流入口边界。

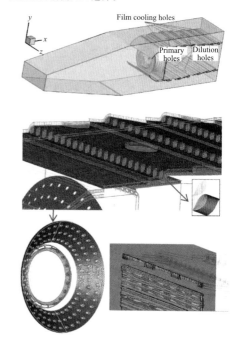

Fig. 1　Schematic of the high fidelity mesh

Table 1　Flow condition

p/kPa	T/K	$\rho/(\text{kg}\cdot\text{m}^{-3})$	$U/(\text{m}\cdot\text{s}^{-1})$
555	860	2.23	50.85

模型空气入口到出口流向距离约 560mm，则一个流动特征时间长度按入口速度估算为 11ms。为更精细地反映流场启动的非定常发展过程，时间步长在前 4 个特征时间长度即 44ms 内取为 0.5μs。此后增加到 1μs 以尽快加速流场计算达到统计定常状态。在 40 个 Intel Xeon X5670CPU 共 480 个计算核心上并行计算 20 万步，物理时间上推进 155.65ms，总耗时约 600h。

3.1　流场变量统计

在空间关键位置设置了监控点以跟踪流场发展情况。以旋流器出口平面圆心为直角坐标原点 $A(0, 0, 0)$，坐标轴方向见图 1，计算得到该点瞬时轴向速度 u 随时间变化如图 2。观察发现约经过 50ms 之后速度脉动趋于统计定常。为保证流场充分发展到统计定常状态，取 7 个特征时间长度 77ms 开始对变量作

时间平均。该点轴向速度 u 的时间序列显示旋流器出口处为旺盛的湍流流动，时间平均的速度值约为 -16m/s，表明该处存在较强的回流，根据其脉动速度均方根值计算得到湍流强度为 59.7%。图 3 给出了对点 A 及其下游 35mm 主燃孔所在截面中心点 B $(35,0,0)$ 处 u 的脉动量进行傅里叶分析得到的能谱，其幅值随频率的变化与湍流经典理论中惯性子区 $-5/3$ 律基本相符，在频率超过 10^4Hz 后也出现了与经典理论曲线一致的迅速衰减的耗散区，表明本文所用网格和计算较好地分辨出了燃烧室内湍流脉动。图 4 显示了旋流器出口轴线上沿途各点 u 的功率谱，点 B 曲线存在 3 个相对明显的峰值，分别位于约 76、130、192Hz 处，体现了一级旋流、二级旋流及下游主燃孔横向射流三股主要流动的叠加对该处流场产生的显著影响。其余各点的功率谱结构与其相似，但特征峰随着测点位置向上游移动而衰减，且高频峰值衰减得更快。这种变化趋势也是对流场结构特点的一种自然反映：越接近旋流器出口的流场受主燃孔横向射流影响越小，以旋流杯内低频振荡为主，而主燃孔正下方流场受到该射流的显著影响，增加了高频

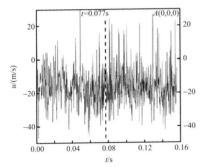

Fig. 2 Instantaneous axial velocity on point A

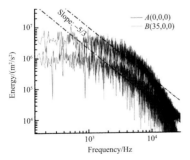

Fig. 3 Energy spectrum of the axial fluctuation velocity on point A, B

不稳定性。测点相似的特征频率分布表明，它们的空间位置可能位于同一个大尺度流动结构中，结合下文中截面上平均流场的流线分布情况（见节 4）证实了测点处于中心主回流区。

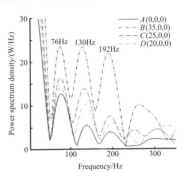

Fig. 4 Power spectrum of the instantaneous axial velocity on probe points

3.2 旋流器内部流场结构

旋流器采用了径向两级反向旋流结构设计。其中二级旋流方向完全垂直于 x 轴方向，而一级旋流通道为斜切式，与 y-z 平面存在一定角度。两者通过一个典型的分层旋流杯结构嵌套在一起。

图 5 给出了旋流杯内部及其出口附近瞬时轴向速度 $u=0$ 的等值面并以涡量值染色显示。流场启动过程中，由一级旋流器进入旋流杯内的流体经过收缩-扩张段后，首先在轴心附近膨胀形成蘑菇状轴向速度为零的等值面。由于旋流通道是斜切式，这种几何限制诱导流体在该型面上形成了显著的螺旋形条带状涡量分布，表明该面上存在强烈的流体旋转，且涡量强度随着流体向下游运动而沿途逐渐衰减，最终在出口膨胀为喇叭状并产生褶皱、破碎而变得不连续。该等值面随时间发展脉动，但一直维持由一级旋流器通道几何限制诱导的螺旋结构。

由于旋转周向速度的存在，径向上产生压力梯度，流体旋转会产生局部低压区。图 6 显示了不同时刻旋流杯内部压强增量 $\Delta p=0$ 的等值面变化情况。结合旋流杯等 x 截面上瞬时压强和流线分布可发现，该型面形成了一对交叉缠绕的内部为负压的双螺旋涡管结构，并随时间推进而周期性地旋转。管状结构自身形状也受流场脉动影响而呈间歇性收缩-膨胀变形。然而一级旋流出口形成的这种螺旋涡管结构向下游运动的过程中，遭遇由二级旋流通道进入的反向旋流后，强烈的逆向剪切作用使得该结构破裂，出现明显的截断。观察二级旋流通道下游截面

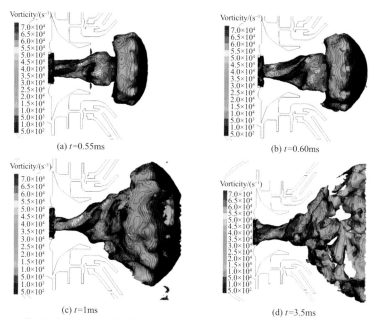

(a) t=0.55ms

(b) t=0.60ms

(c) t=1ms

(d) t=3.5ms

Fig. 5   Instantaneous distribution of axial velocity (iso-surface: u=0, colored by vorticity)

上的流动可以看出流线在两级旋流交汇处发生弯折，在改变旋向的同时还在交界处剪切出一对新的逆时针旋向小涡。

3.3　燃烧室主流场结构

在流场从静止开始启动的过程中，来流空气会分别经头部两级旋流通道、旋流杯层间小孔、主燃孔/掺混孔以及火焰筒上多排气膜冷却孔进入燃烧室内部，多股气流耦合与相互作用产生了复杂的流场结构。图7显示了 t=0.6ms 时刻流场瞬时压强等值面。火焰筒上与机匣之间的流体经完整的主燃孔/掺混孔及半孔横向进入主流域，在孔口形成完整的环管或半环管型低压区旋涡结构。图8显示了该时刻过主燃孔和掺混孔的瞬态流线分布情况。分级进入旋流器的流体在旋流杯出口边缘交汇并产生剧烈混合，而火焰筒内原有静止气体在上下主燃孔进入主燃区的横向射流与主旋流的垂直挤压及旋转剪切作用下，形成了两处明显的螺旋型流线。进一步观察图9瞬时涡量分布可发现，受固壁面限制，火焰筒上两侧的半孔入流产生了更大的涡量，且等值面沿壁面维持了一个相对稳定的形态。而完整圆孔入流的涡量较小，在穿透燃烧室约1/4深度时其等值面即开始破碎。所有孔口涡结构均随时间发展向流场中心运动深入，拉升出颈部结构并逐渐脱落。

脱落的环管型涡结构随后在燃烧室中部相遇，激烈碰撞后破碎成不完整的小涡。与此同时，流经旋流器的流体通过旋流杯内收缩-扩张段结构后，在出口膨胀出气泡状涡旋并随时间发展而周期性地生成-破碎。

4　计算结果对比分析

图10给出了过主燃孔中心 z=0 截面上 LES 最终得到的时间平均流场及标准 $k-\varepsilon$ 模型计算得到的 RANS 流场的流线分布。RANS 模拟的上下两个主回流区几乎完全对称，在主燃孔下游直至出口的流线是平滑的。而 LES 时均结果则不完全对称，上回流区相对略大，涡核位置更接近主燃孔。下涡核的位置相比 RANS 向下游轻微移动了约10mm。这种不对称性的原因之一可能在于观察剖面的选取。尽管燃烧室矩形腔体关于中心截面是对称的，然而旋流器的两级旋流通道却呈轴对称或近似轴对称分布，与中心截面相交的构型剖面并非完全上下对称。

图11进一步展示了中心回流区上半区局部流场细节。与已有PIV试验测量结果相比，两种计算得到的涡核位置 a 差别不大，但 LES 的回流区尺寸更大，形状与试验结果更接近，且准确捕捉到主燃孔下游

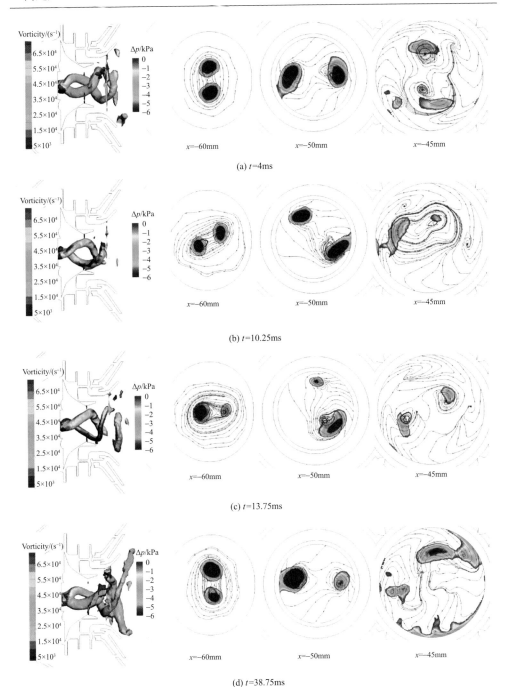

(a) t=4ms

(b) t=10.25ms

(c) t=13.75ms

(d) t=38.75ms

Fig. 6　Instantaneous distribution of pressure increment
〔Column 1: iso-surface Δp=0, colored by vorticity; column 2~4: profiles along x direction〕

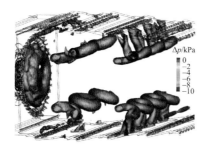

Fig. 7 *t*=0.6ms, low pressure region
(iso-surfaces: Δ*p*=−100Pa & −1000Pa)

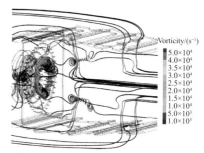

Fig. 8 *t*=0.6ms, streamlines across primary and
dilution holes

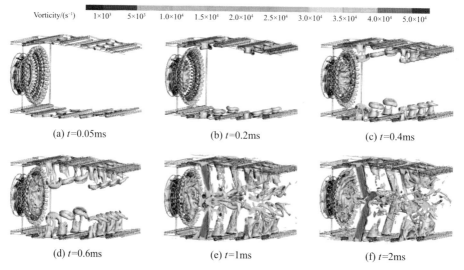

(a) *t*=0.05ms　　　　　　(b) *t*=0.2ms　　　　　　(c) *t*=0.4ms

(d) *t*=0.6ms　　　　　　(e) *t*=1ms　　　　　　(f) *t*=2ms

Fig. 9 Instantaneous distribution of vorticity during flowfield startup (iso-surfaces: $3.5×10^4s^{-1}$, $5×10^4s^{-1}$)

区域 *b* 中已经有所成型的二次小涡,而这是 RANS 计算完全没有模拟出的结构。此外 RANS 在 *c* 处得到的局部高速区形状也较为扁平,相比而言 LES 的速度云图与 PIV 吻合更好。

为进一步量化两种计算对 *a*,*b* 两处区域中大尺度涡旋结构的模拟情况,取多个测点计算值与试验值进行了比较。测点的坐标及数值见表 2。对照结果表明 LES 准确地模拟出两处回流区驻涡结构,对涡核位置的捕捉明显好于 RANS 结果。LES 计算得到的燃烧室出口总压恢复系数为 0.965,RANS 为 0.970,两者差别不大。

Table 2　Axial velocity on some probe points (profile *z*=0)

(*x*, *y*)/mm	PIV/(m/s)	LES/(m/s)	RANS/(m/s)
(35, 10)	−3.02	−4.68	−4.78
(35, 15)	−2.11	−2.62	1.25
(35, 20)	−1.08	−0.90	5.25
(25, 30)	−0.63	−1.35	6.79
(20, 30)	−0.65	−1.40	5.75
(30, 30)	−0.23	0.04	8.71
(60, 25)	−0.24	0.09	1.59
(60, 30)	0.11	0.65	1.01

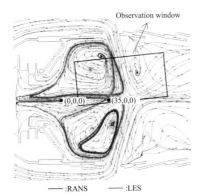

— :RANS   — :LES

Fig. 10　Streamlines on the profile z=0

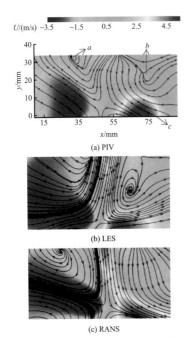

(a) PIV

(b) LES

(c) RANS

Fig. 11　Local flow field in central recirculation zone

5　结　论

本文针对一种航空涡轮发动机径向两级反向旋流燃烧室,开展了冷态流场的大涡模拟研究,结果表明:

(1)模拟获得了燃烧室内流场从静止启动到统计定常的完整非定常发展过程,清楚地捕捉到流动过程中旋流器、主燃孔/掺混孔及主流场中回流区、旋涡生成和破碎等大尺度结构。监测点脉动速度能谱

分布与湍流经典理论相符,表明本文所用网格和计算较好地分辨出了燃烧室内湍流脉动。

(2)旋流器出口中心点至下游35mm处,沿途各点的轴向速度功率谱表明它们位于同一处大尺度流动结构,旋流器出口流场受到下游主燃孔横向射流影响明显,两者耦合共同作用形成了燃烧室内中心主回流区。

(3)一级旋流通道的斜切构型使得流经其中的流体同时具有轴向和旋转速度,并在旋流杯内形成了呈鲜明的螺旋条带状涡量分布的轴向速度零值面,以及交叉缠绕的双螺旋涡管型压强增量零值面。在二级旋流通道流入流体施加的强烈反向旋转剪切作用下,上述型面沿旋流杯扩张段周期性地旋转、断裂并破碎。

(4)中心截面上流场大涡模拟结果表明除大尺度中心回流区外,主燃孔射流下游邻近流体与火焰筒壁面之间还存在一个二次回流小涡,而RANS计算完全没有反映出该区域流场结构。与该模型已有PIV测量结果和RANS结果的对比,验证了本文计算程序对燃烧室高仿真构型复杂流场的大涡模拟能力。

在后续工作中,将在此基础上进一步开展加入燃油喷注的燃烧流场的大涡模拟研究。

参考文献:

[1]　Sayop Kim, Suresh Menon. Large-Eddy Simulation of a High-Pressure Single-Element Lean Direct-Injected Gas-Turbine Combustor[R]. AIAA 2014-0131.

[2]　Bulat G, Jones W P, Navarro-Martinez S. Large Eddy Simulations of Isothermal Confined Swirling Flow in an Industrial Gas-Turbine [R]. *International Journal of Heat and Fluid Flow*, 2015, 51: 50-64.

[3]　Filosa A, Noll B E, Di Domenico M, et al. Numerical Investigations of a Low Emission Gas Turbine Combustor Using Detailed Chemistry[R]. AIAA 2014-3916.

[4]　Widenhorn A, Noll B, Aigner M. Numerical Study of a Non-Reacting Turbulent Flow in a Gas Turbine Model Combustor[R]. AIAA 2009-647.

[5]　Yee Chee See, Matthias Ihme. LES Investigation of Flow Field Sensitivity in a Gas Turbine Model Combustor [R]. AIAA 2014-0621.

[6]　Jong-Chan Kim, Hong-Gye Sung. Large Eddy Simulation of the Turbulent Flow Field in a Swirl Stabilized Annular Combustor[R]. AIAA 2009-645.

[7]　吴治永, 林宇震, 刘高恩, 等. 基于CFD分析改进三旋流燃烧室头部设计[J]. 推进技术, 2009, 30(5):

533-537. (WU Zhi-yong, LIN Yu-zhen, LIU Gao-en, et al. Dome Structure Modification of a High Temperaturerise Combustor Based on CFD Simulation [J]. *Journal of Propulsion Technology*, 2009, 30(5): 533-537.)

[8] 金　义，何小民，彭春梅，等. 驻涡燃烧室驻涡区三维冷态流动特性数值研究[J]. 南京航空航天大学学报，2014，46(2)：272-279.

[9] 胡　斌，黄　勇，王　方，等. 基于冷态数值模拟的航空发动机燃烧室贫油熄火预测[J]. 推进技术，2012，33(2)：232-238. (HU Bin, HUANG Yong, WANG Fang, et al. Lean Blow-Out Prediction of Aero-Engine Combustor Based on Cold Flow Field Numerical Simulation[J]. *Journal of Propulsion Technology*, 2012, 33(2):232-238.)

[10] 韩吉昂，李晓东，钟兢军，等. 环形中心钝体驻涡燃烧室进气方式研究[J]. 推进技术，2015，36(8)：1206-1214. (HAN Ji-ang, LI Xiao-dong, ZHONG Jing-jun, et al. Study of Air Injection Mode in an Annular Central Bluff-Body Trapped Vortex Combustor[J]. *Journal of Propulsion Technology*, 2015, 36(8): 1206-1214.)

[11] 王铮钧，索建秦，黎　明，等. 基于LPP的多点喷射低污染燃烧室头部方案优化研究[J]. 推进技术，2015，36(7)：1036-1045. (WANG Zheng-jun, SUO Jian-qin, LI Ming, et al. Optimized Study of a Dome for Low Emission Combustor with Multipoint Injection Based on LPP[J]. *Journal of Propulsion Technology*, 2015, 36(7): 1036-1045.)

[12] 徐宝鹏，曾佑杰，马宏宇，等. 燃气轮机模型燃烧室的大涡模拟[J]. 航空发动机，2014，40(3):14-18.

[13] 张济民，张宏达，韩　超，等. 分层旋流燃烧器冷态流场的大涡模拟[J]. 航空动力学报，2014，29(10): 2369-2376.

[14] 颜应文，赵坚行，张靖周，等. 模型燃烧室湍流亚网格尺度模型[J]. 南京航空航天大学学报，2009，41(4):461-465.

[15] Yee Chee See, Matthias Ihme. Large Eddy Simulation of a Gas Turbine Model Combustor [R]. *AIAA* 2013-0172.

轴向三级旋流燃烧室流场结构
大涡模拟

周　瑜[1]，乐嘉陵[1,2]，陈柳君[1]，黄　渊[2]

（1. 西北工业大学 动力与能源学院，西安 710072；

2. 中国空气动力研究与发展中心 超高速空气动力研究所

高超声速冲压发动机技术重点实验室，四川 绵阳 621000）

摘　　要：为深入了解真实航空发动机燃烧室内部复杂流场结构，在自有 CFD 平台上采用动态亚网格模型对一种轴向 3 级旋流燃烧室的单个头部矩形试验模型 0.5 MPa 下冷态流场结构进行了 LES（大涡模拟）。为避免试验模型简化误差，对包括火焰筒上约 2 000 个气膜孔在内的燃烧室所有精细结构进行了完全仿真。计算模拟了燃烧室内复杂流场从静止启动到统计定常状态的完整非定常发展过程。成功捕捉到主旋流与横向对冲射流相互影响作用及涡旋破碎等细观结构，获得的测点湍动能谱与湍流经典理论中惯性子区—5/3 规律一致，LES 时间平均流场结构与已有 PIV（particle image velocimetry）试验结果吻合，表明所建立的高仿真网格与 LES 方法可进一步用于真实航空发动机环形燃烧室流场数值模拟。

关　键　词：航空发动机；单头部燃烧室；3 级旋流器；大涡模拟；动态亚网格模型；粒子成像速度仪

中图分类号：V231.3　　文献标志码：A

Large eddy simulation of flow structure in combustor with
axial triple swirler

ZHOU Yu[1]，LE Jia-ling[1,2]，CHEN Liu-jun[1]，HUANG Yuan[2]

（1. School of Power and Energy，

Northwestern Polytechnical University，Xi'an 710072，China；

2. Science and Technology on Scramjet Laboratory，

Hypersonic Aerodynamics Research Institute，

China Aerodynamics Research and Development Center，Mianyang Sichuan 621000，China）

Abstract：To gain an improved understanding of the complex flow field structures in practical aero-engine combustor，LES（large eddy simulation）with dynamic Smagorinsky sub-grid model on self-developed CFD platform was made to explore the cold flow field in single rectangle dome of triple swirler under 0.5 MPa. The complex geometric configuration including all about 2 000 film cooling holes on liner was fully simulated without any conventional simplification in order to reduce the modeling errors. The capability of the CFD code to simulate very complicated flow was validated by the calculation. Unsteady process of the turbulent swirling flow developing from static to statistically stationary status was totally reproduced. Interaction between mainstream and transverse jet was investigated and swirling

周谕，乐嘉陵，陈柳君，黄渊. 轴向三级旋流燃烧室流场结构大涡模拟. 航空动力学报，2017，32(4): 917-925.

flow characteristics like vortex breakdown were well captured. Turbulent kinetic energy spectrum predicted by calculation was in reasonable agreement with the −5/3 law in classic turbulence theory, while time-averaged flow field predicted by LES fitted well with PIV (particle image velocimetry) measurement. It shows that the high fidelity mesh and large eddy simulation can be applied in the simulation of full annular combustor of aero-engine.

Key words: aero-engine; single dome combustor; triple swirler; large eddy simulation; dynamic Smagorinsky sub-grid model; particle image velocity

现代航空发动机为达到更高的推质比，燃烧室设计在向着高温升方向发展。更高的油气比意味着燃烧室需要在非常宽广的油气比范围内正常工作，其头部进气量分配细节和流动结构成为设计关键和研究难点。为获得更高的气动性能，进一步增强燃气混合，提高燃烧稳定性，满足高温升设计需求，目前燃烧室旋流器已逐渐从单级发展到双级，以及近年来逐渐投入实践应用的 3 级旋流器。如文献[1-5]针对 3 旋流燃烧室，在燃烧效率、稳定工作范围和性能等各方面开展了试验研究。由于航空发动机试验存在消耗资源大，难以获得内部流动细节等局限，数值模拟逐渐成为当前燃烧室设计的一个重要研究和发展方向[6-8]。特别是随着计算机性能的快速进步，LES 以其能分辨流场细观结构，计算量远小于直接数值模拟的优点，被越来越多应用于燃烧室研究[9-13]。如 Iudiciani 等[14]针对一种同轴 3 级旋流燃烧室开展了流动特性、网格和亚网格模型敏感性的 LES 研究，并应用本征正交分解分析了螺旋涡结构对进动涡核的影响。

由于真实航空发动机燃烧室结构十分复杂，对其进行准确建模、网格划分和数值求解难度较大，而 LES 相比常用的基于雷诺平均的湍流模型方法，计算量普遍高出一个量级，因此现有研究工作大多针对较为简单的工业构型或者各种实验室研究用简化燃烧器，对真实燃烧室构型的 LES 研究还不多。文献[15-17]分别对一个单头部矩形燃烧室模型冷态流场和两相喷雾燃烧流场进行了 LES，计算与 PIV 试验测量基本吻合，但由于该模型网格数量只有 134 万，流场细节不够显著。文献[18]指出，真实燃烧室复杂构型和工况的 LES 计算迄今还未被充分广泛应用的一大难点就在于其几何复杂度和高质量贴体网格生成，及其相应带来的计算复杂度和对程序计算能力的较高要求。

本文针对一种轴向 3 级旋流燃烧室的单头部构型，采用 LES 方法研究其冷态流场结构，为尽量减少模拟误差，对模型包括火焰筒多斜孔冷却壁上约 2000 个气膜孔在内的所有精细结构进行了完全仿真，在深入了解获得真实燃烧室多旋流湍流流场结构的同时也进一步验证在自有 CFD 平台 AHL-3D 上开发的基于非结构有限体积离散的不可压流动计算程序面对高度复杂流场的模拟能力。

1 研究对象

本文研究对象为航空发动机燃烧室的单头部矩形试验模型，沿流向包括一个扩压器、轴向 3 级旋流器、火焰筒和机匣。火焰筒采用了多斜孔冷却壁，其上下壁面连同头部挡溅板及旋流器端面上一共布置了约 2000 个气膜孔。在实际冷流试验条件下对该模型流场进行了 LES，其入口流动参数见表 1，边界条件为：壁面无滑移条件，出口变量梯度为零，入口按试验流量给定法向速度均值 U，不施加人工扰动以观察来流经过一个较长的扩压段后自然转捩发展过程。

表 1 来流条件
Table 1 Flow condition

p_∞/MPa	T_∞/K	ρ_∞/(kg/m³)	U/(m/s)
0.5	300	5.81	38.74

对图 1(a)中的燃烧室模型进行了结构/非结构混合网格划分，其中火焰筒及机匣空间采用图 1(b)所示各向同性正四面体非结构网格过渡，而每级旋流叶片之间的通道则采用如图 1(c)所示六面体纯结构网格，气膜孔采用图 1(d)所示近壁面结构/中心非结构网格。在开展正式模拟前，通过多次小规模试算对网格进行了反复迭代调整，保证气膜孔及壁面首层网格位于程序计算判断的边界层黏性子层以内，且多处监控点流动变量的变化趋于网格无关。最终每个气膜孔包含约 450 个网格单元，火焰筒内主流区为 600 万网格单元，

模型整体网格总量为 1626 万.

(a) 燃烧室模型整体视图

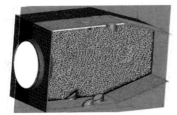

(b) 火焰筒中心截面上网格分布

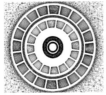

(c) 过2,3级旋流叶片通道　　(d) 火焰筒壁面局部
剖面上网格分布　　　　　气膜孔网格

图 1　燃烧室高仿真计算网格示意图

Fig. 1　Schematic of high fidelity mesh for the combustor

2　数值模拟方法

对 N-S 方程进行过滤得到不可压缩冷态湍流场的 LES 控制方程:

$$\frac{\partial \tilde{\boldsymbol{u}}_i}{\partial \boldsymbol{x}_i} = 0 \qquad (1)$$

$$\frac{\partial \tilde{\boldsymbol{u}}_i}{\partial t} + \frac{\partial (\tilde{\boldsymbol{u}}_i \tilde{\boldsymbol{u}}_j)}{\partial \boldsymbol{x}_j} = \frac{1}{\rho} \left(\frac{\partial^2 \tilde{\boldsymbol{u}}_i}{\partial \boldsymbol{x}_j \partial \boldsymbol{x}_j} - \frac{\partial \tilde{p}}{\partial \boldsymbol{x}_i} - \frac{\partial \boldsymbol{\tau}_{ij}}{\partial \boldsymbol{x}_j} \right) (2)$$

其中亚网格应力 $\boldsymbol{\tau}_{ij} = \widetilde{\boldsymbol{u}_i \boldsymbol{u}_j} - \tilde{\boldsymbol{u}}_i \tilde{\boldsymbol{u}}_j$, 采用涡黏模型封闭有

$$\boldsymbol{\tau}_{ij} = -2\mu_t \tilde{\boldsymbol{S}}_{ij} + \frac{1}{3}\boldsymbol{\tau}_{kk}\boldsymbol{\delta}_{ij} \qquad (3)$$

其中 $\tilde{\boldsymbol{S}}_{ij} = \frac{1}{2}\left(\frac{\partial \tilde{\boldsymbol{u}}_i}{\partial \boldsymbol{x}_j} + \frac{\partial \tilde{\boldsymbol{u}}_j}{\partial \boldsymbol{x}_i} \right)$ 为可解尺度的变形率张量. 黏度 μ_t 需要采用亚网格模型进行模化. 标准的 Smagorinsky 模型为

$$\mu_t = \rho (C_s \hat{\Delta})^2 \sqrt{2\bar{\boldsymbol{S}}_{ij}\bar{\boldsymbol{S}}_{ij}} \qquad (4)$$

式中 C_s 为 Smagorinsky 常数; $\hat{\Delta}$ 为过滤尺度. $\hat{\Delta} = 2|\Omega|^{1/3}$, $|\Omega|$ 为网格单元体积. 航空发动机燃烧

室内流场十分复杂. 采用统一的 C_s 对全局亚网格应力进行模化的标准模型往往会出现耗散过大等问题. 而本文采用的动态亚网格模型 (dynamic Smagorinsky sub-grid model, DSM) 则考虑了可解尺度脉动的局部特性, 应用两种尺度对方程进行两次过滤. 把湍流局部结构信息引入亚网格应力从而在计算过程中动态地确定模型系数, 获得更合理的取值. 其缺点是动态计算系数过程中可能出现负值, 需要人工修正. 动态亚网格模型假设两次过滤尺度均在惯性子区范围内, 两次过滤的亚网格应力系数相等, 则应用 Germano 等式

$$\boldsymbol{L}_{ij} = (\boldsymbol{u}_i)_{1,2}(\boldsymbol{u}_j)_{1,2} - \left[(\boldsymbol{u}_i)_1 (\boldsymbol{u}_j)_1 \right]_2 = (\boldsymbol{\tau}_{ij})_{1,2} - \left[(\boldsymbol{\tau}_{ij})_1 \right]_2 \qquad (5)$$

得到

$$\boldsymbol{L}_{ij} - \frac{1}{3}\boldsymbol{L}_{kk}\boldsymbol{\delta}_{ij} = C_s \boldsymbol{M}_{ij}$$

$$\boldsymbol{M}_{ij} = 2\left[\Delta_2^2 |\bar{\boldsymbol{S}}|\bar{\boldsymbol{S}}_{ij} - \Delta_1^2 \widehat{|\bar{\boldsymbol{S}}|\bar{\boldsymbol{S}}_{ij}} \right] \qquad (6)$$

其中上标 "—" 为尺度 Δ_1 的过滤, 上标 "~" 为尺度 Δ_2 的过滤. 最终得到动态确定的 Smagorinsky 常数 C_s

$$C_s = \frac{\langle \boldsymbol{L}_{ij}\boldsymbol{M}_{ij} \rangle}{\langle \boldsymbol{M}_{ij}\boldsymbol{M}_{ij} \rangle} \qquad (7)$$

计算采用基于非结构有限体积离散的 SIMPLE (semi-implicit method for pressure linked equations) 算法求解, 时间 2 阶隐式, 空间 2 阶迎风格式.

3　燃烧室冷态流场 LES

燃烧室试验模型从扩压段空气入口到火焰筒出口整体长度约为 $775\,\mathrm{mm}$, 一个流动特征时间长度按入口速度估算为 $20\,\mathrm{ms}$. 由于完全仿真了直径约 $1\,\mathrm{mm}$ 的气膜孔, 网格中存在的大量体积极小的单元限制了时间步长无法取较大的数值. 在正式计算前, 同样通过多次试算获得 CFL (Courant-Friedrichs-Lewy) 数分布并验证计算稳定性. 最后综合考虑计算效率, 启动过程的精细模拟以及流场的时间分辨精度. 时间步长在流场启动初期取为 $1 \times 10^{-7}\,\mathrm{s}$. 流场初步建立后再逐渐增大到 $5 \times 10^{-7}\,\mathrm{s}$ 以尽快促进流动发展到统计定常状态. 计算在 40 个 Intel Xeon X5670 CPU 共 480 个计算核心上并行开展, 一共计算了 40 万步, 物理时间上推进 $195\,\mathrm{ms}$. 消耗 CPU 机时约 30 万 (核·小时).

3.1　流场变量统计

以旋流器出口 $z = 0\,\mathrm{mm}$ 截面圆心为坐标原点 $A(0,0,0)$ (单位: mm), 在燃烧室空间设置了多处监控点以观察流场变化情况. 图 2 给出了旋流器出口下游 $60\,\mathrm{mm}$ 处前排主燃孔所在截面中心点 $B(0,0,-60)$ 瞬时流向速度 w 随时间的变化. 该点空间位置同时受到多级旋流及主燃孔横向射流影响, 流动情况比较复杂. 观察发现经过初期一个比较明显的启动发展过程后, 在 $40\sim50\,\mathrm{ms}$ 后速度脉动逐渐趋于稳定, 开始进入统计定常状态. 因此计算中在超过 2 个特征时间长度之后从 $48\,\mathrm{ms}$ 开始对流动变量作统计和时间平均处理. 该点统计的速度时间平均值为 $-38.31\,\mathrm{m/s}$, 脉动速度方均根值为 $10.81\,\mathrm{m/s}$, 湍流强度达到 28.22%, 位于旺盛的湍流区域.

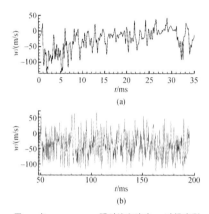

图 2   点 $B(0,0,-60)$ 瞬时流向速度 w 时间序列

Fig. 2   Time-series of instantaneous flow velocity w on probe point $B(0,0,-60)$

为获得燃烧室内部空间关键位置更详细的流场脉动信息, 进一步取火焰筒后排中心掺混孔所在截面中点 $C(0,0,-95)$、1 级旋流器与 2 级旋流器出口交汇处点 $D(0,8,0)$ 及 3 级旋流器所在径向 $35\,\mathrm{mm}$ 处点 $E(0,35,0)$ 的流向速度信息进行了研究. 图 3 显示了各点流向脉动速度的能谱, 湍动能随频率的变化与湍流经典理论中惯性子区的 $-5/3$ 律基本相符, 在频率超过 $10^4\,\mathrm{Hz}$ 后也出现了与湍流经典理论曲线一致的迅速衰减的耗散区, 表明本文所用网格和计算较好地分辨出了燃烧室内湍流脉动. 其中 B、C 两点由于分别同时受到主旋流和前后排主燃孔/掺混孔横向射流影响, 其湍动能数值相比旋流器出口平面上的 D、E 两点高出约一个量级. 图 4 给出了上述各点瞬时速度 w 的功率谱信息, 并增加了旋流器出口沿 x 方向半径 $8\,\mathrm{mm}$ 和 $35\,\mathrm{mm}$ 处的两个监控点 $F(8,0,$

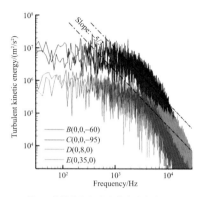

图 3   监控点流向脉动速度湍动能谱

Fig. 3   Turbulent kinetic energy spectrum of flow velocity on probe points

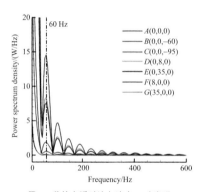

图 4   监控点瞬时流向速度 w 功率谱

Fig. 4   Power spectrum density of flow velocity w on probe points

$0)$ 和 $G(35,0,0)$ 作为对照. 可以看出所有测点的功率谱密度曲线结构非常相似, 均存在一个显著的对应频率约 $60\,\mathrm{Hz}$ 的最大特征峰, 以及随着频率增长而迅速衰减的一串较小的顶峰, 表明流场中可能存在一个以该特征频率进行脉动的主要含能模态. 旋流器中轴线上 A、B、C 三点的特征峰随测点位置沿流向移动而大幅衰减, 体现了其出口流动对下游流场的影响随距离的增长而显著弱化. 点 D 和点 F 均位于坐标原点径向 $8\,\mathrm{mm}$ 处 1 级和 2 级旋流器出口剪切层, 因此其功率谱曲线几乎完全一致. 由于火焰筒为宽高不等的矩形框体, 其几何结构限制了旋流器出口附近流场不可能为理想的轴对称, 从而导致两者存在微小差异. 由于所处空间位置存在旋流器端面和壳体的遮挡, 点 E 和点 G 处流动十分平缓, 其功率密度分

布近似为一条贴近 O 轴的直线.

3.2 燃烧室主流场结构

来流空气在模型燃烧室内的流动经过一个较长的扩压段发展后,分别经轴向3级旋流通道、主燃孔/掺混孔以及火焰筒壁上数千个气膜孔进入燃烧室内部,多股气流的耦合与相互作用产生了复杂的流场结构.此外,在流场启动初期,来流还会首先与模型内部存在的静止空气发生相互作用,在挤压、排气的流动过程中形成一些特殊的瞬态结构.图5给出了0.85 ms时刻燃烧室内瞬态三维流线分布及旋流器局部流场.图5(a)中旋流器出口环状涡量等值面上流向速度 w 的分布表明,此刻经1级和2级旋流通道流出的高速流体在出口附近卷起了一个圆环型局部回流区,该区域的流线由内向外翻卷成一种类似弹簧圈型的分布(见图5(b)).而3级旋流受旋流器端面的阻挡,在叶片尾部与端面之间形成了如图5(c)所示的局部回流并被迫偏转沿径向空间运动.由于流体的旋转运动会在径向上产生压力梯度从而形成局部低压区,因此在某些特定时刻流场低压区域分布可以体现旋流结构.图6(a)给出了不同时刻燃烧室内 $-1\ kPa$ 和 $-5\ kPa$ 的压强增量等值面分布.3级旋流通道流出的空气沿径向空间运动的过程中,其旋转离心作用以及火焰筒壁面限制作用共同诱导形成了上下两条显著的旋臂型压强等值面.结合图6(b)对应时刻中心截面 $x=0\ mm$ 上低压区及流线分布可以看出,该等值面与 $x=0\ mm$ 平面相交的圆形截面上存在明显的径向压力梯度,其流线形状表现为一个局部回流区,表明该

旋臂结构为典型的内部负压的管状涡旋.在实际计算过程中,这种低压区结构会随时间发展而间歇性地出现—消失,但拐角处的局部回流区一直保持存在状态,同时,1级与2级旋流器出口流体的旋转剪切相互作用也形成了局部低压区,里面包含了一对随时间发展的对称旋涡,这对旋涡向下游运动的过程中在主燃孔射流的横向剪切作用下产生回流,在 $t=6.9\ ms$ 时刻已扩展为充满整个火焰筒的两个大涡,并最终形成如图7所示时间平均的速度矢量分布以及等值线 $w=0\ m/s$ 所勾勒出的回流区结构.3级旋流在火焰筒壁面限制下形成了固定的角回流区,同时还与1,2级旋流器出口的主旋流耦合作用在旋流器端面附近产生了次级回流区.这两处回流结构有利于燃油充分雾化、提升燃烧性能,并对主旋流张角、下游回流区尺寸等形成了一定的调节控制作用.而主旋流与3级旋流交汇后,在主燃孔射流横向剪切作用下回转形成了包含一对大尺度旋涡的中心主回流区,有利于火焰稳定.

图8展示了流场启动发展过程中瞬时涡量场结构的非定常变化.受叶片几何形状的限制,来流经过旋流器各级旋流通道后产生周向旋转运动,如图8(a)、图8(b)所示,3级旋流器每个通道出口均形成了与叶片形状对应的整齐排列的涡量等值面.随着时间发展,初始型面在维持了约0.5 ms后,于流体自身旋转剪切作用及挡溅板上气膜孔出流的影响下从图8(c)开始逐渐脱落并破碎.而1,2级旋流首先在旋流器内部经历了相互混合与剪切作用再由共同的出口流出,因此其出口涡旋未能形成规则的拟序结构.火焰筒与外机匣之间

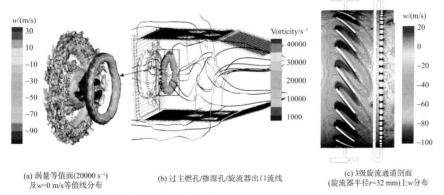

(a) 涡量等值面(20000 s⁻¹)
及w=0 m/s等值线分布　　　(b) 过主燃孔/掺混孔/旋流器出口流线　　　(c) 3级旋流通道剖面
(旋流器半径r=32 mm)上w分布

图5 $t=0.85\ ms$ 时刻燃烧室瞬态流场

Fig.5　Instantaneous flow field in the combustor at $t=0.85\ ms$

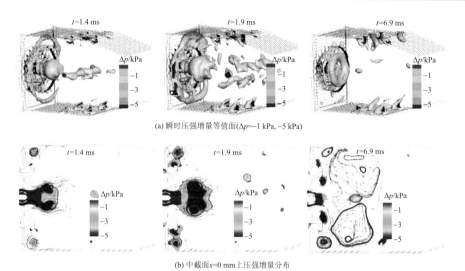

(a) 瞬时压强增量等值面(Δp=—1 kPa, –5 kPa)

(b) 中截面x=0 mm上压强增量分布

图 6　燃烧室内流场瞬时低压区结构

Fig. 6　Distribution of instantaneous low pressure region in the combustor

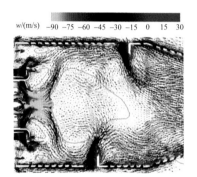

图 7　中截面 $x=0$ mm 上时间平均的速度矢量分布
（等值线 $: w=0$ m/s）

Fig. 7　Distribution of time-averaged velocity vector
on central profile $x=0$ mm (isoline $: w=0$ m/s)

的气流则通过各个主燃孔/掺混孔横向挤压进入主流区,在孔口膨胀形成如图 8(a)所示环管型旋涡结构,这种结构在向流场深入运动的过程中逐渐拉伸(图 8(b)),脱落(图 8(c)),约 1.0 ms 时在火焰筒中部相遇(图 8(d)),在强烈的相互碰撞与剪切作用下破碎成图 8(e)中所示大量的细小涡旋,随后进入图 8(f)所示无序脉动变化状态。

　　为确认计算得到的燃烧室内部流场结构,图 9 和图 10 分别给出了外侧 $x=25$ mm 和内侧 $x=$ -25 mm 截面上平面速度值 V_{yz} 的 LES 时间平均结果与相同来流条件下该燃烧室模型已有 PIV 试验测量结果的比较。其中试验的矩形观测窗口

开设在侧壁中间,尺寸约为 $110\,\text{mm}\times72\,\text{mm}$,观测范围基本覆盖了从旋流器出口到后排掺混孔在内的核心流动区域。图中可以看出上下孔两股射流的错位碰撞剪切作用使得两孔之间的流线发生扭曲转向,在旋流器出口主流的影响下,主燃孔上游流线也出现了扭转形成回流小涡。整体而言计算与试验基本一致,LES 得到的两个截面的上下孔射流强度、穿透深度对称性更好一些,试验测得的外侧截面上孔射流较强而下孔则相对稍弱,因此图 9(a)中由下孔进入的流线水平夹角相比计算结果要小一些。而在内侧截面上,图 10(a)测得的下孔射流强度与计算相当,上孔射流强度较弱导致其与上游主流耦合形成的一个局部小回流区维持其旋涡形态,而非如计算所示被射流挤压成大幅度扭曲变向的流线。这种差别一个可能的原因在于本文的 LES 没有在入口添加额外的人工扰动,计算等同于模拟了一个自然转捩的流动状态,同时实际试验中入口来流存在一定振荡和不均匀性,不利于 PIV 测量获得准确的平均流场。

　　等 x 截面上的流场反映了火焰筒上下孔对冲射流的相互作用影响。为继续考察旋流器对主旋流场结构形成的诱导作用,取其出口附近两个等 z 截面进行了对比研究。图 11 为旋流器下游 $z=-30$ mm 截面上的平面速度值 V_{xy} 及流线分布,其出口气流在此截面上旋转形成了一个充满整个框体的大涡,计算与试验基本一致,差别主要

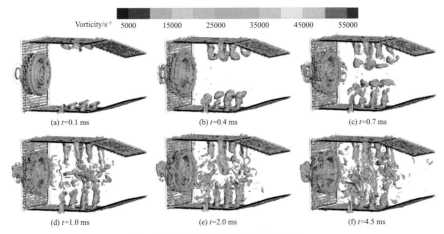

(a) $t=0.1$ ms (b) $t=0.4$ ms (c) $t=0.7$ ms

(d) $t=1.0$ ms (e) $t=2.0$ ms (f) $t=4.5$ ms

图 8　流场发展过程中瞬时涡量等值面分布

Fig. 8　Distribution of instantaneous vorticity in flow field developing process

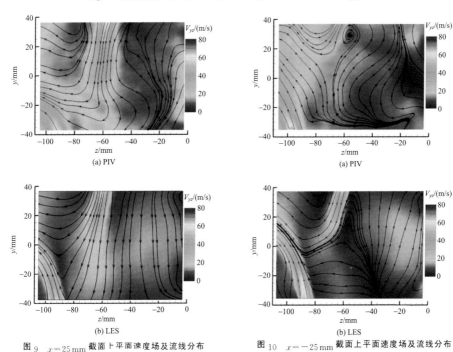

(a) PIV

(b) LES

图 9　$x=25$ mm 截面上平面速度场及流线分布

Fig. 9　Distribution of planar velocity and streamlines on profile $x=25$ mm

(a) PIV

(b) LES

图 10　$x=-25$ mm 截面上平面速度场及流线分布

Fig. 10　Distribution of planar velocity and streamlines on profile $x=-25$ mm

在于图 11(a) 中试验测得的旋涡中心与 x 轴零点存在约 10 mm 的偏移，流线轨迹更接近于圆形，而图 11(b) 中计算模拟的涡心严格落在坐标原点、旋流器中轴线上，流线轨迹近似呈现为一个倾斜的椭圆形，其长轴方向与截面左下到右上的对

角线基本一致。由于火焰筒剖面为上下较高左右较窄的矩形，且旋流方向为顺时针，因此该截面的流线轨迹形状应当体现出壁面几何的这种不对称约束对旋流的影响，即 LES 计算的平均场流线更符合该截面构型在理论上应有的分布。图 12 为

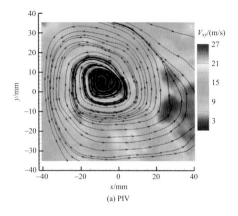

(a) PIV

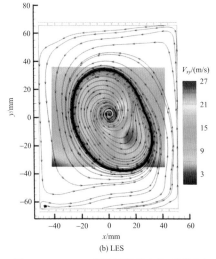

(b) LES

图 11　$z=-30\,\mathrm{mm}$ 截面上平面速度场及流线分布

Fig. 11　Distribution of planar velocity and streamlines on profile $z=-30\,\mathrm{mm}$

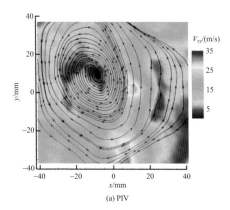

(a) PIV

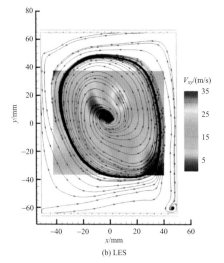

(b) LES

图 12　$z=-45\,\mathrm{mm}$ 截面上平面速度场及流线分布

（后附彩图）

Fig. 12　Distribution of planar velocity and streamlines on profile $z=-45\,\mathrm{mm}$

$z=-45\,\mathrm{mm}$ 截面上的平面速度值及流线分布, 可以看出与图 11 类似, 试验测得的大涡中心在 x 方向相对原点存在约 $10\,\mathrm{mm}$ 的偏移, 而计算得到的旋涡中心继续保持在截面中点上. 由于进一步远离了旋流器出口, 该截面上旋流强度减弱导致流线轨迹的椭圆化和偏斜程度不如 $z=-30\,\mathrm{mm}$ 截面显著.

4　结　论

本文对一种带轴向 3 级旋流器的航空发动机单头部燃烧室进行了高仿真建模, 并采用 LES 方法对其冷态流场进行了数值计算, 获得了燃烧室从静止启动发展到统计定常状态过程中关键位置多处测点、截面与三维空间的非定常流动特性及流场结构.

1) 第 3 级旋流分别与火焰筒壁面和主旋流耦合作用形成了角回流区及旋流器端面回流区, 此种设计可促进燃油进一步充分雾化, 提高燃烧性能. 火焰筒中心仍存在与传统双旋流燃烧室类似的由主旋流和横向射流诱导的主回流区, 有利于火焰稳定.

2) 计算的测点湍动能谱符合湍流经典理论规律, 时间平均流场与 PIV 测量结果相吻合, 表明本文所建立的高仿真计算网格与 LES 方法可进一步拓展应用于航空发动机完整环形燃烧室复杂流场数值模拟, 为燃烧室设计提供可靠依据.

参考文献:

[1] 刘伟,林宇震,刘高恩,等. 低压条件下复合式多级旋流杯燃烧室燃烧效率研究[J]. 航空动力学报,2004,19(5):650-655.
LIU Wei,LIN Yuzhen,LIU Gaoen,et al. Research of combustion efficiency of combustors using hybrid airblast atomizer under low pressure[J]. Journal of Aerospace Power,2004,19(5):650-655. (in Chinese)

[2] 袁怡祥,林宇震,刘高恩. 三旋流器头部燃烧室拓宽燃烧稳定工作范围的研究[J]. 航空动力学报,2004,19(1):142-147.
YUAN Yixiang,LIN Yuzhen,LIU Gaoen. Combustor dome design with three swirlers for widening the operation stability range[J]. Journal of Aerospace Power,2004,19(1):142-147. (in Chinese)

[3] 丁国玉,何小民,赵自强,等. 油气比及进口参数对三级旋流器燃烧室性能的影响[J]. 航空动力学报,2015,30(1):53-58.
DING Guoyu,HE Xiaomin,ZHAO Ziqiang,et al. Effect of fuel-air ratio and inlet parameters on performance of triple swirler combustor[J]. Journal of Aerospace Power,2015,30(1):53-58. (in Chinese)

[4] 王成军,江平,辛欣,等. 基于PIV技术对三级旋流杯燃烧流场的测量[J]. 航空动力学报,2015,30(5):1032-1039.
WANG Chengjun,JIANG Ping,XIN Xin,et al. Measurement of triple-stage swirler cup combustor flow field based on PIV technology[J]. Journal of Aerospace Power,2015,30(5):1032-1039. (in Chinese)

[5] 彭云晖,林宇震,刘高恩. 三旋流燃烧室出口温度分布的初步试验研究[J]. 航空动力学报,2007,22(4):554-558.
PENG Yunhui,LIN Yuzhen,LIU Gaoen. A preliminary experimental study of pattern factor for a triple swirler combustor[J]. Journal of Aerospace Power,2007,22(4):554-558. (in Chinese)

[6] 李锋,郭瑞卿,尚守堂,等. 高温升三旋流燃烧室与双旋流燃烧室的性能对比[J]. 航空动力学报,2015,30(1):10-15.
LI Feng,GUO Ruiqing,SHANG Shoutang,et al. Performance comparison between high temperature rise triple-swirler combustor and double-swirler combustor[J]. Journal of Aerospace Power,2015,30(1):10-15. (in Chinese)

[7] 吴治永,林宇震,刘高恩,等. 基于CFD分析改进三旋流燃烧室头部设计[J]. 推进技术,2009,30(5):533-537.
WU Zhiyong,LIN Yuzhen,LIU Gaoen,et al. Dome structure modification of a high temperature rise combustor based on CFD simulation[J]. Journal of Propulsion Technology,2009,30(5):533-537. (in Chinese)

[8] 吴振宇,王成军,王丹丹. 三级旋流器的设计及其流场模拟[J]. 沈阳航空工业学院学报,2010,27(5):38-41.
WU Zhenyu,WANG Chengjun,WANG Dandan. Triple swirler design and fluid numerical simulation[J]. Journal of Shenyang Institute of Aeronautical Engineering,2010,27(5):38-41. (in Chinese)

[9] Tucker P G. Computation of unsteady turbomachinery flows:Part 2 LES and hybrids[J]. Progress in Aerospace Sciences,2011,47(7):546-569.

[10] Kim S,Menon S. Large-eddy simulation of a high-pressure single-element lean direct-injected gas-turbine combustor[R]. AIAA-2014-0131,2014.

[11] Filosa A,Noll B E,Di Domenico M,et al. Numerical investigations of a low emission gas turbine combustor using detailed chemistry[R]. AIAA-2014-3916,2014.

[12] Yee C S,Ihme M. LES investigation of flow field sensitivity in a gas turbine model combustor[R]. AIAA-2014-0621,2014.

[13] Ruggles A,Kelman J. A gas turbine combustor for instability research and LES validation:methods and mean results[J]. Combustion Science and Technology,2014,186(3):313-331.

[14] Iudiciani P,Duwig C,Hosseini S M,et al. LES investigation and sensitivity analysis of the flow dynamics in a gas turbine swirl combustor[R]. AIAA-2011-65,2011.

[15] 颜应文,赵坚行,张靖周,等. 模型燃烧室满流亚网格尺度模型[J]. 南京航空航天大学学报,2009,41(4):461-465.
YAN Yingwen,ZHAO Jianxing,ZHANG Jingzhou,et al. Sub-grid scale model for model combustor[J]. Journal of Nanjing University of Aeronautics and Astronautics,2009,41(4):461-465. (in Chinese)

[16] 颜应文,赵坚行,张靖周,等. 大涡模拟模型环形燃烧室污染特性[J]. 航空动力学报,2008,23(7):1161-1167.
YAN Yingwen,ZHAO Jianxing,ZHANG Jingzhou,et al. Large eddy simulation of pollution formation in model annular combustor[J]. Journal of Aerospace Power,2008,23(7):1161-1167. (in Chinese)

[17] 何跃龙,邓远灏,颜应文,等. 大涡模拟模型燃烧室燃烧性能计算[J]. 航空动力学报,2012,27(9):1939-1947.
HE Yuelong,DENG Yuanhao,YAN Yingwen,et al. Large-eddy simulation of two-phase reacting flows and combustion performance in model combustor[J]. Journal of Aerospace Power,2012,27(9):1939-1947. (in Chinese)

[18] See Y C,Ihme M. Large eddy simulation of a gas turbine model combustor[R]. AIAA-2013-0172,2013.

Design and analysis osculating general curved cone waverider

Xuzhao He, Jialing Le and Si Qin

Science and Technology on Scramjet Laboratory, Hypervelocity Aerodynamics Institute of CARDC, Mianyang, China

Abstract

Purpose – Waverider has high lift to drag ratio and will be an idea aerodynamic configuration for hypersonic vehicles. But a structure permitting aerodynamic like waverider is still difficult to generate under airframe's geometric constrains using traditional waverider design methods. And furthermore, traditional waverider's aerodynamic compression ability cannot be easily adjusted to satisfy the inlet entrance requirements for hypersonic air-breathing vehicles. The purpose of this paper is to present a new method named osculating general curved cone (OCC) method aimed to improve the shortcomings of traditional waveriders.

Design/methodology/approach – A basic curved cone is, first, designed by the method of characteristics. Then the waverider's inlet captured curve and front captured tube are defined in the waverider's exit plane. Osculating planes are generated along the inlet captured curve and the designed curved cone is transformed to the osculating planes. Streamlines are traced in the transformed curved cone flow field. Combining all streamlines which have been obtained, OCC waverider's compression surface is generated. Waverider's upper surface uses the free stream surface.

Findings – It is found that OCC waverider has good volumetric characteristics and good flow compression abilities compared with the traditional osculating cone (OC) waverider. The volume of OCC waverider is 25 per cent larger than OC waverider at the same design condition. Furthermore, OCC waverider can compress incoming flow to required flow conditions with high total pressure recovery in the waverider's exit plane. The flow uniformity in the waverider exit plane is quite well.

Practical implications – The analyzed results show that the OCC waverider can be a practical high performance airframe/forebody for hypersonic vehicles. Furthermore, this novel waverider design method can be used to design a structure permitting aerodynamic like waverider for a practical hypersonic vehicle.

Originality/value – The paper puts forward a novel waverider design method which can improve the waverider's volumetric characteristics and compression abilities compared with the traditional waverider design methods. This novel design approach can extend the waverider's applications for designing hypersonic vehicles.

Keywords General curved cone, Method of characteristics, Osculating method, Streamline tracing, Waverider

Paper type Research paper

Nomenclature

Symbols

ϕ	= Ratio of long to short axis of super ellipse [-];
n	= exponent of super ellipse [-];
θ	= Quadrant angle of super ellipse °;
a, b, c, d, e and f	= quintic curve's coefficients [-];
H	= Height between ICC and FCT at symmetry plane [m];
L	= Straight line's length in FCT curve [m];
R_A	= Axisymmetric radius at point A [m];
V	= Waverider's volume [m³]; and
S	= waverider's wet surface area [m²].

Definitions, acronyms and abbreviations

OCC	=	Osculating general curved cone;
OC	=	Osculating cone;
ICC	=	Inlet captured curve;
FCT	=	Front captured tube;
MOC	=	Method of characteristic; and
AHL3D	=	Air-breathing hypersonic lab three dimensional.

Introduction

Waverider has high lift to drag ratio which can be used as hypersonic vehicles' airframe or forebody. Waverider forebody has many advantages compared with the traditional lift forebodies (Duveau *et al.*, 1999) for air-breathing hypersonic vehicle, such as high flow capture ratio and high flow uniformity. A lot of studies have been conducted to study the waverider configurations and their applications. Bowcutt *et al.*

Xuzhao He, Jialing Le, Si Qin. Design and analysis osculating general curved cone waverider. Aircraft Engineering and Aerospace Technology, 2017, 89(6): 797-803.

(1987) and Corda and Anderson (1988) studied waverider configuration's optimization problems considering the viscous effect. Lewis (1991) used the waverider as vehicle's forebody to design the air-breathing hypersonic vehicle. Mazhul (2010) studied the waverider's performances under off-design conditions. Takashima and Lewis (1992, 1994a, 1994b) studied and optimized the waverider's performance under viscous conditions using the viscous Navier–Stokes computation tools. Takashima et al. (1996) designed the hypersonic dual fuel vehicle using the waverider configurations.

Previous waverider studies as introduced above are all based on the traditional waverider configuration design method. As we know, the waverider concept was proposed by Nonweiler (1959) in 1950s first, and the original concept gradually evolved four kinds of waverider design methods. The first kind of waverider design method was introduced by Nonweiler (1963). The streamline tracing technique and two-dimensional inviscid supersonic wedge flow filed were used to design wedge-based waveriders. The second kind of waverider design method is derived from supersonic flow passing outer or inner cones, which have axisymmetric shock wave and flow field structures, such as Goonko et al.'s (2000) introduction. The third kind of waverider design method is derived from supersonic wedge and cone flow fields and elliptic cones flow fields, such as Takashima and Lewis (1994a, 1994b) and Rasmussen's (1980) introduction. The final but most popular waverider design method is introduced by Sobieczky et al. (1990), called osculating cone (OC) method, which uses supersonic outer cone flow field and osculating method to generate waverider. Later, Sobieczky et al. (1997) used curved shockwave and osculating method to design waverider. Rodi (2005) extends the OC method to osculating flow field method which uses power law bodies to generate the basic flow field and osculating method to design waverider, but Rodi (2005) did not analysis the waverider's flow field and also the design method was not verified in the paper.

As shown above, many waverider design methods have been put forward presently, but there are still some difficulties for using a waverider in an utility vehicle. The first obstacle is geometric structure constrain. Using present waverider design methods, a structure permitting aerodynamic-like waverider cannot be generated under airframe's geometric constrains. The designed waveriders are too bend to satisfy the structure or volume requirements. For this reason, the practical hypersonic waverider vehicles do not appear up to now.

The second obstacle for traditional waveriders is that the waverider's aerodynamic compression ability cannot satisfy the inlet entrance requirements. For a 10 degree wedge or cone under Mach 6 incoming flow, only Mach number 4.7 or 5.0 can be obtained when the flow passes through the leading edge shock compression. The compressed Mach number is too high to fit the inlet inner compression part's requirement. If the wedge or cone angle is increased, then the compressed Mach number will decrease but the total pressure recovery will decrease too. This kind of waverider is obviously not suitable for airframe propulsion integration.

A new design method named osculating general curved cone (OCC) method is presented in this paper for practical waverider design which can satisfy the geometric needs and inlet requirements of the future hypersonic air-breathing vehicles. The purpose to use "general curved cone" is to refer to a kind of cone with arbitrary curved surface, but it has a straight leading edge shock. The curved cone surface ensures that the aerodynamic compression ability and volume characteristics of the OCC waverider can be adjusted to fit with the requirements of the potential air-breathing hypersonic vehicle. The aerodynamic characteristics and flow field structures are simulated and analyzed in detail. The theoretical designing results are compared with the inviscid CFD simulation results. The comparison study between OCC and OC waveriders is also presented, which shows that OCC waverider will be a better aerodynamic configuration for hypersonic vehicle than OC waverider.

General curved cone design

General curved cone is used as the basic flow flied for OCC waverider. Curved cone is designed using the method of characteristic (MOC) for axisymmetric flow. Details of the MOC have been introduced by Zucrow and Hoffman (1977).

Figure 1 shows the curved cone flow field structure. The curved cone includes three sections. The first section (OP) is a straight cone which generates a straight initial shock wave and finishes the initial compression. Taylor–Maccoll equations (Zucrow and Hoffman, 1977) were used to solve the flow field variables in this part. The second part is a curve (section PQ) which is at a tangent with section OP at point P. It can be any curve which continuously compresses the incoming flow to the required flow conditions and at the same time, do not disturb the initial straight shock AO. For instance, PQ's shape can be defined first and then MOC tools (Zucrow and Hoffman, 1977) are used to solve the flow field which is determined by curve PQ and flow variables on AP. AP is a characteristic line and its flow variables are obtained by Taylor–Maccoll equations. For a special case, PQ can be an isotropic compression section whose isotropic compression waves meet at point A. The methods introduced by Anderson (1969) can be used to design the isotropic compression curve. The third section QC is a curve which is at a tangent with section PQ at point Q. Its shape can be adjusted to fit the specific requirements for inlet entrance conditions. Shape of QC can be defined by cubic curve, and for the present case, it is a straight line tangent with curve PQ. Flow field determined by transition section QC and characteristic line QA is calculated by MOC.

Figure 2 shows the specific designing result of a curved cone flow field, for which incoming free stream Mach number is 6.0. The curved cone has a 10 degree initial straight cone, and then an isotropic compression curve PQ connects to the straight cone. The Mach number at point A at the end of

Figure 1 General basic curved cone flow field structure

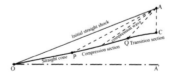

Figure 2 MOC design result of a Mach 6 to Mach 4 curved cone

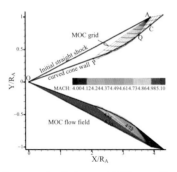

isotropic compression fan is set to 4.0. The transition section QC for this specific design case is a straight line tangent with the isotropic compression section. The curved cone flow field's Mach number contour and MOC design grid are shown in Figure 2.

Osculating general curved cone waverider design method

Many articles have introduced OC waverider design method (Sobieczky *et al.*, 1990; Chauffor *et al.*, 2004). In OC method, the waverider's inlet captured curve (ICC) and front captured tube (FCT) are defined first in the waverider's exit plane. Second, osculating planes are generated along ICC. Third, in an osculating plane, streamline is traced in a straight conical flow field. Combining all streamlines which have been obtained in the osculating plane, OC waverider's compression surface is generated. Normally, waverider's upper surface uses the free stream surface. The difference between OCC and OC methods is that OCC method uses the designed curved cone as the osculating flow field. And the similarity law for the inviscid basic flow field is used in different osculating planes.

For the OCC waverider design method, ICC and FCT curves are defined in the OCC waverider's exit plane first. Actually, any second derivatives continuity curves can be used as the ICC. For a specific example, ICC is defined as a super elliptic curve. The definition of ICC is shown as Equation (1) and ICC's shape is shown in Figure 3:

$$x = \phi(\cos(\theta))^{2/n} \quad y = (\sin(\theta))^{2/n} \tag{1}$$

ϕ is the ratio of long to short axis and n is exponent of super ellipse. A reasonable shock shape curve can generated by

Figure 3 Sketch map of ICC and FCT curves in waverider exit plane

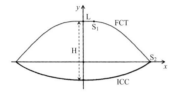

properly adjusting ϕ, n and θ. The parameters of the ICC curve are defined as $\phi = 2$, $n = 2$ and $\theta = 0.6$ presently.

The FCT line uses a straight line connected to a quintic curve to generate. Its shape is shown in Figure 3 and the formulas are shown below:

$$y = H \; x \leq L \; in \; DC \; section \tag{2}$$

$$y = ax^5 + bx^4 + cx^3 + bx^2 + ex + f \; in \; CB \; section \tag{3}$$

Where a, b, c, d, e and f are quintic curve's coefficients which can be determined by geometric information at point S_1 and S_2. H is the height between ICC and FCT at symmetry plane. The straight line's length L is set to 0.1 H. The quintic has zero first-order derivation and zero curvature at point S_1. Its first-order derivation and curvature at point S_2 are -1.0 and 0.1, respectively.

Then, the curvature center of an arbitrary point, such as point A, on ICC can be found, which is A'. An osculating plane AA' can be generated, as shown in Figure 4. Osculating plane is a vertical plane which is perpendicular with the waverider's exit plane. The osculating plane AA' intersects with the FCT curve at point D in the waverider's exit plane.

Figure 5 shows the design procedures in the osculating plan. In the osculating plane AA', the designed curved cone is used as the osculating flow field. The inviscid similarity law is used to scale the designed curved cone flow field to match with the geometrical size which is defined in the osculating plane. Such as the corresponding points A-B-C-D-A' must match with each other in Figures 4 and 5. Then, as shown in Figure 5, a horizontal line stretches out from point D and intersects with the initial straight shock OA at D'. After that, a streamline is traced from the leading edge point D' until the exit of the

Figure 4 Design sketch map in the waverider's exit plane

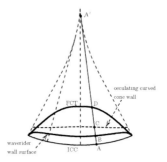

Figure 5 Design sketch map in the osculating plane AA'

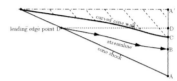

curved cone flow field. This streamline forms the waverider's lower surface in the osculating plan AA'.

Repeating the previous procedures along ICC point by point, the entire OCC waverider's compression surface is generated. The free stream surface is used as the upper surface of OCC waverider. Figure 6 shows the three side views of a designed OCC waverider.

Flow field structure analysis of osculating general curved cone waverider

CFD software AHL3D (He *et al.*, 2006) is used to simulate the designed waverider's inviscid aerodynamic characteristics. AHL3D can simulate two-dimensional, axisymmetric or three-dimensional Euler or Navier–Stocks equations on structured multi-block grids. A variety of robust upwind schemes can be used for inviscid flux construction. The working fluid can be simulated as a calorically perfect single component gas or a mixture of thermally perfect gas.

The simulation is conducted at design condition, which is Mach number 6, zero angle of attack. Simulation results will be compared with design ones to verify OCC waverider's design method. Figure 7 shows the CFD simulation grids. Structured grids with four blocks and totally 3.2 million points are used in present simulation.

In Figure 8, pressure contour in the OCC waverider's exit plane is shown. Figure 9 shows the waverider's three-dimensional shock wave structure which is generated by the iso-pressure surface of CFD simulation result. Above results show that the three-dimensional shock wave attaches with the OCC waverider's side edge closely. The shock wave attachment characteristic of the waverider is show clearly from those two figures.

Figure 10 is the Mach number contour in OCC waverider's symmetric plane. The flow field structure in symmetric plane

Figure 6 Three sides view of the designed OCC waverider

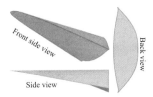

Figure 7 CFD calculation grid of the designed OCC

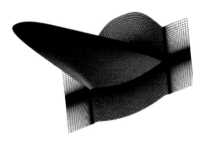

Figure 8 pressure contour in exit plane of the designed OCC waverider

Figure 9 The OCC waverider and its three dimensional shock wave

Figure 10 Mach number contour in symmetry plane of the designed OCC waverider

is identical with the basic curved cone's flow field structure. The initial compression shock and isentropic compression wave can be seen clearly in waverider's symmetry plane.

Figure 11 shows the pressure contour (p/p_∞) comparison between the CFD simulation and theoretical design results on the OCC waverider's compression surface. The pressure

Figure 11 Comparison of the waverider's compression surface pressure contour (p/p_∞) between the CFD simulation and theoretical design result (top view)

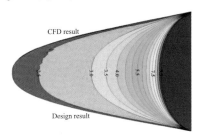

contours of the design result are interpolated from the traced streamlines in the basic curved cone's flow field. The design result is almost identical with the CFD simulation result. From Chauffor and Lewis (2004), the azimuthal pressure gradients along the waverider's osculating plane are negligible at high Mach number. Even in low supersonic speed, osculating method still can generate a general waverider which can isolate the high pressure gas in waverider's compression surface. Figure 12 is the back view of the comparison results between design and simulation pressure contours (p/p_∞) in OCC waverider's exit plane and compression surface. The shock wave's size and structure, the pressure contours' distribution and value are almost identical in design and simulation condition.

Figure 13 is the front view of wall surface streamlines and osculating planes in the waverider's compression surface. The solid lines in this figure indicate the streamlines drawing from the CFD computational result, while the dashed lines are the theoretical streamlines which overlap with the osculating planes. From Figure 11, the pressure distribution on waverider's compression surface decreases from center to side edge. So small azimuthal pressure gradients will appear on waverider's compression surface, and consequently, the streamlines move slightly outward to the side edge. Actually, the flow is not totally axisymmetric in the osculating plane and the streamlines will not stay in the osculating plane exactly. Instead, the streamlines stray away from the high pressure region into the low pressure region; however, the deflection of the streamlines from the osculating planes is small. All those above results have proven that present OCC waverider design method is theoretically correct.

Comparison between osculating cone and osculating general curved cone waverider

Having introduced an innovative waverider design method above, we want to know what are the differences between

OCC and OC waveriders. For comparing OC and OCC waverider, the same ICC and FCT curves are used which have been defined in Section 3. A straight cone with 10 degree cone angle is used as the basic flow field for the OC waverider design, which has the same cone angle as the straight part of the curved cone which has been defined in Section 2. The OCC waverider studied in the present paper is used as the counterpart for comparison.

First, from the geometrical point of view, as shown in Figure 14, the OCC waverider is more plump than OC waverider. The OCC waverider has a convex compression surface and its structure is realizable. The OC waverider has an inner concave surface and its structure is difficult to realize. OCC waverider looks more like a practical vehicle's airframe or forebody than OC waverider.

Figure 15 is the Mach number contour's comparison between OC and OCC waverider on their compression surfaces and exit planes. From this figure, OC waverider can compress incoming flow from Mach number 6 to Mach number 5.0 ~ 5.1. The compressed Mach number varies from 5.0 to 5.1. The OCC waverider compresses the Mach number 6 incoming flow to Mach number 4.0 ~ 3.98 at its exit. Its compressed Mach number varies from 3.98 to 4.0. Those results show that the OCC waverider has better compression ability and flow uniformity than OC waverider.

Figure 16 is the comparison between the OC and OCC waverider's pressure contour (p/p_∞) on their compression surface and exit plane. We can see that OC waverider can produce around 2.5 times static pressure rise in its exit plane. But the OCC waverider has around 10 times static pressure rise at the same place. Obviously, OCC waverider has stronger flow compression ability than OC wavrider, and it can provide suitable compressed incoming flow for inlet's inner compression section. OCC waverider will be a suitable forebody for waverider inlet integration.

Figure 17 is the comparison between OC and OCC waverider's Mach number contour in their symmetry plane. Their symmetry plane's flow field structures give clear

Figure 12 Comparison of exit plane and waverider compression surface pressure contour (p/p∞) between the CFD simulation and theoretical design results (back view)

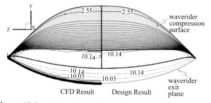

Figure 13 Front view of streamlines and osculating planes along the lower surface of the designed OCC waverider

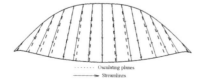

Figure 14 Three dimensinal view of the OC and OCC waverider

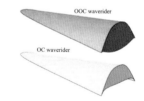

Figure 15 Comparison of mach number contour on waveriders low surface and exit

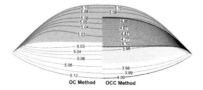

Figure 16 Comparison of pressure contour (p/p∞) on waveriders low surface and exit plane

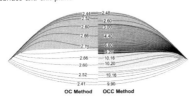

OC Method   OCC Method

Figure 17 Comparison of Mach number contour on OC and OCC waveriders' symmetry plane

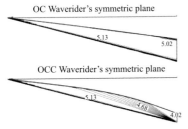

OC Waverider's symmetric plane

OCC Waverider's symmetric plane

explanations for their different flow characteristics. OCC waverider's symmetry plane has shock wave and isotropic compression waves, which continuously compress the incoming flow to the defined flow condition. In the OC waverider's flow field, the compression mainly relies on the initial compression shock, which will cause huge pressure recovery loss if low Mach number is needed in waverider's exit plane. On waverider's exit plane, OCC waverider's Mach number is lower than OC waverider's, but OCC waverider's pressure recovery ratio is same as OC waverider's. This is thanks to the isotropic compression section used in the OCC waverider's design.

When we talk about the waverider's lift to drag ratio, it directly relates to the waverider's volume characteristic. Although the OCC waverider's lift to drag ratio can be increased by minimizing the basic curved cone's drag under specific volume constrains, we only focus on the novel OCC waverider design method without making any optimization for present demonstration case.

The volume parameters and lift to drag ratios at design condition for OCC and OC waverider are shown in Table I. The OCC waverider's volume is 25 per cent larger than OC waverider's and its lift to drag ratio is a little bit lower than OC waverider's. As mentioned above, if a high lift to drag ratio OCC waverider is wanted, then it can be generated using an optimized low drag basic curved cone.

Table I The comparison of volume and lift to drag ratio characteristics between OCC and OC waverider

Waverider types	$V^{2/3}/S$	Lift to drag ratio
OCC waverider	0.0985	4.1
OC waverider	0.0788	5.0

Conclusion

The design method of OCC waverider has been introduced in present papers. OCC waverider uses general curved cone as the osculating flow field. Osculating and streamline tracing methods are used to generate the OCC waverider when the inlet capture curve and front capture tube are defined. Based on the study of present paper, there are several advantages for this kind of waverider: 1. OCC waverider has good volumetric characteristics. Using the same inlet capture curve and front capture tube curves, the OCC waverider has bigger volume than OC waverider. The OCC waverider looks more like a structure realizable waverider. 2. OCC waverider has good flow compression abilities. OCC waverider can compress incoming flow to required flow conditions with high total pressure recovery in the waverider's exit plane. The flow uniformity in the waverider exit plane is quite well. The high and qualified flow compression ability can provide qualified compressed gas for hypersonic air-breathing vehicle's inlet. From above indications, we can draw a conclusion that OCC waverider can be a practical high performance airframe/forebody for hypersonic vehicles in the near future.

References

Anderson, B.H. (1969), "Design of supersonic inlets by a computer program incorporating the method of characteristics", NASA TN D-4960 Lewis Research Center, Cleveland, OH.

Bowcutt, K.G., Anderson, J.D. and Capriotti, D. (1987), "Viscous optimized hypersonic wa-veriders", AIAA Paper 87-0272.

Chauffor, M.L. and Lewis, M.J. (2004), "Corrected Waverider Design for Inlet applications", AIAA 2004-3405.

Corda, S. and Anderson, J.D. (1988), "Viscous optimized hypersonic waveriders designed from axisymmetric flow fields", AIAA Paper 1988-0369.

Duveau, P.H., Hallard, R., Novelli, P.H. and Eggers, T.H. (1999), "Aerodynamic performance analysis of the hypersonic airbreathing vehicle Japhar", ISABE 1999-7286.

Goonko, Y.P., Mazhul, I.I. and Markelov, G.N. (2000), "Convergent-flow-derived waveriders", *Journal of Aircraft*, Vol. 37 No. 4, pp. 697-709.

He, X.Z., Zhao, H.Y. and Le, J.L. (2006), "Application of wall function boundary condition considering heat transfer and compressibility", *Journal of ACTA Aerodynamic SINICA*, Vol. 24 No. 4, pp. 450-453.

Lewis, M.J. (1991), "Application of waverider-based configurations to hypersonic vehicle design", AIAA Paper 91-3304.

Mazhul, I.I. (2010), "Off-design regimes of flow past waveriders based on isentropic compression flows", *Fluid Dynamics*, Vol. 45 No. 2, pp. 271-280.

Nonweiler, T.R.F. (1959), "Aerodynamic problems of manned space vehicles", *Journal of the Royal Aeronautical Society*, Vol. 63, pp. 521-530.

Nonweiler, T.R.F. (1963), "Delta wings of shape amenable to exact shock wave theory", *Journal of Royal Aeronautical Society*, Vol. 67 No. 1, pp. 39-44.

Rasmussen, M.L. (1980), "Waverider configurations derived from inclined circular and elliptic cones", *Journal of Spacecraft and Rockets*, Vol. 17 No. 6, pp. 537-545.

Rodi, P.E. (2005), "The Osculating Flowfield method of Waverider geometry generation", AIAA Paper 2005-0511.

Sobieczky, H., Dougherty, F.C. and Jones, K. (1990), "Hypersonic waverider design from given shockwaves", *Proceedings of the 1st International Waverider Symposium*, University of Maryland, Maryland, 17-19 October.

Sobieczky, H., Zores, B., Wang, Z. and Qian, Y.J. (1997), "High speed flow design using osculating Axisymmetric flows", *Proceedings of the 3th Congress of PICAST, Xi'an*, 1-5 September.

Takashima, N. and Lewis, M.J. (1992), "Navier–Stokes computation of a viscous optimized waverider", AIAA Paper 92-0305.

Takashima, N. and Lewis, M.J. (1994a), "Waverider configurations based on non-axisymmetric flow fields for engine-airframe integration", AIAA 94-0380.

Takashima, N. and Lewis, M.J. (1994b), "Navier–Stokes computation of a viscous optimized waverider", *Journal of Spacecraft and Rockets*, Vol. 31 No. 3, pp. 383-391.

Takashima, N., Lewis, M.J., Lockwood, M.K., Bogar, T. and Johnson, D. (1996), "Waverider configuration development for the dual fuel vehicle", AIAA Paper 96-4593.

Zucrow, M.J. and Hoffman, J.D. (1977), *Gas Dynamics*, John Wiley and Sons, New York, NY.

附录 乐嘉陵院士发表的论文

一、高超声速非平衡流动研究

1. 期刊论文

1979 年

[1] 乐嘉陵. 高温空气总辐射强度的测定. 气动研究与发展, 1979, 3: 1-10.

1984 年

[2] 乐嘉陵, 杜易鑫. $2H_2+O_2$ 爆轰波后壁面热交换测量. 空气动力学学报. 1984, 3: 71-74.

1990 年

[3] 乐嘉陵, 王晓栋. 用随机选取法(RCM)计算高温平衡气体的激波管流动. 空气动力学学报, 1990, 8(4): 445-451.

1993 年

[4] 乐嘉陵, 董维中. 高温高压喷管化学和热力学非平衡流计算. 气动实验与测量控制, 1993, 7(3): 30-37.

[5] 胡光初, 乐嘉陵, 曹文祥. 二维非定常化学非平衡流动高精度数值模拟及其与实验的比较. 气动实验与测量控制, 1993, 7(3): 94-100.

[6] 乐嘉陵, 曹文祥, 叶希超, 吴兴源, 孙启明, 丁慧芳. 双马赫反射的数值模拟与光学干涉定量测量的比较(Ms=6.42). 气动试验与测量控制. 1993, 7(1): 23-30.

[7] 李超, 乐嘉陵. 激波绕拐角运动流场的数值模拟. 气动实验与测量控制. 1993, 7(3): 1-4.

[8] 胡光初, 曹文祥, 乐嘉陵. 正激波后平衡高温混合气体(空气加氩)电子密度及热力特性计算. 宇航学报, 1993, 14(1): 28-36.

1994 年

[9] 胡光初, 乐嘉陵, 曹文祥. 二阶迎风 TVD 数值格式在非平衡高超声速钝体绕流中的应用. 力学学报, 1994, 26(2): 139-148.

[10] 胡光初, 曹文祥, 乐嘉陵. TVD 数值技术在弱电离化学非平衡斜激波反射流场计算中的应用. 计算物理, 1994, 11(2): 195-202.

1995 年

[11] 乐嘉陵, 曹文祥, 雒朝富. 再入钝头体近尾流动计算, 宇航学报, 1995, 16(3): 69-75.

1996 年

[12] 乐嘉陵, 倪鸿礼, 胡光初. 高超声速马赫反射. 气动实验与测量控制. 1996, 10(2): 1-8.

2000 年

[13] 董维中, 乐嘉陵, 刘伟雄. 驻点壁面催化速率常数确定的研究. 流体力学实验与测量, 2000, 14(3): 1-6.

2002 年

[14] 董维中, 乐嘉陵, 高铁锁. 钝体标模高焓风洞试验和飞行试验相关性的数值分析. 流体力学实验与测量, 2002, 16(2): 1-8,20.

[15] 柳军, 乐嘉陵, 杨辉. 高超声速圆球模型飞行流场的数值模拟和实验验证. 流体力学实验与测量, 2002, 16(1): 67-79.

[16] Le Jialing, V L Ganimedov, M I Muchnaja, V N Vetlutsky. The calculations of aerodynamic heating and viscous friction forces on the surface of hypersonic flight vehicle. Experiments and Measurements in Fluid Mechanics, 2002, 16(1): 8-20. 乐嘉陵, 詹妮迈德芙, 曼彻娜娅·维特拉斯夫. 高超声速飞行器表面气动热和粘性摩擦力计算. 流体力学实验与测量, 2002, 16(1): 8-20.

[17] 柳军, 曾明, 赵慧勇, 乐嘉陵. AUSMPW+格式在高超声速热化学非平衡流数值模拟中的应用. 国防科技大学学报, 2002, 24(6): 6-10.

2003 年

[18] 柳军, 刘伟, 曾明, 乐嘉陵. 高超声速三维热化学非平衡流场的数值模拟. 力学学报, 2003, 35(6): 730-734.

2004 年

[19] 许勇, 乐嘉陵. 基于 CFD 的电磁散射数值模拟. 空气动力学学报, 2004, 22(2): 185-189.

2005 年

[20] 岳斌, 乐嘉陵, 鲍伟仪. 高稳定性电弧等离子体超声速喷管流动装置. 实验流体力学, 2005, 19(2): 45-48.

2006 年

[21] 张若凌, 乐嘉陵, 王苏, 崔季平. 强激波阵面的非平衡结构研究. 实验流体力学, 2006, 20(2): 36-40, 49.

[22] 柳军, 石安华, 乐嘉陵. 弹道靶中球模型非平衡可见光辐射的测量. 实验流体力学, 2006, 20(2): 58-62.

2007 年

[23] 贺旭照, 乐嘉陵. 空间推进方法求解抛物化 Navier-Stokes 方程及其验证. 空气动力学学报, 2007, 25(2): 189-193.

2009 年

[24] 乐嘉陵, 张若凌, 王苏, 崔季平. 强激波阵面的非平衡特性研究. 空气动力学学报, 2009, 27(增刊): 1-8.

2. 国际会议

1991 年

[1] Le jialing. Some progress of wind tunnel testing technique and aerodynamic research in CARDC, Proceedings of the 1th International Conference on Experimental Fluid Mechanics. Chengdu, China, 1991.

1992 年

[2] Le Jialing, Dong Weizhong, Han Longheng. Studies of non-equilibrium properties in the nozzle flow. The Asia-Pacific Conference on Plasma Science and Technology (APCPST). Nanjing, China, 1992: 36-40.

[3] Le Jialing, Dong Weizhong. Calculations of thermo-chemical non-equilibrium nozzle flow for high reservoir pressures. Annual Fluid, Gas and Plasma-dynamics (FGPD) School-workshop. Tsagi, Moscow, 1992.

[4] Le Jialing, Hu Guangchu, Cao Wenxiang. Pseudo stationary oblique shock wave reflection with real gas effects. International Conference on Methods of Aerophysical Research, Novosibirsk, 1992.

[5] Le Jialing, Hu Guangchu, Cao Wenxiang. A high-resolution calculation of reflection of oblique shock wave from a compression corner in chemical nonequilibrium gas flow. International Conference on Methods of Aerophysical Research, Novosibirsk, 1992.

1993 年

[6] Le Jialing, Dong Weizhong. Studies of nonequilibrium effects in nozzle flow under high reservoir conditions. Proceedings of the 19th International Symposium on Shock Waves. Marseille, France, 1993: 307-312.

1994 年

[7] Le Jialing, Li Chao, Wu Xingyan, Yang Hui, Ye Xichao. Propagation of shock wave over a wedge. Proceedings of the 2nd Intenational Conference on Experimental Fluid Mechanics, Torino, Italy, 1994.

1995 年

[8] Le Jia-Ling, Cao Wen-Xiang, Hu Guang-Chu. Numerical calculation of hypersonic chemical nonequilibrium near wake flow. First Asian Computational Fluid Dynamics Conference. Hongkong, China. 1995.

1996 年

[9] Dong WeiZhong, Le JiaLing. Numerical studies of non-equilibrium heat transfer testing in the shock tube and extrapolation to flight. Second Asia workshop on CFD. Tokyo, 1996: 219-225.

2004 年

[10] Zhang R L, Le J L, Cui J P, Yu F M, Han L H, Wang S, Liu J. Measurement and calculation of vibrational temperature behind strong shock waves. Proceedings of the 24th International Symposium on Shock Waves. Beijing, China, 2004: 131-136.

[11] Liu J, Zeng M, Shi A H, Le J L, Qu Z H. Numerical calculation of visible spectral radiation of sphere model in ballistic range. Proceedings of the 24th International Symposium on Shock Saves. Beijing, China, 2004: 233-238.

[12] Wang S, Cui J P, Fan B C, He Y Z, Zhang R L, Han L H, Yu F M, Le J L. Measurement of electron density profile behind strong shock waves with a Langmuir probe. Proceedings of the 24th International Symposium on Shock Waves. Beijing, China, 2004: 269-274.

3. 国内会议

1993 年

[1] 乐嘉陵. 高超声速流动中的一些试验问题. 第四届全国试验流体力学学术会议论文集. 北京, 1993.

1995 年

[2] 董维中, 乐嘉陵. 化学振动非平衡流数值模拟的热流初步分析. 第七届全国激波管与激波

学术交流会论文集. 河南洛阳, 1995: 294-296.

1996 年

[3] 董维中, 乐嘉陵. 激波管中壁面材料催化速率常数确定的研究. 第一届全国航空航天空气动力学前沿问题学术讨论会文集. 山东威海, 1996.

[4] 董维中, 乐嘉陵. 热化学非平衡效应对典型高超声速流动影响的数值计算. 西安全国物理力学会议, 1996.

1998 年

[5] 董维中, 乐嘉陵, 刘维维. 高温气体流动中壁面催化速率常数确定的研究. 中国空气动力学物理气体动力专业委员会第八次学术年会. 四川绵阳, 1998.

2002 年

[6] 张若凌, 乐嘉陵, 崔季平. 高温高压准一维喷管流动的计算.第十届全国激波与激波管学术讨论会. 安徽黄山, 2002: 264-270.

[7] 赵慧勇, 张若凌, 乐嘉陵. AUSMPW+格式在热化学非平衡流计算中的应用.第十届全国激波与激波管学术讨论会. 安徽黄山, 2002: 277-283.

[8] 王苏, 崔季平, 范秉诚, 何宇中, 张若凌, 韩隆恒, 俞富明, 乐嘉陵. 电子探针法测定强激波后电子密度分布.第十届全国激波与激波管学术讨论会. 安徽黄山, 2002: 182-188.

[9] 柳军, 乐嘉陵. 高超声速三维热化学非平衡流计算程序及应用.第十届全国激波与激波管学术讨论会论文集. 安徽黄山, 2002: 65-72.

[10] 邢建文, 乐嘉陵, 郑忠华.Scramjet 燃烧室流场的二维模拟及试验比较. 计算流体力学研究进展. 第十一届全国计算流体力学会议论文集. 河南洛阳, 2002: 313-318.

2004 年

[11] 贺旭照, 乐嘉陵. 求解 PNS 方程的流向通量分裂法. 第十一届全国激波与激波管学术会议论文集. 四川绵阳, 2004.

[12] 赵慧勇, 乐嘉陵. 三维化学非平衡整体发动机的计算. 第十一届全国激波与激波管学术会议论文集. 四川绵阳, 2004.

[13] 张若凌, 乐嘉陵. 强正激波后非平衡流动的计算方法. 第十一全国激波与激波管学术讨论会.四川绵阳 2004.

二、激波与流动显示研究

1. 期刊论文

1992 年

[1] 李超, 乐嘉陵, 王泽刚. 管道中复杂马赫反射流场的高精度数值模拟. 空气动力学学报, 1992, 10(3): 332-338..

1993 年

[2] 乐嘉陵, 高铁锁, 曹文祥. 有限空间中激波运动的数值模拟. 航空学报, 1993, 14(1): 1-6.

[3] 乐嘉陵, 曹文祥, 叶希超等. 双马赫反射的数值模拟与光学干涉定量测量的比较(M_s=6.42). 气动试验与测量控制, 1993, 7(1): 22-31.

[4] 李超, 乐嘉陵. 激波绕拐角运动流场的数值模拟. 气动实验与测量控制, 1993, 7(3): 1-4.

[5] 何晓东, 乐嘉陵. 考虑粘性时激波反射及壁面热流计算. 气动实验与测量控制, 1993, 7(3):

10-14.

[6] 乐嘉陵, 曹文祥. 双马赫反射的数值模拟与光学干涉定量测量的比较. 气动实验与测量控制, 1993, 7(1): 22-31.

1996 年

[7] 李超, 乐嘉陵, 叶希超. 激波运动与反射流场的 ENO 算法. 气动实验与测量控制, 1996, 10(2): 34-40.

1997 年

[8] Ye Xichao, Yang Hui, Le Jialing. Flow field of shock wave through curved channel. Experiments and Measurements in Fluid Mechanics, 1997, 11(1): 50-55.

1998 年

[9] 乐嘉陵, 倪鸿礼. 激波(爆炸波)与物体相互作用的数值模拟. 流体力学试验与测量. 1999, 13(3): 1-9.

[10] 乐嘉陵, 陈阳生. 运动激波绕尖劈流动的研究. 流体力学实验与测量. 1998, 12(1): 29-37.

1999 年

[11] Le Jia ling, Ni Hong li. Numerical simulation of shock (blast) wave interaction with bodies. Communications in nonlinear science and numerical simulation, 1999, 13(3): 1-9. 乐嘉陵, 倪鸿礼. 激波(爆炸波)与物体相互作用的数值模拟. 流体力学实验与测量. 1999, 13(3): 1-9.

2000 年

[12] 杨辉, 岳茂雄, 郭自力, 乐嘉陵, 张力虎. 激波过弯道绕山坡对三维物体的冲击效应研究. 流体力学实验与测量. 2000, 14(2): 31-34.

[13] 周璐, 乐嘉陵, 李晓梅. 计算流动显示-概念、原理及实现. 计算机工程与科学, 2000, 22(1): 7-9.

2001 年

[14] Le Jialing, Wu Yingchuan, Ni Hong li, Wang Hui-ling. Computational interferometry for three-dimensional flow. Experiments and Measurements in Fluid Mechanics, 2001, 15(2): 1-9. 乐嘉陵, 倪鸿礼, 王惠玲. 三维流场的计算干涉方法. 流体力学实验与测量. 2001, 15(2): 1-9.

[15] Ni Hongli, Le Jialing. Research on blast wave interaction with body traveling supersonically. Experiments and Measurements in Fluid Mechanics. 2001, 15(2): 10-15. 倪鸿礼, 乐嘉陵. 爆炸波与超声速飞行物体相互作用研究. 流体力学实验与测量, 2001, 15(2): 10-15.

2002 年

[16] 毛雄兵, 乐嘉陵. 激波管内边界层数值模拟. 航空计算技术, 2002, 32(2): 7-8,14.

[17] 吴颖川, 乐嘉陵, 贺安之. 彩色计算干涉技术及应用. 流体力学实验与测量. 2002, 16(1): 80-86,93.

2003 年

[18] 吴颖川, 乐嘉陵, 贺安之. 轴对称流场的计算流动显示算法. 中国激光, 2003, 30(8): 721-725.

[19] 郭隆德, 周肇飞, 杨辉, 乐嘉陵, 杨建军. 物理靶上自由飞流场全息干涉诊断. 光电工程, 2003, 30(3): 40-42.

2004 年

[20] 贺安之, 姚红兵, 吴颖川, 乐嘉陵. 复杂流场全息干涉图的计算干涉编码处理. 光电子激光, 2004, 15(3): 348-351.

[21] 姚红兵, 贺安之, 吴颖川, 乐嘉陵. 虚拟和真实实验相融合的三维非完全数据层析. 光学学报, 2005, 25(4): 479-484.

2006 年

[22] 王惠玲, 乐嘉陵, 吴颖川. 激波绕楔形物体的流场分析. 实验流体力学 2006, 20(1): 31-35.

[23] 王惠玲, 李玉亮, 倪鸿礼, 乐嘉陵. 爆炸波与超声速运动物体相互作用的数值研究. 实验流体力学, 2006, 20(2): 1-6,17.

2008 年

[24] 赵慧勇, 乐嘉陵. 双时间步方法的应用分析. 计算物理, 2008, 25(3)：253-258.

2. 国际会议

1994 年

[1] Le Jialing, Li Chao, Wu Xingyan, Yang Hui, Ye Xichao. Propagation of shock wave over a wedge. Proceedings of the 2nd International Conference on Experimental Fluid Mechanics, Torino, Italy, 1994.

1995 年

[2] Le Jialing, Ni Hongli, Hu Guangchu. Mach reflection flow at hypersonic speeds. 20th International Symposium on Shock Waves. 1995.

[3] Le Jialing, Ye Xichao, Wu Xingyan, Yang Hui. Study the propagation of shock wave bodies with holographic interferometer. 20th International Symposium on Shock Waves. 1995.

1996 年

[4] Jialing Le, Xichao Ye, Hui Yang. Experimental study on the propagation of shock wave over cylinders with a holographic interference. Proceedings of SPIE (the International Society for Optical Engineering), 1996, 2866: 286-290.

[5] Le Jialing, Ni Hongli. Study on the propagation of shock wave over cylinders in a channel bend. International Conference on Methods of Aerophysical Research. Novosibirsk, Russia, 1996.

1997 年

[6] Le Jialing, Ni Hongli. Study on propagation of blast wave over cylinder in a channel bend. 21th International Symposium on Shock Waves. Great keppel island, Australia, 1997.

[7] Ni Hongli, Le Jialing. The numerical study of shock-on-shock interaction. 21th International Symposium on Shock Waves. Great keppel island, Australia, 1997.

[8] Ni Hongli, Le Jialing. The numerical study of shock-on-shock interaction. 21th International Symposium on Shock Waves. Great keppel island, Australia, 1997.

1998 年

[9] Le Jialing. Shock wave interaction with bodies. Sino-Russian Hypersonic Flow Conference, 1998.

1999 年

[10] Le Jialing, Ni Hongli, Wang Huiling. Research on the motion of shock wave over bodies in a channel bend. Proceedings of he Eighth Asian Congress of Fluid Mechanics. 1999.

2002 年

[11] Yingchuan Wu, Jialing Le, Anzhi He. Parallel volume rendering algorithms for computational flow imaging. Proceedings of International Conference on the Methods of Aerophysical

Research, Novosibirsk, Russia, 2002.

[12] Zheng Zhonghua, Le Jialing. Parallel modeling of three-dimensional scramjet combustor and comparisons with experiment's results. Proceedings of International Conference on the Methods of Aerophysical Research, Novosibirsk, Russia, 2002.

2003 年

[13] Jialing Le, Yingchuan Wu, Hongli Ni, Hui Yang, Huiling Wu. Some research in application of holographic interferometry and computational flow imaging. Proceedings of Spie - the International Society for Optical Engineering, 2003.

3. 国内会议

1995 年

[1] 乐嘉陵, 叶希超. 二维密度场定量测量技术进展. 第五届全国实验流体力学学术会议论文集. 四川松潘县, 1995: 11-17.

1998 年

[2] 乐嘉陵. 爆炸波与物体相互作用研究进展. 第八届全国激波与激波管学术讨论会. 北京, 1998: 1-2.

2002 年

[3] 郭隆德, 周肇飞, 杨辉, 乐嘉陵, 杨建军. 物理靶上自由飞流场诊断. 第五届全国流动显示学术会议论文集. 山东, 烟台, 2002: 182-188.

[4] 乐嘉陵. 光学流动显示和 CFI 技术及其若干应用. 第五届全国流动显示学术会议论文集. 山东, 烟台, 2002: 1-3.

[5] 贺安之, 乐嘉陵, 吴颖川. 计算流动显示技术的研究与应用. 第五届全国流动显示学术会议论文集. 山东, 烟台, 2002: 329-336.

[6] 倪鸿礼, 王惠玲, 乐嘉陵. 坑道内爆炸冲击波载荷数值研究. 第十届全国激波与激波管学术讨论会论文集. 安徽黄山, 2002.

[7] 黄为民, 乐嘉陵, 倪鸿礼, 岳茂雄, 杨辉. 爆炸波及其绕流流场研究. 第十届全国激波与激波管学术讨论会论文集. 安徽黄山, 2002.

2006 年

[8] 杨顺华, 乐嘉陵, 刘鑫, 陆林生. 滑移网格技术及其在高速列车穿越隧道问题中的应用. 第十二届全国激波与激波管学术会议论文集. 河南洛阳, 2006: 348-353.

三、高超声速技术研究

1. 期刊论文

1997 年

[1] 乐嘉陵, 刘陵. 高超声速飞行器的碳氢燃料双模态超燃冲压方案研究. 流体力学实验与测量, 1997, 11(2): 1-13.

2000 年

[2] 乐嘉陵, 胡欲立, 刘陵. 双模态超燃冲压发动机研究进展. 流体力学实验与测量, 2000, 14(1): 1-12.

[3] 乐嘉陵. 高超声速技术及其在军事上的应用. 现代军事, 2000, 24(6): 10-12.

[4] Wang Xiaodong, Le Jialing. Computations of inlet/isolator for scramjet engine. Journal of thermal science, 2000, 9(4): 334-338.

2001 年

[5] J.L. Le, Z.C. Zhang, H.C. Bai, M.A. Goldfeld, R.V. Nestoulia, A.V. Starov. Preliminary investigation of full model of two – mode scramjet. Journal of thermal science, 2001, 10(2): 97-102.

[6] He Yuanyuan, Le Jialing. Numerical simulation of integrative flow field for hypersonic vehicle, Journal of thermal science, 2001, 10(2): 103-108.

2002 年

[7] 郑忠华, 乐嘉陵. Scramjet 燃烧室流场的三维并行数值模拟及试验比较. 流体力学实验与测量, 2002, 16(2): 9-15.

[8] 白菡尘, 刘伟雄, 贺伟, 曾来荣, 谭宇, 李向东, 乐嘉陵. 氢燃料双模态冲压模型发动机 M6 的试验研究. 流体力学实验与测量. 2002, 16(4): 1-6.

[9] 王晓栋, 乐嘉陵. 前缘对进气道性能影响的数值模拟. 推进技术, 2002, 23(6): 460-462.

[10] 王晓栋, 乐嘉陵. 入口温度剖面对喷管流场结构的影响. 推进技术, 2002, 23(4): 283-286.

2003 年

[11] Le Jialing. Combustion of liquid and gaseous fuels in a supersonic combustor. Combustion explosion and shock waves, 2003, 39(3): 292-299.

[12] 白菡尘, 刘开胜, 苟永华, 焦伟, 乐嘉陵. M6 双模态冲压模型发动机氢燃料的燃烧试验研究. 南京航空航天大学学报, 2003, 35(1): 53-57.

[13] 乐嘉陵, 白菡尘. Исследование горения жидкого и газообразного топлив в сверхзвуковой камере сгорания. 爆轰与燃烧（俄罗斯）, 2003,（3）.

2004 年

[14] 王晓栋, 乐嘉陵, 宋文艳. 冲压燃烧室内的燃料扩散性能研究. 空气动力学学报, 2004, 22(2): 147-150.

[15] 王晓栋, 乐嘉陵, 宋文艳. 带支板的冲压燃烧室的燃烧性能研究. 空气动力学学报, 2004, 22(3): 274-278.

2005 年

[16] 乐嘉陵, 刘伟雄, 贺伟, 谭宇, 白菡尘. 脉冲燃烧风洞及其在火箭和超燃发动机研究中的应用. 实验流体力学, 2005, 19(1): 1-10. Le Jialing, et al. Impulse Combustion wind tunnel and its application in rocket and scramjet research. Experiments & measurements in fluid mechanics, 2005, 19(1): 1-10.

2006 年

[17] 肖保国, 钱炜祺, 杨顺华, 乐嘉陵. 甲烷点火燃烧的简化学反应动力学模型研究. 推进技术, 2006, 27(2): 101-105.

[18] 贺旭照, 赵慧勇, 乐嘉陵. 考虑可压缩与热传导的壁面函数边界条件及其应用. 空气动力学学报, 2006, 24(4): 450-453.

[19] 蒋劲, 张若凌, 乐嘉陵. 超燃冲压发动机再生冷却热结构设计的计算工具. 实验流体力学, 2006, 20(3): 1-7.

2007 年

[20] 宋文艳, 王靛, 陈亮, 乐嘉陵. 纯净空气来流下的超声速燃烧实验装置及其初步实验结果.

实验流体力学, 2007, 21(1): 1-6.

[21] 贺元元, 乐嘉陵, 倪鸿礼. 吸气式高超声速机体/推进一体化飞行器数值和试验研究. 实验流体力学, 2007, 21(2): 29-34.

[22] 贺元元, 倪鸿礼, 乐嘉陵. 一体化高超声速飞行器气动-推进性能评估. 实验流体力学, 2007, 21(2): 63-67.

[23] 刘伟雄, 谭宇, 毛雄兵, 乐嘉陵. 一种新运行方式脉冲燃烧风洞研制及初步应用. 实验流体力学, 2007, 21(4): 59-64.

[24] 钱炜祺, 杨顺华, 肖保国, 乐嘉陵. 用准稳态方法建立碳氢燃料点火燃料的简化化学反应动力学模型. 空气动力学学报, 2007, 25(1): 12-18.

[25] 余安远, 乐嘉陵, 郭荣伟. 隐身外形飞行器用埋入式进气道的设计与风洞实验研究. 空气动力学学报, 2007, 25(2): 150-156.

[26] 钱炜祺, 杨顺华, 肖保国, 乐嘉陵. 碳氢燃料点火燃烧的简化化学反应动力学模型. 力学学报, 2007, 39(1): 37-44.

[27] 贺旭照, 张勇, 汪广元, 倪鸿礼, 乐嘉陵. 高超声速飞行器单壁膨胀喷管的自动优化设计. 推进技术, 2007, 28(2): 148-151,224.

[28] 乐嘉陵. CARDC 吸气式高超声速技术研究进展. 现代炮兵学报. 2007, 23(3): 4-8.

2008 年

[29] 邢建文, 乐嘉陵. 火焰面模型在超燃冲压发动机数值模拟中的应用. 实验流体力学, 2008, 22(2): 40-45.

[30] 赵慧勇, 贺旭照, 乐嘉陵. 一种新的壁面距离计算方法 – 循环盒子法. 计算物理, 2008, 25(4): 427-430.

[31] 杨顺华, 乐嘉陵. 超声速气流中液体燃料雾化数值模拟. 推进技术, 2008, 29(5): 519-522.

[32] 王兰, 邢建文, 郑忠华, 乐嘉陵. 超燃冲压发动机内流性能的一维评估. 推进技术, 2008, 29(6): 641-645.

[33] 杨阳, 邢建文, 乐嘉陵, 王金诺. 湍流燃烧模型对氢燃料超燃室流场模拟的影响. 航空动力学报, 2008, 23(4): 605-610.

[34] 张小庆, 乐嘉陵. 脉冲式燃烧风洞起动特性数值研究. 航空动力学报, 2008, 23(9): 1568-1572.

[35] 宋文艳, 肖隐利, 乐嘉陵. 电阻加热器在超声速燃烧研究中的应用. 航空动力学报, 2008, 23(12): 2268-2273.

[36] 刘伟雄, 贺伟, 李宏斌, 李象远, 乐嘉陵. 污染组分对氢燃料发动机燃烧动力学的影响. 科学通报, 2008, 53(18): 2257-2260.

[37] 杨阳, 刘伟雄, 乐嘉陵, 王金诺. 直连式脉冲燃烧高温风洞的设计. 西南交通大学学报, 2008, 43(3): 387-391.

[38] 张小庆, 乐嘉陵, 许明恒. 超声速脉冲风洞起动过程数值模拟, 西南交通大学学报, 2008, 43(6): 751-755.

[39] 贺旭照, 赵慧勇, 乐嘉陵. 吸气式高超声速飞行器气动力气动热的数值模拟方法及应用. 计算物理. 2008, 25(5): 555-560.

2009 年

[40] 赵慧勇, 雷波, 乐嘉陵. 非对称交叉激波和湍流边界层相互作用的数值研究. 航空动力学报, 2009, 24(10): 2183-2188.

[41] 蒋劲, 张若凌, 乐嘉陵. 再生冷却超燃冲压发动机整机稳态热分析. 航空动力学报, 2009,

24(12): 2649-2654.

[42] 贺旭照, 乐嘉陵, 宋文艳. 二维带动力吸气式高超声速飞行器绕流的 PNS-NS 混合求解. 航空动力学报, 2009, 24(12): 2741-2747.

[43] 杨顺华, 乐嘉陵, 赵慧勇, 郑中华. 煤油超燃冲压发动机三维大规模并行数值模拟. 计算物理, 2009, 26(4): 534-539.

[44] 王兰, 吴颖川, 乐嘉陵. 氢燃料超燃冲压发动机燃烧室非结构网格数值模拟. 空气动力学学报, 2009, 27(3): 308-313.

[45] 肖隐利, 杨顺华, 宋文艳, 乐嘉陵. 超声速凹槽大涡数值模拟研究. 空气动力学学报, 2009, 27(4): 469-473.

[46] 周正, 倪鸿礼, 贺旭照, 乐嘉陵. 基于 Rao 方法的二维单壁膨胀喷管优化设计. 推进技术, 2009, 30(4): 451-456.

[47] 贺旭照, 倪鸿礼, 周正, 乐嘉陵, 宋文艳. 吸气式高超声速飞行器三维后体尾喷管优化设计. 推进技术, 2009, 30(6): 687-716.

[48] 刘伟雄, 毛雄兵, 谭宇, 乐嘉陵. 挤压燃料供应技术在脉冲设备中的应用研究. 四川大学学报(工程科学版), 2009, 41(5): 202-205.

[49] 刘伟雄, 杨阳, 邵菊香, 宋文艳, 李象远, 乐嘉陵. 空气污染组分 H2O 和 CO2 对乙烯燃烧性能的影响. 物理化学学报, 2009, 25(8): 1618-1622.

[50] 王兰, 吴颖川, 乐嘉陵. Scramjet 整机流场的非结构网格并行数值模拟. 推进技术, 2009, 30(1): 34-40.

[51] 张小庆, 杨富荣, 鲍伟仪, 乐嘉陵. 直连式脉冲燃烧风洞起动过程研究. 实验流体力学, 2009, 23(2): 63-67.

2010 年

[52] 赵慧勇, 雷波, 黄为民, 乐嘉陵. TNT k-omega 湍流模型的隐式解法及其在超声速流动模拟中的应用. 数值计算与计算机应用, 2010, 31(2): 108-115.

[53] 贺旭照, 乐嘉陵, 宋文艳, 赵志. 基于轴对称喷管的三维内收缩进气道的设计与初步评估. 推进技术, 2010, 31(2): 147-152.

[54] 赵慧勇, 雷波, 乐嘉陵. 对称交叉激波和湍流边界层相互作用的数值研究. 推进技术, 2010, 31(4): 406-411.

[55] 肖隐利, 杨顺华, 宋文艳, 乐嘉陵. 超声速燃烧室凹槽动态特性试验研究. 实验流体力学, 2010, 24(5): 7-12.

[56] 贺旭照, 乐嘉陵, 宋文艳. 超声速化学反应流动的 LU-SGS 伪时间迭代空间推进求解. 航空动力学报, 2010, 25(5): 1043-1048.

[57] 肖保国, 杨顺华, 赵慧勇, 钱炜祺, 乐嘉陵. RP-3 航空煤油燃烧的详细和简化化学动力学模型. 航空动力学报, 2010, 25(9): 1948-1955.

2011 年

[58] 赵慧勇, 雷波, 乐嘉陵. 非线性 EASM 模型在激波与湍流边界层相互作用中的数值应用. 航空动力学报, 2011, 26(4): 860-866.

[59] 蒋劲, 张若凌, 乐嘉陵, 刘伟雄, 伍军, 赵国柱. 燃油冷却面板传热特性试验与计算分析研究. 实验流体力学, 2011, 25(1): 400-411.

[60] 陈亮, 曹娜, 乐嘉陵, 宋文艳, 杨顺华. 超声速冷态流场液体射流雾化实验研究. 实验流体力学, 2011, 25(2): 29-34.

[61] 乐嘉陵. 吸气式高超声速技术研究进展. 西南科技大学学报, 2011, 26(4): 1-9.

2012 年

[62] 王西耀, 杨顺华, 乐嘉陵. 油气比对超燃冲压发动机点火过程的影响. 推进技术, 2012, 33(4), 522-529.

[63] 张弯洲, 乐嘉陵, 田野, 杨顺华, 邓维鑫, 程文明. 超燃发动机混合效率评估方法探讨. 航空动力学报, 2012, 27(9): 1958-1966.

2013 年

[64] 邓维鑫, 乐嘉陵, 杨顺华, 张弯洲, 周华波, 许明恒. 脉冲风洞发动机试验多油位多时序高精度燃料供应系统. 实验流体力学, 2013, 27(3): 70-76.

[65] 田野, 乐嘉陵, 杨顺华, 张弯洲. 空气节流对超燃燃烧室流场结构和燃料混合影响的数值研究. 推进技术, 2013, 34(1): 54-61.

[66] 张磊, 乐嘉陵, 张若凌, 张香文, 景凯, 高洋. 超临界压力下湍流区碳氢燃料传热研究. 推进技术, 2013, 34(2): 225-229.

[67] 王西耀, 杨顺华, 乐嘉陵. 超燃冲压发动机带凹槽的燃烧室流场振荡研究. 推进技术, 2013, 34(5): 651-657.

[68] 田野, 乐嘉陵, 杨顺华, 张弯洲, 邓维鑫. 空气节流对超燃燃烧室火焰稳定影响的数值研究. 推进技术, 2013, 34(6): 795-801.

[69] 邓维鑫, 乐嘉陵, 杨顺华, 田野, 张弯洲, 许明恒. M3 条件下乙烯燃料超燃冲压发动机空气节流点火试验. 推进技术, 2013, 34(9): 1240-1247.

[70] 张弯洲, 乐嘉陵, 杨顺华, 程文明, 邓维鑫. 马赫数 4 下氢气自燃辅助乙烯点火实验研究. 推进技术, 2013, 34(12): 1628-1635.

[71] 邓维鑫, 乐嘉陵, 王西耀, 杨顺华, 张弯洲, 许明恒. 空气节流对超燃发动机燃烧性能的影响. 航空动力学报. 2013, 28(2): 316-323.

[72] 张弯洲, 乐嘉陵, 杨顺华, 程文明, 邓维鑫, 周化波. 马赫数为 4 的超燃发动机碳氢燃料点火试验. 航空动力学报, 2013, 28(4): 800-806.

[73] 邓维鑫, 乐嘉陵, 杨顺华, 田野, 张弯洲, 许明恒. 注油方式对超燃冲压发动机燃烧性能的影响, 航空动力学报, 2013, 28(7): 1449-1457.

[74] 田野, 乐嘉陵, 杨顺华, 张弯洲, 邓维鑫. 空气节流对超燃冲压发动机燃烧室起动点火影响的数值研究. 航空动力学报. 2013, 28(7): 1495-1502.

[75] 吴颖川, 贺元元, 余安远, 乐嘉陵. 展向截断曲面乘波压缩进气道气动布局. 航空动力学报, 2013, 28(7): 1570-1575.

[76] L. Zhang, R.L. Zhang, S.D. Xiao, J. Jiang, J.L. Le. Experimental investigation on heat transfer correlations of n-decane under supercritical pressure. International journal of heat and mass transfer, 2013, 64: 393-400.

2014 年

[77] 赵国柱, 宋文艳, 张若凌, 乐嘉陵. 超临界压力下正十烷流动传热的数值模拟. 推进技术, 2014, 35(4): 537-543.

[78] Jin Jiang, Ruoling Zhang, Jialing Le, et al. Regeneratively cooled scramjet heat transfer calculation and comparison with experimental data. Journal of aerospace engineering, 2014, 778(8): 1227-1234.

[79] 吴颖川, 贺元元, 贺伟, 乐嘉陵. 基于密切曲锥的乘波构型一体化飞行器设计方法研究. 空气动力学学报, 2014, 32(1): 8-13.

[80] 张磊, 张若凌, 肖世德, 张香文, 乐嘉陵. 超临界压力正十烷对流传热实验及计算研究. 实验流体力学, 2014, 28(2): 14-20.

[81] 贺旭照, 周正, 毛鹏飞, 乐嘉陵. 密切曲面内锥乘波前体进气道设计和试验研究. 实验流体力学, 2014, 28(3): 39-44.

2015 年

[82] Zhang Ruoling, Le Jialing. Laminar-to-turbulent transition flow inside a heated circular tube. Modern applied science, 2015, 9(9): 205-219

[83] 田野, 乐嘉陵, 杨顺华, 邓维鑫, 张弯洲. 空气节流对乙烯燃料超燃冲压发动机流场结构影响研究. 推进技术, 2015, 36(4): 481-487.

[84] 田野, 乐嘉陵, 杨顺华, 肖保国. 乙烯燃料超燃冲压发动机流场振荡及其控制研究. 推进技术, 2015, 36(7): 961-967.

[85] Tian Ye, Yang Shunhua, Le Jialing. Study on the effect of air throttling on flame stabilization of an ethylene fueled scramjet combustor. International journal of aerospace engineering, 2015, 1-10.

[86] 田野, 杨顺华, 肖保国, 乐嘉陵. 空气节流对煤油燃料超燃燃烧室燃烧性能影响. 宇航学报, 2015, 36(12): 1421-1427.

[87] 吴颖川, 贺元元, 贺伟, 乐嘉陵. 吸气式高超声速飞行器机体推进一体化技术研究进展. 航空学报. 2015, 36(1): 245-260.

2016 年

[88] Tian Ye, Yang Shunhua, Le Jialing. Numerical study on effect of air throttling on combustion mode formation and transition in a dual-mode scramjet combustor. Aerospace science and technology, 2016, 52: 173–180.

[89] Tian Ye, Yang Shunhua, Le Jialing. Study on flame stabilization of a hydrogen and kerosene fueled combustor. Aerospace science and technology, 2016, 59: 183-188.

[90] Tian Ye, Yang Shunhua, Le Jialing, Su Tie, Yue Maoxiong, Zhong Fuyu, Tian Xiaoqiang. Investigation of combustion and flame stabilization modes in a hydrogen fueled scramjet combustor. international journal of hydrogen energy, 2016,41:19218-19230.

[91] 易森荣, 赵慧勇, 乐嘉陵. 强制转捩对高超声速进气道性能影响. 航空动力学报, 2016, 31(8): 1380-1387.

[92] 熊模友, 乐嘉陵, 黄渊, 宋文艳, 杨顺华, 郑忠华. 采用火焰面/反应进度变量方法模拟湍流燃烧. 航空动力学报, 2016, 31(11): 2604-2612.

[93] 田野, 杨顺华, 肖保国, 乐嘉陵. 双模态冲压发动机燃烧性能初步研究.航空动力学报, 2016, 31(12): 2921-2927.

[94] 刘日超, 乐嘉陵, 杨顺华, 郑忠华, 宋文艳, 黄渊. 直流喷射首次破碎的形变过程研究. 推进技术, 2016, 37(7): 1334-1340.

[95] 周正, 贺旭照, 卫锋, 乐嘉陵. 密切曲内锥乘波前体进气道低马赫数性能试验研究. 推进技术, 2016, 37(8): 1455-1460.

[96] 刘日超, 乐嘉陵, 杨顺华, 郑忠华, 宋文艳, 黄渊. 亚声速横向气流中液体射流破碎过程的直接模拟. 推进技术, 2016, 37(11): 2135-2141.

[97] 张弯洲, 乐嘉陵, 杨顺华, 程文明, 邓维鑫. Ma4 下超燃发动机乙烯点火及火焰传播过程试验研究. 实验流体力学, 2016, 30(3): 40-46.

[98] Xuzhao He, Zheng Zhou, Si Qin, Feng Wei, Jialing Le. Design and experimental study of a

practical osculating inward cone waverider inlet. Chinese Journal of Aeronautics, 2016, 29(6): 1582-1590.

2017 年

[99] 宋文艳, 王靛, 陈亮, 乐嘉陵. 纯净空气来流下的超声速燃烧室装置及其初步试验结果. 实验流体力学, 2017, 31(1): 1-6.

[100] 贺旭照, 秦思, 卫锋, 乐嘉陵. 比热比和压比对高超飞行器尾喷流影响的实验研究. 实验流体力学, 2017, 31(1): 13-19.

[101] 王宇辉, 乐嘉陵, 杨洋, 谭宇. 旋转爆轰发动机燃烧室的燃烧与流动特性研究. 实验流体力学, 2017, 31(1): 32-38.

[102] 张若凌, 乐嘉陵. 电加热圆管内流动的自然转捩过程研究. 实验流体力学, 2017, 31(2): 51-60.

[103] 武龙, 王锋, 乐嘉陵. 脉冲燃烧风洞测力系统动态标定方法. 实验流体力学, 2017, 31(4): 51-58.

[104] 熊模友, 乐嘉陵, 杨顺华. 采用 CARS 试验技术与 UFPV 数值方法研究航空发动机燃烧室. 实验流体力学, 2017, 31(5): 15-23.

[105] 刘日超, 乐嘉陵, 陈柳君, 杨顺华, 宋文艳. 双旋流燃烧室两相喷嘴试验和数值研究. 实验流体力学, 2017, 31(5): 24-31.

[106] 岳茂雄, 苏铁, 杨顺华, 袁强, 乐嘉陵. 燃烧场波系显示及燃烧区域诊断研究. 红外与激光工程, 2017, 46(2): 14-19.

[107] Tian Ye, Yang Shunhua, Le Jialing, Zhong Fuyu, Tian Xiaoqiang. Investigation of combustion process of a kerosene fueled combustor with air throttling. Combustion and Flame. 2017, 179: 74-85.

[108] Tian Ye, Yang ShunHua, Le JiaLing. Investigation of the effects of fuel injector iocations on ignition and flame stabilization in a kerosene fueled scramjet combustor. Aerospace Science and Technology, 2017, 70: 310–316.

[109] Weixin Deng, Jialing Le, Shunhua Yang, Wanzhou Zhang, Ye Tian. Experimental research of air-throttling ignition for scramjet at Ma 6.5. Chinese Journal of Aeronautics, 2017, 30(3): 932-938.

[110] 熊模友, 乐嘉陵, 黄渊, 宋文艳, 杨顺华, 郑忠华. 基于 LES 的非稳态火焰面/反应进度变量方法模拟部分预混抬举火焰. 推进技术, 2017, 38(1): 148-157.

[111] 邓维鑫, 杨顺华, 张弯洲, 王西耀, 田野, 乐嘉陵. 高超声速流动的气体吹除控制方法研究. 推进技术, 2017, 38(4): 759-763.

[112] 周瑜, 乐嘉陵, 陈柳君, 黄渊. 径向双旋流燃烧室流场结构大涡模拟研究. 推进技术, 2017, 38(4): 909-917.

[113] 贺旭照, 乐嘉陵. 曲外锥乘波体进气道实用构型设计和性能分析. 航空学报, 2017, 38(6): 9-19.

[114] 卫锋, 贺旭照, 陈军, 吴颖川, 乐嘉陵. 微修形异型转圆内转式进气道的设计与试验研究. 推进技术, 2017, 38(6): 1218-1225.

[115] 熊模友, 乐嘉陵, 黄渊, 宋文艳, 杨顺华, 郑忠华. 采用基于火焰面的燃烧模型研究部分预混燃烧. 推进技术, 2017, 38(7): 1459-1467.

[116] 周瑜, 乐嘉陵, 黄渊. 同轴射流燃烧室非预混湍流燃烧流场特性大涡流模拟研究. 推进技术, 2017, 38(7): 1514-1522.

[117] 刘日超, 乐嘉陵, 杨顺华. KH-RT 模型在横向来流作用下射流雾化过程的应用. 推进技术, 2017, 38(7): 1595-1602.

[118] 贺旭照, 秦思, 卫锋, 乐嘉陵. 吸气式高超声速飞行器非均匀尾喷流试验. 航空学报, 2017, 38(3): 35-42.

[119] 周瑜, 乐嘉陵, 陈柳君, 黄渊. 轴向三级旋流燃烧室流场结构大涡模拟. 航空动力学报, 2017, 32(4): 917-925.

[120] 邓维鑫, 乐嘉陵, 杨顺华, 田野, 王西耀. 空气/煤油火炬点火器设计及试验研究. 航空动力学报, 2017, 32(7): 1554-1560.

[121] 秦思, 贺旭照, 曾学军, 乐嘉陵. 喷流落压比对高超飞行器尾喷管内外流干扰的实验. 航空动力学报, 2017, 32(10): 2491-2497.

[122] Xuzhao He, Jialing Le, Si Qin. Design and analysis osculating general curved cone waverider. Aircraft Engineering and Aerospace Technology, 2017, 89(6): 797-803.

2. 国际会议

1996 年

[1] Le Jialing. Short duration techniques for simulating exhaust plumes at high altitudes. The Second Sino-Russia Hypersonic flow Conference. Moscow, Russia, 1996.

1999 年

[2] Le Jialing, Liu Ling. Numerical simulation of dual mode in combustion chamber. Proceedings of the Eighth Asian Congress of Fluid Mechanics. Shenzhen, Chian, 1999.

2002 年

[3] J.L. Le, Z.C. Zhang, H.C. Bai, M.A. Goldfeld, R.V. Nestoulia, A.V. Starov. Methodical aspects of investigation of kerosene ignition and combustion in scramjet model. International Conference on Methods of Aerophysical Research. Novosibirsk, Russia, 2002.

[4] Le Jialing, Liu Weixiong, He Wei, Tan Yu. Pulse combustion facility and its preliminary application in scramjet research. International Conference on Methods of Aerophysical Research. Novosibirsk, Russia, 2002.

[5] Zheng Zhonghua, Le Jialing. Parallel modeling of three-dimensional scramjet combustor and comparisons with experiments results, International Conference on Methods of Aerophysical Research. Novosibirsk, Russia, 2002.

[6] He Yuanyuan, Le Jialing. Numerical research of airframe/engine integrative hypersonic vehicle, International conference on methods of aerophysical research. Novosibirsk, Russia, 2002, 94-100.

2003 年

[7] He Wei, Tan Yu, Liu Weixiong, Le Jialing. Performance study of model scramjet with fuel of kerosene in pulse facility. 12th AIAA International Space Planes and Hypersonic Systems and Technologies. Norfolk, Virginia, 2003.

[8] Le Jialing, Liu Weixiong, He Wei. Research progress in CARDC's short duration test techniques. 41th AIAA Aerospace sciences Meeting and Exhibition. Reno, Nevada 6-9 Jan, 2003.

2004 年

[9] W. He, Y. Tan, X.D. Li, W.X. Liu, J.L. Le. Impulse combustion tunnel and its application in

experimental research of scramjet. Proceedings of the 24th International Symposium on Shock Waves, Beijing, China, 2004: 401-406.

[10] Zheng Zhonghua, Le Jialing. Massively parallel computation of three dimension scramjet combustor. Proceedings of the 24th International Symposium on Shock Waves, Beijing, China, July 11-16, 2004, pp. 897-902.

[11] Le J L, Liu W X, Tan Y, He W. Time requirements for scramjet performance study with fuel of kerosene. Proceedings of the 24th International Symposium on Shock Waves. Beijing, China, 2004: 947-952.

[12] Jialing Le. Investigation of kerosene combustion in a scramjet model at artificial and self-ignition. International Conference on the Methods of Aerophysical Research. 2004.

[13] Zheng Zhonghua, Le Jialing, Zhao Huiyong. Massive parallel computation of 3d hydrocarbon fueled scramjet engine. International Conference on the Methods of Aerophysical Research. 2004.

2005 年

[14] Yang Shunhua, Le Jialing. Computational analysis of kerosene fueled scramjet. ISABE-2005-1195, 18th International Symposium on Air Breathing Engines. 2005: 1-12.

[15] Le Jialing. Massively parallel simulations of kerosene-fueied model scramjet and comparisons with experiments. AIAA/CIRA 13th International Space Planes and Hypersonics Systems and Technologies Conference. Capua, Italy, 2005.

[16] Le Jialing, Yang Shunhua, Liu Weixiong, Xing Jianwei. Massively parallel simulations of kerosene-fueled model scramjet. AIAA/CIRA 13th International Space Planes and Hypersonics Systems and Technologies Conference. Capua, Italy, 2005.

[17] Zhao Huiyong, Le Jialing. Numerical investigation on flow field in scramjet engine. Eighth International Symposium on Fluid Control, Measurement and Visualization. Chengdu, China, 2008.

[18] He Xuzhao, Le Jialing. Unsteady flow simulation using newly developed parallel cfd program with pseudo-time sub-iteration method. The 8th International Symposium on Fluid Control, Measurement and Visualization. Chengdu China, 2005.

2006 年

[19] Yang Shunhua, Le Jialing. Fuel atomization and droplet breakup models for parallel spray combustion CFD code, Asian joint conference on propulsion and power. 2006.

[20] Yang Shunhua, Le Jialing, Zhao Huiyong, Zheng Zhonghua. Parallel numerical investigation of fuel atomization and combustion in a scramjet. 14th AIAA/AHI Space Planes and Hypersonic Systems and Technologies Conference. Canberra, Australia, 2006.

[21] Zheng Zhonghua, He Yuanyuan, He Wei, Liu Weixiong, Le Jialing. Preliminary study of integrated aero-propulsive performance of vehicle with CFD and experiments. 14th AIAA/AHI Space Planes and Hypersonic Systems and Technologies Conference. Canberra, Australia, 2006.

2008 年

[22] Zhang Ruoling, Jiang Jin, Le Jialing. The simulation of endothermic fuel flow in cooling channels of scramjet. 14th International Conference on Methods of Aerophysical Research. Akademgorodok, Novosibirsk, Russia, 2008.

[23] Xing Jianwen, Le Jialing. Application of flamelet model for the numerical simulation of

turbulent combustion in scramjet. International Conference on Methods of Aerophysical Reseach, Akademgorodok, Novosibirsk, Russia, 2008.

[24] Le Jialing. Preliminary aero-propulsive performance study for vehicle in short duration facility. International conference on Methods of Aerophysical Reseach, Akademgorodok, Novosibirsk, Russia, 2008.

2009 年

[25] Le Jialing, Liu Weixiong, Song Wenyan, Xing Jianwen, Yang Yang. Experimental and numerical investigation of air vitiation effects on scramjet test performance. 16th AIAA/DLR/DGLR International Space Planes and Hypersonic Systems and Technologies Conference. AIAA 2009-7344, 2009.

[26] J Le, B Xiao, S Yang. Implementation of reduced chemical kinetics and ISAT in modeling of supersonic combustion. 19th International Symposium on Air Breathing Engines. 2009.

[27] W He, J Le, W Liu, S Yang. Investigation of ignition characteristics for kerosene fueled scramjet. 19th International Symposium on Air Breathing Engines. 2009.

2011 年

[28] Zhang Ruoling, Jiang Jin, Liu Weixiong, Le Jialing, Zhao Guozhu, Yang Yang. The study on coolant flow and heat transfer along the cooling channels in scramjet. 20th International Symposium on Air Breathing Engines. Goteborg, Sweden, 2011, ISABE-2011-1518.

[29] Le Jialing, Yang Shunhua, Wang Xiyao. Numerical investigations of unsteady spray combustion in a liquid kerosene fueled scramjet. 20th International Symposium on Air Breathing Engines. Goteborg, Sweden, 2011, ISABE-2011-1524.

[30] Zhang Ruoling, Zhang Lei, Zhang Xiangwen, Le Jialing. Researches on heat transfer correlations of hydrocarbon fuel under supercritical condition. The 8th Sino-Russia Hypersonic Flow Conference. Shanghai, 2011.

2012 年

[31] Le Jialing, S Yang, Li Hongbin. Analysis and correlation of flame stability limits in supersonic flow with cavity flameholder. 18th AIAA/3AF International Space Planes and Hypersonic Systems and Technologies Conference. Tours, France, 2012.

[32] Zhang Ruoling, Jiang Jin, Yang Yang, Le Jialing, Zhang Lei, Zhang Xiangwen. Researches on heat transfer correlations of hydrocarbon fuel under supercritical pressure. 18th AIAA/3AF International Space Planes and Hypersonic Systems and Technologies Conference. Tours, France, 2012.

2013 年

[33] Weixin Deng, Jialing Le,Shunhua Yang, Ye Tian, Wanzhou Zhang. High speed efficient and reliable air-breathing hypersonic propulsion system. Experiment Computation & Analysis. International conference on transportation engineering. Chengdu, 2013: 2207-2220.

[34] Ruoling Zhang, Jin Jiang, Jialing Le, Yang Yang, Lei Zhang, Guozhu Zhao. The computation of heat transfer in a regeneratively cooled scramjet. 21th International Symposium on Air Breathing Engines, Korea, ISABE-2013-1629.

2014 年

[35] Deng Weixin, Le Jialing, Yang Shunhua, Zhang Wanzhou, Tian Ye. A mathematical model of scramjet ignition. Asian Joint Conference on Propulsion and Power. Jejuna island, Korea, 2014.

2015 年

[36] Yang Shunhua, Liu Weixiong, Le Jialing, Wei Feng, Zhang Wanzhou and Chen Jun. Experimental testing of a hypersonic inward turning inlet with water-drop like shape to circular shape transition. 20th AIAA International Conference on Spaceplanes, Hypersonic Systems and Technologies. Glasgow, AIAA 2015-3620.

[37] Ruoling Zhang, Guozhu Zhao, Jialing Le, Weimin Huang, Zerun Tong, Xiaojian Xu. Numerical study on heat transfer of hydrocarbon fuel with thermal cracking. 20th AIAA International Conference on Spaceplanes, Hypersonic Systems and Technologies. Glasgow, AIAA 2015-3621.

2016 年

[38] Tian Ye, Yang ShunHua, Le JiaLing, Su Tie, Yue Maoxiong, Zhong Fuyu, Tian Xiaoqiang. Investigation of combustion and flame stabilization modes in a hydrogen fueled scramjet combustor. International Journal of Hydrogen Energy, 2016: 19218　19230.

[39] Zhang Ruoling, Le Jialing. Natural laminar-to-turbulent transition inside an electrically heated circular tube. International Conference on the Methods of Aerophysical Research (ICMAR 2016), AIP Conference Proceedings 1770, 2016, 030035; doi: 10.1063/1.4963977.

2017 年

[40] Ruoling Zhang, Shuai Zheng, Jialing Le. The influence of reynolds stress on the iaminar heat transfer characteristics of hydrocarbon fuel. 21st International Space and Hypersonic Systems and Technology Conference. Xiamen, 2017:AIAA 2017-2101.

[41] Miaorong Yi, Huiyong Zhao, Jialing Le. Hypersonic natural and forced transition simulation by correlation-based interminably model. 21st AIAA International Space Planes and Hypersonics Technologies Conference: Xiamen.2017.

3. 国内会议

1996 年

[1] 乐嘉陵. 脉冲式试验技术在空气动力学研究中的应用. 纪念北航院长沈元院士 80 寿辰报告. 四川绵阳, 1996.

[2] 乐嘉陵, 张志成. 关于发展我国超然冲压发动机的建议. 21 世纪初我国航天高技术发展研讨会.四川绵阳, 1996.

2002 年

[3] 王晓东, 乐嘉陵. 冲压燃烧室内部流场的数值模拟研究. 第十一全国计算流体力学会议论文集. 河南洛阳, 2002.

[4] 邢建文, 乐嘉陵, 郑忠华. Scramjet 燃烧室流场的三维并行数值模拟及试验比较. 第十一全国计算流体力学会议论文集. 河南洛阳, 2002.

[5] 贺伟, 刘伟雄, 白菡尘, 丛京伟, 李向东, 乐嘉陵. 脉冲燃烧风洞及其在超燃发动机试验研究中的应用. 第十届全国激波与激波管学术讨论会论文集. 安徽黄山, 2002.

[6] 杨顺华, 乐嘉陵. 燃烧加热风洞中有限速率的水蒸气的凝结. 第十届全国激波与激波管学术讨论会论文集. 安徽黄山, 2002.

[7] Jialing Le, Weixiong Liu. Resaerch progress on shock duration test techniques and its application. Proceedings of the 10th Chinese symposium on shock wave. Anhui, huangshan. 2002, 14-20.

2004 年

[8] 钱炜祺, 肖保国, 乐嘉陵. 复杂化学反应动力学模拟的简化方法. 第十一全国激波与激波管学术讨论会. 河南洛阳, 2004.

[9] 乐嘉陵, 杨顺华. 煤油超燃冲压发动机数值模拟. 第十一全国激波与激波管学术讨论会. 四川绵阳, 2004.

[10] 王兰, 吴颖川. 非结构网格复杂超声速流场的数值模拟. 第十一全国激波与激波管学术讨论会. 四川绵阳, 2004.

[11] 吴颖川, 王兰, 乐嘉陵. 非结构网格数值计算并行处理的重排序算法. 第十一全国激波与激波管学术讨论会. 四川绵阳, 2004.

[12] 贺元元, 乐嘉陵. 机体/推进一体化高超声速飞行器气动-推进性能数值研究. 第十一届全国激波与激波管学术会议论文集. 四川绵阳, 2004.

[13] 邢建文, 乐嘉陵. 应用两方程湍流模型模拟超声速流动中的横向射流. 第十一届全国激波与激波管学术会议. 四川绵阳, 2004.

2006 年

[14] 杨顺华, 乐嘉陵. 超声速横向气流中液体射流雾化研究. 第十二届激波与激波管学术讨论会. 河南洛阳, 2006: 70-75.

[15] 王兰, 吴颖川, 乐嘉陵. 非结构混合网格超声速化学反应流场的数值模拟. 第十二届激波与激波管学术讨论会. 河南洛阳, 2006: 342-347.

[16] 贺旭照, 乐嘉陵, 赵慧勇. 高超声速流动中物面摩阻及热流的 CFD 验证. 第十二届全国激波与激波管学术会议论文集. 河南洛阳, 2006: 336-341.

[17] 黄挺, 乐嘉陵, 余安远. 二维高超声速进气道起动特性研究. 第十二届全国激波与激波管学术会议论文集. 河南洛阳, 2006: 42-46.

[18] 肖保国, 杨顺华, 钱炜祺, 乐嘉陵. 乙烯预混层流稳态火焰简化化学反应动力学模型研究. 全国激波与激波管学术会议. 河南洛阳, 2006: 60-64.

[19] 杨阳, 毛雄兵, 李宏斌, 刘伟雄, 乐嘉陵. 脉冲燃烧风洞甲烷加热器试验研究与初步计算分析. 第十二届全国激波与激波管学术会议论文集. 河南洛阳, 2006: 93-98.

[20] 余安远, 乐嘉陵. 一种降低进气道起动马赫数的方法——内置横向隔板法的初步数值研究. 第十二届全国激波与激波管学术会议论文集. 河南洛阳, 2006: 76-81.

[21] 张小庆, 毛雄兵, 刘伟雄, 乐嘉陵. 扩压器在脉冲燃烧风洞中的应用研究. 第十二届全国激波与激波管学术会议论文集. 河南洛阳, 2006: 99-103.

[22] 郑忠华, 贺伟, 邢建文, 乐嘉陵. 煤油燃料单模块发动机性能评估. 第十二届全国激波与激波管学术会议论文集. 河南洛阳, 2006: 82-87.

[23] 王兰, 吴颖川, 乐嘉陵. 基于非结构、混合网格的超燃冲压发动机整机化学反应流数值模拟. 中国近代空气动力学与气动热力学会议. 四川绵阳, 2006.

[24] 乐嘉陵, 刘伟雄. CARDC 高超声速吸气式推进技术近期进展. 第十二届全国激波与激波管学术会议论. 四川绵阳, 2006: 1-2.

[25] 余安远, 乐嘉陵, 郭荣伟. 开口于平表面的埋入式进气道流场的非结构网格数值模拟研究. 中国第一届近代空气动力学与气动热力学会议论文集. 四川绵阳, 2006: 514-520.

2008 年

[26] 赵慧勇, 乐嘉陵. 高超声速进气道四种 $k-\omega$ 湍流模型的比较. 第十三届全国激波与激波管学术会议文集. 湖南长沙, 2008: 101-106.

[27] 贺旭照, 乐嘉陵, 宋文艳. 采用空间推进和时间迭代结合的并行 CFD 方法求解超燃冲压发

动机整机流动. 第十三届全国激波与激波管学术会议论文集. 湖南长沙, 2008: 32-38.

[28] 黄挺, 余安远, 乐嘉陵. 二维高超声速进气道起动马赫数下性能数值研究. 第十三届全国激波与激波管学术会议论文集. 湖南长沙, 2008: 45-50.

[29] 焦伟, 刘伟维, 乐嘉陵, 卢传喜, 蒲旭阳. 一种新型壁面燃料喷注方式在大尺度超燃发动机研究中的初步应用. 第十三届全国激波与激波管学术会议论文集. 湖南长沙, 2008: 112-116.

[30] 倪鸿礼, 贺伟, 贺元元, 乐嘉陵, 刘伟雄. 带动力缩比机体/推进一体化飞行器推阻特性试验研究. 第十三届全国激波与激波管学术会议论文集. 湖南长沙, 2008: 20-25.

[31] 杨富荣, 杨阳, 陈力, 鲍伟义, 苏铁, 乐嘉陵. 脉冲燃烧风洞起动过程光电方法测量. 第十三届全国激波与激波管学术会议论文集. 湖南长沙, 2008: 458-463.

[32] 钱炜祺, 肖保国, 乐嘉陵, 杨顺华. 化学动力学在超燃冲压发动机数值模拟中的应用. 第十三届全国激波与激波管学术会议论文集. 湖南长沙, 2008: 88-93.

[33] 余安远, 黄挺, 倪鸿礼, 乐嘉陵. FL-23 风洞用超燃冲压进气道试验平台的设计与初步应用. 第十三届全国激波与激波管学术会议论文集. 湖南长沙, 2008: 94-100.

[34] 张小庆, 乐嘉陵. 脉冲式燃烧风洞起动过程的数值模拟. 第十三届全国激波与激波管学术会议论文集. 湖南长沙, 2008: 510-515.

[35] 赵慧勇, 乐嘉陵. 高超声速进气道四种 $k-\omega$ 湍流模型的比较. 第十三届全国激波与激波管学术会议论文集. 湖南长沙, 2008: 101-106.

[36] 刘伟雄, 贺伟, 李宏斌, 乐嘉陵. 污染组分对氢燃料发动机燃烧动力学影响的初步研究. 第一届全国高超声速科技学术会议论文集. 云南丽江, 2008: 1-5.

2009 年

[37] 余安远, 刘伟雄, 郭荣伟, 乐嘉陵. 脉冲风洞中超燃冲压进气道起动试验研究. 第二届近代实验空气动力学会议论文集. 吉林延吉. 2009: 213-219.

2010 年

[38] 乐嘉陵. CARDC 脉冲燃烧风洞的研究的历史回顾及发展. 第十四届全国激波与激波管学术会议论文集. 安徽黄山, 2010.

[39] 王兰, 吴颖川, 乐嘉陵. Scramjet 带凹槽燃烧室流场数值模拟. 第十四届全国激波与激波管学术会议论文集. 安徽黄山, 2010.

[40] 王西耀, 乐嘉陵, 杨顺华. 超燃发动机燃烧室内两相燃烧的数值模拟. 第十四届全国激波与激波管学术会议论文集. 安徽黄山, 2010.

[41] 余安远, 乐嘉陵, 倪鸿礼. 高超声速二元进气道起动性能的定常数值评估方法研究. 第十四届全国激波与激波管学术会议论文集. 安徽黄山, 2010.

[42] 余安远, 吴颖川, 贺旭照, 乐嘉陵. 两种乘波前体-进气道性能的数值计算与分析. 第十四届全国激波与激波管学术会议论文集. 安徽黄山, 2010.

2012 年

[43] 乐嘉陵, 刘伟雄. 吸气式高超声速地面设备研究进展及其应用. 第十五届全国激波与激波管学术会议论文集. 浙江杭州. 2012: 1-2.

[44] 贺旭照, 乐嘉陵, 周正, 倪鸿礼. 乘波前体进气道的一体化设计方法及性能分析. 第十五届全国激波与激波管学术会议论文集. 浙江杭州, 2012: 25-32.

[45] 吴颖川, 贺元元, 余安远, 乐嘉陵. 展向截断曲面乘波压缩前体进气道气动布局研究. 第十五届全国计算流体力学会议论文集. 山东烟台, 2012: 722-726.

2014 年

[46] 卫锋, 贺旭照, 乐嘉陵. 内外锥混压的乘波前体进气道一体化设计初探. 第十六届全国激波与激波管学术会议. 河南洛阳, 2014.

[47] 余安远, 杨大伟, 吴杰, 倪鸿礼, 乐嘉陵. 高超声速进气道超声速时激波振荡现象预测与验证. 第十六届全国激波与激波管学术会议. 河南洛阳, 2014.

2015 年

[48] 邓维鑫, 乐嘉陵, 杨顺华, 张弯洲. 超燃发动机燃料当量比与燃烧压力数学模型. 第十七届中国空气动力学物理气体动力学学术交流会. 四川成都, 2015.

[49] 邓维鑫, 蒋劲, 乐嘉陵, 杨顺华, 张磊. 超燃冲压发动机试验模型设计关键技术研究. 第五届冲压发动机技术交流会. 福建厦门, 2015.

[50] 乐嘉陵, 倪鸿礼, 余安远, 杨大伟, 吴杰. 高超声速二元进气道起动性能数值模拟技术研究与试验验证. 2015 第二届中国航空科学技术大会论文集. 北京. 2015: 341-347.

[51] 田野, 乐嘉陵, 杨顺华. 空气节流对双模态冲压发动机燃烧模态形成与转换影响研究. 第八届全国高超声速会议. 黑龙江哈尔滨, 2015.

2016 年

[52] 白菡尘, 刘伟雄, 乐嘉陵, 毛雄兵, 谭宇. 半自由射流冲压发动机试验方法初步探索. 中国第一届近代空气动力学与气动热力学会议论文集. 四川绵阳, 2016: 108-112.

[53] 贺旭照, 乐嘉陵. 高超声速流动中气动力气动热的 CFD 验证及应用. 中国第一届近代空气动力学与气动热力学会议论文集. 四川绵阳, 2016: 185-190.

[54] 田野, 邓维鑫, 乐嘉陵等. Ma 6.5 条件下超燃冲压发动机模型点火特性. 第十七届激波与激波管学术会议. 四川成都, 2016.

[55] 张若凌, 乐嘉陵. 电加热圆管内流动的自然转捩过程研究. 第十七届激波与激波管学术会议. 四川成都, 2016.

2017 年

[56] 郑榆山, 乐嘉陵, 王宇辉. 入口流量对旋转爆震发动机性能和流场结构的影响. 第十七届全国计算流体力学会议. 浙江杭州, 2017.

[57] 邓维鑫, 乐嘉陵, 任虎, 杨顺华, 张弯洲. 超燃冲压发动机隔离段入口气流参数控制技术研究. 中国航天第三专业信息网第三十八届技术交流会暨第二届空天动力联合会议. 辽宁大连, 2017.

[58] 邓维鑫, 乐嘉陵, 任虎, 卫峰, 杨顺华, 张弯洲. 超燃冲压发动机隔离段入口气流畸变问题研究综述及进展. 第六届冲压发动机技术交流会. 海南文昌, 2017.

[59] 田野, 杨顺华, 乐嘉陵. 空气节流对煤油燃料超燃燃烧室火焰稳定影响研究. 第六届冲压发动机会议. 海南文昌, 2017.

[60] 郑榆山, 王超, 乐嘉陵. 旋转爆震三维非预混数值模拟研究. 第五届爆震与新型推进学术研讨会. 湖南长沙. 2017.

专著

乐嘉陵, 高铁锁, 曾学军. 再入物理. 北京. 国防工业出版社, 2005.